Plant Response to Air Pollution

Plant Response to Air Pollution

Editor:

Dr. Nisha Dahiya, Ph.D.

Delve Publishing
www.delvepublishing.com

Plant Response to Air Pollution

Editor: Dr. Nisha Dahiya, Ph.D.

708 3rd Avenue, 6th Floor
New York, NY 10017
United States of America

www.delvepublishing.com

ISBN: 978-1-68095-743-3

Library of Congress Control Number: 2016951502

About the Editor

Dr. Nisha Dahiya, Ph.D.

Dr. Nisha Dahiya is serving as a Research Scientist at the Indian Council of Medical Research (ICMR), New Delhi, India. She completed her PhD in 2014 from University of Camerino, Italy, under the Malaria and Human Development Program, which was supported by World Health Organization, Global Malaria Program. Her main interests are drug discovery for infectious diseases and surveillance and molecular epidemiology of rising drug resistant pathogens. Currently, she is assisting in the coordination of national network for surveillance of anti microbial resistance through out the country.

Contents

List of Contributors

JiazhiZhang
State Key Laboratory of Food Science and Technology, School of Environment and Civil Engineering, Jiangnan University, Wuxi 214122, China
Jiangsu Coorperative Innovation Center of Technology and Material of Water Treatment, Suzhou University of Science and Technology, Suzhou 215009, China

Xingyi Li
State Key Laboratory of Food Science and Technology, School of Environment and Civil Engineering, Jiangnan University, Wuxi 214122, China
Jiangsu Coorperative Innovation Center of Technology and Material of Water Treatment, Suzhou University of Science and Technology, Suzhou 215009, China

Li Zhou
State Key Laboratory of Food Science and Technology, School of Environment and Civil Engineering, Jiangnan University, Wuxi 214122, China
Jiangsu Coorperative Innovation Center of Technology and Material of Water Treatment, Suzhou University of Science and Technology, Suzhou 215009, China

LihongWang
State Key Laboratory of Food Science and Technology, School of Environment and Civil Engineering, Jiangnan University, Wuxi 214122, China
Jiangsu Coorperative Innovation Center of Technology and Material of Water Treatment, Suzhou University of Science and Technology, Suzhou 215009, China

Qing Zhou
State Key Laboratory of Food Science and Technology, School of Environment and Civil Engineering, Jiangnan University, Wuxi 214122, China
Jiangsu Coorperative Innovation Center of Technology and Material of Water Treatment, Suzhou University of Science and Technology, Suzhou 215009, China

Xiaohua Huang
Jiangsu Collaborative Innovation Center of Biomedical Functional Materials, Jiangsu Key Laboratory of Biomedical Materials, College of Chemistry and Materials Science, Nanjing Normal University, Nanjing 210046, China

Marcello Iriti
Università degli Studi di Milano, Dipartimento di Produzione Vegetale, Sezione di Patologia Vegetale, Via Celoria 2, 20133 Milano, Italy

Franco Faoro
Università degli Studi di Milano, Dipartimento di Produzione Vegetale, Sezione di Patologia Vegetale, Via Celoria 2, 20133 Milano, Italy

S. Deb Roy
Indian Institute of Tropical Meteorology, 411008 Pune, India

G. Beig
Indian Institute of Tropical Meteorology, 411008 Pune, India

Sachin D. Ghude
Indian Institute of Tropical Meteorology, 411008 Pune, India

Zhenzhu Xu
State Key Laboratory of Vegetation and Environmental Change, Institute of Botany, Chinese Academy of Sciences, Beijing, China

Yanling Jiang
State Key Laboratory of Vegetation and Environmental Change, Institute of Botany, Chinese Academy of Sciences, Beijing, China

Guangsheng Zhou
State Key Laboratory of Vegetation and Environmental Change, Institute of Botany, Chinese Academy of Sciences, Beijing, China
Chinese Academy of Meteorological Sciences, Beijing, China

Andréia Caverzan
Programa de Pós-Graduação em Genética e Biologia Molecular, Departamento de Genética, Universidade Federal Rio Grande do Sul, Porto Alegre, RS, Brazil

Gisele Passaia
Programa de Pós-Graduação em Genética e Biologia Molecular, Departamento de Genética, Universidade Federal Rio Grande do Sul, Porto Alegre, RS, Brazil

Silvia Barcellos Rosa
Programa de Pós-Graduação em Biologia Celular e Molecular, Centro de Biotecnologia, Universidade Federal Rio Grande do Sul, Porto Alegre, RS, Brazil

Carolina Werner Ribeiro
Programa de Pós-Graduação em Biologia Celular e Molecular, Centro de Biotecnologia, Universidade Federal Rio Grande do Sul, Porto Alegre, RS, Brazil

Fernanda Lazzarotto
Programa de Pós-Graduação em Genética e Biologia Molecular, Departamento de Genética, Universidade Federal Rio Grande do Sul, Porto Alegre, RS, Brazil

Márcia Margis-Pinheiro
Programa de Pós-Graduação em Genética e Biologia Molecular, Departamento de Genética, Universidade Federal Rio Grande do Sul, Porto Alegre, RS, Brazil
Programa de Pós-Graduação em Biologia Celular e Molecular, Centro de Biotecnologia, Universidade Federal Rio Grande do Sul, Porto Alegre, RS, Brazil

G. U. Chibuike
Department of Soil Science, University of Nigeria, Nsukka, Nigeria

S. C. Obiora
Department of Geology, University of Nigeria, Nsukka, Nigeria

Hyacinthe Le Gall
EA3900-BIOPI, Biologie des Plantes et Innovation, Université de Picardie Jules Verne, 80039 Amiens, France

Florian Philippe
EA3900-BIOPI, Biologie des Plantes et Innovation, Université de Picardie Jules Verne, 80039 Amiens, France

Jean-Marc Domon
EA3900-BIOPI, Biologie des Plantes et Innovation, Université de Picardie Jules Verne, 80039 Amiens, France

Françoise Gillet
EA3900-BIOPI, Biologie des Plantes et Innovation, Université de Picardie Jules Verne, 80039 Amiens, France

Jérôme Pelloux
EA3900-BIOPI, Biologie des Plantes et Innovation, Université de Picardie Jules Verne, 80039 Amiens, France

Catherine Rayon
EA3900-BIOPI, Biologie des Plantes et Innovation, Université de Picardie Jules Verne, 80039 Amiens, France

Misa Takahashi
Department of Mathematical and Life Sciences, Graduate School of Science, Hiroshima University, Higashi-Hiroshima, Japan

Hiromichi Morikawa
Department of Mathematical and Life Sciences, Graduate School of Science, Hiroshima University, Higashi-Hiroshima, Japan

Fernanda Freitas Caregnato
Centro de Estudos em Estresse Oxidativo, Departamento de Bioquímica, Instituto de Ciências Básicas da Saúde, Universidade Federal do Rio Grande do Sul (UFRGS), Av. Ramiro Barcelos, 2600, Anexo, CEP 90035-003, Porto Alegre, RS, Brazil

Rafael Calixto Bortolin
Centro de Estudos em Estresse Oxidativo, Departamento de Bioquímica, Instituto de Ciências Básicas da Saúde, Universidade Federal do Rio Grande do Sul (UFRGS), Av. Ramiro Barcelos, 2600, Anexo, CEP 90035-003, Porto Alegre, RS, Brazil

Armando Molina Divan Junior
Laboratório de Bioindicação Vegetal, Centro de Ecologia, Instituto de Biociências, Universidade Federal do Rio Grande do Sul (UFRGS), Av. Bento Gonçalves, 9500, Prédio 43411, CEP 91501-970, Porto Alegre, RS, Brazil

José Cláudio Fonseca Moreira
Centro de Estudos em Estresse Oxidativo, Departamento de Bioquímica, Instituto de Ciências Básicas da Saúde, Universidade Federal do Rio Grande do Sul (UFRGS), Av. Ramiro Barcelos, 2600, Anexo, CEP 90035-003, Porto Alegre, RS, Brazil

Sowbiya Muneer
Department of Horticulture, College of Agricultural Life Sciences, Chonnam National University, 300 Young Bong-Dong Buk-Gu, Gwangju 500-757, Republic of Korea

Tae Hwan Kim
Department of Animal Science, Institute of Agricultural Science and Technology, College of Agriculture & Life Science, Chonnam National University, 300 YoungBong-Dong Buk-Gu, Gwangju 500-757, Republic of Korea

Byung Chul Choi
School of Mechanical Systems Engineering, College of Engineering, Chonnam National, University, 300 Young Bong-Dong Buk-Gu, Gwangju 500-757, Republic of Korea

Beom Seon Lee
Dayung GS Co., Ltd., Damyang, Jeonnam 517-922, Republic of Korea

Jeong Hyun Lee
Department of Horticulture, College of Agricultural Life Sciences, Chonnam National University, 300 Young Bong-Dong Buk-Gu, Gwangju 500-757, Republic of Korea

Seyyednejad, S.M.
Department of Biology, Faculty of Sciences, Shahid Chamran University of Ahvaz, Ahvaz, Iran

Koochak, H.
Department of Biology, Faculty of Sciences, Shahid Chamran University of Ahvaz, Ahvaz, Iran

Vaezi, J
Department of Biology, Faculty of Sciences, Shahid Chamran University of Ahvaz, Ahvaz, Iran

Otniel Freitas-Silva
Institute for Biotechnology and Bioengineering IBB, Centre of Biological Engineering, Universidade do Minho, Campus de Gualtar, 4710-057 Braga, Portugal
EMBRAPA Food Technology, Avenida das Américas 29501, 23020-470 Rio de Janeiro, RJ, Brazil

Héctor Morales-Valle
Institute for Biotechnology and Bioengineering IBB, Centre of Biological Engineering, Universidade do Minho, Campus de Gualtar, 4710-057 Braga, Portugal

Armando Venâncio
Institute for Biotechnology and Bioengineering IBB, Centre of Biological Engineering, Universidade do Minho, Campus de Gualtar, 4710-057 Braga, Portugal

Oleg Blum
M. M. Gryshko National Botanical Garden, National Academy of Sciences of Ukraine, Timiryazevs'ka St., 1, 01014 Kyiv, Ukraine

Nataliya Didyk
M. M. Gryshko National Botanical Garden, National Academy of Sciences of Ukraine, Timiryazevs'ka St., 1, 01014 Kyiv, Ukraine

Nataliya Pavluchenko
M. M. Gryshko National Botanical Garden, National Academy of Sciences of Ukraine, Timiryazevs'ka St., 1, 01014 Kyiv, Ukraine

Barbara Godzik
W. Szafer Institute of Botany, Polish Academy of Sciences, Lubicz 46, 31-512 Krakow, Poland

Marta Wilton Pereira Leite de Vasconcelos
Universidade Católica Portuguesa, Centro de Biotecnologia e Química Fina (CBQF), Laboratório Associado, Escola Superior de Biotecnologia, Rua Arquiteto Lobão Vital, Apartado 2511, 4202-401 Porto, Portugal

Paloma Koprovski Menguer
Universidade Federal do Rio Grande do Sul, 91501-970 Porto Alegre, RS, Brazil

Yanbo Hu
College of Life Science, Northeast Forestry University, No. 26 Hexing Road, Xiangfang District, Harbin 150040, China

Luis Fernando Revers
EMBRAPA Uva e Vinho, Rua Livramento 515, 95701-008 Bento Gonçalves, RS, Brazil

Raul Antonio Sperotto
Centro de Ciências Biológicas e da Saúde (CCBS), Programa de Pós-Graduação em Biotecnologia (PPGBiotec), Centro Universitário UNIVATES, Rua Avelino Tallini 171, 95900-000 Lajeado, RS, Brazil

Alain Vian
Université d'Angers, Campus du Végétal, UMR 1345 IRHS, CS 60057, SFR 4207 QUASAV, 49071 Beaucouzé Cedex, France

Eric Davies
Department of Plant and Microbial Biology, North Carolina State University, Raleigh, NC 27695, USA

Michel Gendraud
Université Blaise Pascal, 24 avenue des Landais, 63177 Aubière Cedex, France

Pierre Bonnet
Institut Pascal, Université Blaise Pascal, BP 10448, 63000 Clermont-Ferrand, France
CNRS, UMR 6602, 63171 Aubière, France

Abida Begum
Department of Chemistry, P.E.S School of Engineering, Bangalore-560100, India
Lecturer, Department of Chemistry, Shirdi Sai Engineering College Anekal, Bangalore, India

S. Harikrishna
Department of Chemistry, P.E.S School of Engineering, Bangalore-560100, India
Lecturer, Department of Chemistry, Shirdi Sai Engineering College Anekal, Bangalore, India

Vijeta Verma
Department of Botany, University of Lucknow, Lucknow 226007, India

Neelam Chandra
Department of Botany, University of Lucknow, Lucknow 226007, India

Hussani Mubarak
Department of Environmental Engineering, School of Metallurgy and Environment, Central South University, Changsha, Hunan 410083, China
National Engineering Research Center for Control and Treatment of Heavy Metal Pollution, Changsha 410083, China

Nosheen Mirza
Department of Environmental Engineering, School of Metallurgy and Environment, Central South University, Changsha, Hunan 410083, China
National Engineering Research Center for Control and Treatment of Heavy Metal Pollution, Changsha 410083, China
Department of Environmental Sciences, COMSATS Institute of Information Technology, Abbottabad 22060, Pakistan

Li-Yuan Chai
Department of Environmental Engineering, School of Metallurgy and Environment, Central South University, Changsha, Hunan 410083, China
National Engineering Research Center for Control and Treatment of Heavy Metal Pollution, Changsha 410083, China

Zhi-Hui Yang
Department of Environmental Engineering, School of Metallurgy and Environment, Central South University, Changsha, Hunan 410083, China
National Engineering Research Center for Control and Treatment of Heavy Metal Pollution, Changsha 410083, China

Wang Yong
Department of Environmental Engineering, School of Metallurgy and Environment, Central South University, Changsha, Hunan 410083, China
National Engineering Research Center for Control and Treatment of Heavy Metal Pollution, Changsha 410083, China

Chong-Jian Tang
Department of Environmental Engineering, School of Metallurgy and Environment, Central South University, Changsha, Hunan 410083, China
National Engineering Research Center for Control and Treatment of Heavy Metal Pollution, Changsha 410083, China

Qaisar Mahmood
Department of Environmental Sciences, COMSATS Institute of Information Technology, Abbottabad 22060, Pakistan

Arshid Pervez
Department of Environmental Sciences, COMSATS Institute of Information Technology, Abbottabad 22060, Pakistan

Umar Farooq
Department of Chemistry, COMSATS Institute of Information Technology, Abbottabad 22060, Pakistan

Shah Fahad
College of Plant Science and Technology, Huazhong Agricultural University, Wuhan, Hubei 430070, China

Wajid Nasim
Department of Environmental Sciences, COMSATS Institute of Information Technology (CIIT), Vehari, Pakistan

Kadambot H. M. Siddique
The UWA Institute of Agriculture, The University of Western Australia, Perth, WA 6009, Australia

Ademir Sérgio Ferreira Araújo
UESPI - Campus de Parnaíba, Av. Nossa Senhora de Fátima, s/n, 64202-220 - Parnaíba, PI - Brasil

Regina Teresa Rosim Monteiro
USP/CENA - Lab. Ecotoxicologia, C.P. 96 - 13400-970 - Piracicaba, SP – Brasil

Elizabeth Olivares
Centro de Ecología, Instituto Venezolano de Investigaciones Científicas, Apartado 21827, Caracas 1020 A, Venezuela

Preface

Air, water and soil pollution influences vegetation. Primary air pollutants such as reactive compounds of sulphur (sulphur oxides), nitrogen (nitrogen oxides), carbon monoxide and hydrocarbons, are emitted from fossil fuel combustion. Secondary pollutants, primarily ground level zone have severe implications on plants. These pollutants enter through stomata and adversely affect plant functions. On exposure to the automobile pollution, Sida cordifolia L. and Catharanthus roseus L. showed significant increase in the levels of catalase and peroxidase and evident alteration in chlorophyll, carotenoid and phaeophytin. The stomatal indices, number of epidermal cells per unit area and the number of stomata reportedly decreased due to exposure, while increase in the size of stomata and epidermal cells was found. Bisphenol A, another common industrial pollutant induced dose dependent and stage specific response in soyabean roots, causing only membrane peroxidation at low doses. On the contrary, at higher doses, in addition to membrane peroxidation, the BPA also activated antioxidant systems and enhanced reactive oxygen species levels. Low level exposure did not affect membrane peroxidation at growth stages other than seedling stage.

Plant response to pollutants varies with species, age, variety and other abiotic factors. Ozone, among all the pollutats, has been documented to have the most detrimental effects on the plants. Ozone causes damage to almost all levels of plant functions, such as leaf injury, stomatal conductance, photosynthesis, carbon allocation and respiration. Phaseolus vulgaris varieties Irai and Fepargo were studied to demonstrate the effect of high concentration of ozone on enzymatic and non-enzymatic antioxident system of the plant. Ozone increased total thiol levels in Irai while the levels were unchanged in Fepargo 26. Chlorophyll a and b levels were fund to be reduced along with reduction in protein concentration indicating compromised photosynthetic apparatus. Irai had higher levels of reactive oxygen species (ROS) and ascorbate (AsA) concentration, but lower total thiol levels and catalase content. Contrarily, aqueous ozone may be used for the control of fungal pathogens especially against the conidia of A. flavus with effective dose depending upon initial spore inoculum. Certain catagories of agrochemicals such as strobilurin group fungicides, growth regulators such as "Emistym C" demonstrate protective effects against ozone injury in ozone sensitive cultivars of clover and tobacco. Such agrochemicals may possible be used for increased ozone tolerance in sensitive crops. Elevated levels of CO2, however, increase photosynthetic rate and demonstrate positive impact on crop productivity. Elevation of CO2 levels in atmosphere stimulates the light-saturated photosynthetic CO2 assimilation rate in C3 and C4 plants with varying degrees ranging from maximum stimulation for trees and C3 grasses, moderate for shrubs, C3, C4 crops and legumes, and negative for C4 grass. Under elevated CO2, carbohydrate accumulations, such as starch size and number of chloroplasts, can be enhanced, partially due to the carbon substrate increase. However, the excessive carbohydrate accumulation may cause feedback inhibition or physical damage at the chloroplast level, reducing the photosynthetic capacity. Certain measures taken to reduce emission of primary pollutants, such as, using fuels low in sulphur content, introduction of catalytic converters to reduce exhaust pollutants, reducing application of nitrogen-based fertilizers, and switching to cleaner energy sources, may reduce air pollution and its deleterious effects on vegetation. High Frequency non-ionizing Electromagnetic Fields emitted from electronic devices were found to cause altered gene expression in plants. In a study, metabolic activities such as reactive oxygen species metabolism, α- and β-amylase, Krebs cycle, pentose phosphate pathway, chlorophyll content, terpene emission, etc. were reportedly modified as a result of the exposure. Heavy metal pollution causes growth reduction, poor performance and low yield in plants growing on heavy metal polluted soil. In hydroponics culture, Boehmeria nivea L. at 20 mg/l showed changes in chlorophyll, electrolyte leakage and hydrogen peroxide levels. Certain plants such as Boehmeria nivea L. and micro-organisms accumulate heavy metals and are suitable for bioremediation which has been accepted as an effective method of treating soils carrying heavy metal pollution. Boehmeria nivea L. can uptake various heavy metals, such as Sb, Cd, Pb, and Zn. The plant can accumulate arsenic catalase and malondialdehyde in plant parts, which indicated that it may be used for phytoremediation. In plants, antioxidant defenses such as ascorbate-glutathione cycle, which is one of the major hydrogen peroxide detoxifying system in plant cells, protect plants from the reactive oxygen species (ROS), produced on exposure to pollutants or environmental stress. Ascorbate peroxidase enzymes, present in chloroplasts, mitochondria, peroxisomes and cytosol, catalyze conversion of H2O2 into water using ascorbate as a specific electron donor.

Editor
Dr. Nisha Dahiya, Ph.D.

INTRODUCTION

Studies of the response of vegetation to air pollution stress have shifted in emphasis from descriptions of symptoms to physiological explanations and ecological ramifications. Physiological analysis of the effects of pollution has focused on suspected direct and indirect alterations of normal plant structure and function, and the results show a variety of effects which range from alterations in organismal physiology to changes in cellular biochemistry. A common objective of all research is the attempt to understand how plant and environmental factors affect susceptibility of vegetation to air pollution. An organizational framework for the analysis of the response of vegetation to any environmental stress is provided by Levitt's study of stress phenomena in plants. His ideas have been applied in investigations of the response of plants to high saline conditions, water stress, low temperatures, heavy metal toxicity and resistance to sculpture dioxide in lichens. His analysis is applicable also to the response of vascular plants to gaseous air pollutants. This approach provides a coherent, logical scheme of alternative factors affecting the sensitivity of foliage to air pollution. Some of the terminology and concepts advanced by Levitt have permeated the stature, but often their application has been either incorrect or inappropriate. Consequently the objective of this paper is to present finically Levitt's proposal and to use its provisions in the analysis of the physiological response of foliage to gaseous air pollution.

The rapid growth of industries. Urbanization, and transportation and their injudicious planning without due regard for sustainable development are leading to serious environmental hazards. The destruction of ecological balance as a consequence has a significant unfavorabk effect on the biosphere. and the entire life-support system is rapidly facing danger. The significant contribution of air pollution to the problems of ill health of human beings, loss of agricultural productivity. Forest decline, etc. has been a cause of increasing public concern throughout the world. It has also become a topic of intense scientific, industrial, and government interest. The environment constitutes the largest portion of the biosphere and is one of the vital components on which the life depends. Biosphere contributes significantly to the chemical composition of the atmosphere and its properties and, vice versa, changes in atmospheric composition have a strong feedback effect on ecosystem functioning and biodiversity. Air quality is critical to plant health.

Earth's atmosphere is finite and its capacity to cleanse itself is limited. Earth has evolved in an atmosphere of life-supporting chemical constituents and therefore has a limited ability for homeostasis and resilience against changes in air quality.' A wide array of air pollutants including particulates, liquids, and gases is being emitted both from natural and anthropogenic sources. After emission, air pollutants are subjected to physical. chemical. and photochemical transformations. which ultimately decide the fate and atmospheric concentrations of the pollutants. The application of abatement technology. Increased urbanization, population growth especially in developing countries, and the shift to large fossil-fueled electric generating plants have altered the nature of air quality problems and the focus of air pollution research during recent years. The problems of air pollution that were thought to be concentrated around large uncontrolled point sources have been supplanted by those associated with regional-scale elevated concentrations of pollutants. Air pollutants do not remain confined near the source of emission, but spread over far off distances. transcending natural and political boundaries depending upon topography and meteorological conditions, especially wind direction and speed and vertical and horizontal thermal gradients. The topic of the regional impact of air pollution on different ecosystems is one of the major ecological, scientific, and political issues of the modern industrialized world. Adverse effects of air pollutants on biota and ecosystems have been documented worldwide!" Since humans live in an extremely holocenotic environmental complex where all factors are interrelated, a slight change in any one factor may produce numerous and involved alterations in other factors. In the last three decades, environmental scientists have begun to appreciate the complex interactions of air pollutants with abiotic and biotic components of an ecosystem. Tingey et al." have emphasized the consideration of ecological impacts of air pollution in developing ambient air quality standards.

Causes of Biodiversity Loss

The continued functioning of ecosystems is dependent on the constituent species and their distribution, as well as on genetic variation within those species and the dynamics of the interactions that exist between different species and the physical environment. Ecosystems have characteristic biodiversity at different levels of organization depending upon the abiotic and biotic factors." Both genetic and natural selection processes are responsible for the diversity of an arc. Biodiversity originates at the gene level and extends through species, populations. Communities, to ecosystems. Any kind of disturbance is a key factor in analyzing the dynamics of communities and its role in the functioning and maintenance of the ecosystem. In response to environmental stimuli. Biodiversity constantly changes, leading to loss of genes. Species. Populations. And habitats.' "Humans now have the dominant influence on biodiversity. Humans have made changes in our landscapes due to agriculture. Silviculture. Road building, industrial development, urbanization. And pollution — the major causes of biodiversity loss at the global scale. The issue of biodiversity loss has raised an alarm throughout the world. The present species extinction rate has surpassed even the massive extinction reported during the geologic past. Habitat fragmentation and destruction, species overexploitation, introduction of exotic species, and release of toxic chemicals into air, water, and soil are implicated as major causes of loss of biodiversity."" The first two stressors attained immediate importance as they are directly related to species toss, but the release of toxic chemicals has a subtle effect on biological diversity because of the adverse effects of pollution on sensitive species. As plants live outdoors and are continuously exposed to air pollutants, they show the effects of pollutants in due course of time. An air pollutant may affect species differently in a community, where some species exhibit no response and some may be unfavorably affected. After several generations, the sensitive species may be progressively eliminated leading to species loss from the community. This response to low-intensity stressors for a prolonged duration is micro-evolution leading to a shift in genetic structure toward greater resistance." Genetic diversity is found responsive to anthropogenic stresses."'" Regional-scale impact on vegetation may be ascribed to secondary air pollutants formed as a result of different transformations of primary pollutants present in air =Ma traveling long distances. Airborne pollutants such as ozone (Os) and peroxyacetyl nitrate (PAN) and acid depositions are the problems of regional scale affecting plant species directly and indirectly through interactions with other natural stresses such as water and nutrient deficiencies. Temperature stress, and insect and pathogen infestations."01ohal-level effects of changes in atmospheric composition on biodiversity are due to increases in persistent pesticides and trace metals, greenhouse gases, and stratospheric of depleting gases, etc. The depletion of stratospheric 0, with the consequence of increasing UV-8 radiation will have direct effects on biodiversity and ecosystem functioning by impairment of photosynthesis, nitrogen metabolism, and orientation of marine phytoplankton, which may affect dimethyl sulfide (DMS) emission and thus affect cloud formation." Increased UV-I3 not only affects primary producers. But the next tropic levels as well?' Disruption of the carbon cycle due to increase in the concentration of CO_2 in atmosphere and the nitrogen cycle due to human-induced nitrogen fixation leading to increase in nitrous oxide emissions and higher levels of N depositions may cause drastic changes in species diversity." Plant species richness has been shown to reduce in short grass steppe, tall grass prairie, tundra, and deciduous forests upon N fertilization and depositions.s." CO_2 enhancement experiments suggest changes in plant-plant interactions and alterations of competitive balances among species,*$ which might lead to a decrease in plant species diversity. Polley et al° correlated the increase in woody Cs Prosopis glanduloxa in the C, grasslands in the southwestern U.S. to a 27% rise in CO_2 concentration. Vine growth in tropical forest has been found to be favored. Leading to increase in tree mortality due to increased CO_2. Increase of some insects, plant pathogens, and weeds has been implicated to the increase in CO: and consequent temperature." These perturbations, corn only termed as global change. Can reduce species diversity. Which in turn will affect ecosystem tanctioning. IIA change in climate faster than the migration of most species will reduce the suitable area for a large number of species, leading to drastic reduction in global species diversity. A big pole ward shift of vegetation patterns is predicted under a double-CO: climate's the biosphere is a sink of atmospheric pollutants. Depositions of atmospheric S and N com-pounds of anthropogenic origin affect the ecosystem functioning in temperate and boreal forests and grasslands, as well as in lakes, and rivers, by acidifying the water and soil through N and S enrichment. These perturbations have been found to affect biodiversity severely." Depositions of atmospheric pollutants in the biosphere is an important ecosystem function. Highly dependent on biodiversity. There is a great deal of concern about the effects of air pollutants on biodiversity, but there has been relatively less known about the effects of air pollutants on species of natural and semi natural communities. The chemical composition of the atmosphere has changed on an unprecedented scale over the past decades. Air pollution thus constitutes a major evolutionary challenge for biodiversity in managed and natural ecosystems. This chapter is concerned with the effects of air pollutants on plant diversity at different levels of organization.

Air Pollution and Plant Responses

The plant response to air pollutants depends upon the chemical toxicity and exposure pattern of the pollutant and the sensitivity level of the species. Plant species differ in their perception and response to air pollutants due to differential expression under the influence of genetic and/or environmental factors leading to constitutive and inducible differences among the plants within and between the populations to the same air pollutant.'" Pollutants may influence plant growth and development through multiple pathways and mechanisms over widely varying timescales. Effects may be direct due to deposition on plant surfaces. Resulting in impaired physiology and metabolism of foliar surfaces. Or indirect, as a consequence of altered nutrients or toxic metals availability or due to altered susceptibility to other stresses. Air pollutants in combination may induce additive, antagonistic, or synergistic impacts on plants depending upon the specific pollutants, their concentration and exposure duration, and environmental conditions, Lower-level chronic exposures are generally more representative of toady's regional-scale air pollution situations.

Variability in species response has been suggested to be governed by a sequence of events. Which either reduces penetration of pollutant in leaf interior (pollutant avoidance)" and/or enhances the ability of plant tissue to withstand the pollutant and its products after reaching the target (pollutant tolerance)."." These sequences of events initially depend upon constitutive resource thereafter regulation of gene expression and acceleration of secondary metabolite production may modify the response pattern:"

Air pollutants enter the plant body through stomata' pores in the leaves. Darrall' has reviewed the stomata! Response of plants to a variety of air pollutants as of partial stomata! Closure, wider opening, or no response depending upon pollutant concentration and exposure duration. Environmental conditions. And plant specks and their cultivars. Some plants are known to exhibit intrinsically higher stomata' conductance and tend to be more susceptible to damage than others." Implications of stomata] responses to air pollutants also depend on responses of other critical physiological processes such as carbohydrate assimilation and partitioning!" Variability in inter-and intraspecific response to air pollutants with respect to photosynthetic partitioning may alter the nature of competition within a plant community.'.'" After entering into the leaf interior, air pollutants may be metabolized. Sequestered, accumulated, or excreted. Detoxification of pollutants and their products at the cell surface and interior has been widely documented in the literature."

Impact of Air Pollution on Plant Diversity at Different Levels of Organization

Biological diversity. Or in shortened fonts biodiversity. Means the variability among living organisms inhabiting the Earth. In a broader perspective, biodiversity is ecological diversity referring to the number of species in a given area, the ecological roles of the species, and the changes in composition across a region, and their grouping together with processes and interactions within and between systems.

Three attributes of biodiversity are suggested: composition (number of alleles or species): structure (physical arrangement of biotic components and biomass distribution): and function (natural processes of energy flow and nutrient cycling)." It is well established that the stresses from regionally transported gaseous pollutants and acidic deposition can cause structural and functional changes in plant communities."`: leading to subtle degradation of ecosystems.* Air pollutants can affect various levels of ecosystem organization. Starting from an individual leaf to an entire ecosystem." Although the species-level diversity is the most-studied element of biological diversity in relation to atmospheric pollutants, genetic- and community-level diversities arc also very important.

Genetic Diversity

This element of biodiversity is the foundation of all other aspects of diversity. Loss of plant species of small population sizes can lead to reduced genetic diversity, towered fitness, and increased extinction risk. Intrinsic variations within populations serve as the immediate source of genetic diversity. Taylor and Murdyic demonstrated plant-to-plant variation of foliar injury in the herbaceous annual Geranium cans'inianum L due to SOI. Exposure. Even after a 30-year period. The populations continued to exhibit a high degree of variability. Bell et al."' have also shown variations across the taxa for most physiological and growth parameters in response to air pollution. Evolution of SO; resistance in grassland species in industrialized regions of the U.K. has been repotted." The ecological and biological issues of the genetic variations are physiological mechanisms of resistance, speed and frequency with which adaptation rises, and the role of microevolution. Since plants follow dissimilar mechanisms of resistance to air pollutants and consequently have their own fitness costs, a shift in competitive interactions may favor the species with minimum cost of resistance. When the cost of resistance is very high and the rate of selection against resistant individuals is low, species replacement takes place.'"

Dunn" had suggested long ago that ambient levels of 0, in the Los Angeles basin were high enough to drive the selection of resistant genotypes of Lapinus bicolor. Recent studies on Populus sremuloides. Trifolium repens, and Plamago major have provided variations in genetic diversity among populations due to evolution of resistance to ambient 0, under field conditions.'° Inherent resistance of populations to 0, is correlated with the 0, level at the source of the plant material. Direct involvement of 0, in this genetic diversity other than any environmental factors has been proved as the percentage of resistant individuals increased in populations exposed to elevated levels of 0, over long periods of time.""'

Genetic diversity has profound implications on community stricture. Reduction in population sire and restriction of exchange of alleles among populations due to air pollution effects will increase the chances of spatial isolation of the species. ft has been hypothesized that the evolution of resistance due to air pollution contributes to the loss of genetic diversity within plant species."" Development of resistant populations of glasses and herbs growing in the vicinity of coal-fired plants has been reported." A reduction in genetic variability of populations and the evolution of resistance and adaptation to gaseous air pollutants have been observed by Ayzaloo and Bell" and Coleman and Mooney." Genetic diversity can be affected at the community level by mortality or reduced growth and reproduction of sensitive species which are subsequently replaced by more-resistant species. Parsons and Pitelka" were of the view that significant losses of genetic diversity may only occur due to population extinction or genetic drift at restricted areas experiencing intense pollution episodes"" The chronic level of pollution in recent years should not actually be a cause of great concern for genetic diversity.

Community Divinity

Plant community structure depends on various environmental factors. Air quality is an important factor which has changed substantially due to rapid industrialization and urbanization in the recent past. Air pollutants have not been normal constituents of the environment in which plants have evolved over the centuries. Thus. Nature has not provided a continuous selection pressure to isolate genes for resistance. However, genetic modification and recombination may provide morphological and physiological systems that exclude and tolerate excesses of the pollutants!"

Methodological Approach for Studying Community Response

There arc a variety of research approaches used for studying the community response to air pollution. Among various approaches. controlled environment growth chamber. Greenhouse, and open-top chamber (OTC) studies have limitations for studying the community-level diversity response to air pollution." Duchelle et al." have used OTC to investigate the response of small samples of herbaceous communities to 0,. Open-air fumigation though an expensive methodology provided direct measures of community response on a long-term basis."'" Annentano and Menges" used statistical approaches to relate symptoms to influencing factors under native conditions. Mathematical models to evaluate the long-term behavior of communities subjected to chronic air pollution were also tried." The success of models depends upon the state of the art of pollution effect research based on fumigation studies!' Quasi-experimental approach depends upon the gradient of pollutant concentration in an area under natural conditions with other factors more or less uniform." This approach can be successfully used in the field, where the response to air pollutants has clear-cut spatial and temporal variations and influencing factors are known through previous studies." Long-term studies are required to elucidate the subtle effects of air pollution and its by-products at statistically acceptable levels" Field studies of natural communities around pollution stress regimes are the most-used approach for studying changes in community diversity in response to air pollution.

Can Studios

The differential susceptibility of plant specks to air pollutants may be reflected in the altered pattern of species composition and structure at the community level. Smith" classified the relationship between air pollutants and communities into three categories. Class I depicts the relationship where the vegetation acts only as a sink for the pollutants. Class II relationships occur when the pollutants have subtle deleterious effects on the plant community. Class Ill relationships represent acute morbidity or mortality of the members of the plant community. Murray" documented information for class II relationships, where changes in community structure were studied in an area around an aluminum smelter in New South Wales. Australia from immediate proximity of the source in a sequence of increasing distance. It was reported that specks susceptible to air pollution stress are removed and others which are more tolerant become prevalent. It may be suggested that variation in air quality taxes the adaptive ability of organisms with the result that only those species which have already adapted to these conditions or those which can acclimatize to the new conditions may participate in community formation.

The impact of air pollution on plant diversity in a community may be viewed in terms of changes in ecological function and/or a loss of ecological integrity. Diversity is principally a mechanism that generates community stability. The more stable the community. The greater the species diversity. Air pollutants diminish the stability by increasing the mortality of the mast-sensitive species. In a study to assess the effect of coal-fired power plant emission on the vegetation of a mixed oak forest with white pine. Rosenberg et al" found that the important value index (IV!) of certain species and species diversity were inversely related to the distance from the emission source. An obvious community simplification around an iron smelter near Wawa. Ontario has been reported.' Wood and Nash" and Freedman and Hutchinson"' have reported low-diversity communities with low standing crop productivity and nutrient status with a high proportion of weedy herbaceous species with high tolerance to stress around metal smelters.

A decline in the abundance of lichens has been suggested to be an early response to the unfavorable effect atmospheric pollutants on an ecosystem?"" Das" observed that along the Red Road of Calcutta. India with a traffic load of about 50 to 70 and sometimes 100 motor vehicles per minute. No lichen on tree trunks was found, whereas along Rallygunge circular mad with 5 to 10 motor vehicles per minute. some foliose lichens were present. The diversity and abundance of lichens were found to be reduced in the presence of atmospheric S and N compounds."

Photochemical oxidants are known to cause species-specific damage in agro ecosystems and monoculture." Diversity and functioning of microbial populations in the soil have been found to be unfavorably altered as a feedback mechanism to atmospheric depositions."'" Microbial diversity is known to be sensitive to acid alkaline depositions"" and heavy metals."'

In a detailed study conducted around two coal tired power plants situated in Sonhhadra District of Uttar Pradesh. India to analyze the impact of emission on structure of herbaceous communities, the number of individuals and species richness were found to increase with decreasing pollution load, The species richness values ranged from 2.65 at a heavily polluted site to 9.31 at the less-polluted site during the rainy season.

CHAPTER 1

EFFECTS OF ENVIRONMENTAL POLLUTANTS ON ANTIOXIDANT SYSTEMS IN PLANTS AT DIFFERENT GROWTH STAGES

JiazhiZhang, Xingyi Li, Li Zhou, LihongWang, Qing Zhou & Xiaohua Huang

INTRODUCTION

Bisphenol A [BPA; 2,2-bis (4-hydroxyphenyl)] is a chemical intermediate for synthesizing polycarbonate and epoxy[1]. It is widely applied in the production of daily necessities, such as baby bottles, food packaging, thermal paper, electronic equipment and medical facilities[2,3]. Recently, it was reported that the global annual output of BPA is up to 6.8 million tons[4]. The production and consumption of BPA are fairly stable in Europe, but in Asia, especially China, they are increasing gradually[3]. Due to the massive use and emissions of BPA products, BPA is widespread in the global environment. It was reported that BPA concentrations in surface water in the Netherlands and Germany are up to $21\,\mu g \cdot L^{-1}$[1] and $0.41\,\mu g \cdot L^{-1}$[5], respectively. In Japan, the concentration of BPA in garbage leachate is $17.2\,mg \cdot L^{-1}$[6]. Toxicological studies on animals showed that BPA has estrogen activity that can disrupt physiological functions of animals and humans, especially in the reproductive and endocrine systems[7,8,9].

Plants are the primary producers in ecological system, and can provide organic materials and energy for secondary consumers. Through food chains in ecological systems, the hazards of BPA can extend to animals, even humans[10]. The European Union has released a risk assessment report of BPA on terrestrial ecosystems. The draft of the report requires researchers to focus on BPA toxicity on plants to further investigate the potential risks of BPA on plants[11]. Recently, the effects of BPA on algae have been frequently reported[12,13], but the effects of BPA on terrestrial plants are less studied[14], and the microscopic mechanism of BPA on plants remains unclear[15]. Previous studies showed that certain doses of BPA exposure could promote or inhibit the growth[15,16], germination[17], pollen tube elongation[18], photosynthesis[19], and hormone contents[20] in plants. As a vital organ of plants in long-term adaptations to land, roots can absorb water and nutrients from soil for plant growth and development[21], and roots also have the ability to absorb, fix and transfer pollutants. Compared to the aerial parts of plants, roots are directly exposed to BPA in the soil; thus, it is important to study the microscopic mechanisms of BPA on plant roots. Previous studies showed that $1.5\,mg \cdot L^{-1}$ BPA could promote glutamine synthetase/glutamate synthetase cycles and glutamate dehydrogenase pathways in soybean roots for the synthesis of amino acids and proteins, thereby promoting root growth[22]. However, 17.2

Citation: Zhang, J. et al. effects of environmental pollutants on antioxidant systems in plants at different growth stages; doi: 10.1038/srep23782 (2016).

and 50.0 mg·L^{-1} BPA inhibit glutamine synthetase/glutamate synthetase cycles and promote glutamate dehydrogenase pathways, leading to the production of excessive amino acids and proteins, thus inhibiting root growth22.

Reactive oxygen species (ROS) are a type of oxidants, which include superoxide anion (O_2^-), hydrogen peroxide (H_2O_2) and hydroxyl radical (·OH). They have active chemical properties and strong oxidizability23·24. ROS are produced in aerobic metabolism in plants24. As signaling molecules, ROS can regulate metabolism in plant roots, thus affecting plant growth and development25. Plants have formed complete antioxidant systems (antioxidant enzymes and non-enzymatic substances) during long-term evolution processes to regulate ROS levels and to resist adversity stress. The antioxidant enzymes mainly include superoxide dismutase (SOD), peroxidase (POD) and catalase (CAT)26, and antioxidant non-enzymatic substances mainly include glutathione (GSH), proline (Pro), and ascorbic acid (AsA)27. Therefore, studying the effects of BPA on ROS levels and antioxidant systems in plant roots is important for understanding root function, growth, and even the entire life status of plants.

Plant life cycles consist of different growth stages. The effects of the combined stress of soil acidification and lead ion on the antioxidant systems in soybean roots varies at different growth stages28. Unfortunately, the indicator systems used in previous studies on the environmental toxicology of BPA was relatively simple, or previous studies focused mainly on single growth stages in the plant life cycle15·16·22·28. Therefore, previous studies could not comprehensively, systematically, and accurately reflect plant responses to BPA exposure, thus the results could mislead scientists working to objectively evaluate the ecological risk of BPA. In this study, the effects of BPA on ROS levels, membrane lipid fatty acids, membrane lipid peroxidation, and antioxidant enzyme activities and non-enzymatic substance contents in soybean roots at different growth stages (seedling stage, flowering and podding stage, seed-filling stage) were investigated to evaluate whether there are differences among different growth stages. The results will provide new information to evaluate the microscopic mechanism of BPA on plants at all growth stages.

RESULTS

Effects of BPA on ROS levels and Membrane Lipid Peroxidation in Soybean Roots at Different Growth Stages

Figure 1 shows the effects of BPA on ROS levels and membrane lipid peroxidation in soybean roots at different growth stages. Compared to the control, after 7 d of 1.5 mg·L^{-1} BPA exposure, the H_2O_2 content at the seedling stage increased ($p<0.05$), but the O_2^- and malondialdehyde (MDA) contents and membrane permeability did not change ($p>0.05$), and those indices at both flowering and podding stages and seed-filling stages did not change ($p>0.05$). With an increasing BPA dose (6.0 and 12.0 mg·L^{-1}), the O_2^- content, MDA content, and membrane permeability significantly increased ($p<0.05$) at the three growth stages. Moreover, the increase degree of the O_2^- content and membrane permeability were positively related to the BPA dose, but negatively related to the growth stage. Moreover, the O_2^- content, MDA content, and membrane permeability at the three growth stages did not change ($p>0.05$) after withdrawal of 1.5 mg·L^{-1} BPA exposure, but increased ($p<0.05$) after withdrawal of 6.0 and 12.0 mg·L^{-1} BPA exposure.

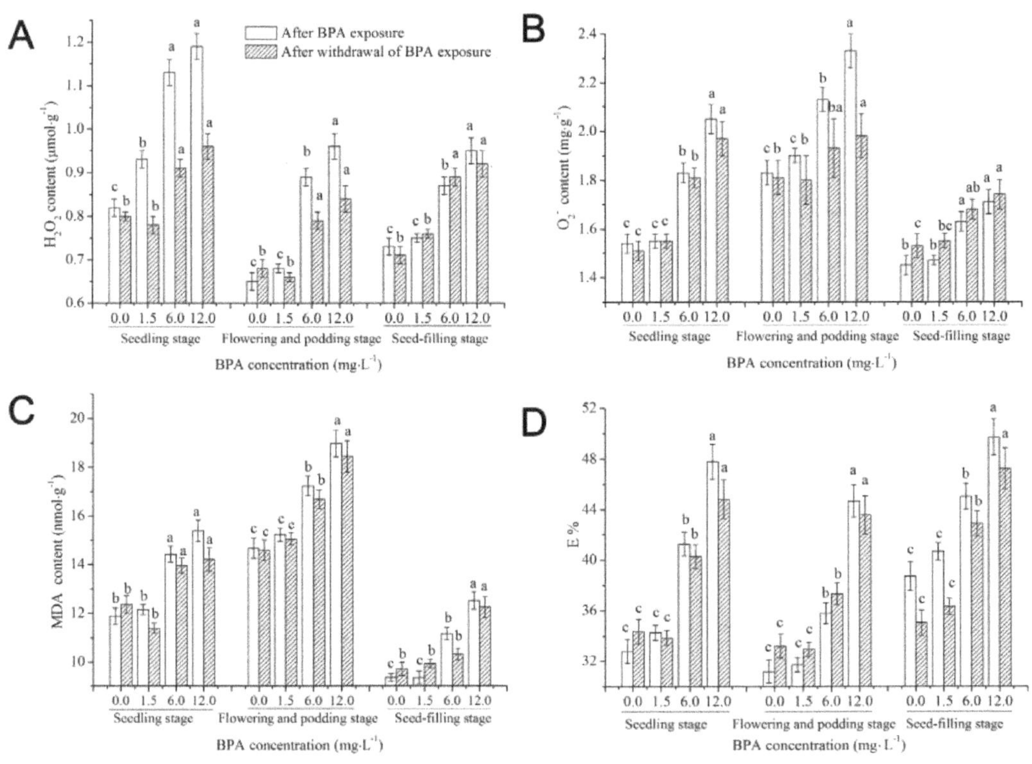

Figure 1: Effects of BPA on the contents of hydrogen peroxide (H_2O_2), superoxide anion (O_2^-), malondialdehyde (MDA) and membrane permeability (E%) in soybean roots at different growth stages following BPA exposure and withdrawal of BPA exposure. Significant differences at $p<0.05$ are denoted with different letters at different growth stages.

Effects of BPA on the Composition of Membrane Lipid Fatty Acids in Soybean Roots at Different Growth Stages

Tables 1, 2, and 3 show the effects of BPA on the composition of membrane lipid fatty acids in soybean roots at different growth stages. Compared to the control, after 7 d of 1.5 mg·L^{-1} BPA exposure, the composition of fatty acids at all growth stages did not change ($p>0.05$), but the percentage content of different fatty acids changed ($p<0.05$). The total contents of saturated fatty acids (SFA) increased ($p<0.05$), whereas the total contents of unsaturated fatty acids (UFA) and the index of unsaturated fatty acids (IUFA) decreased ($p<0.05$). After 7 d of 6.0 mg·L^{-1} BPA exposure, the SFA contents at the seedling stage as well as at the flowering and podding stage increased ($p<0.05$), whereas the UFA content and IUFA decreased ($p<0.05$). Moreover, only the IUFA at the seed-filling stage decreased ($p<0.05$). After 12.0 mg·L^{-1} BPA exposure at different growth stages, the changes in all indices were similar to those after 6.0 mg·L^{-1} BPA exposure, and the change degrees were positively related to the BPA dose but negatively related to the growth stage. After withdrawal of BPA exposure, the SFA contents increased ($p<0.05$) at the seedling stage, while the UFA content and IUFA decreased ($p<0.05$); the SFA and UFA content as well as the IUFA at the flowering and podding stages did not change obviously ($p>0.05$). Interestingly, the SFA content at the seed-filling stage decreased ($p<0.05$).

Table 1: Effects of BPA on the contents of saturated fatty acids (SFA), containing myristic acid (14:0), palmitic acid (16:0), stearic acid (18:0) and arachidic acid (20:0), the contents of unsaturated fatty acids (UFA), containing palmitoleic acid (16:1), oleic acid (18:1), linoleic acid (18:2) and linolenic acid (18:3) and the index of unsaturated fatty acids (IUFA) in the membrane lipid of soybean roots at the seedling stage following BPA exposure and withdrawal of BPA exposure. Values are means± standard deviation (n= 6). Values in parentheses are the percentage of treatments in control. Significant differences at p< 0.05 are shown with different letters in the same column following BPA exposure and withdrawal of BPA exposure.

	Composition of fatty acids										
BPA (mg·L^{-1})	14:00	16:00	16:01	18:00	18:01	18:02	18:03	20:00	SFA	UFA	IUFA
After BPA exposure											
0.0	0.58 ± 0.01c (100.00)	25.79 ± 0.42d (100.00)	0.63 ± 0.01b (100.00)	10.31 ± 0.17d (100.00)	2.71 ± 0.05a (100.00)	24.79 ± 0.40a (100.00)	33.01 ± 0.54a (100.00)	2.19 ± 0.04bc (100.00)	38.86 ± 0.63c (100.00)	61.14 ± 1.00a (100.00)	151.95 ± 2.47a (100.00)
1.5	0.80 ± 0.03b (137.93)	28.39 ± 0.98c (110.08)	0.62 ± 0.03b (98.41)	18.06 ± 0.62b (175.17)	2.19 ± 0.08b (80.81)	21.12 ± 0.73b (85.20)	26.12 ± 0.91b (79.13)	2.69 ± 0.09a (122.83)	49.95 ± 1.730b (128.54)	50.05 ± 1.73b (81.86)	123.42 ± 4.28b (81.22)
6.0	0.96 ± 0.02a (165.52)	31.78 ± 0.73b (123.23)	0.82 ± 0.04a (130.16)	15.69 ± 0.36c (152.18)	2.02 ± 0.06b (74.54)	22.26 ± 0.51b (89.79)	24.42 ± 0.57bc (73.98)	2.06 ± 0.05c (94.06)	50.49 ± 1.17b (129.93)	49.51 ± 1.14b (80.98)	120.60 ± 2.78b (79.37)
12.0	0.90 ± 0.03a (155.17)	35.94 ± 1.04a (139.36)	0.80 ± 0.02a (126.98)	23.33 ± 0.68a (226.29)	2.04 ± 0.05b (75.28)	15.42 ± 0.44c (62.20)	19.29 ± 0.55c (58.44)	2.28 ± 0.06b (104.11)	62.44 ± 1.80a (160.68)	37.56 ± 1.09c (61.43)	119.25 ± 3.44b (78.48)
After withdrawal of BPA exposure											
0.0	0.77 ± 0.02c (100.00)	27.95 ± 0.81c (100.00)	0.71 ± 0.02c (100.00)	10.86 ± 0.31b (100.00)	2.72 ± 0.08c (100.00)	25.18 ± 0.73a (100.00)	29.19 ± 0.84a (100.00)	2.61 ± 0.08b (100.00)	42.19 ± 1.22c (100.00)	57.81 ± 1.67a (100.00)	141.37 ± 4.08a (100.00)
1.5	0.83 ± 0.01bc (107.79)	31.86 ± 0.55b (113.99)	0.74 ± 0.01c (104.23)	11.48 ± 0.20b (105.71)	2.62 ± 0.05c (96.32)	22.54 ± 0.39b (89.52)	27.32 ± 0.47a (93.59)	2.61 ± 0.05b (100.00)	46.78 ± 0.81b (110.88)	53.22 ± 0.92b (92.06)	130.40 ± 2.26b (92.24)
6.0	0.91 ± 0.02b (118.18)	38.18 ± 0.88a (136.60)	1.00 ± 0.02b (140.85)	12.94 ± 0.30a (119.15)	3.39 ± 0.08b (124.63)	19.92 ± 0.46c (79.11)	20.01 ± 0.46b (68.55)	3.65 ± 0.10a (139.85)	55.69 ± 1.29a (132.00)	44.31 ± 1.02c (76.65)	104.25 ± 2.41c (73.74)
12.0	1.48 ± 0.04a (192.21)	37.10 ± 1.07a (132.74)	2.19 ± 0.06a (308.45)	13.62 ± 0.39a (125.41)	4.90 ± 0.14a (180.15)	17.20 ± 0.50d (68.31)	19.99 ± 0.58b (68.48)	3.52 ± 0.09a (134.87)	55.71 ± 1.61a (132.05)	44.29 ± 1.28c (76.61)	101.46 ± 2.93c (71.77)

Table 2: Effects of BPA on the contents of saturated fatty acids (SFA), containing myristic acid (14:0), palmitic acid (16:0), stearic acid (18:0) and arachidic acid (20:0), the contents of unsaturated fatty acids (UFA), containing palmitoleic acid (16:1), oleic acid (18:1), linoleic acid (18:2) and linolenic acid (18:3) and the index of unsaturated fatty acids (IUFA) in the membrane lipid of soybean roots at the flowering and podding stage following BPA exposure and withdrawal of BPA exposure. Values are means± standard deviation (n= 6). Values in parentheses are the percentage of treatments in control. Significant differences at p< 0.05 are shown with different letters in the same column following BPA exposure and withdrawal of BPA exposure.

	Composition of fatty acids										
BPA (mg·L^{-1})	14:00	16:00	16:01	18:00	18:01	18:02	18:03	20:00	SFA	UFA	IUFA
After BPA exposure											
0.0	0.68 ± 0.01b (100.00)	29.84 ± 0.49b (100.00)	1.07 ± 0.02a (100.00)	12.29 ± 0.20b (100.00)	4.96 ± 0.08a (100.00)	19.11 ± 0.31a (100.00)	29.75 ± 0.49a (100.00)	2.32 ± 0.04ab (100.00)	45.13 ± 0.74b (100.00)	54.87 ± 0.89ab (100.00)	133.47 ± 2.17ab (100.00)
1.5	0.60 ± 0.02c (88.24)	30.15 ± 1.05b (101.04)	0.92 ± 0.03b (85.98)	11.63 ± 0.40b (94.63)	3.36 ± 1.12b (67.74)	19.54 ± 0.68a (102.25)	31.69 ± 1.10a (106.52)	2.11 ± 0.08b (90.95)	44.49 ± 1.54b (98.58)	55.51 ± 1.92a (101.17)	138.43 ± 4.80a (103.72)
6.0	0.66 ± 0.02bc (97.06)	34.35 ± 0.79a (115.11)	0.82 ± 0.02c (76.64)	11.81 ± 0.27b (96.09)	3.43 ± 0.08b (69.15)	20.00 ± 0.46a (104.66)	26.47 ± 0.61b (88.97)	2.46 ± 0.06a (106.03)	49.28 ± 1.12a (109.20)	50.72 ± 1.17bc (92.44)	123.66 ± 2.86bc (92.65)
12.0	0.75 ± 0.02a (110.29)	33.52 ± 0.97a (112.33)	0.66 ± 0.02d (61.68)	14.15 ± 0.41a (115.13)	3.20 ± 0.09b (64.52)	20.50 ± 0.59a (107.27)	24.83 ± 0.72b (83.46)	2.38 ± 0.07a (102.59)	50.80 ± 1.47a (112.56)	49.20 ± 1.42c (89.67)	119.36 ± 3.45c (89.43)
After withdrawal of BPA exposure											
0.0	1.26 ± 0.03b (100.00)	34.61 ± 1.00ab (100.00)	1.48 ± 0.04a (100.00)	11.26 ± 0.32b (100.00)	5.69 ± 0.16b (100.00)	19.15 ± 0.55b (100.00)	23.17 ± 0.67a (100.00)	3.38 ± 0.10a (100.00)	50.51 ± 1.46ab (100.00)	49.49 ± 1.43ab (100.00)	114.98 ± 3.32a (100.00)
1.5	1.62 ± 0.03a (128.57)	36.61 ± 0.64a (105.78)	1.31 ± 0.02b (88.51)	11.67 ± 0.20b (103.64)	4.22 ± 0.08c (74.17)	19.80 ± 0.34ab (103.39)	22.21 ± 0.39a (95.86)	2.55 ± 0.05b (75.44)	48.46 ± 0.84b (95.94)	51.54 ± 0.89a (104.14)	119.84 ± 2.08a (104.23)
6.0	1.18 ± 0.03b (93.65)	37.13 ± 0.86a (107.28)	1.49 ± 0.03a (100.68)	12.68 ± 0.29a (112.61)	6.35 ± 0.14a (111.60)	18.51 ± 0.43b (96.66)	20.04 ± 0.46b (86.49)	2.62 ± 0.06b (77.51)	53.61 ± 1.24a (106.14)	46.39 ± 1.07b (93.74)	104.97 ± 2.42b (91.29)
12.0	0.87 ± 0.02c (69.05)	33.51 ± 0.97a (96.82)	1.01 ± 0.03c (68.24)	11.61 ± 0.33b (103.11)	5.72 ± 0.17b (100.53)	21.32 ± 0.62a (111.33)	23.49 ± 0.68a (101.38)	2.47 ± 0.07b (73.08)	52.44 ± 1.51ab (103.82)	47.56 ± 1.37b (96.10)	111.80 ± 3.23ab (97.23)

Table 3: Effects of BPA on the contents of saturated fatty acids (SFA), containing myristic acid (14:0), palmitic acid (16:0), stearic acid (18:0) and arachidic acid (20:0), the contents of unsaturated fatty acids (UFA), containing palmitoleic acid (16:1), oleic acid (18:1), linoleic acid (18:2) and linolenic acid (18:3) and the index of unsaturated fatty acids (IUFA) in the membrane lipid of soybean roots at the seed-filling stage following BPA exposure and withdrawal of BPA exposure. Values are means± standard deviation (n= 6). Values in parentheses are the percentage of treatments in control. Significant differences at p< 0.05 are shown with different letters in the same column following BPA exposure and withdrawal of BPA exposure.

BPA ($mg{\cdot}L^{-1}$)	Composition of fatty acids								SFA	UFA	IUFA
	14:00	16:00	16:01	18:00	18:01	18:02	18:03	20:00			
After BPA exposure											
0.0	1.64 ± 0.03c (100.00)	31.58 ± 0.51b (100.00)	2.28 ± 0.04d (100.00)	9.84 ± 0.16c (100.00)	6.91 ± 0.11c (100.00)	25.05 ± 0.41a (100.00)	19.26 ± 0.31a (100.00)	3.44 ± 0.06a (100.00)	46.50 ± 0.75a (100.00)	53.50 ± 0.87a (100.00)	117.08 ± 1.90a (100.00)
1.5	2.20 ± 0.08b (134.15)	34.81 ± 1.21a (110.23)	3.73 ± 0.13c (163.60)	11.96 ± 0.42a (121.54)	4.35 ± 0.15d (62.95)	22.37 ± 0.77b (89.30)	18.24 ± 0.63a (94.70)	2.34 ± 0.08c (68.02)	47.31 ± 1.64a (101.74)	52.69 ± 1.82a (98.49)	107.54 ± 3.72a (91.85)
6.0	2.95 ± 0.07a (179.88)	32.12 ± 0.74ab (101.71)	5.64 ± 0.12b (247.37)	10.41 ± 0.24bc (105.79)	12.48 ± 0.29a (180.61)	22.34 ± 0.51b (89.18)	11.11 ± 0.25b (57.68)	2.95 ± 0.07b (85.76)	48.43 ± 1.12a (104.15)	51.57 ± 1.19a (96.39)	96.13 ± 2.22b (82.11)
12.0	3.10 ± 0.09a (189.02)	33.49 ± 0.96ab (106.05)	4.21 ± 0.13a (184.65)	11.12 ± 0.32ab (113.01)	11.01 ± 0.32b (159.33)	22.35 ± 0.65b (89.22)	11.26 ± 0.32b (58.46)	3.45 ± 0.10a (100.29)	48.72 ± 1.48a (104.77)	48.83 ± 1.41a (91.27)	93.70 ± 2.70b (80.03)
After withdrawal of BPA exposure											
0.0	1.13 ± 0.03bc (100.00)	30.97 ± 0.89a (100.00)	1.43 ± 0.04c (100.00)	9.89 ± 0.28b (100.00)	4.28 ± 0.12a (100.00)	27.03 ± 0.78a (100.00)	22.75 ± 0.66a (100.00)	2.51 ± 0.08c (100.00)	44.51 ± 0.89a (100.00)	55.49 ± 1.60a (100.00)	128.02 ± 3.70a (100.00)
1.5	1.46 ± 0.02a (129.20)	29.62 ± 0.51a (95.64)	1.64 ± 0.03b (114.69)	9.16 ± 0.16c (92.62)	8.40 ± 0.14a (196.26)	25.28 ± 0.44b (93.53)	22.27 ± 0.39a (97.89)	2.17 ± 0.04d (86.45)	42.41 ± 0.85b (95.28)	57.59 ± 1.00a (103.78)	127.41 ± 2.21a (99.52)
6.0	1.11 ± 0.02c (98.23)	31.39 ± 0.73a (101.36)	1.51 ± 0.03bc (105.59)	11.45 ± 0.27a (115.77)	4.86 ± 0.11b (113.55)	24.13 ± 0.56b (89.27)	22.75 ± 0.53a (100.00)	2.81 ± 0.06b (111.95)	46.75 ± 0.94a (105.03)	53.25 ± 1.23a (95.96)	122.87 ± 2.83a (95.98)
12.0	1.21 ± 0.03b (107.08)	29.06 ± 0.84a (93.83)	3.38 ± 0.10a (236.36)	3.38 ± 0.10d (34.18)	3.38 ± 0.10d (78.97)	3.38 ± 0.10c (12.50)	3.38 ± 0.10b (14.86)	3.38 ± 0.10a (134.66)	37.48 ± 0.74c (84.21)	53.52 ± 1.55a (96.45)	123.66 ± 3.57a (96.59)

Effects of BPA on the Activities of Antioxidant Enzymes in Soybean Roots at Different Growth Stages

Figure 2 shows the effects of BPA on the activities of antioxidant enzymes in soybean roots at different growth stages. Compared to the control, after 1.5 $mg{\cdot}L^{-1}$ BPA exposure, the activities of SOD, POD, and CAT at three growth stages did not change significantly ($p>0.05$). After 6.0 $mg{\cdot}L^{-1}$ BPA exposure, the above indices increased ($p<0.05$) at the seedling stage, and the SOD and CAT activities at the flowering and podding stage and the seed-filling stage increased ($p<0.05$). After 12.0 $mg{\cdot}L^{-1}$ BPA exposure, the activities of the three antioxidant enzymes at the three growth stages increased significantly ($p<0.05$). After withdrawal of 1.5 $mg{\cdot}L^{-1}$ BPA exposure, the activities of the antioxidant enzymes at different growth stages did not change ($p>0.05$); however, after withdrawal of 6.0 and 12.0 $mg{\cdot}L^{-1}$ BPA exposures, the activities of antioxidant enzymes at all growth stages increased ($p<0.05$), except for the POD activity at the flowering and podding stage after withdrawal of 6.0 $mg{\cdot}L^{-1}$ BPA exposure.

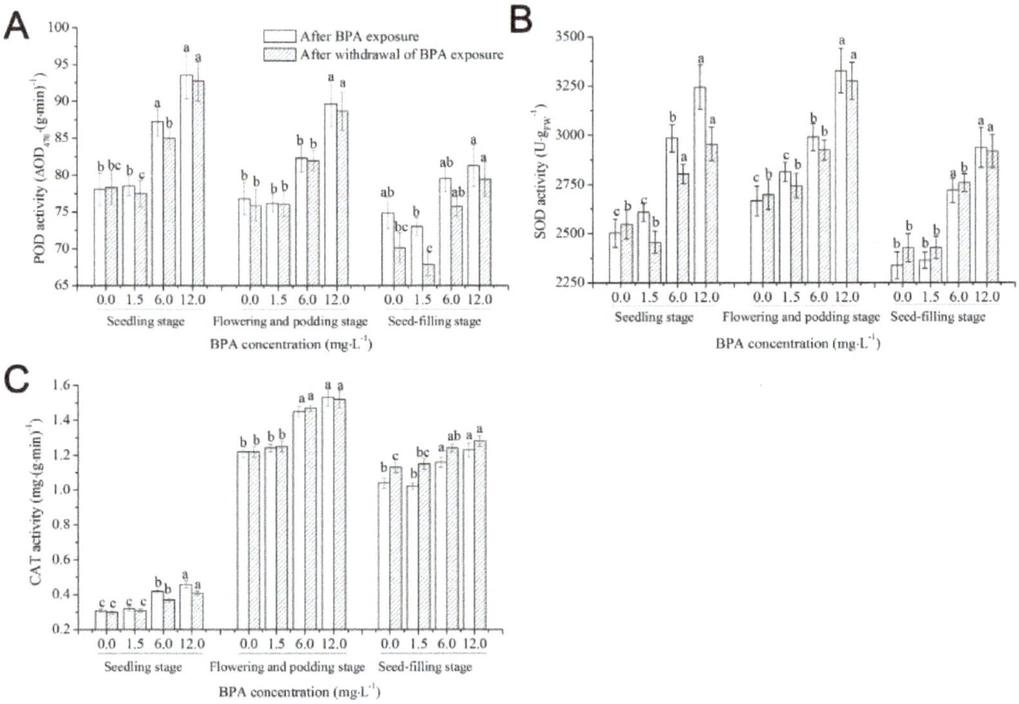

Figure 2: Effects of BPA on the activities of peroxidase (POD), superoxide dismutase (SOD) and catalase (CAT) in soybean roots at different growth stages following BPA exposure and withdrawal of BPA exposure.

Significant differences at $p<0.05$ are denoted with different letters at different growth stages.

Effects of BPA on the Contents of Antioxidant Substances in Soybean Roots at Different Growth Stages

Figure 3 shows the effects of BPA on the contents of antioxidant substances in soybean roots at different growth stages. Compared to the control, after 1.5 mg·L^{-1} BPA exposure, the Pro content at seedling stage increased ($p<0.05$), and the contents of Pro, AsA, and GSH at the flowering and podding stage and the seed-filling stage did not change ($p>0.05$). With an increase in the BPA dose (6.0 and 12.0 mg·L^{-1}), the contents of all antioxidant substances at different growth stages increased ($p<0.05$). After withdrawal of 1.5 mg·L^{-1} BPA exposure, the changes in the content of antioxidant substances at the seedling stage were similar with those after this dose of BPA exposure. The Pro content recovered, but did not recover to the control levels; however, the content of antioxidant substances at the latter two growth stages had no significant changes ($p>0.05$). After withdrawal of 6.0 and 12.0 mg·L^{-1} BPA exposures, the content of the antioxidant substances at each growth stage increased ($p<0.05$).

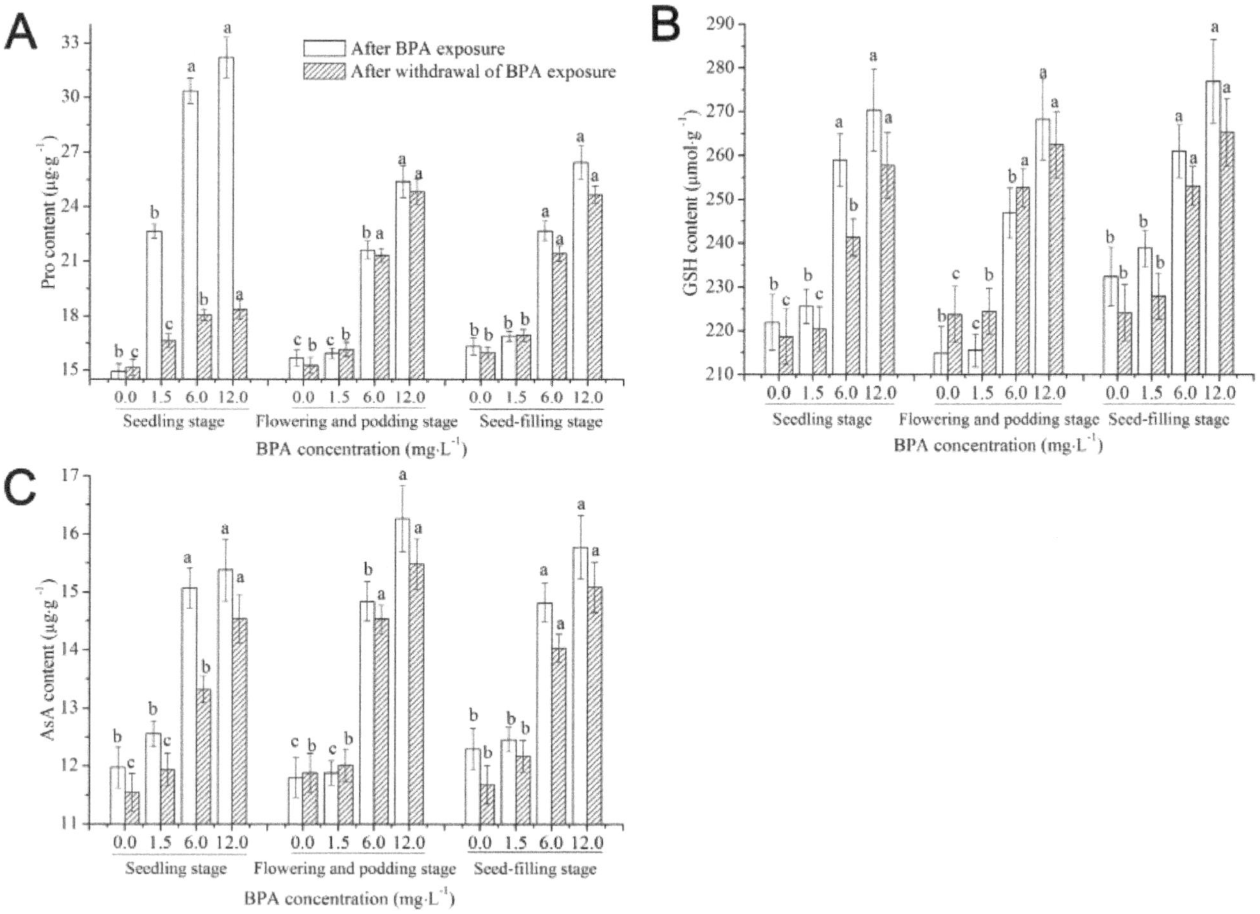

Figure 3: Effects of BPA on the contents of proline (Pro), glutathione (GSH) and ascorbic acid (AsA) in soybean roots at different growth stages following BPA exposure and withdrawal of BPA exposure. Significant differences at $p<0.05$ are denoted with different letters at different growth stages.

Correlation Analysis

Table 4 shows the relationship between antioxidant systems and ROS levels as well as membrane lipid peroxidation in soybean roots at different growth stages. At the seedling stage, after BPA exposure, $H_2O_2$and MDA contents were positively correlated with antioxidant levels and O_2^- content and membrane permeability were positively related to the activities of antioxidant enzymes and GSH content, but the IUFA showed no correlation with antioxidant levels. After withdrawal of BPA exposure, the correlations between ROS levels and antioxidant levels were positive, except for the Pro content; the MDA content was positively correlated with the SOD activity; and the IUFA was negatively correlated with the CAT activity and Pro content. At the flowering and podding stage, after BPA exposure, those correlation relationships were similar to those at the seedling stage, but the IUFA showed a negative correlation with most antioxidant system indexes. After withdrawal of BPA exposure, the correlations were similar to those after BPA exposure. At the seed-filling stage, after BPA exposure, the correlations became more significant than those at the flowering and podding stage, but the IUFA still showed no correlation with antioxidant system levels. After withdrawal of BPA exposure, the correlations were similar to those after BPA exposure, except that the MDA content showed no correlation with antioxidant levels.

Table 4: Relationships between antioxidant systems and ROS levels as well as membrane lipid peroxidation in soybean roots at different growth stages following BPA exposure and withdrawal of BPA exposure.

	SOD activity	POD activity	CAT activity	AsA content	Pro content	GSH content		SOD activity	POD activity	CAT activity	AsA content	Pro content	GSH content
	After BPA exposure							After withdrawal of BPA exposure					
Seedling stage													
H_2O_2 content	0.977*	0.952*	0.976*	0.990*	0.989*	0.979*	H_2O_2 content	0.997**	0.974*	0. 982*	0.969*	0.863	0.983*
O_2^- content	0.992**	1.000**	0.988*	0.948	0.890	0.983*	O_2^- content	0.972*	0.980*	1.000**	0.997**	0.922	0.997**
MDA content	0.995**	0.992**	0.999**	0.999**	0.987**	0.963*	MDA content	0.973*	0.898	0.901	0.873	0.753	0.903
E%	0.996**	0.997**	0.984*	0.948	0.910	0.980*	E%	0.989*	0.995**	0.992**	0.988*	0.865	0.997**
IUFA	−0.728	−0.653	−0.692	−0.759	−0.905	−0.701	IUFA	−0.902	−0.884	−0. 956*	−0.946	−0.982*	−0.936
Flowering and podding stage													
H_2O_2 content	0.939	0.945	0. 999**	0. 995**	0.988*	0.986*	H_2O_2 content	0.930	0.968*	0. 979*	0.992**	0.981*	0.993**
O_2^- content	0.985*	0.978*	0.985*	0.988*	0.994**	0. 993**	O_2^- content	0.934	0.970*	0.987*	0.997**	0.988*	0. 998**
MDA content	0.985*	0.979*	0.986*	0.989*	0.995**	0.994**	MDA content	0.988*	0.994**	0.958*	0.975*	0. 995**	0.972*
E%	0.979*	0.990*	0.919	0.937	0.959*	0.961*	E%	0.995**	0.997**	0.915	0.947	0.972*	0.944
IUFA	−0.856	−0.940	−0.959*	−0.968*	−0.960*	−0.962*	IUFA	−0.433	−0.593	0.740	−0.711	−0.622	−0.723
Seed-filling stage													
H_2O_2 content	0.999**	0.953*	0.987*	0.995**	1.000**	0.999**	H_2O_2 content	0.970*	0.914	0.990**	0.986*	0.973*	0.984*
O_2^- content	0.998**	0.963*	0.989*	0.999**	0.998**	0.996**	O_2^- content	0.997**	0.966*	0. 999**	0. 998**	0.996**	1.000**
MDA content	0.998**	0.965*	0.994**	0.989*	0.998**	0.991**	MDA content	0.886	0.871	0.871	0.892	0.917	0.882
E%	0.988*	0.912	0. 963*	0.976*	0.990**	0.996**	E%	0.995**	0.964*	0.996**	0.999**	1.000**	0.998**
IUFA	−0.924	−0.840	−0.883	−0.938	−0.926	−0.948	IUFA	−0.919	−0.877	−0.938	−0.921	−0.897	−0.929

DISCUSSION

The production and removal of ROS in plants maintain a dynamic balance under normal conditions25. Under adversity, however, a large number of ROS (H_2O_2 and O_2^-) will be produced in plant cells, thus breaking the balance29. Our results showed that low dose of BPA (1.5 mg·L^{-1}) exposure caused higher production of H_2O_2 than removal in soybean roots (Fig. 1) at the seedling stage, and this imbalance was aggravated as BPA exposure dose increased (Fig. 1). After withdrawal of BPA exposure, the imbalance did not return to the control levels (Fig. 1). Excessive ROS can attack the polyunsaturated fatty acid in the cell membrane, causing membrane lipid peroxidation30 and an increase in membrane permeability. Moreover, excessive ROS can be used to generate more toxic ·OH by promoting the Fenton reaction31, affecting plant physiological activity and inhibiting plant growth and development29. MDA is a product of membrane lipid peroxidation; thus, the MDA content is an important index to evaluate the degree of membrane lipid peroxidation32. UFA is the main component of cell membrane lipids in organisms, and it can regulate a variety of abiotic stresses in plants33. Low dose of BPA (1.5 mg·L^{-1}) exposure led to excessive production of ROS in soybean roots at the seedling stage, causing oxidative stress to the UFA (Table 1), rather than membrane lipid peroxidation (Fig. 1). Higher doses of BPA exposure (6.0 and 12.0 mg·L^{-1}) caused excessive accumulation of ROS, leading to the self-catalytic effect of free radical chain reactions aggravating the oxidation of the UFA (Table 1), finally causing decreases in membrane fluidity and membrane lipid peroxidation (Fig. 1). SOD is the first line of defense in plants to disproportionate O_2^- to H_2O_2. POD is an important enzyme in the cell walls or cytosol of plants, and it can catalyze reactions from H_2O_2 to H_2O to achieve detoxification34. CAT is in the peroxisome or mitochondria, and its function is similar to that of POD, though CAT has a low substrate affinity34. AsA is an effective scavenger of free radicals and has antioxidant functions35, and the decrease in its levels indicates a decline in the total antioxidant capacity. As an antioxidant, GSH can directly remove ROS, and at the same time, GSH itself can be oxidized to oxidized glutathione36. Pro regulates the osmosis of cell membranes and coordinates intracellular metabolic processes37. In this work, we found that low dose (1.5 mg·L^{-1}) of BPA exposure caused oxidative stress to soybean roots at the seedling stage (Fig. 1) and the accumulation of Pro in cells to regulate cell membrane permeability and reduce the absorption of BPA by roots37, but this did not cause membrane lipid peroxidation or activate the antioxidant systems (Figs 1, 2, 3). High doses of BPA exposure activated the antioxidant enzyme systems to remove excess ROS (Table 1), while membrane lipid peroxidation was not avoided (Fig. 1 and Table 1). After withdrawal of high doses of BPA exposure, membrane lipid peroxidation did not disappear, indicating that the structure of soybean roots at the seedling stage did not develop well and stress resistance and self-healing abilities were relatively low38; thus, BPA exposure damaged the membrane structure of soybean seedling roots.

To objectively evaluate whether the effects of BPA on antioxidant systems in soybean roots varies at different growth stages, we also studied the effects of BPA exposure on antioxidant systems in roots at the flowering and podding stage and the seed-filling stage. The flowering and podding stage is the important period for soybean to alternately grow and reproduce, as well as a period for forming dry matter. Our results showed that low-dose BPA (1.5 mg·L^{-1}) exposure did not significantly affect ROS levels, membrane lipid peroxidation or the antioxidant systems in soybean roots at the flowering and podding stage (Figs 1, 2, 3 and Table 2). This result indicates that the growth and metabolism of soybean at the flowering and podding stage are relatively strong and able to accumulate more biomass, and the structure and function of roots are robust, which enables resilience to also be relatively strong37. The 6.0 mg·L^{-1} BPA exposure increased ROS levels, and the increased ROS attacked polyunsaturated fatty acids in roots' cell membranes, causing decreases in the UFA content (Table 2). Meanwhile, the activities of antioxidant enzymes and the levels of non-enzymatic substances increased to remove excess ROS (Figs 2 and 3). However, the increased

ROS in soybean root cells was beyond the elimination abilities of antioxidant enzymes and non-enzymatic substances, leading to excessive accumulation of ROS and membrane lipid peroxidation (Fig. 1)30. Moreover, after withdrawal of BPA exposure, membrane lipid peroxidation did not disappear (Fig. 1), indicating that high-dose BPA exposure caused irreversible damage to soybean roots at the flowering and podding stage. At the same time, the antioxidant ability and ROS levels were positively correlated (Table 4), indicating that high doses of BPA exposure did not cause fatal damage to the antioxidant systems of soybean roots.

The seed-filling stage is the key period for the yield formation of soybean39. Previous studies showed that the growth, photosynthesis, nitrogen transfer and assimilation of soybean declined after entering the seed-filling stage40, but soybean roots still showed higher activities at late growth stages41. Our data suggested that the effects of BPA exposure on ROS levels, membrane lipid peroxidation and antioxidant systems in soybean roots at the seed-filling stage were similar to those at the flowering and podding stage (Figs 1, 2, 3 and Table 3), and the H_2O_2 content was higher than the O_2^- content (Fig. 1). We speculate that SOD constantly disproportionated O_2^- to $H_2O_2$42 and that H_2O_2 is relatively stable and with greater longevity. Compared to the three growth stages, with the developing of soybean root structure and function, and the increasing of biomass, BPA stress on unit biomass of soybean roots was decreased37. Previous studies showed that low-dose (1.5 mg·L^{-1}) BPA exposure can promote roots to absorb and make use of microelements (such as Mg and Mn)43, and microelements can be used to synthesize antioxidants to resist ROS oxidative stress44. The resistance of soybean roots to BPA stress at three growth stages followed the order: seed-filling stage>flowering and podding stage>seedling stage. The effects of BPA on antioxidant systems at different growth stages in the entire soybean plant, including leaves, should be an area of future investigation.

In conclusion, the effects of BPA on ROS levels, membrane lipid peroxidation and antioxidant systems in soybean roots were aggravated with increasing BPA exposure dose but weaker with prolonging the growth stages. After withdrawal of BPA exposure, these effects became weaker and the recovery degree was negatively related to BPA exposure dose and positively related to growth stage.

MATERIALS AND METHODS

Preparation of BPA Solution

Given current global BPA pollution situation, especially in developing countries6·45, previous studies about the effects of BPA on plants and animals46·47·48, and high environmental levels of BPA released by pollution accident, three BPA doses (1.5, 6.0, and 12.0 mg·L^{-1}) were selected. Of these concentrations, 1.5 mg·L^{-1} BPA is assigned by the United States Environmental Protection Agency as a safe dose for drinking water and the upper safety limit for individuals49, and this dose is often used to investigate the effects of BPA on plants17·47·49; 6.0, and 12.0 mg·L^{-1} are the BPA concentrations in soil, river sediment, and hazardous landfill leachates50·51·52, and these two doses are also used to study the effects of BPA on plants53. Different doses of BPA solutions were prepared by dissolving suitable amounts of BPA in one-half strength Hoagland solution (pH 7.0).

Plant Culture and BPA Exposure

Soybean (*Glycine max*, Zhonghuang 25) seeds were germinated in accordance with previously reported methods19. Thirty days after germination, when there were both flowers and young raised pods (approximately 2 cm) in the soybean plants, the soybean plants were transplanted into BPA solutions at different doses (0.0, 1.5, 6.0, and 12.0 mg·L^{-1}). The control plants were cultured in one-half strength Hoagland nutrient solution (pH 7.0) without BPA. Each treatment was performed 6 times. The BPA solution and the nutrient solution were renewed every 3 d. The plants were exposed to BPA solution for 7 d and then moved to one-half strength Hoagland solution (pH 7.0) without BPA for 7 d, based on our previous pre-experiment (in which the results showed that the effect of BPA on soybean seedling growth became steady after BPA exposure for 7 d and withdrawal of BPA exposure for another 7 d). After 7 d of BPA exposure, followed by 7 d of BPA withdrawal, the roots were collected for measurements of test indices.

Determination of ROS Content and Membrane Lipid Peroxidation

The H_2O_2 content in soybean roots was determined using a modification of a method described in previous reports54. Fresh roots (0.5 g) were homogenized in 3 mL of 50 mM potassium phosphate buffer (pH 6.5) at 4 °C and the homogenate was centrifuged at 11,500 × *g* for 15 min. The reaction mixture contained 3 mL supernatant and 1 mL of 20% H_2SO_4 (containing 0.1% $TiCl_4$). The absorbance of the reaction mixture was determined at 410 nm, and the H_2O_2 content was calculated by a standard curve.

The O_2^- content was determined according to previous methods37. Roots (2 g) were homogenized in 3 mL of 3% trichloroacetic acid, and then the homogenate was centrifuged at 12,000 × *g* for 15 min. The reaction mixture contained 1 mL supernatant and 1 mL of 50 mM potassium phosphate buffer (pH 7.0, containing 1 mM hydroxylamine hydrochloride). The absorbance of the reaction mixture was recorded at 530 nm, and the O_2^- content was calculated by the standard curve.

The MDA content was determined according to previous reports36. Roots (0.5 g) were collected and homogenized in 3 mL 1% trichloroacetic acid, and the homogenate was centrifuged at 11,500 × *g* for 10 min. The reaction mixture contained 1 mL supernatant and 4 mL of 0.5% thiobarbituric acid, and it was reacted in a boiling water bath for 30 min then quickly cooled in an ice bath. The reaction mixture was centrifuged again at 11,500 × *g* for 15 min. The absorbance of the supernatant was recorded at 450 nm, 532 nm and 600 nm. The concentration of MDA was calculated according to the following equation: Concentration (μmol·L^{-1}) = $6.45 \times (OD_{532} - OD_{600}) - 0.56 \times OD_{450}$, where OD stands for optical density. The MDA content was expressed as nmol of MDA per g of fresh weight.

Membrane permeability was determined in accordance with previous methods, with slight modification55. Root fragments (0.5 g) were rinsed with deionized water and then put in glass tubes, which contained 10 mL deionized water.

The roots were pumped in a vacuum environment for 20 min to translucent. The conductivity of the solution was measured with a conductivity meter (L_1), and then the glass tubes were put in a boiling water bath to heat for 3–5 min. After being cooled to room temperature, the conductivity of the solution was measured again (L_2). Membrane permeability was calculated using the followed formula: membrane permeability (E%) = $(L_1/L_2) \times 100\%$.

Determination of Fatty Acid Composition

The extraction of total membrane lipids in soybean root cells was based on previous reports56. Soybean roots (1.0 g) were baked at 105 °C for 5 min, then the samples were homogenized in a mixture of chloroform and methanol (1:2, V: V) and the homogenate was centrifuged at 1,000 × *g* for 10 min. The mixture of the supernatant and 0.76% NaCl was oscillated for 15 min and then concentrated by nitrogen rotary evaporator to obtain total membrane lipids. Fatty acid composition was determined by gas chromatography57. Methanol (2 mL) and concentrated sulfuric acid (4–5 drops) were added into the total membrane lipid. The mixture was put in a water bath (50–60 °C) for 10 min then added into n-hexane (1–2 mL). After shock, the mixture stood for 15 min, and 2 mL of distilled water were then added. The solvent in the supernatant was evaporated using nitrogen, after which the residues were analyzed using a Shimadzu GC-2010 gas chromatograph to automatically determine the sample. Standard substances of fatty acids were purchased from Sigma Company. The fatty acid content was obtained through comparisons with the peak area of the standard samples using quantitative analysis.

Determination of Antioxidant Enzymes Activity

Fresh soybean roots (0.5 g) were collected and homogenized in a 50 mM phosphate buffer (pH 7.8), which contained 5 mM ascorbic acid, 5 mM dithiothreitol, 5 mM EDTA, and 2% (v/v) polyvinylpyrrolidone. The homogenates were centrifuged at 15,000 × *g* for 15 min, and the supernatant was used to determine the antioxidant enzyme activity.

The SOD activity was determined using a modified method based on previous reports36. The mixture reacted under fluorescent lights for 30 min. One unit of enzyme activity was defined as the quantity of SOD required for 50% inhibition of nitroblue tetrazolium reduction at 560 nm.

The POD activity was determined in accordance with previously reported methods with slight modification58. The reaction mixture (3 mL) contained 1 mL 25 mM phosphate buffer (pH 6.8), 5 µL 28 mM guaiacol and 15 µL enzyme extract. A few drops of H_2O_2 were added to initiate reaction, and one unit of enzyme activity was defined as the change of absorbance at 470 nm within 1 min.

The determination of the CAT activity was based on previous reports59. The reaction mixture (3 mL) contained 100 mM phosphate buffer (pH 7.0) and 50 µL of enzyme extract. The 15 mM H_2O_2 (a few drops) was added into the mixture to initiate reaction. One unit of enzyme activity was defined as the decomposition of H_2O_2 (mg) from a fresh sample (g) at 240 nm within 1 min.

Determination of the Contents of Non-Enzymatic Antioxidant Substance

The AsA content was determined according to previous reports37. Roots (0.5 g) were homogenized in 5% trichloroacetic acid. Then, the homogenate was centrifuged at 4,000 × *g* for 10 min. The reaction mixture contained 0.2 mL supernatant, 0.5 mL of 150 mM phosphate buffer (pH 7.4, containing 5 mM EDTA) and 0.2 mL deionized water. The reaction mixture was incubated at 40 °C for 40 min and the absorbance was recorded at 532 nm. The AsA content was calculated by the standard curve.

The determination of the Pro content was carried out according to previous reports with slight modifications36. Roots (0.5 g) were collected and homogenized in 10 mL of 3% sulfosalicylic acid. Then, the homogenate was centrifuged at 1,000 × *g* for 10 min. The mixture (containing 2 mL supernatant, 2 mL ice acetic acid, and 4 mL 2.5% ninhydrin) underwent a reaction in a 100 °C water bath for 40 min and then was rapidly cooled to terminate the reaction. The absorbance was recorded at 520 nm and the Pro content was calculated by the standard curve.

The GSH content was determined in accordance with previously reported methods36. Roots (0.5 g) were collected and homogenized in 5 mL of 0.1 M HCl (pH 2.0), and the homogenate was centrifuged at 10,000 × *g* at 4 °C for 10 min. The reaction mixture contained 200 µL supernatant, 800 µL of 0.5 M KH_2PO_4/K_2HPO_4 buffer (pH 8.0), a few drops of 1 mM EDTA, and 100 µL of 6 mM 5, 5-dithiobis-(2-nitrobenzoic acid) (Sigma, USA). The mixture was reacted at 30 °C in a water bath for 15 min. The absorbance was determined at 412 nm, and the GSH content was calculated by standard curve.

Statistical Analysis

Each treatment group was set up 6 times, and all data for the three independent experiments of the mean value ± standard error (mean ± SD). The significant differences ($p < 0.05$) between the different treatments were analyzed using an LSD test with SPSS 17.0 software. The symmetric quantitative variables were used to calculate the Pearson's correlation coefficient and to perform correlation analysis among test indices.

REFERENCES

1. Belfroid, A., van Velzen, M., van der Horst, B. & Vethaak, D. Occurrence of bisphenol A in surface water and uptake in fish: evaluation of field measurements. Chemosphere 49, 97–103 (2002).
2. Staples, C. A., Dorn, P. B., Klecka, G. M., O'Block, S. T. & Harris, L. R. A review of the environmental fate, effects, and exposures of bisphenol A. Chemosphere 36, 2149–2173 (1998).
3. Huang, Y. Q. et al. Bisphenol A (BPA) in China: a review of sources, environmental levels, and potential human health impacts. Environ. Int. 42, 91–99 (2012).
4. Jandegian, C. M. et al. Developmental exposure to bisphenol A (BPA) alters sexual differentiation inpainted turtles (Chrysemys picta). Gen. Comp. Endocrinol. 216, 77–85 (2015).

5. Fromme, H. et al. Occurrence of phthalates and bisphenol A and F in the environment. Water Res. 36, 1429–1438 (2002).
6. Yamamoto, T., Yasuhara, A., Shiraishi, H. & Nakasugi, O. Bisphenol A in hazardous waste landfill leachates. Chemosphere 42, 415–418 (2001).
7. Crain, D. A. et al. An ecological assessment of bisphenol-A: evidence from comparative biology. Reprod. Toxicol. 24, 225–239 (2007).
8. Liu, Y. et al. Acute toxicity of nonylphenols and bisphenol A to the embryonic development of the abalone Haliotis diversicolor supertexta. Ecotoxicology 20, 1233–1245 (2011).
9. Rhee, J. et al. Bisphenol A modulates expression of sex differentiation genes in the self-fertilizing fish, Kryptolebias marmoratus. Aquat. Toxicol. 104, 218–229 (2011).
10. Jondeau-Cabaton, A. et al. Characterization of endocrine disruptors from a complex matrix using estrogen receptor affinity columns and high performance liquid chromatography-high resolution mass spectrometry. Environ. Sci. Pollut. R. 20, 2705–2720 (2013).
11. European Commision. Updated European risk assessment report 4,4′ isopropylidenediphenol (bisphenol-A). Brussels (2008). Available at: http://publications.jrc.ec.europa.eu/repository/handle/JRC59980 (Accessed: 4th November 2010).
12. Ji, M. et al. Biodegradation of bisphenol A by the freshwater microalgae Chlamydomonas mexicana and Chlorella vulgaris. Ecol. Eng. 73, 260–269 (2014).
13. Zhang, W. et al. Acute and chronic toxic effects of bisphenol A on Chlorella pyrenoidosa and Scenedesmus obliquus. Environ. Toxicol. 29, 714–722 (2014).
14. Zhang, J. et al. Effects of bisphenol A on chlorophyll fluorescence in five plants. Environ. Sci. Pollut. R. 22, 17724–17732 (2015).
15. Qiu, Z., Wang, L. & Zhou, Q. Effects of bisphenol A on growth, photosynthesis and chlorophyll fluorescence in above-ground organs of soybean seedlings. Chemosphere 90, 1274–1280 (2013).
16. Sun, H., Wang, L. & Zhou, Q. Effects of bisphenol A on growth and nitrogen nutrition of roots of soybean seedlings. Environ. Toxicol. Chem. 32, 174–180 (2013).
17. Dogan, M., Korkunc, M. & Yumrutas, O. Effects of bisphenol A and tetrabromobisphenol A on bread and durum wheat varieties. Ekoloji 21, 114–122 (2012).
18. Speranza, A., Crosti, P., Malerba, M., Stocchi, O. & Scoccianti, V. The environmental endocrine disruptor, bisphenol A, affects germination, elicits stress response and alters steroid hormone production in kiwifruit pollen. Plant Biol. 13, 209–217 (2011).
19. Jiao, L. et al. Effects of bisphenol A on chlorophyll synthesis in soybean seedlings. Environ. Sci. Pollut. R. 22, 5877–5886 (2015).
20. Wang, S. et al. Effects of bisphenol A, an environmental endocrine disruptor, on the endogenous hormones of plants. Environ. Sci. Pollut. R. 22, 17653–17662 (2015).
21. Wang, H., Inukai, Y. & Yamauchi, A. Root development and nutrient uptake. Crit. Rev. Plant Sci. 25, 279–301 (2006).
22. Sun, H., Wang, L. H., Zhou, Q. & Huang, X. H. Effects of bisphenol A on ammonium assimilation in soybean roots. Environ. Sci. Pollut. R. 20, 8484–8490 (2013).
23. Cordeiro, R. M. Molecular dynamics simulations of the transport of reactive oxygen species by mammalian and plant aquaporins. BBA-Gen. Subjects. 1850, 1786–1794 (2015).
24. Karkonen, A. & Kuchitsu, K. Reactive oxygen species in cell wall metabolism and development in plants. Phytochemistry 112, 22–32 (2015).
25. Gao, Y.-B. et al. Low temperature inhibits pollen tube growth by disruption of both tip-localized reactive oxygen species and endocytosis in Pyrus bretschneideri Rehd. Plant Physiol. Bioch. 74, 255–262 (2014).
26. Wang, Z. et al. The effect of exogenous salicylic acid on antioxidant activity, bioactive compounds and antioxidant system in apricot fruit. Sci. Hortic-Amsterdam. 181, 113–120 (2015).
27. Ramakrishna, B. & Rao, S. S. R. Foliar application of brassinosteroids alleviates adverse effects of zinc toxicity in radish (Raphanus sativus L.) plants. Protoplasma 252, 665–677 (2015).
28. Wang, L., Wang, Q., Zhou, Q. & Huang, X. Combined effect of soil acidification and lead ion on antioxidant system in soybean roots. Chem. Ecol. 31, 123–133 (2015).
29. Ahmad, P., Sarwat, M. & Sharma, S. Reactive oxygen species, antioxidants and signaling in plants. J. Plant Biol. 51, 167–173 (2008).
30. Parlak, K. U. & Yilmaz, D. D. Response of antioxidant defences to Zn stress in three duckweed species. Ecotox. Environ. Safe. 85, 52–58 (2012).
31. Smith, B. A., Teel, A. L. & Watts, R. J. Identification of the reactive oxygen species responsible for carbon tetrachloride degradation in modified Fenton's systems. Environ. Sci. Technol. 38, 5465–5469 (2004).
32. Jemec, A., Tisler, T., Erjavec, B. & Pintar, A. Antioxidant responses and whole-organism changes in Daphnia magna acutely and chronically exposed to endocrine disruptor bisphenol A. Ecotox. Environ. Safe. 86, 213–218 (2012).
33. Sui, N. & Han, G. Salt-induced photoinhibition of PSII is alleviated in halophyte Thellungiella halophila by increases of unsaturated fatty acids in membrane lipids. Acta. Physiol. Plant 36, 983–992 (2014).
34. Xu, C., Natarajan, S. & Sullivan, J. H. Impact of solar ultraviolet-B radiation on the antioxidant defense system in soybean lines differing in flavonoid contents. Environ. Exp. Bot. 63, 39–48 (2008).
35. Liu, F. et al. Higher transcription levels in ascorbic acid biosynthetic and recycling genes were associated with higher ascorbic acid accumulation in blueberry. Food Chem. 188, 399–405 (2015).
36. Du, S. et al. Atmospheric application of trace amounts of nitric oxide enhances tolerance to salt stress and improves nutritional quality in spinach (Spinacia oleracea L.).

Food Chem. 173, 905–911 (2015).

37. Wang, Q. et al. Effects of bisphenol A on antioxidant system in soybean seedling roots. Environ. Toxicol. Chem. 34, 1127–1133 (2015).

38. Malekzadeh, P. Influence of exogenous application of glycinebetaine on antioxidative system and growth of salt-stressed soybean

39. seedlings (Glycine max L.). Physiol. Mol. Biol. Plants 21, 225–232 (2015).

40. Kumudini, S. V., Pallikonda, P. K. & Steele, C. Photoperiod and e-genes influence the duration of the reproductive phase in soybean. Crop Sci. 47, 1510–1517 (2007).

41. Kaschuk, G., Hungria, M., Leffelaar, P. A., Giller, K. E. & Kuyper, T. W. Differences in photosynthetic behaviour and leaf senescence of soybean (Glycine max L. Merrill) dependent on N2 fixation or nitrate supply. Plant Biol. 12, 60–69 (2010).

42. Afza, R., Hardarson, G., Zapata, F. & Danso, S. K. A. Effects of delayed soil and foliar N fertilization on yield and N2 fixation of soybean. Plant Soil 97, 361–368 (1987).

43. Ahuja, N., Singh, H. P., Batish, D. R. & Kohli, R. K. Eugenol-inhibited root growth in Avena fatua involves ROS-mediated oxidative damage. Pestic. Biochem. Phys. 118, 64–70 (2015).

44. Nie, L. et al. Effects of bisphenol A on mineral nutrition in soybean seedling roots. Environ. Toxicol. Chem. 34, 133–140 (2015).

45. Sun, Z., Wang, L., Zhou, Q. & Huang, X. Effects and mechanisms of the combined pollution of lanthanum and acid rain on the root phenotype of soybean seedlings. Chemosphere 93, 344–352 (2013).

46. Goodson, A., Summerfield, W. & Cooper, I. Survey of bisphenol A and bisphenol F in canned foods. Food. Addit. Contam. 19, 796–802 (2002).

47. Mandich, A. et al. In vivo exposure of carp to graded concentrations of bisphenol A. Gen. Comp. Endocr. 153, 15–24 (2007).

48. Mihaich, E. M. et al. Acute and chronic toxicity testing of bisphenol A with aquatic invertebrates and plants. Ecotoxicol. Environ. Safe 72, 1392–1399 (2009).

49. Saiyood, S., Inthorn, D., Vangnai, A. S. & Thiravetyan, P. Phytoremediation of bisphenol A and total dissolved solids by the mangrove plant, Bruguiera gymnorhiza. Int. J. Phytoremediat. 15, 427–438 (2013).

50. Geens, T., Goeyens, L. & Covaci, A. Are potential sources for human exposure to bisphenol-A overlooked? Int. J. Hyg. Environ. Health. 214, 339–347 (2011).

51. Yamada, K., Urase, T., Matsuo, T. & Suzuki, N. Constituents of organic pollutants in leachates from different types of landfill sites and their fate in the treatment processes. J. Jpn. Soc. Phys. Water Environ. 22, 40–45 (1999).

52. Yasuhara, A. et al. Determination of organic components in leachates from hazardous waste disposal sites in Japan by gas chromatography–mass spectrometry. J. Chromatogra. A 774, 321–332 (1997).

53. Sun, W. L., Ni, J. R., O'Brien, K. C., Hao, P. P. & Sun, L. Y. Adsorption of bisphenol A on sediments in the Yellow River. Water Air Soil Poll. 167, 353–364 (2005).

54. Li, R. et al. Physiological responses of the alga Cyclotella caspia to bisphenol A exposure. Bot. Mar. 51, 360–369 (2008).

55. Hossain, M. A., Hasanuzzaman, M. & Fujita, M. Up-regulation of antioxidant and glyoxalase systems by exogenous glycinebetaine and proline in mung bean confer tolerance to cadmium stress. Physiol. Mol. Biol. Plants 16, 259–272 (2010).

56. Alpaslan, M. & Gunes, A. Interactive effects of boron and salinity stress on the growth, membrane permeability and mineral composition of tomato and cucumber plants. Plant Soil 236, 123–128 (2001).

57. Bligh, E. G. & Dyer, W. J. A rapid method of total lipid extraction and purification. Can. J. Biochem. Phys. 37, 911–917 (1959).

58. Harwood, J. Gas chromatography and lipids: a practical guide: by W. W. Christie, the oily press, Ayr, Scotland. Phytochemistry 28, 3251–3252 (1989).

59. Saba, M. K., Arzani, K. & Barzegar, M. Postharvest polyamine application alleviates chilling injury and affects apricot storage ability. J. Agr. Food. Chem. 60, 8947–8953 (2012).

60. Shi, S. Y., Wang, G., Wang, Y. D., Zhang, L. G. & Zhang, L. X. Protective effect of nitric oxide against oxidative stress under ultraviolet-B radiation. Nitric. Oxide-Biol. Ch. 13, 1–9 (2005).

CHAPTER 2

CHEMICAL DIVERSITY AND DEFENCE METABOLISM: HOW PLANTS COPE WITH PATHOGENS AND OZONE POLLUTION

Marcello Iriti and Franco Faoro

Keywords: Stress physiology; secondary metabolism; phytochemicals; phytoalexins; phytoanticipins; tropospheric ozone pollution; volatile organic compounds

INTRODUCTION

In their ecosystem, plants have to cope with a plethora of potentially unfavourable conditions. Stress factors affecting plant fitness not only derive from natural sources, such as adverse temperature fluctuations (heating, chilling, freezing), high irradiation (photoinhibition, photooxidation), osmotic imbalance (salinity, drought), hypoxia/anoxia (flooding), mineral (macro- and micronutrient) deficiency, wounding, phytophagy and pathogen attack, but also from anthropogenic activities. The latter include xenobiotics employed in agriculture (pesticides), environmental (air, soil and water) pollutants and increased UV radiation. In any case, regardless of natural or anthropogenic stress factor, plants have to cope with their stressors. From a pathophysiological point of view, a plant may avoid or adapt to a particular stress with a dose-dependent mechanism (Figure 1). Under a certain threshold, a mild stress may be compensated by the plant, whereas, at higher levels, the detrimental effects of a severe stress may cause irreversible damages, according to the stressor dose-stress effect relationship [1]. Furthermore, the stress tolerance threshold depends not only on the type of stressor and exposure time, but also on the plant stress-coping capacity. In this view, the shift between normal and stress metabolism represents a fundamental trait in plant acclimation (short-term) and adaptation (long-term) strategies, although is almost impossible to define exactly the threshold between them [2].

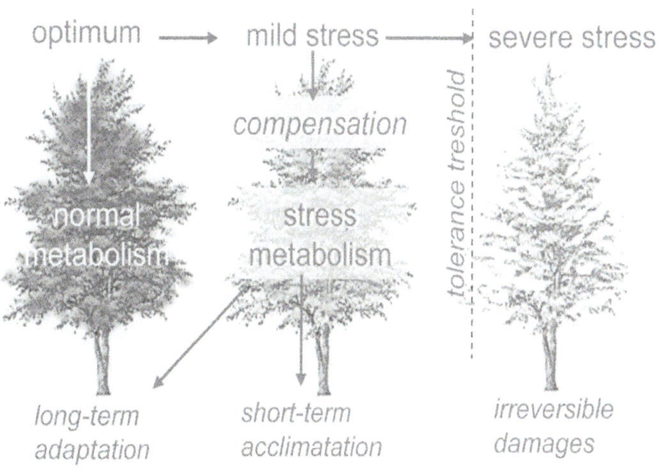

Figure 1. Stressor dose-stress effect relationships in plants.

Citation: Marcello Iriti and Franco Faoro, Chemical Diversity and Defence Metabolism: How Plants Cope with Pathogens and Ozone Pollution, doi:10.3390/ijms10083371.

In this survey, we deal with the secondary metabolites involved in plant resistance against pathogens and tolerance to ozone, a widespread atmospheric pollutant, emphasising on metabolic fluxes between primary and secondary metabolism induced by both biotic and abiotic stresses.

PLANT SECONDARY METABOLISM

In living organisms, secondary metabolites are usually considered not essential for growth and development, unlike the products of primary metabolism. This does not mean they are not important, as they relate plants with the components of their ecosystem, i.e., the physic environment (biotope) and the living community (biocenosis), thus resulting indispensable for the survival of the species. Moreover, it must be pointed out that the widely accepted division of metabolism in "primary" and "secondary" is somewhat unsatisfactory. In fact, some major chemical groups still classified as secondary metabolites, such as lipids, are required by all living organisms, as are some metabolites arising from a secondary pathway (i.e., giberellin A_1 from the isoprenoid pathway) that can have a determinant endogenous regulator role [3] Additionally, stress metabolism could be regarded as a particular expression of secondary metabolism, when stressful conditions, of both biotic and abiotic nature, change the dynamic equilibrium of the ecosystem. As an instance, phytoalexins are compounds synthesized *ex novo* or whose synthesis increase after pathogen challenging, raising their tissue concentration [4].

In plants, chemical diversity has determined their evolutionary success. Because of their sessile habitus, plants are unable to avoid worsening environmental conditions, and they cannot escape the plethora of previously mentioned biotic stresses. Consequently, unlike animals, plants have evolved an enormous number of secondary metabolites to overcome any danger, though most of them are species-specific. This implies that the actual number of secondary metabolites within a given species is much more limited. The functional role of these phytochemicals range from the ecology to defence, improving protection against both biotic and abiotic stresses, besides being involved in ecological roles as attractants or repellents, for pollinators and phytophagy respectively, colours and scents of reproductive organs (flowers and fruits).

Generally, precursors of secondary metabolic pathways are products of the primary metabolism. Therefore, a severe or long-lasting stress factor could induce an excessive shift between primary and secondary metabolism and, consequently, a diversion of essential available resources from growth to defence. To a large extent, secondary metabolites derive from three biosynthetic routes, namely the phenylpropanoid, isoprenoid and alkaloid pathways. Phytochemicals arising from these pathways include not only compounds with a broad-spectrum antibiotic activity, but also powerful antioxidants able to counteract oxidative stress [5–7].

Phenylpropanoid Pathway

Phenylpropanoids are a class of phenylalanine derivatives with a basic C_6–C_3 (phenyl-propane) skeleton (Scheme 1). In turn, the essential amino acid phenylalanine arises from the shikimate pathway, as well as the other aromatic amino acids tyrosine and tryptophan. Precursors of this pathway are phosphoenolpyruvate, from glycolysis, and erythrose 4-phosphate from pentose phosphate pathway, leading to two important intermediates, shikimic and chorismic acid. In further steps, after a branch point, phenylalanine and tyrosine are synthesized from prephenic and arogenic acid, whereas tryptophan from anthranilic acid [8].

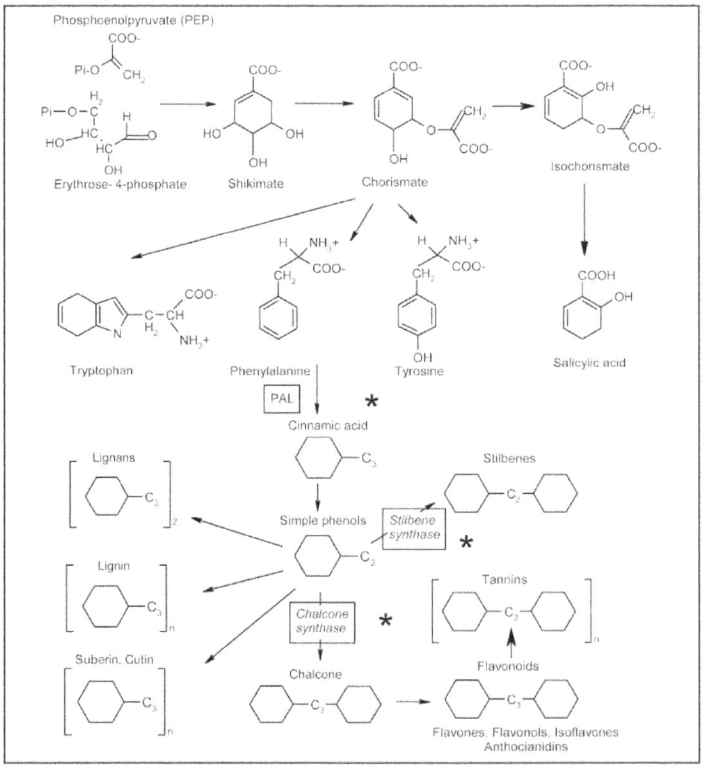

Scheme 1. Aromatic amino acid biosynthesis from shikimate and phenyl-alanine derivatives of the phenylpropanoid pathway. Depending on the species and the conditions (i.e., during pathogen attack) salicylic acid can also be synthesized from chorismate.

The removal of an amino group from phenylalanine (deamination) via the step catalyzed by the enzyme phenylalanine ammonia-lyase (PAL), leads to cinnamic acid and, in turn the precursor of hydroxycinnamates after a series of hydroxylation of the benzene ring (Scheme 2). These compounds, including coumaric, ferulic and sinapic acids are reduced to the corresponding alcohols via aldehyde intermediates, namely coumaryl, coniferyl and sinapyl alcohol, collectively termed monolignols.

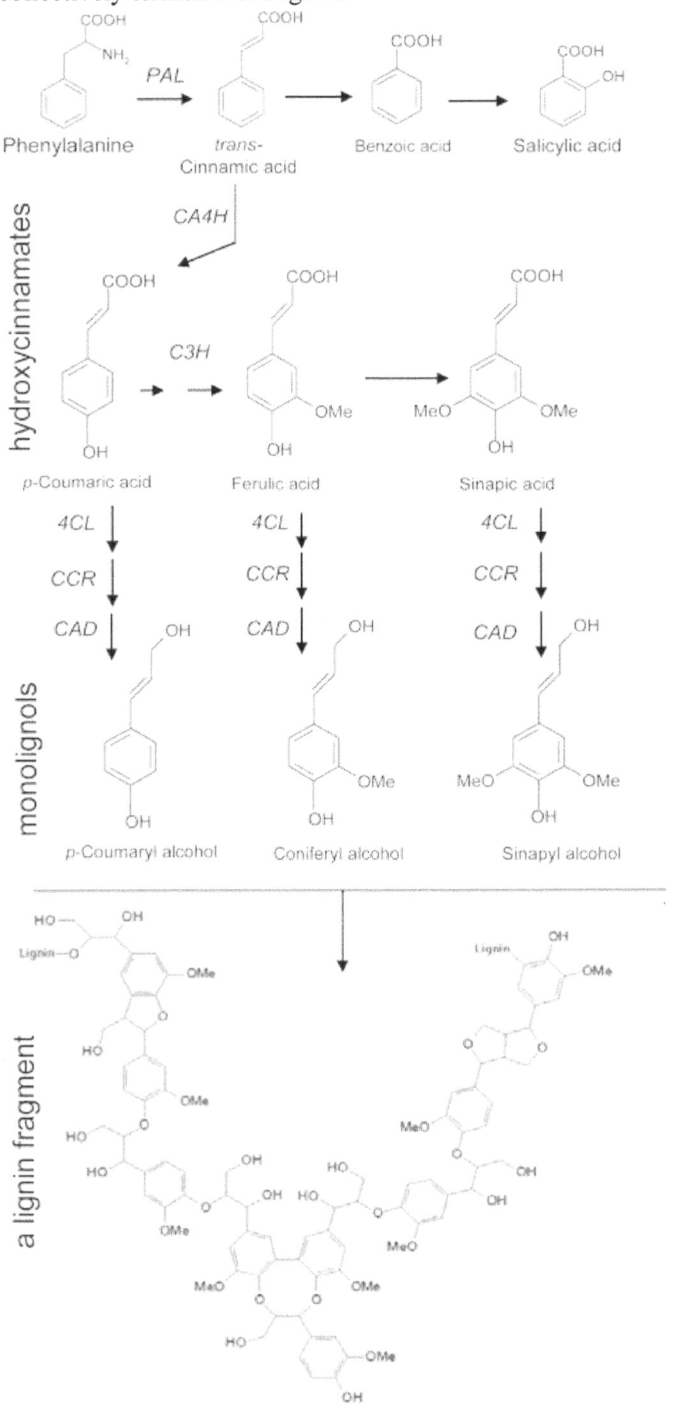

Scheme 2. Main steps of the phenylpropanoid pathway leading to benzoic acids, hydroxycinnamates (coumaric, ferulic and sinapic acids) and lignin, (CA4H, cinnamate-4-hydroxylase; C3H, coumarate-3-hydroxylase; CAD, cinnamyl alcohol dehydro-genase CCR, cinnamoyl: CoA reductase; CL, 4-Coumarate: CoA ligase; PAL, phenylalanine ammonia-lyase.

Dimerization or polymerization of monolignols leads to lignans and lignin, respectively (Scheme 2). Lignification is a complex reaction in which peroxidases catalyze the polymerization of lignin units, consuming H_2O_2. Apposition of lignin in plant cell wall is a process occurring during the development of particular tissues, as well as in plant defence response, in order to strengthen the cell wall and protect the plasmalemma. Benzoic and hydroxybenzoic acids (C_6–C_1), such as salicylic acid (SA), a molecule importantly involved in systemic acquired resistance (SAR), represent another group of cinnamic acid derivatives, formed by cleavage of a C_2 fragment from the phenylpropane structure (Scheme 2) [7,9]. However, it must be pointed out that, depending on the species and the conditions (i.e., during pathogen attack), salicylic acid can also be synthesized directly from chorismate-isochorismate by isochorismate synthase [10].

The metabolites mentioned till now can be collectively named simple phenols, to differentiate them from polyphenolic compounds. The latter have an additional benzene ring, arising from the cyclization of three malonic acid (C_2) residues via stilbene synthase (STS) or chalcone synthase (CHS), to give stilbenes (C_6–C_2–C_6) or flavonoids (C_6–C_3–C_6), respectively (Scheme 1). Finally proanthocyanidins (PA), or condensed tannins, include oligo- and polymeric flavonoid derivatives (from flavanols catechins), with polymerisation degree ranging from 2 to 17 and more (Scheme 1). The numerous hydroxyl groups available in these molecules promote the formation of complexes with macromolecules, such as proteins and polysaccharides or metal ions [7–9].

Isoprenoid Pathway

Isoprenoids, also named terpenoids, represent the chemically and functionally most diversified class of low molecular mass lipids in plants, both primary and secondary metabolites [11]. They include electron carriers (quinones), membrane constituents (sterols), vitamins (A, D, E and K), plant hormones (side chain of cytokinins, abscisic acid, gibberellins and brassinosteroids), photosynthetic pigments (chlorophyll, phytol and carotenoids) and essential oils [12].

Acetyl coenzyme A (CoA) represents the precursor for the isprenoid biosynthesis (Scheme 3). Firstly, two molecules of acetyl CoA react to give acetoacetyl CoA and, then, with a further acetyl CoA to produce β-hydroxy-β-methylglutaryl(HMG)-CoA. In plants, the same enzyme, HMG-CoA synthase, catalyses both reactions. The conversion of HMG-CoA into mevalonate, via HMG-CoA reductase, is the rate limiting enzyme of this pathway [13]. Mevalonate kinase and mevalonate phosphate kinase phosphorylate, respectively, mevalonate and, then, mevalonate 5-phosphate, yielding mevalonate 5-diphosphate. Afterwards, mevalonate diphosphate decarboxylation, via mevalonate diphosphate decarboxylase, produces isopentenyl diphosphate (IPP), the five-carbon building block for the formation of isoprenoid chains. The enzyme IPP:dimethylallyl-PP isomerase converts IPP into dimethylallyl diphosphate (DMAPP), the acceptor for successive transfers of isopentenyl residues [13,14].

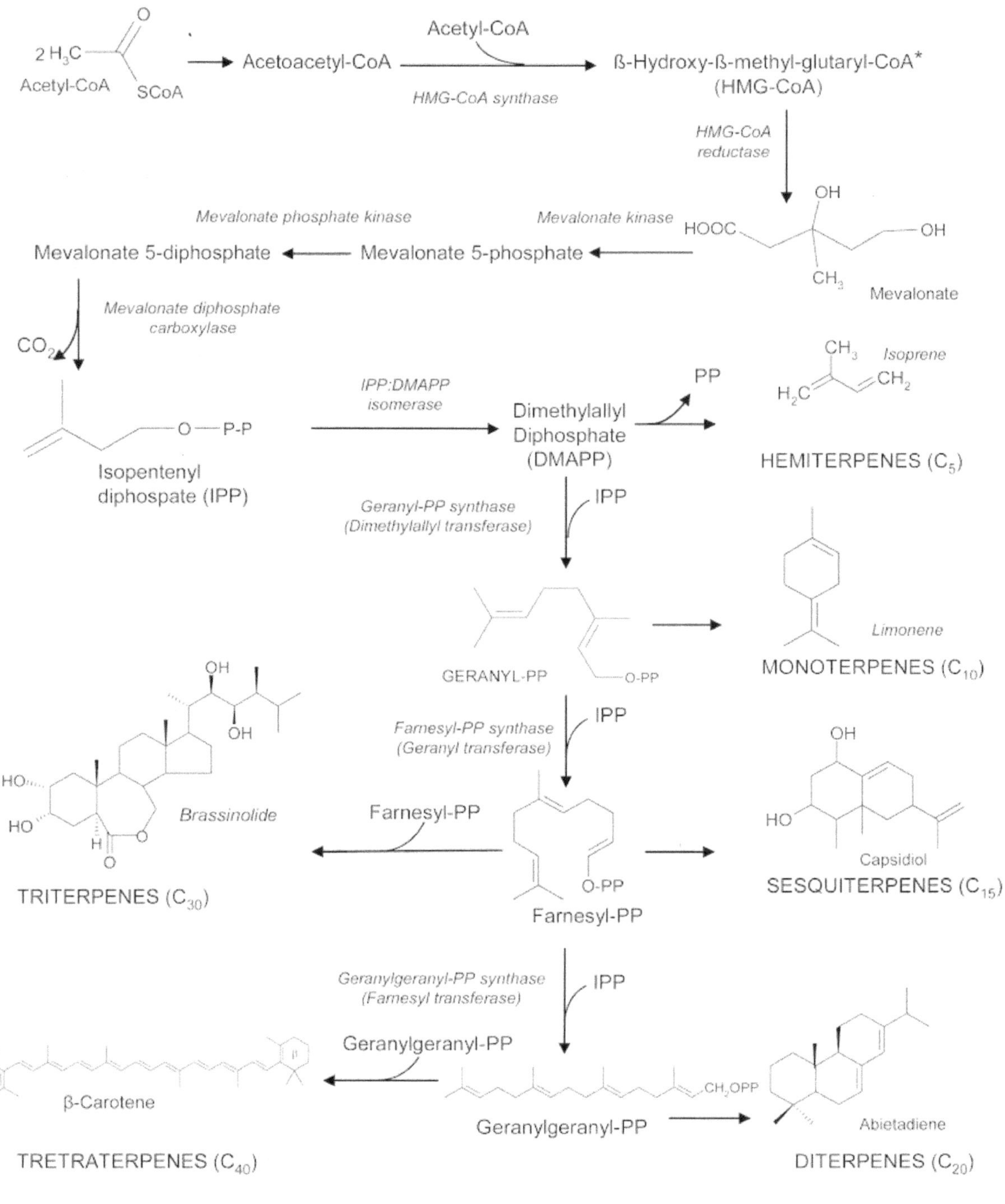

Sckeme 3. Isoprenoid pathway from acetyl-CoA. The enzyme with an asterisk is reported to be influenced by ozone.

Hemiterpenes (C_5), such as isoprene, originate from dimethylallyl-PP, upon the release of diphosphate. Differently, dimethylallyl-PP can condense with IPP, to form geranyl-PP via geranyl-PP synthase. In the same way, further chain elongation is attained by head to tail condensation of geranyl-PP to IPP, to produce farnesyl-PP via farnesyl-PP synthase. Analogously, geranylgeranyl-PP synthase catalyses the head to tail condensation of farnesyl-PP to IPP, thus yielding geranylgeranyl-PP [15]. Geranyl-PP is the precursor for the formation of monoterpenes (C_{10}) or essential oils, including highly volatile open chain and cyclic compounds, such as menthol, limonene, geraniol, linalool and pinene. They are active in plant-microbe, plant-pronubi, plant-phytophagous and plant-plant interactions, due to their attractiveness or repulsiveness [16]. Farnesyl-PP is the precursor for the synthesis of open chain and cyclic sesquiterpenes (C_{15}), the largest group of isoprenoids, including essential oils and antibiotic compounds (phytoalexins) [17].

Diterpenes (C_{20}) derive from geranylgeranyl-PP, consisting of phytoalexins, plant hormones, the phytol side chain of chlorophylls, tocopherols and phylloquinone [18].

Additionally, triterpenes (C_{30}) are synthesized from two molecules of farnesyl-PP (C_{15}), by a reductive head to head condensation. The triterpene squalene is the precursor for sterols, important membrane constituents, via squalene synthase [19]. Analogously, head to head condensation of two molecule of geranylgeranyl-PP (C_{20}) leads to tetratepenes (C_{40}), such as the carotenoids (carotene, lycopene) and xanthophylls (lutein, zeaxanthin, violaxanthin). Besides, isoprenoids are involved in protein prenylation, that is the synthesis of variously lengthened isoprenoid chains anchoring proteins in membranes, such as G proteins, ubiquinone, plastoquinone and cytochrome-a. Finally, natural rubber is a polyterpen, composed of over 1,000 isoprene units and deriving from polymerisation of geranylgeranyl-PP units [20].

Alkaloid Pathways

A common pathway for alkaloid biosynthesis does not exist. A majority of alkaloids are amino acid derivatives grouped on the basis of their precursor and chemical structure. Therefore, the main groups include alkaloids arising from ornithine, leucine, lysine, tyrosine, tryptophan, histidine and phenylalanine, in addition to alkaloids arising from nicotinic (pyridine alkaloids) and anthranilic acid, acetate, isoprenoids and purine (Figure 2). In turn, these classes include some minor divisions. Accordingly, pyrrole alkaloids arise from leucine; pyrrolidine, tropane and pyrrolizidine alkaloids from ornithine; piperidine, quinolizidine and indolizidine alkaloids derive from lysine; catecholamines, isoquinoline, tetrahydroisoquinoline and benzyltetrahydroisoquinoline alkaloids originate from tyrosine; indolamines, indole, carboline, quinoline, pyrrolindole and ergot alkaloids come from tryptophan and imidazole alkaloids from histidine. Anthranilic acid is the precursor of quinazoline, quinoline and acridine alkaloids, whereas isoprenoid alkaloids include mono- (geraniol), di- (geranylgeranyl-PP) and triterpene (cholesterol) derivatives (Figure 2) [5,21,22].

Precursors	Alkaloid classes	Examples
Ornithine 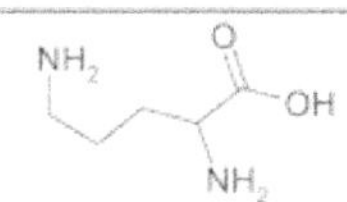	Pyrrolidine[1], Tropane, Pyrrozidine, Polyamines[2]	1. Cocaine, Atropine 2. Putrescine, Spermine, Spermidine
Leucine 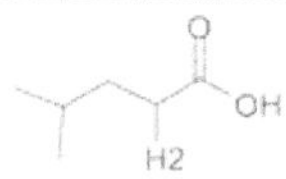	Pyrrole	
Lysine 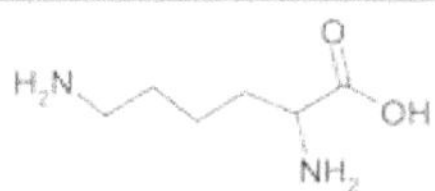	Pyperidine, Quinolizidine, Indolizidine	
Tyrosine 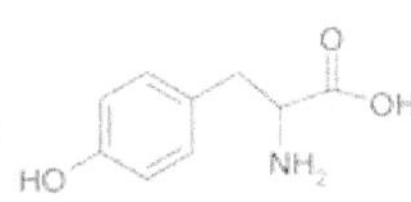	Isoquinoline Tetrahydroisoquinoline, Benzyltetrahydroiso-quinoline[1], Catechol-amines[2]	1. Morphine, Curarines, Papaverine 2. Noradrenaline, Adrenaline
Tryptophan 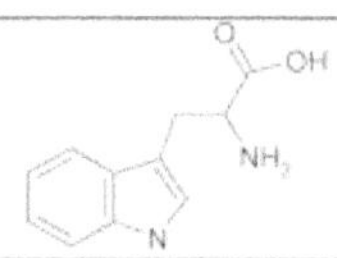	Indole[1], Carbolines, Quinoline[2], Pyrrolindole, Indolamines[3]	1. Vindoline,Catharantine 2. Quinine, Capthotecin 3. Melatonin, Serotonin
Histidine 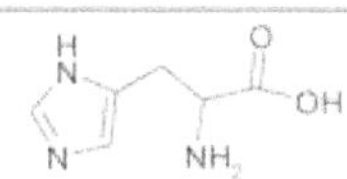	Imidazole	
Phenylalanine 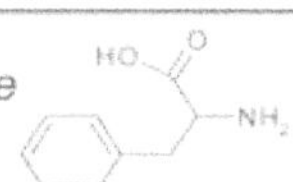		Ephedrine, Capsaicin
Anthranilic acid 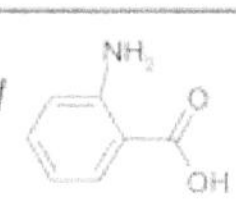	Quinazoline, Qunoline,Acridine	
Purine 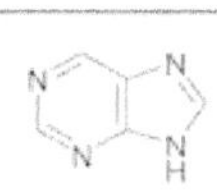		Theobromine, Theophylline, Caffeine
Geranylgeranyldiphosphate 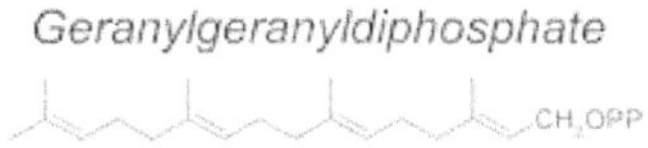	Terpenoidic	
Cholesterol 	Steroidal	Solanin
Acetate 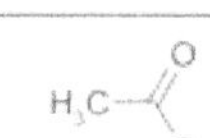	Piperidine	
Nicotinic acid 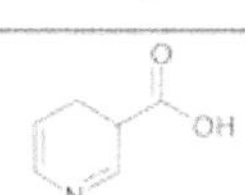	Pyridine	Nicotine

Figure 2. Main classes of alkaloid precursors and derivatives (the example number refers to the corresponding alkaloid class).

Alkaloids constitute an enormous number of phytochemicals of toxicological, pharmacological, nutritional and cosmetic interest, and of ecological importance for plants. For instance, tropane alkaloids include cocaine and atropine, nicotine is a pyridine alkaloid, noradrenaline (or norepinephrine), adrenaline (epinephrine), papaverine, curarines and morphine arise from tyrosine. Melatonin and serotonin are indolamines, vindoline, catharantine (and their derivatives vincristine and vinblastine) are indole alkaloids, quinine and capthotecin are quinoline alkaloids and lysergic acid diethylamide (LSD) is an ergot alkaloid, all these arising from tryptophan. Histamine is an imidazole alkaloid, ephedrine and capsaicin derive from phenylalanine, solanin is a steroidal glycoalkaloid from cholesterol. Finally, purine alkaloids include theophylline, theobromine and caffeine, found in tea, cacao and coffee, respectively [5,21,22].

SECONDARY METABOLITES INVOLVED IN PLANT RESISTANCE AGAINST PATHOGENS

Higher plants are regularly exposed to microorganisms, at the level of both epigeous and hypogeous organs. However, only a few of them causes diseases, because of an efficient plant innate immune system. Therefore, as for animals, compatibility of micro-organisms with a susceptible plant represents an exception in nature, while incompatibility with a resistant host is the rule. The latter condition depends on the plant defence machinery, a complex array of both physical and chemical preformed and inducible defences [23]. Plant antimicrobial compounds are broadly classified into two categories: phytoanticipins and phytoalexins. The former are low molecular weight compounds present in plant before micro-organism challenge, or produced after infection from pre-existing precursors, while the latter are both synthesized and accumulated in plant after exposure to micro-organisms or abiotic agents [24].

Phytoanticipins

Plant preformed antimicrobial compounds (phytoanticipins) include three important classes: saponins, cyanogenic glycosides and glucosinolates [25].

Saponins are glycosylated compounds widely distributed among plant families and divided into three major groups, depending on the structure of the aglycone, which may be a triterpenoid, a steroid or a steroidal glycoalkaloid. The genus*Avena* is particularly rich in these metabolites, with oat containing two different families of saponins, the triterpenoid avenacins, in roots, and the steroid avenacosides in leaves, both groups mainly concentrated in the epidermal layers (Scheme 4). In turn, avenacosides are closely related to the steroidal glycolakaloid α-tomatine, occurring in tomato (Scheme 4). Saponins destabilize plasmalemma integrity by forming complexes with membrane sterols and causing pore formation. Plants avoid these toxic effects by compartmentalizing them in the cell vacuole or in other organelles, whose membranes are protect because of a low or different sterol composition [26–28]. Resistance of oat to the root-infecting fungus*Gaeumannomyces graminis* pv tritici, the causal agent of the take all disease in wheat and barley, has been attributed to the presence of avenacins in oat roots [29–32]. Conversely, the oat-infecting *G. graminis* pv avenae is able to tolerate and detoxify avenacins by secreting extracellular glycosidases (avenacinases) which remove the terminal sugar from their carbohydrate chain, thus altering the amphiphatic properties of these saponins [31,33,34]. In tomato, the content of α-tomatine has been correlated with the resistance to the tracheomycosis (vascular diseases) caused by *Fusarium oxysporum* pv lycopersici and*Verticillium albo-atrum*, as well as with the foliar pathogen *Cladosporium fulvum* [35–37].

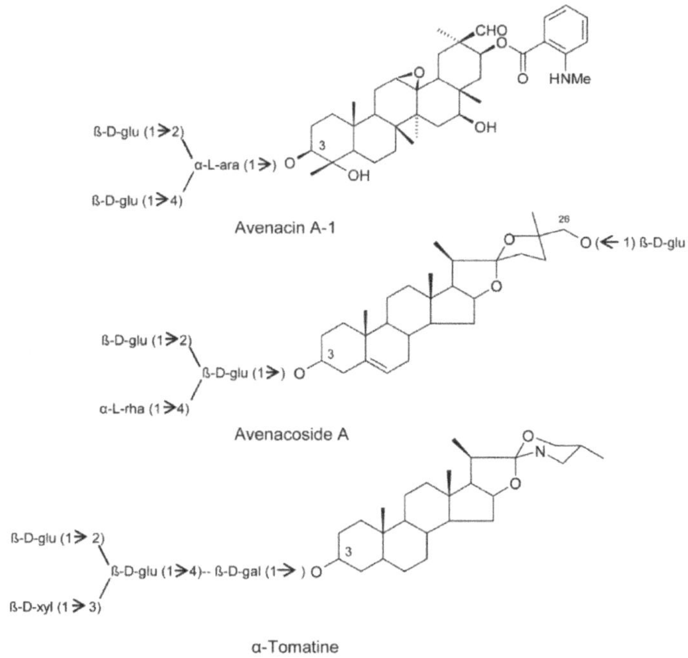

Scheme 4. Molecular structures of the major oat and tomato saponins (adapted from Osbourn, [32]).

Cyanogenic glucosides are phytoanticipins known to be present in more than 2,500 plant species. They are regarded as having an important role in plant defense against herbivores due to bitter taste and release of toxic hydrogen cyanide (HCN) upon tissue disruption (reviewed in [38]). In fact cyanogenic glycosides are precursors of HCN, a respiratory poison released after the cleavage of the carbohydrate moiety by a β-glucosyl hydrolase (or simply a β-glucosidase; Scheme 5).

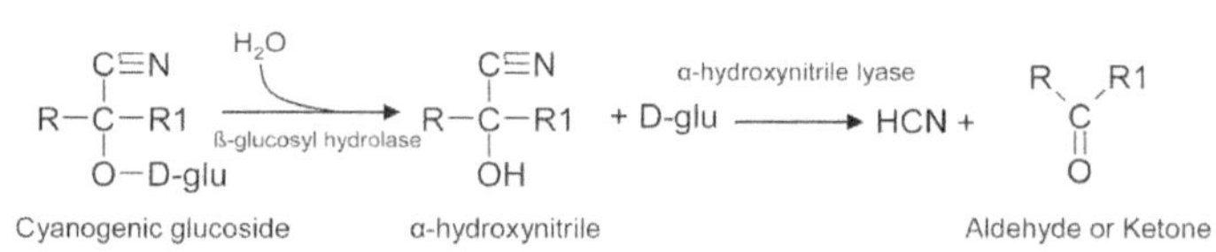

Scheme 5. Generation of hydrocyanic acid from a cyanogenic (HCN) glucoside precursor. The α-hydroxynitrile intermediate is then converted by α-hydroxynitrile lyase to HCN and aldehyde or ketone.

The corresponding α-hydroxynitrile intermediate is then converted to HCN by an α-hydroxynitrile lyase, associated with defence against pathogens and phytophagy (Scheme 5) [39]. Healthy cyanogenic plants do not contain detectable amounts of HCN, because, in their tissues, cyanogenic glycosides are normally compartmentalized, i.e., spatially separated from the enzymes that catalyse HCN production. Important cyanogenic glycosides include dhurrin and amygdalin, from sorghum and almond respectively [40–42]. Pathogens of cyanogenic plants produce the cyanide-inducible enzyme cyanide hydratase, which detoxify HCN by converting it to formamide [43]. Glucosinolates are sulphur-containing glucosides widespread in members of the family Cruciferae, including the genus*Brassica*, of relevant agronomic interest, and the weed Arabidopsis, a model plant in molecular genetic and biology. They represent a large chemical family including over 120 different compounds with known defensive properties against herbivores [44,45]. Glucosinolates can be subdivided into three major classes, depending on their side chain, which may be derived from aliphatic (methionine), indolyl (tryptophan) or aralkyl (phenylalanine) α-amino acids. Within individual plants, their distribution is tissue specific, as in oilseed rape, where aliphatic glucosinolates abound in leaves, whereas indolyl and phenylethyl glucosinolates predominate in roots and stems. Like cyanogenic glycosides, these compounds are also activated in response to tissue damage by the action of myrosinase (thioglucoside glucohydrolase or simply thioglucosidase), an enzyme that, in healthy plants, is separated from its glucosinolate substrates by subcellular compartmentalization (Scheme 6). The unstable aglycone thus generated may then form various products, including nitriles, thiocyanates and volatile isothiocyanates, being the latter the major and the most fungitoxic glucosinolate breakdown products (Scheme 6) [39,46,47]. Therefore, high levels of glucosinolates have also been associated with plant (*Brassica*) resistance to pathogens, such as the blackleg fungus *Leptosphaeria maculans* and the oomycete *Peronospora parasitica*, causal agents of stem canker and downy mildew, respectively [48,49]. Inhibition of ascospores of *Mycosphaerella brassicae* has been early reported, in both cabbage and cauliflower leaves [50] Interestingly, indolyl glucosinolates may be used as precursors for the biosynthesis of auxin, a indole/tryptophan derivative, by fungi that cause hyperplasic and hypertrophic tissue growth, such as the root gall fungus *Plasmodiophora brassicae* [37].

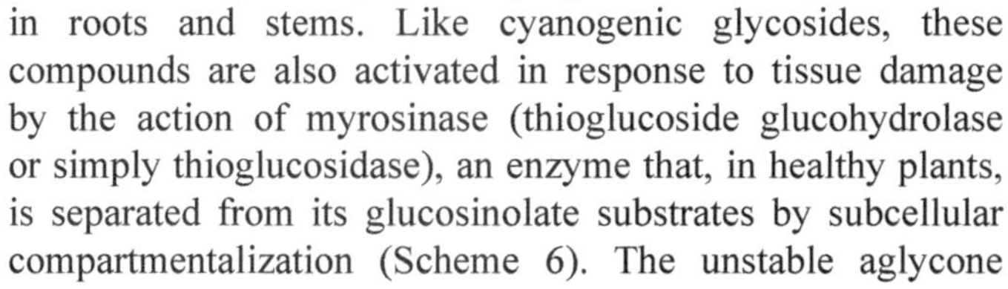

Scheme 6. Hydrolysis of glucosinolates and degradation of the unstable aglycone to isothiocyanates, nitriles and athiocyanates.

Phytoalexins

The recognition of elicitors by the plant's own perception repertoire represents a crucial stimulus to initiate phytoalexin synthesis. Elicitors are (exogenous) signal molecules of microbial origin, or generated endogenously from plant cell wall fragments produced at the host-pathogen interface. In any case, plant necessitates of receptors in order to recognize the pathogen and mount its own defence reaction [23]. Interestingly, phytoalexins accumulate both in resistant and susceptible hosts at the same concentrations, though with a different kinetic, thus pointing out that their efficacy strictly depends on the timing of their synthesis at the infection site [51].

Activation of plant defence response generally induces an intense, though transient, diversion of normal (primary) metabolism into synthesis of phytoalexins, secondary metabolites *de novo* synthesized from essential substrates such as phenylalanine, acetyl-CoA, malonyl-CoA, mevalonic acid or other amino acids. Phytoalexins have been identified in the majority of plant families, and members of a plant family normally produce similar types of phytoalexins, though phenylpropanoids include several classes of well-studied phytoalexins, for instance isoflavonoids from the Leguminoseae family [52,53].

The *Phaseolus vulgaris-Colletotrichum lindemuthianum* pathosystem has provided a good model for studying the role of phytoalexins in plant resistance against pathogen attack. *C. lindemuthianum*, the causal agent of bean anthracnose, is a hemibiotrophic fungus whose colonization is restricted, in resistant hosts, because of the production of isoflavonoid phytoalexins, including phaseollin, phaseollin isoflavan and kievitone [28,54]. Similar examples include medicarpin and pisatin, two isoflavonoid phytoalexins from alfalfa (*Medicago sativa*) and pea (*Pisum sativum*), respectively [55,56]. However, broad bean (*Vicia faba*) provides a notable exception. Like most other legumes, it produces isoflavonoid phytoalexins, but the principal induced antimicrobial compounds are furanoacetylenic wyerone derivatives [28,52].

In the Vitaceae family, the phytoalexins which have been well characterized constitute a rather restricted group of molecules belonging to the stilbene family, synthesized as a general response to fungal attack. These compounds possess the skeleton based on the trans-resveratol (3,5,4′-trihydroxy-stilbene) structure, including piceids, pterostilbenes and viniferins, that are, respectively glucosides, dimethylated derivatives and oligomers of resveratrol [57]. In grapevine (*Vitis vinifera*), the activities of CHS and STS are differentially regulated, according to the plant developmental stage (see section 2.1). During the initial phase of berry ripening (véraison), resveratol accumulation in cells of berry skin declines, while anthocyanin synthesis increases, due to the competition between the two branches of the same pathway. As a consequence, since véraison, anthocyanin accumulation confers colour to berry skin, whereas the lowering levels of the powerful phytoalexin resveratrol make bunches more susceptible to *Botrytis cinerea* infections (grey mould) [59]. Interestingly, open field treatments with plant activators (compounds able to activate the plant immune system) can reverse, to a certain degree, the inverse relationship between resveratrol and anthocyanin content, reducing the CHS and STS competition for the substrate and avoiding the metabolic switch from one to the other pathway. Thus, higher levels of resveratrol protect grape from grey mould after véraison, without hampering the colouring phase, which is an important qualitative trait [59].

In the model plant Arabidopsis, camalexin (3-thiazol-2′-yl-indole) (Scheme 7) represents the main phytoalexin, involved in inducible defence mechanisms against a variety of pathogens, such as the bacterium *Pseudomonas syringae* and the fungus*Alternaria brassicicola*. Camalexin is a N-and S- containing indole phytoalexin synthesized from tryptophan via indole-3-acetaldoxime, a branch point metabolite that also leads to the biosynthesis of the previously described glucosinolates, besides the plant hormone indole acetic acid (IAA) and melatonin (Scheme 7) [60].

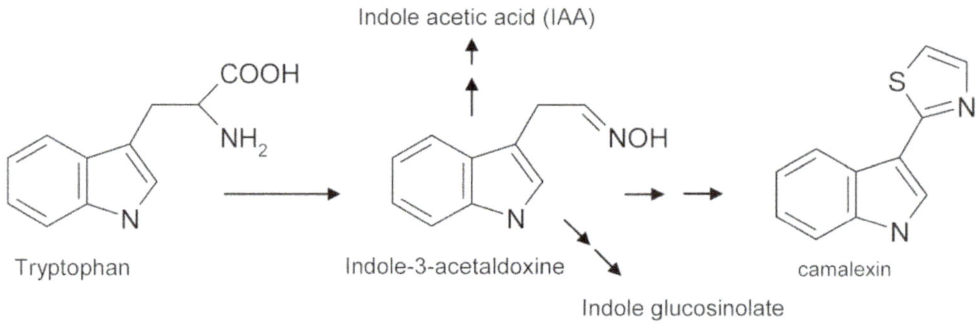

Scheme 7. Schematic biosynthetic pathway of camalexin via indole-3-acetoaldoxime. Other important indolic compounds arise from the same route.

Finally, it is important to underline that phytoalexin (and phytoanticipin) synthesis is only a component of the plant co-ordinated defence strategy, contributing to restrict the pathogen spreading together with structural barriers, oxidative burst, hypersensitive response (HR) and pathogenesis related (PR) proteins [23].

TROPOSPHERIC OZONE AND PLANT HEALTH

In the atmosphere, two different pools of O_3 exists, the beneficial and the detrimental one (Figure 3). In the stratosphere (the higher atmosphere, ranging approximately from 15 to 40 km in altitude), the ozone layer absorbs the harmful UV-B and UV-C radiations, thus screening the living organisms [61,62]. Conversely, in the troposphere (the lower part of the atmosphere, approximately from the Earth surface to 10-12 km in altitude), ozone is regarded as a pollutant [63]. Tropospheric ozone is a secondary pollutant produced through reactions between primary pollutants, emitted directly into the air (mainly nitric oxides, sulphur oxides, carbon oxides and hydrocarbons), catalyzed by sunlight (Figure 3) [64]. Consequently, ozone is produced on bright sunny days over areas with intense primary pollution, due mainly to vehicle exhausts, fossil fuel burning and industrial processes, in the so-called photochemical cycle [65,66].

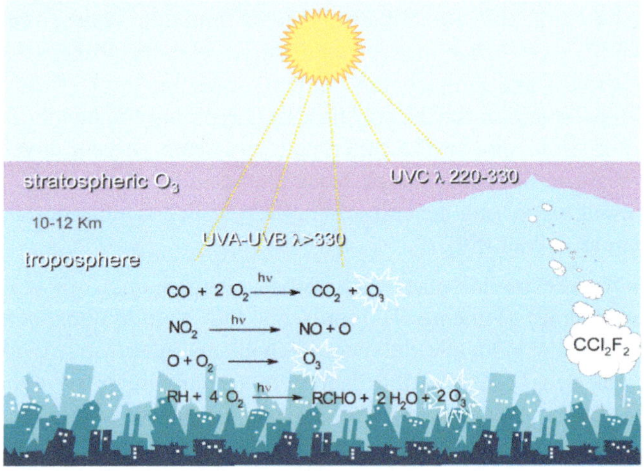

Figure 3. Beneficial and detrimental pools of ozone in the stratosphere and troposphere, respectively. In the stratosphere this pool is depleted by fluorochlorocarbons (CCl_2C_2) while in the troposphere it is raised by photochemical smog.

Ozone enters the leaf tissues through the stomata and plants are exposed to either acute and chronic ozone doses, according to the gas concentration and exposure time. Because of its strong oxidizing potential (+ 2.07 eV), O_3 is a powerful oxidizing agent capable of reacting with virtually any biomacromolecule, including lipids, proteins, nucleic acids and carbohydrates, although it is neither a radical species nor a ROS [67,68]. In cell membrane, polyunsaturated fatty acids represent the primary target for ozone, stimulating lipid peroxidation and impairing membrane fluidity [69]. Modification of proteins can occur, both in their structure and activity, and DNA damage can also be produced, as shown in animals by the increased activity of poly(ADP-ribose) synthetase (PARP), a chromatin-bound enzyme promoting damaged DNA repair [70,71]. Though the increase of PARP activity has not been so far investigated in plants after ozone exposure, it is likely to occur as a general defence response, similar to that observed in Arabidopsis after pathogen attack [72].

Symptoms of injury are generally unifacial and visible in the adaxial (upper) leaf surface, depending on leaf morphology. The most common symptoms due to chronic injury include chlorosis (yellowing due to the chlorophyll breakdown, often distributed in spots over the leaf) and bronzing (red-brown pigmentation caused by phenylpropanoid accumulation), while lesions attributable to acute exposures are much more diversified. In broad-leaved plants, they include bleaching (small unpigmented necrotic spots), flecking (small brown necrotic areas fading to grey or white), stippling (small punctuate spots, white, black or red in colour) and tipburn (dying tips, firstly reddish, later turning brown) specially in conifers [73,74].

At a physiological level, ozone exposure impairs stomatal function and reproductive development, decreases photosynthetic activity and CO_2 photoassimilation, and induces senescence, resulting in dysfunction of transpiration and water use efficiency, reduction of dry matter production and yield losses [75–79].

Ozone and Phenylpropanoids

In plants, phenylpropanoid metabolism is induced as a general response to stress. Therefore, enhancement of key enzyme activities and accumulation of secondary metabolites are events occurring early after challenging, in order to improve the resistance against pathogen attack and/or tolerance to adverse environmental conditions and pollutants. PAL is an extremely sensitive indicator of stress conditions, and it is commonly considered as a biochemical marker indicating the activation of plant defences, including the synthesis of both structural and protective compounds. In particular, ozone exposure elevates the level of flux through the phenylpropanoid pathway, thereby supplying carbon skeletons for secondary metabolites [80].

The enhancement of phenylpropanoid biosynthesis by ozone is well documented. A very early report, dated more than 30 years ago, documented the accumulation of isoflavonoids in soybeans following ozone exposure [81]. Ever since, ozone-stimulated induction of transcripts for defence-related genes, the same induced by pathogens, as well as the increased activity of key enzymes of the phenylpropanoid pathway have been reported in several plant systems. In Arabidopsis (*Arabidopsis thaliana*), PAL mRNA is rapidly and transiently induced within 3 h of ozone treatment (300 ppb daily for 6 h), reaching a 3-fold higher levels than control plants [82]. A similar trend has been reported in parsley (*Petroselinum crispum*) plants, in which ozone treatment (200 nL L^{-1} for 10 h) induced an early 3-fold and 1.2-fold increase of PAL and CHS activity, respectively, followed by a 2-fold increase of total leaf furanocoumarins and flavone glycosides [83]. The content of psoralen, bergapten and other furanocoumarins in celery (*Apium graveolens*) dropped 24 h after ozone (0.2 ppm for 2 h) fumigation, but levels of these chemicals, in treated leaves, increased rapidly at 120 h [84]. Further studies have shown the overlap between patterns of genes induced by ozone exposure and pathogen infection, probably due to the role of ROS as effector molecules involved in transduction pathways activated either by pathogens and ozone. The enhancement of SA content, in plant tissues, due to ozone treatment is well documented [85]. Intriguingly, in tobacco (*Nicotiana tabacum*) plants, a pulse ozone treatment (120–170 nmol mol^{-1} for 5 h) enhanced the emission of methyl salicylate, a volatile SA derivative, to a greater extent in sensitive cv. Bel W3 compared with tolerant cv. Bel B [86]. In Arabidopsis, SA accumulation, necessary for the expression of hypersensitive response to pathogens (HR) and SAR, is also required for the accumulation of some ozone-induced mRNAs, particularly PAL and pathogenesis related protein 1 (PR1) transcripts. Interestingly, ozone-induced PAL mRNA accumulation was unaffected in plants expressing salicylate hydroxylase, demonstrating that the ozone induced accumulation of PAL transcripts does not depend on SA accumulation [88]. These findings demonstrate that ozone activates at least two distinct signaling pathways, including a salicylic acid dependent pathway previously shown to be associated with the activation of pathogen defense reactions, and a second one SA-independent, as a protective response to ozone. [88]. HR is a type of programmed cell death (PCD) triggered at the attempted pathogen penetration site, frequently at the onset of systemic immunity (SAR) [83,84], whereas PR proteins are enzymes induced in plants by pathogen infection, as well as by abiotic and environmental stresses, included ozone [89–93]. In sensitive bean (*Phaseolus vulgaris*) cv. Pinto, ozone exposure (120 nL L^{-1} for 4 h) causes a stimulation of phenylpropanoid route and flavonoid branch, as shown by the increased mRNA accumulation of PAL, CHS and chalcone isomerase (CHI), the latter involved in isoflavonoid biosynthesis [93]. In grapevine (*Vitis vinifera*), STS, the first enzyme of the stilbene branch involved in the synthesis of resveratrol and other stilbenic compounds, is considered the most sensitive ozone-induced biomarker [94]. In grapevine callus, either PAL and STS activity increase, after ozone fumigation (0.3 μmol mol^{-1} for 2 h), unlike CHS [90]. This could be due to the STS and CHS competition for the same substrate (Scheme 1), as it happens during the véraison, previously described. Furthermore, a general drop in the amount of some assayed phenylpropanoids (coumaric acid, ferulic acid, gallic acid and catechin) has been reported, unlike caffeic acid, whose level raised only in one cell line [95]. In 20 soybean (*Glycine max*) cultivars, ozone tolerance has been associated with the presence of kaempferol glycosides, a powerful antioxidant flavonol, as well as tolerance to manganese (Mn) toxicity in one soybean line [96]. In European silver birch (*Betula pendula*) chronically exposed to ozone, a 16.2% increase in totals phenylpropanoids and a corresponding 9.9% increase of 10 compounds, among simple phenols and flavonoids, such as chlorogenic acid and catechin, respectively, has been reported [92]. Interestingly, the combined action of CO_2 and ozone greatly enhances the synthesis of total and polymeric PA, in *Betula* sp. leaf tissues, suggesting an additive effect of these environmental pollutants on phenylpropanoid biosynthesis [98].

Shikimate dehydrogenase (SKDH), a key enzyme of shikimate pathway, PAL and cinnamyl alcohol dehydrogenase (CAD), a key enzyme of lignin biosynthesis which forms monolignols (Scheme 2), have been investigated in poplar (*Populus tremula* x *alba*) leaves. Under ozone exposure (60–120 nL L^{-1}, during the 14 h light period, for 1 month), either CAD activity and transcript levels have been rapidly and strongly stimulated, increasing up to 15-fold and 23-fold the control values, respectively. In contrast, SKDH and PAL activity raised only in old and middle-aged leaves, but not in the youngest ones. Interestingly, the increased activity of these enzymes has been associated with a higher lignin content in ozone-exposed leaves and, additionally, the newly synthesized lignin structurally differed from the control lignin. Particularly, stress lignin appeared more condensed, i.e., enriched in carbon-carbon interunit linkages, in *p*-hydroxyphenyl (H) units and in terminal units with free phenolic groups (Scheme 2) [99]. The same enzymes have been studied in two genotypes of ozone-treated (150 nL L^{-1} for 3 h) tomato (*Lycopersicon esculentum*) plants [100]. However, while SKDH and PAL activity augmented significantly only in one line, CAD activity diminished in both the genotypes, in contrast with results reported by other authors [99,101]. An explanation could reside in the different response of herbaceous and woody plants and in the different acute or chronic ozone dose employed in tomato and poplar, respectively.

To conclude, the importance of phenylpropanoids in plant tolerance against ozone injury is related to their different properties. These compounds include an array of molecules with a plethora of biological activities, besides precursors of structural biopolymer, such as lignin. Particularly, their protective role is mainly inherent to their antioxidant power, i.e., the ability to trap free radicals (ROS), functioning as electron donors. Nonetheless, increased synthesis of lignin, as well as structural modifications of lignin itself, represent another important defence mechanism, in order to protect plasmalemma from the ROS injury, thus preventing membrane damages due to lipid peroxidation [7].

Ozone and Isoprenoids

Strictly speaking, not all the isoprenoids are secondary metabolites, as some primary metabolites, such as sterols, arise from the isoprenoid pathway as well. The effects of the pollutant on either the sterol concentration and composition have been early reported in several plants [102–104]. Sterols are important component of cell membranes, involved in their stabilization. Generally, exposure to high ozone concentrations (>12 μL L^{-1}) results in a decrease of free sterols (FS) and an increase in bound sterols (BS). Conversely, at low ozone concentration (<12 μL L^{-1}) an accumulation of FS occurs, resulting in a decrease of the FS:BS ratio. In this case the enhancement of FS synthesis is followed by a shift towards sterols with a more bulky C-17 side chain, i.e., sitosterol and stigmasterol vs. campesterol. As FS have a much greater capacity to stabilize membranes than BS, the FS:BS ratio has a more determinant effect on membrane permeability than the composition of FS fraction itself. Accordingly, decrease in FS:BS ratio with high ozone concentration results in cell injury and visible damages, whereas modification of FS composition, occurring at lower ozone levels, results neither in permanent membrane injury nor in visible damages [105–107]. In tobacco leaves, ozone fumigation resulted in increased lipid concentrations, but decreased levels of FS and triglycerides [103], whereas a higher amount of total phytosterols has been reported in fumigated plants by other authors [108].

In ozone treated (0.3 μL L^{-1} for 8 h) Scots pine (*Pinus sylvestris*), it has been reported a transient increase of a transcript corresponding to the cytosolic/endoplasmic HMG-CoA synthase, a key enzyme of isoprenoid biosynthesis [109]. The C_5precursor of isoprenoids, IPP, arises either from mevalonate pathway, in the cytosol, as previously described, or from plastidial precursors [110]. In ozone-treated pine seedlings (250 nL L^{-1}, 12 h day^{-1} for 4 days), the biosynthesis of plastidial IPP has been inhibited, differently from the mevalonate synthesis in the cytosol, due to the resource allocation between the two IPP synthesis pathways [111].

Biogenic volatile organic compounds (BVOC) comprise mainly isoprenoids (particularly hemi-, mono- and sesquiterpenes) emitted from plants during cell growth and in response to several kinds of stresses, thus having various ecophysiological functions and mediating plant-arthropod interactions (Figure 4) [112]. BVOC can act as attractants for pollinators or repellents for noxious insects, besides being involved in tritrophic signalling, i.e., the relationship between plant, herbivorous and carnivorous arthropods. In particular, phytophagus feeding can induce BVOC emission from plants, which can function as foraging cues for the recruitment of the natural enemies of herbivores [113], and for the orientation, landing and feeding of other phytophagy [114]. Besides wounding and trithrophic interactions, several environmental factors can affect the BVOC emission from plants, such as light intensity, temperature, water supply and pollutants (Figure 4) [115]. The relationships between ozone and BVOC is somewhat complex, both exerting a mutual influence [116]. On the one hand, the release of either BVOC and volatile organic compounds (VOC) from anthropogenic origin into the atmosphere can constitute a significant input of photochemical oxidant precursors, thus contributing to the regional-scale air pollution; on the other hand BVOC emission can be triggered by the exposure to high ozone concentration [86,117]. Additionally, chronic ozone exposure can modify the composition of the plant BVOC emissions [118–121], thus not only affecting the tritrophic interactions, but also directly weakening plant defence responses against arthropods [122]. Isoprenoids can be synthesized and emitted in order to tolerate the ozone injury. Isoprene, a hemiterpene (Scheme 3), has been reported to reduce ozone damages in leaves, because of its antioxidant activity. This gas protects the photosynthetic apparatus, quenches ozone byproducts and radical species responsible for lipid peroxidation of cell membranes and cell death [123–125]. Besides, monoterpenes may exert an isoprene-like antioxidant activity, too [126,127].

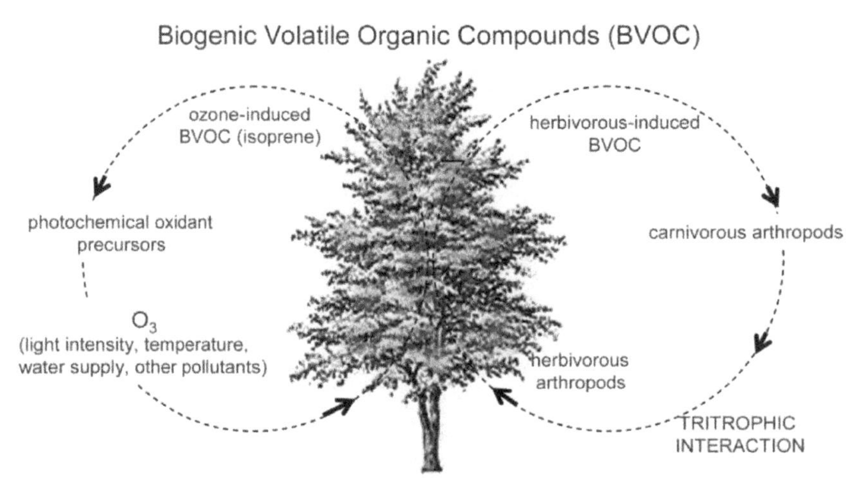

Figure 4. Relationships between ozone and herbivorous-induced BVOCs and their influence on photochemical ozone precursors and on tritrophic interactions, respectively.

Ozone and Alkaloids

The effect of ozone on alkaloid biosynthetic pathways has not been extensively investigated, although a generalized influence of the pollutant on the nitrogen metabolism has been ascertained [128–130]. Generally, ozone exposure reduces the amount of total alkaloids in tobacco plants [128,129], and lower levels of nicotine, a pyridine alkaloid, in ozone exposed plants have been related to increased survival, growth and development of hornworm (*Manduca sexta*) larvae [130]. Several studies reported increased preferences or enhanced survival and fitness of different insect species on ozonated plants, thus pointing out again the detrimental effect of the pollutant on the plant chemical defences [132–136].

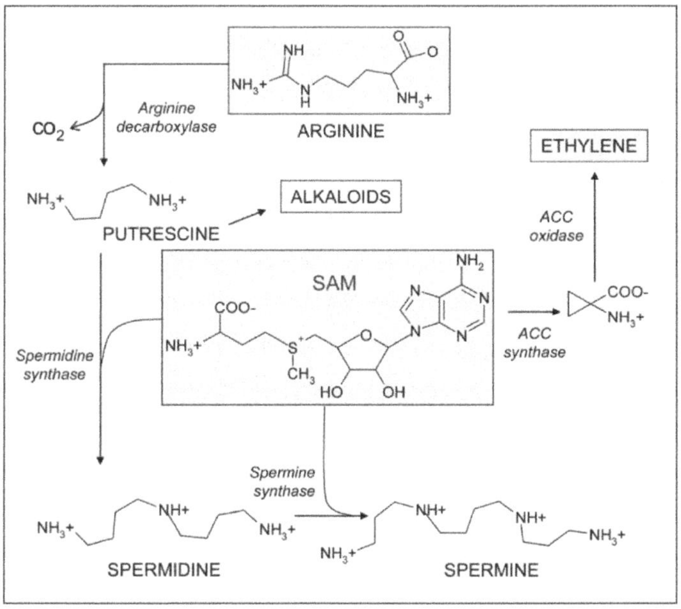

Scheme 8. Involvement of S-adenosylmethionine (SAM) in polyamine and ethylene biosynthesis. The metabolic shift to ethylene or polyamine biosynthesis can enhance ozone susceptibility or tolerance, respectively.

The role of polyamines, important alkaloid precursors (Scheme 8), has been correlated with ozone tolerance. Polyamines are polycationic nitrogenous compounds of low molecular weight ubiquitous in all living organisms. In plants, they function as growth regulators involved in an array of physiological processes, being involved in embryogenesis, cell division, morphogenesis, development, flowering and senescence [137]. In addition, they serve as an integral component of plant response to both biotic and abiotic stresses [138,139]. The most important polyamines are the diamine putrescine (Put), the triamine spermidine (Spd) and the tetramine spermine (Spm), arising directly from the free, non proteinogenic amino acid ornithine, by ornithine decarboxylase (ODC), or indirectly from arginine by arginine decarboxylase (ADC). Further steps include Put conversion into Spd, via spermidine synthase (SPDS), and the similar synthesis of Spm from Spd via spermine synthase (SPMS). Either these enzymes employed the aminopropyl moiety provided by S-adenosylmethionine (SAM) [140]. In plants, ADC activity is enhanced by ozone exposure, whereas the ODC one remaines unchanged [141,142]. Free and conjugated polyamines improve ozone tolerance with two different mechanisms: by inhibiting the ethylene biosynthesis and by direct ROS scavenging, respectively [143,144]. Ethylene and polyamines share the same biosynthetic precursor (SAM) and, thereby, they mutually inhibit their own biosynthesis. In particular, aminocyclopropane carboxylic acid (ACC), a precursor of ethylene, arises from SAM via ACC synthase, a rate-limiting step in the ethylene production. Therefore, this metabolic shift to ethylene or polyamine biosynthesis can enhance ozone susceptibility or tolerance, respectively, due to the correlation between the stress ethylene production and visible ozone injury [142]. Additionally, apoplastic polyamines can form conjugates with hydroxycinnamates and phenolic acid derivatives, effective in ROS detoxification [145]. As reported above, polyamine are precursors of alkaloids, thus, apart they role in ozone tolerance, higher amounts of these compounds, as a consequence of ozone exposure, could induce an increase of pyrrolidine, tropane and pyrrolizidin alkaloids, deriving from ornithine via Put.

CONCLUSIONS

Chemical diversity determined the evolutionary success of plant organisms, favouring their adaptation to a changing environment. Plants cope with a plethora of stressful abiotic and biotic conditions by modifying their secondary metabolic pathways [146]. In this view, chemical diversity improved, during the evolution, the fitness of plant organisms, thus ensuring their evolutionary radiation [147]. Phytochemicals, with their broad spectrum activities, are primarily involved in plant tolerance towards worsening climatic conditions and environmental pollutants, as well as in resistance against pests and pathogens. The metabolic processes activated in these defence responses may be tightly separated or overlapping, according to the stress factor and the plant cultivar, as a result of negative (*trade off*) or positive (*cross resistance*) cross talk, respectively, i.e., the communication between molecular signals and transduction pathways involved in different plant defence responses. A cross tolerance may also occur, in plant, from a cross induction, that is the ability to alert the defence mechanisms against an abiotic stress in consequence of priming with a biotic elicitor and vice versa.

Therefore, the influence of atmospheric pollution in plant-pathogen interactions is quite complex [148,149]. On the one hand, ozone may induce the same sequel of events involved in plant innate immunity [23], i.e., oxidative burst and hypersensitive response at the onset of systemic acquired resistance (SAR) [88], on the other hand plants weakened by ozone stress my be particularly susceptible to infections [149]. Interestingly, wheat rust (*Puccinia recondita* f. sp. *tritici*) was strongly inhibited by ozone, but was unaffected by elevated CO_2, both in presence or absence of ozone stress [150]. Vice versa, a protective effect of rust (*Uromyces fabae*) infection in broad bean (*Vicia faba*) was reported against ozone, sulphur dioxide either alone or combined [151].

Finally, to the other side of the coin, an excessive production of secondary metabolites may be detrimental for the plant's fitness (Figure 5), in terms of allocation costs and autotoxicity [152,153], though it may be favourable, at least in food plants, which could provide foodstuffs and beverages enriched of bioactive phytochemicals [154].

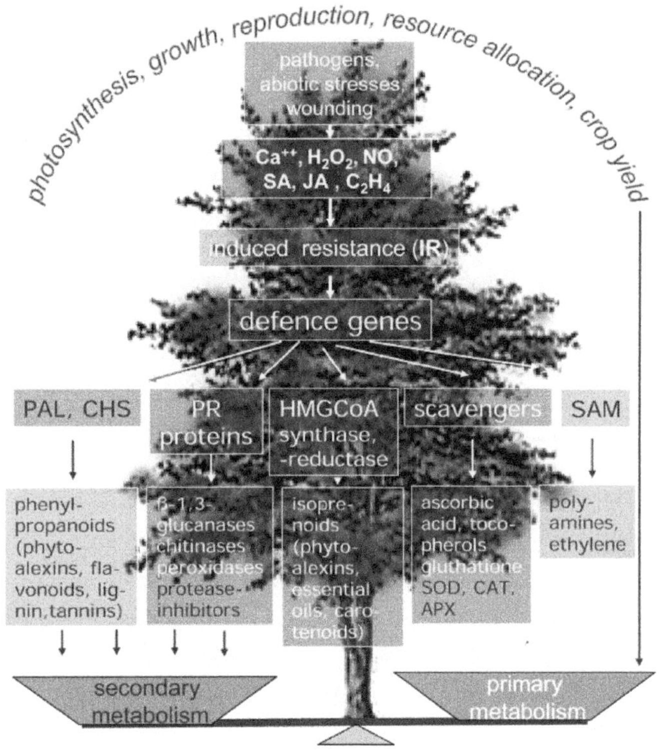

Figure 5. Influence of different stresses on plant metabolism. The activation of induced resistance leads to the accumulation of numerous defence compounds that may unbalance the equilibrium between primary and secondary metabolism, thus resulting in fitness costs for the plant. On the other hand, the synthesis of secondary metabolites due to a stress may protect plant from other different stresses.

REFERENCES AND NOTES

1. Lichtenthaler, HK. The stress concept in plants: An introduction. *Ann. N.Y. Acad. Sci* **1998**, *851*, 187–198.
2. Heiser, I; Elstner, EF. The biochemistry of plant stress and disease. *Ann. N.Y. Acad. Sci* **1998**, *851*, 224–232.
3. Firn, RD; Jones, CG. A Darwinian view of metabolism: Molecular properties determine fitness. *J. Exp. Bot* **2009**, *60*, 719–726.
4. Paiva, N. An introduction to the biosynthesis of chemicals used in plant-microbe communication. *J. Plant Growth Regul***2000**, *19*, 131–143.
5. Facchini, PJ. Alkaloid biosynthesis in plants: Biochemistry, cell biology, molecular regulation and metabolic engineering. *Annu. Rev. Plant Physiol. Plant Mol. Biol* **2001**, *52*, 29–66.
6. Holstein, SA; Hohl, RJ. Isoprenoids: Remarkable diversity of form and function. *Lipids* **2004**, *34*, 293–309.
7. Iriti, M; Faoro, F. Plant defense and human nutrition: the phenylpropanoids on the menu. *Curr. Top. Nutrac. Res* **2004**,*2*, 47–95.
8. Weaver, LM; Hermann, KM. Dynamics of the shikimate pathway in plants. *Tends Plant Sci* **1997**, *2*, 346–351.
9. Hahlbrock, K; Scheel, D. Physiology and molecular biology of phenylpropanoid metabolism. *Annu. Rev. Plant Physiol. Plant Mol. Biol* **1989**, *40*, 347–369.
10. Wildermuth, MC; Dewdney, J; Wu, G; Ausubel, FM. Isochorismate synthase is required to synthesize salicylic acid for plant defence. *Nature* **2001**, *414*, 562–565.
11. Iriti, M; Faoro, F. Lipid biosynthesis in Spermatophyta. In *Floriculture, Ornamental and Plant Biotechnology*; Teixeira da Silva, JA, Ed.; Global Science Books: London, UK, 2006; Volume 1, Chapter 40; p. 359.
12. Sacchettini, JC; Poulter, CD. Creating isoprenoid diversity. *Science* **1997**, *277*, 1788–1789.
13. Chappell, J. The biochemistry and molecular biology of isoprenoid metabolism. *Plant Physiol* **1995**, *107*, 1–6.
14. Bach, TJ. Synthesis and metabolism of mevalonic acid in plants. *Plant Physiol. Biochem* **1987**, *25*, 163–178.
15. Poulter, CD; Rilling, HC. The prenyl transferase reaction, enzymatic and mechanistic studies of the 1′-4 coupling reaction in the terpene biosynthetic pathway. *Acc. Chem. Res* **1978**, *1*, 307–313.
16. Croteau, R. Biosynthesis and catabolism of monoterpenoids. *Chem. Rev* **1987**, *87*, 929–954.
17. Cane, DE. Enzymatic formation of sesquiterpenes. *Chem. Rev* **1990**, *90*, 1089–1103.
18. Dogbo, A; Camara, B. Purification of isopentenyl pyrophosphate isomerase and geranylgeranyl pyrophosphate synthase from *Caspicum* chromoplasts by affinity chromatography. *Biochim. Biophys. Acta* **1987**, *92*, 140–148.
19. Abe, I; Rohmer, M; Prestwich, GD. Enzymatic cyclization of squalene and oxidosqualene to sterols and triterpens.*Chem. Rev* **1993**, *93*, 2189–2206.
20. Wendt, KU; Schulz, GE. Isoprenoid biosynthesis: manifold chemistry catalysed by similar enzymes. *Curr. Biol* **1998**, *6*, 127–133.
21. Cordell, GA; Quinn-Beattie, ML; Farsnsworth, NR. The potential of alkaloids in drug discovery. *Phytot. Res* **2001**, *15*, 183–205.
22. Hughes, EH; Shanks, JV. Metabolic engineering of plants for alkaloid production. *Metab. Engin* **2002**, *4*, 41–48.
23. Iriti, M; Faoro, F. Review on innate and specific immunity in plants and animals. *Mycopathologia* **2007**, *164*, 57–64.
24. VanEtten, HD; Mansfield, JW; Bailey, JA; Farmer, EE. Two classes of plant antibiotics: Phytoalexins versus 'phytoanticipins'. *Plant Cell* **1994**, *6*, 1191–1192.
25. Osbourn, AE. Preformed antimicrobial compounds and plant defence against fungal attack. *Plant Cell* **1996**, *8*, 1821–1831.
26. Price, KR; Johnson, IT; Fenwick, GR. The chemistry and biological significance of saponins in food and feedingstuffs.*Crit. Rev. Food Sci. Nutr* **1987**, *26*, 27–133.
27. Osbourn, AE. Saponins and plant defence-A soap story. *Trends Plant Sci* **1996**, *1*, 4–9.
28. Mansfield, JW. Antimicrobial Compounds. In *Mechanisms of Resistance to Plant Diseases*; Slusarenko, AJ, Fraser, RSS, van Loon, LC, Eds.; Kluwer Academic Press: Dordrecht, The Netherlands, 2000.
29. Turner, EM. The nature of resistance of oats to the take-all fungus. *J. Exp. Bot* **1953**, *4*, 264–271.
30. Crombie, L; Crombie, WML; Whiting, DA. Structures of the four avenacins, oat root resistance factors to take-

all disease. *J. Chem. Soc., Chem. Commun* **1984**, *4*, 246–248.

31. Crombie, WML; Crombie, L. Distribution of avenacins A-1, A-2, B-1 and B-2 in oat roots: Their fungicidal activity toward 'take-all' fungus. *Phytochemistry* **1986**, *25*, 2069–2073.
32. Osbourn, AE; Clarke, BR; Lunness, P; Scott, PR; Daniels, MJ. An oat species lacking avenacin is susceptible to infection by *Gaeumannomyces graminis* var. *tritici*. *Physiol. Mol. Plant Pathol* **1994**, *45*, 457–467.
33. Osbourn, AE; Clarke, BR; Dow, JM; Daniels, MJ. Partial characterization of avenacinase from *Gaeumannomyces graminis* var.*avenae*. *Physiol. Mol. Plant Pathol* **1991**, *38*, 301–312.
34. Osbourn, AE. Preformed antimicrobial compounds and plant defense against fungal attack. *Plant Cell* **1996**, *8*, 1821–1831.
35. Dow, JM; Callow, JA. A possible role for α-tomatine in the varietal-specific resistance of tomato to *Cladosporium fulvum*. *Phytopathol. Z* **1978**, *92*, 211–216.
36. Smith, CA; MacHardy, WE. The significance of tomatine in the host response of susceptible and resistant tomato isolines infected with two races of *Fusarium oxysporum* f. sp. *lycopersici*. *Phytopathology* **1982**, *72*, 415–419.
37. Pegg, GF; Woodward, S. Synthesis and metabolism of α-tomatine in tomato isolines in relation to resistance to*Verticillium albo-atrum*. *Physiol. Mol. Plant Pathol* **1986**, *28*, 187–201.
38. Zagrobelny, M; Bak, S; Møller, BM. Cyanogenesis in plants and arthropods. *Phytochemistry* **2008**, *69*, 1457–1468.
39. Fenwick, GR; Heaney, RK; Mullin, WJ. Glucosinolates and their breakdown products in food and food plants. *Crit. Rev. Food Sci. Nutr* **1983**, *18*, 123–301.
40. Nahrstedt, A. Cyanogenic compounds as protecting agents for organisms. *Plant Syst. Evol* **1985**, *150*, 35–47.
41. Poulton, JE; Li, CP. Tissue level compartmentation of (*R*)-amygdalin and amygdalin hydrolase prevents large-scale cyanogenesis in undamaged *Prunus* seeds. *Plant Physiol* **1994**, *104*, 29–35.
42. Swain, E; Poulton, JE. Utilization of amygdalin during seedling development of *Prunus serotina*. *Plant Physiol* **1994**,*106*, 437–445.
43. Wang, P; Matthews, DE; VanEtten, HD. Purification and characterization of cyanide hydratase from the phytopathogenic fungus *Gloeocercospora sorghi*. *Arch. Biochem. Biophys* **1992**, *298*, 569–575.
44. Benderoth, M; Textor, S; Windsor, AJ; Mitchell-Olds, T; Gershenzon, J; Kroymann, J. Positive selection driving diversification in plant secondary metabolism. *Proc. Nat. Acad. Sci. USA* **2006**, *109*, 9118–9123.
45. Bidart-Bouzat, MG; Kliebenstein, DJ. Differential levels of insect herbivory in the field associated with genotypic variation in glucosinolates in *Arabidopsis thaliana*. *J. Chem. Ecol* **2008**, *34*, 1026–1037.
46. Mithen, R. Leaf glucosinolate profile and their relationship to pest and disease resistance in oilseed rape. *Euphytica***1992**, *63*, 71–83.
47. Lenman, M; Falk, A; Xue, J; Rask, L. Characterization of a *Brassica napus* myrosinase pseudogene: myrosinases are members of the BGA family of β-glycosidases. *Plant Mol. Biol* **1993**, *21*, 463–474.
48. Greenhalgh, JG; Mitchell, ND. The involvement of flavour volatiles in the resistance to downy mildew of wild and cultivated forms of *Brassica oleracea*. *New Phytol* **1976**, *77*, 391–398.
49. Mithen, R; Magrath, R. Glucosinolates and resistance to *Leptosphaeria maculans* in wild and cultivated *Brassiaca* species.*Plant Breed* **1992**, *108*, 60–68.
50. Harthill, WFT; Sutton, PG. Inhibition of *Mycosphaerella brassicola* ascospores on young cabbage and cauliflower leaves.*Ann. Appl. Biol* **1980**, *96*, 153–161.
51. Mert-Türk, F. Phytoalexins: Defence or just a response to stress? *J. Cell Mol. Biol* **2002**, *1*, 1–6.
52. Grayer, RJ; Harborne, JJ. A survey of antifungal compounds from higher plants. *Phytochemistry* **1994**, *37*, 19–42.
53. Dixon, RA; Paiva, NL. Stress-induced phenylpropanoid metabolism. *Plant Cell* **1995**, *7*, 1085–1097.
54. Bailey, JA; Rowell, PM; Arnold, GM. The temporal relationship between infected cell death, phytoalexin accumulation and the inhibition of hyphal development during resistance of *Phaseolus vulgaris* to *Colletotrichum lindemuthianum*.*Physiol. Plant Pathol* **1980**, *1*, 435–449.
55. Blount, JW; Dixon, RA; Paiva, NL. Stress responses in alfalfa (*Medicago sativa* L.) XVI. Antifungal activity of medicarpin and its biosynthetic precursors: Implications for the genetic manipulation of stress metabolites. *Physiol. Mol. Plant Pathol* **1993**, *41*, 333–349.
56. DiCenzo, GL; VanEtten, HD. Studies on the late steps of (+) pisatin biosynthesis: Evidence for (−) enantiomeric intermediates. *Phytochemistry* **2006**, *67*, 675–683.
57. Jeandet, P; Douillet-Breuil, AC; Bessis, R; Debord, S; Sbaghi, M; Adrian, M. Phytoalexins from the Vitaceae: Biosynthesis, phytoalexin gene expression in transgenic plants, antifungal activity and metabolism. *J. Agric. Food Chem* **2002**, *50*, 2731–2741.
58. Jeandet, P; Sbaghi, M; Bessis, R; Meunier, P. The potential relationship of stilbene (resveratrol) synthesis to anthocyanin content in grape berry skins. *Vitis* **1995**, *34*, 91–94.
59. Iriti, M; Rossoni, M; Borgo, M; Faoro, F. Benzothiadiazole enhances resveratrol and anthocyanin biosynthesis in grapevine, meanwhile improving resistance to *Botrytis cinerea*. *J. Agric. Food Chem* **2004**, *52*, 4406–4413.
60. Glawischnig, E; Hansen, BG; Olsen, CE; Halkier, BA. Camalexin is synthesized from indole-3-acetaldoxime, a key branching point between primary and secondary metabolism in Arabidopsis. *Proc. Nat. Acad. Sci. USA* **2004**, *101*, 8245–8250.
61. Dutsch, HV. Vertical ozone distribution on a global scale. *Pure Appl. Geophys* **1978**, *116*, 511–529.
62. Kerr, JB; McElroy, CT. Evidence for large trend of ultraviolet-B radiation linked to ozone depletion. *Science* **1993**, *262*, 1032–1034.
63. Logan, JA. Tropospheric ozone: seasonal behaviour,

trends and anthropogenic influences. *J. Geophys. Res* **1985**, *90*, 10463–10482.

64. Crutzen, PJ; Lelieveld, J. Human impacts on atmospheric chemistry. *Annu. Rev. Earth Planet. Sci* **2001**, *29*, 17–45.
65. Fowler, D; Cape, JN; Coyle, M; Smith, RI; Hjellbrekke, A-G; Simpson, D; Derwent, RG; Johnson, CE. Modelling photochemical oxidant formation, transport, deposition and exposure of terrestrial ecosystems. *Environ. Pollut* **1999**,*100*, 43–55.
66. Kley, D; Kleinmann, M; Sandermann, H; Krupa, S. Photochemical oxidants: State of the science. *Environ. Pollut* **1999**,*100*, 19–42.
67. Mustafa, MG. Biochemical basis of ozone toxicity. *Free Radical Biol. Med* **1990**, *9*, 245–265.
68. Kelly, FJ; Mudway, I; Krishna, MT; Holgate, ST. The free radical basis of air pollution: Focus on ozone. *Resp. Med* **1995**,*89*, 647–656.
69. Pryor, WA; Das, B; Church, DF. The ozonation of unsaturated fatty acids: Aldehydes and hydrogen peroxide as products and possible mediators of ozone toxicity. *Chem. Res. Toxicol* **1991**, *4*, 341–348.
70. Mudd, JB; Leavitt, R; Ongun, A; McManus, TT. Reaction of ozone with amino acids and proteins. *Atmos. Environ* **1969**,*3*, 669–682.
71. Hussain, MZ; Mustafa, MG; Ghani, QP; Bhatnagar, RS. Stimulation of poly(ADP-ribose) synthetase activity in the lungs of mice exposed to a low level of ozone. *Archiv. Biochem. Biophys* **1985**, *241*, 477–485.
72. Adams-Phillips, L; Wan, J; Tan, X; Dunning, FM; Meyers, BC; Michelmore, RW; Bent, AF. Discovery of ADP-Ribosylation and other plant defense pathway elements through expression profiling of four different *Arabidopsis–Pseudomonas R-avr* interactions. *Mol. Plant-Microbe Interact* **2008**, *21*, 646–657.
73. Bergmann, E; Bender, J; Weigel, J. Ozone threshold doses and exposure-response relationships for the development of ozone injury symptoms in wild plant species. *New Phytol* **1999**, *144*, 423–435.
74. Krupa, S; McGrath, MT; Andersen, CP; Booker, F; Burkey, KO; Chappelka, AH; Chevone, BI; Pell, EJ; Zilinskas, BA. Ambient ozone and plant health. *Plant Dis* **2000**, *85*, 4–12.
75. Torsethaugen, G; Pell, EJ; Assmann, SM. Ozone inhibits guard cell K+ channels implicated in stomatal opening. *Proc. Nat. Acad. Sci. USA* **1999**, *96*, 13577–13582.
76. Black, VJ; Black, CR; Roberts, JA; Stewart, CA. Impact of ozone on the reproductive development of plants. *New Phytol***2000**, *147*, 421–447.
77. Enyedi, AJ; Eckardt, NA; Pell, EJ. Activity of ribulose bisphosphate carboxylase/oxygenase from potato cultivars with differential response to ozone stress. *New Phytol* **1992**, *122*, 493–500.
78. Reddy, GN; Arteca, RN; Dai, YR; Flores, HE; Negram, FB; Pell, EJ. Changes in ethylene and polyamines in relation to mRNA levels of the large and small subunits of ribulose bisphosphate oxygenase in ozone stressed potato foliage.*Plant Cell Environ* **1993**, *120*, 819–826.
79. Guidi, L; Degl'Innocenti, E; Martinelli, F; Piras, M. Ozone effects on carbon metabolism in sensitive and insensitive*Phaseolus vulgaris* cultivars. *Environ. Exp. Bot* **2009**, *66*, 117–125.
80. Toumainen, J; Pellinen, R; Roy, S; Kiiskinen, M; Eloranta, T; Karjalainen, R; Kangasjärvi, J. Ozone affect birch (*Betula pendula* Roth) phenylpropanoid, polyamine and reactive oxygen detoxifying pathways at biochemical and gene expression levels. *J. Plant Physiol* **1996**, *148*, 179–188.
81. Keen, NT; Taylor, OC. Ozone injury in soybeans: Isoflavonoid accumulation is related to necrosis. *Plant Physiol* **1975**,*55*, 731–733.
82. Sharma, YK; Davis, KR. Ozone-induced expression of stress-related genes in *Arabidopsis thaliana*. *Plant Physiol* **1994**,*105*, 1089–1096.
83. Eckey-Kaltenbach, HE; Ernst, D; Heller, W; Sandermann, H. Biochemical plant response to ozone IV. Cross-induction of defensive pathways in parsley (*Petroselinum crispum* L.) plants. *Plant Physiol* **1994**, *104*, 67–74.
84. Dercks, W; Trumble, JT; Winter, C. Impact of atmospheric pollution on linear furanocoumarin in celery. *J. Chem. Ecol***1990**, *16*, 443–454.
85. Rao, MV; Davis, KR. The physiology of ozone-induced cell death. *Planta* **2001**, *213*, 682–690.
86. Heiden, AC; Hoffmann, T; Kahl, J; Kley, D; Klockow, D; Langebartels, C; Mehlhorn, H; Sandermann, H; Schraudner, M; Schuh, G; Wildt, J. Emission of volatile organic compounds from ozone-exposed plants. *Ecolog. Appl* **1999**, *9*, 1160–1167.
87. Sharma, YK; Leon, J; Raskin, I; Davis, KR. Ozone-induced responses in *Arabidopsis thaliana*: The role of salicylic acid in the accumulation of defense-related transcripts and induced resistance. *Proc. Nat. Acad. Sci. USA* **1996**, *93*, 5099–5104.
88. Langebartels, C; Kerner, K; Leonardi, S; Schraudner, M; Trost, M; Heller, W; Sandermann, H. Biochemical plant resonses to ozone I: Differential induction of polyamine and ethylene biosynthesis in tobacco. *Plant Physiol* **1991**, *9*, 882–889.
89. Faoro, F; Iriti, M. Plant cell death and cellular alterations induced by ozone: Key studies in Mediterranean conditions.*Environ. Pollut* **2009**, *157*, 1470–1477.
90. Schraudner, M; Ernst, D; Langebartels, C; Sandermann, H. Biochemical plant responses to ozone III. Activation of the defence-related proteins β-1,3-glucanase and chitinase in tobacco leaves. *Plant Physiol* **1992**, *99*, 1321–1328.
91. Thalmair, M; Bauw, G; Thiel, S; Döhring, T; Langebartels, C; Sandermann, H. Ozone and ultraviolet B effects on the defense-related proteins β-1,3-glucanase and chitinase in tobacco. *J. Plant Physiol* **1996**, *148*, 222–228.
92. Pääkkönen, E; Seppänen, S; Holpainen, T; Kokko, H; Kärenlampi, S; Kärenlampi, L; Kangasjärvi, L. Induction of genes for stress proteins PR-10 and PAL in relation to growth, visible injury and stomatal conductance in birch (*Betula pendula*) clones exposed to ozone and/or

drought. *New Phytol* **1998**, *138*, 295–305.

93. Paolacci, AR; D'Ovidio, R; Marabottini, R; Nali, C; Lorenzini, G; Abenavoli, MR; Badiani, M. Ozone induces a differential accumulation of phenylalanine ammonia-lyase, chalcone synthase and chalcone isomerase RNA transcripts in sensitive and resistant bean cultivars. *Aust. J. Plant Physiol* **2001**, *28*, 425–428.
94. Schubert, R; Fischer, R; Hain, R; Schreier, PH; Bahnweg, D; Ernst, H; Sandermann, H. An ozone-responsive region of the grapevine resveratrol synthase promoter differs from the basal pathogen-responsive sequence. *Plant Mol. Biol***1997**, *34*, 417–436.
95. Sgarbi, E; Baroni Fornasiero, R; Lins, AP; Medeghini Bonatti, P. Phenol metabolism in differentially affected by ozone in two cell lines from grape (*Vitis vinifera* L.) leaf. *Plant Sci* **2003**, *165*, 951–957.
96. Foy, D; Lee, EH; Rowland, R; Devine, TE; Buzzel, RI. Ozone tolerance related to flavonol glycoside genes in soybean.*J. Plant Nutr* **1995**, *18*, 637–647.
97. Saleem, A; Loponen, Y; Pihlaja, K; Oksanen, E. Effects of long-term open-field ozone exposure on leaf phenolics of european silver birch (*Betula pendula* Roth). *J. Chem. Ecol* **2001**, *27*, 1049–1062.
98. Karonen, M; Ossipov, V; Ossopova, S; Kapari, L; Loponen, J; Matsumura, H; Kohno, T; Mikami, C; Sakai, Y; Izuta, T; Pihlaja, K. Effects of elevated carbon dioxide and ozone on foliar proanthocyanidins in *Betula platyphylla*, *Betula ermanii* and *Fagus crenata* seedlings. *J. Chem. Ecol* **2006**, *32*, 1445–1458.
99. Cabané, M; Pireaux, J-C; Léger, E; Weber, E; Dizengremel, P; Pollet, B; Lapierre, C. Condensed lignins are synthesized in poplar leaves exposed to ozone. *Plant Physiol* **2004**, *134*, 586–594.
100. Guidi, L; Degl'Innocenti, E; Genovesi, S; Soldatini, GF. Photosynthetic process and activities of enzymes involved in the phenylpropanoid pathway in resistant and sensitive genotypes of *Lycopersicon esculentum* L. exposed to ozone.*Plant Sci* **2005**, *168*, 153–160.
101. Galliano, H; Heller, W; Sandermann, H. Ozone induction and purification of spruce cinnamyl alcohol dehydrogenase.*Phytochemistry* **1993**, *32*, 557–563.
102. Tomlinson, H; Rich, S. Effect of ozone on sterols and sterol derivatives in bean leaves. *Phytopathology* **1971**, *61*, 1404–1405.
103. Trevathan, LE; Moore, LD; Orcutt, DM. Symptom expression and free sterol and fatty acid composition of flue-cured tobacco plants exposed to ozone. *Phytopathology* **1979**, *69*, 582–585.
104. Grunwald, C; Endress, AG. Foliar sterols in soybeans exposed to chronic levels of ozone. *Plant Physiol* **1985**, *77*, 245–247.
105. Grunwald, C. Effects of free sterols, steryl ester and steryl glycoside in membrane permeability. *Plant Physiol* **1971**, *48*, 653–655.
106. Evans, LS; Ting, IP. Ozone-induced membrane permeability changes. *Am. J. Bot* **1973**, *60*, 155–162.
107. Grunwald, C. Effects of free sterols, steryl ester and steryl glycoside in membrane permeability. *Plant Physiol* **1974**, *54*, 624–628.
108. Menser, HA; Chaplin, JF; Cheng, ALS; Sorokin, T. Polyphenols, phytosterols and reduced sugar in air-cured tobacco leaves injured by ozone air pollution. *Tobacco Sci* **1977**, *21*, 35–38.
109. Wegener, A; Gimbel, W; Werner, T; Hani, J; Ernst, D; Sandermann, H. Molecular cloning of ozone-inducible protein from *Pinus sylvestris* L. with high sequence similarity to vertebrate 3-hydroxy-3-methylglutaryl-CoA-synthase.*Biochim. Biophys. Acta* **1997**, *1350*, 247–252.
110. Cheng, A-X; Lou, Y-G; Mao, Y-B; Lu, S; Wang, L-J; Chen, X-Y. Plant terpenoids: Biosynthesis and ecological functions.*J. Integr. Plant Biol* **2007**, *49*, 179–186.
111. Shamay, Y; Raskin, VI; Brandis, AS; Steinberger, HE; Marder, JB; Schwartz, A. Ozone treatment affects pigment precursor metabolism in pine seedlings. *Physiol. Plant* **2001**, *112*, 285–292.
112. Kesselmeier, J; Staudt, M. Biogenic volatile organic compounds (VOC): An overview on emission, physiology and ecology. *J. Atmos. Chem* **1999**, *33*, 23–88.
113. Dicke, M. Chemical ecology of host-plant selection by herbivorous arthropods: A multitrophic perspective. *Biochem. Syst. Ecol* **2000**, *28*, 601–617.
114. Brilli, F; Ciccioli, P; Frattoni, M; Prestininzi, M; Spanedda, AF; Loreto, F. Constitutive and herbivore-induced monoterpenes emitted by *Populus euroamericana* leaves are key volatiles that orient *Chrysomela populi* beetles. *Plant Cell Environ* **2009**, *32*, 542–552.
115. Peñuelas, J; Llusià, J. The complexity of factors driving volatile organic compound emissions by plants. *Biol. Plant***2001**, *44*, 481–487.
116. Llusià, J; Peñuelas, J; Gimeno, BS. Seasonal and species-specific response of VOC emissions by Mediterranean woody plant to elevated ozone concentrations. *Atmos. Environ* **2002**, *36*, 3931–3938.
117. Grote, R; Niinemets, U. Modeling volatile isoprenoid emissions–A story with split ends. *Plant Biol* **2008**, *10*, 8–28.
118. Roselle, SJ. Effects of biogenic emission uncertainties on regional photochemical modelling of control strategies. *Atmos. Environ* **1994**, *28*, 1757–1772.
119. Pinto, DM; Blande, JD; Nykanen, R; WenXia, D; Ner, AM; Holopainen, JK. Ozone degrades common herbivore-induced plant volatiles: Does this affect herbivore prey location by predators and parasitoids? *J. Chem. Ecol* **2007**, *33*, 683–694.
120. Himanen, SJ; Nerg, AM; Nissinen, AM; Pinto, DM; Stewart, CN, Jr; Poppy, GM; Holopainen, JK. Effects of elevated carbon dioxide and ozone on volatile terpenoid emissions and multitrophic communication of transgenic insecticidal oilseed rape (*Brassica napus*.). *New Phytol* **2009**, *181*, 174–186.
121. Freiwald, V; Haikio, E; Julkunen-Tiitto, R; Holopainen, JK; Oksanen, E. Elevated ozone modifies the feeding behaviour of the common leaf weevil on hybrid aspen through shifts in developmental, chemical, and structural

properties of leaves. *Entomol. Exp. Appl* **2008**, *128*, 66–72.

122. Alstad, DN; Edmunds, GF, Jr; Weinstein, LH. Effects of air pollutants on insect populations. *Ann. Rev. Entomol* **1982**,*27*, 369–384.

123. Loreto, F; Velikova, V. Isoprene produced by leaves protects the photosynthetic apparatus against ozone damage, quenches ozone products and reduces lipid peroxidation of cellular membrane. *Plant Physiol* **2001**, *127*, 1781–1787.

124. Loreto, F; Ferranti, F; Mannozzi, M; Maris, C; Nascetti, C; Pasqualini, S. Ozone quenching properties of isoprene and its antioxidant role in plants. *Plant Physiol* **2001**, *126*, 993–1000.

125. Velikova, V; Fares, S; Loreto, F. Isoprene and nitric oxide reduce damages in leaves exposed to oxidative stress. *Plant Cell Environ* **2008**, *31*, 1882–1894.

126. Loreto, F; Pinelli, P; Manes, F; Kollist, K. Impact of ozone on monoterpene emission and evidence for an isoprene-like antioxidant action of monoterpenes emitted by *Quercus ilex* leaves. *Tree Physiol* **2004**, *24*, 361–367.

127. Vitale, M; Salvatori, E; Loreto, F; Fares, S; Manes, F. Physiological responses of *Quercus ilex* to water stress and acute ozone exposure under controlled conditions. *Water Air Soil Pollut* **2008**, *189*, 113–125.

128. Menser, HA; Chaplin, JF. Effects of ozone air pollution on nitrogen constituents, pH and water soluble ash of air-cured tobacco leaves. *Tobacco Sci* **1975**, *19*, 108–110.

129. Aycock, MK. Phenotypic correlation of weather fleck with certain agronomic and chemical traits in Maryland tobacco.*Tobacco Sci* **1975**, *19*, 96–97.

130. Jackson, DM; Rufty, TW; Heagle, AS; Severson, RF; Eckel, RVW. Survival and development of tobacco hornworm larvae on tobacco plants grown under elevated levels of ozone. *J. Chem. Ecol* **2000**, *26*, 1–19.

131. Menser, HA; Chaplin, JF. Air pollution: Effects of phenol and alkaloid content of cured tobacco leaves. *Tobacco Sci***1969**, *13*, 169–170.

132. Jeffords, MR; Endress, AG. Possible role of ozone in tree defoliation by the gypsy moth (Lepidoptera: Lymantriidae).*Environ. Entomol* **1984**, *13*, 1249–1252.

133. Endress, AG; Post, SC. Altered feeding preferences for Mexican bean beetle *Epilachna varivestis* for ozonated soybean foliage. *Environ. Pollut* **1985**, *39*, 9–16.

134. Trumble, JT; Hare, JD; Musselmann, RC; McCool, PM. Ozone-induced changes in host-plant suitability: interactions of*Keiferia lycopersicella* and *Lycopersicon esculentum*. *J. Chem. Ecol* **1987**, *13*, 203–218.

135. Chappelka, AH; Kraemer, ME; Mebrahtu, T; Rangappa, M; Benepal, PS. Effects of ozone on soybean resistance to Mexican bean beetle (*Epilachna varivestis* Mulsant). *Environ. Exp. Bot* **1988**, *28*, 53–60.

136. Heagle, AS; Branderburg, RL; Burns, JC; Miller, JE. Ozone and carbon dioxide effects on spider mites in white clover and peanut. *J. Environ. Qual* **1994**, *23*, 1168–1176.

137. Martin-Tanguy, J. Metabolism and function of polyamines in plants: Recentdevelopment (new approaches). *Plant Growth Regul* **2001**, *34*, 135–148.

138. Walters, DR. Polyamines in plant-microbe interactions. *Physiol. Mol. Plant Pathol* **2000**, *57*, 137–146.

139. Liu, J-H; Kitashiba, H; Wang, J; Ban, Y; Moriguchi, T. Polyamines and their ability to provide environmental stress tolerance to plants. *Plant Biotechn* **2007**, *24*, 117–126.

140. Bagni, T; Tassoni, A. Biosynthesis oxidation and conjugation of aliphatic polyamines in higher plants. *Amino Acids***2001**, *20*, 301–317.

141. Rowland-Bamford, AJ; Borland, AM; Lea, PJ; Mansfield, TA. The role of arginine decarboxylase in modulating the sensitivity of barley to ozone. *Environ. Pollut* **1989**, *61*, 95–106.

142. Langebartels, C; Wohlgemuth, H; Kschieschan, S; Grün, S; Sandermann, H. Oxidative burst and cell death in ozone-exposed plants. *Plant Physiol. Biochem* **2002**, *40*, 567–575.

143. Ormrod, DP; Beckerson, DW. Polyamines as antiozonant for tomato. *HortScience* **1986**, *21*, 1070–1071.

144. Bors, W; Langebartels, C; Michel, C; Sandermann, H. Polyamines as radical scavengers and protectans agaist ozone damages. *Phytochemistry* **1989**, *28*, 1589–1595.

145. Bouchereau, A; Aziz, A; Larher, F; Martin-Tanguy, J. Polyamines and environmental challenges: Recent development.*Plant Sci* **1999**, *140*, 103–125.

146. Hemantaranjan, A. Plant secondary metabolism in response to the environment. *Adv. Plant Physiol* **2008**, *10*, 31–89.

147. Wink, M. Plant secondary metabolism:diversity, function and its evolution. *Nat. Prod. Comm* **2008**, *3*, 1205–1216.

148. Violini, G. Ozone and plant pathogens: an overview. *Riv. It. Pat. Veg* **1995**, *5*, 113–130.

149. Booker, F; Muntifering, R; McGrath, M; Burkey, K; Decoteau, D; Fiscus, E; Manning, W; Krupa, S; Chappelka, A; Grantz, D. The ozone component of global change: Potential effects on agricultural and horticultural plant yield, product quality and interactions with invasive species. *J. Integrat. Plant Biol* **2009**, *51*, 337–351.

150. von Tiedemann, A; Firsching, KH. Interactive effects of elevated ozone and carbon dioxide on growth and yield of leaf rust-infected versus non-infected wheat. *Environ. Pollut* **2000**, *72*, 205–224.

151. Lorenzini, G; Medeghini-Bonatti, P; Nali, C; Baroni Fornasiero, R. The protective ffect of rust infection against ozone, sulphur dioxide and paraquat toxicity symptoms in broad bean. *Physiol. Mol. Plant Pathol* **1994**, *45*, 263–279.

152. Walters, DR; Boyle, C. Induced resistance and allocation costs: What is the impact of pathogen challenge? *Physiol. Mol. Plant Pathol* **2005**, *66*, 40–44.

153. Walters, DR; Heil, M. Costs and trade offs associated with induced resistance. *Physiol. Mol. Plant Pathol* **2007**, *71*, 3–17.

154. Iriti, M; Faoro, F. Bioactivity of grape chemicals for human health. *Nat. Prod. Comm* **2009**, *4*, 611–634.

CHAPTER 3 | EXPOSURE-PLANT RESPONSE OF AMBIENT OZONE

S. Deb Roy, G. Beig, and Sachin D. Ghude

INTRODUCTION

Near surface ozone is a pollutant of important concern due to its adverse effects on agricultural productivity (Mills et al., 2007) and human health (Avol et al., 1998), and is now a major environmental concern in many regions of the world. Surface ozone is mainly produced through photochemical reactions involving volatile organic compounds and NOx in the presence of sunlight, and also acts as a precursor for the highly reactive hydroxyl radical (Logan et al., 1981; Thompson 1992). Epidemiological and toxicological studies indicate that higher concentration of ozone in the boundary layer is harmful to biological health (Avol et al., 1998). The enhanced levels of ozone close to surface also decrease the yield of agricultural crops (Mills et al., 2007). A number of recent studies from selected sites across South Asia have specifically addressed the sensitivity of locally-grown crops to the ozone levels (Agrawal, 2003; Wahid, 2003). Some studies done elsewhere in the world (Tuovinen et al., 2000) have focused on the application of micrometeorological principles to improve the characterisation of vegetation exposure to ozone and described the transition from the exposure index AOT40 to a flux-based approach for estimating ozone risks. Tuovinen et al. (2002) have discussed about the possibility of estimating the value of the AOT40 exposure index from time-averaged (passive sampling) data by using a simplified statistical method even in the absence of continuous data, provided the time-averaged concentration can be measured reliably. Studies like Keller et al. (1999) have shown that assessment based on stomatal ozone flux instead of AOT40, can be used to differentiate ozone risks between regions and years at a national scale. Grunhage et al. (1999) talked about ¨ improving the usage of European critical levels for ozone by deriving meaningful, reliable and effective dose-response relationships with the application of effective AOT40, which suggests that the estimated crop losses using the effective dose are within the bounds of probability. Few other studies (Fuhrer et al., 1997) provide a critical assessment of the scientific basis of the concepts used to define critical levels for ozone and identifies the key limitations and uncertainties involved. Girgzdiene et al. (2007) have analysed the ground-level ozone data and AOT40 in the rural areas of Lithuania and discussed about the variations and trends in the same. Ghude et al. (2006) have shown that the observed surface ozone levels over Delhi are high enough to exceed "Critical Levels" which

---

Citation: Deb Roy, S., Beig, G., and Ghude, S. D.: Exposure-plant response of ambient ozone doi:10.5194/acp-9-5253-2009, 2009.

are considered to be safe for human health. However, detailed results related to AOT40 for vegetation index based on observational data for the Indian region is sparse. Using surface level measurements from Pune (18.54° N, 73.81° E), Beig et al. (2008) have recently calculated the AOT40 and shown that ozone levels have surpassed the "Critical Levels" in most parts of the year for vegetation and forest. The monitoring station is located at the northwestern parts of the Pune city which is a well industrialized semi urban city of India at an altitude of around 550 m. Surrounded mostly by hilly areas the site is at a considerable distance from the major traffic junctions and hence de- fined as semi-urban rather than urban. This is the only work where detailed study about the AOT40 based on observed data has been reported over any Indian station. In the past few decades, India has been undergoing rapid industrial and economic growth with increasing emissions of trace gases and pollutants (Ghude et al., 2008, 2009). Due to all these reasons, the tropospheric ozone level is an issue of concern in the last few decades. Large variabilities even in the vertical distribution of tropospheric ozone over a subtropical site in the IG area, have been reported and analyzed by Gupta et al. (2007). Necessary and effective emission reduction strategies are required to be developed in order to curb the surface level ozone pollution to protect the vegetation from further damage in India whose economy is highly dependent on agricultural sector. Hence, more efforts are needed to understand the spatial and temporal distribution of cumulative exposure indices and threshold exceedances over the whole domain of India. Presently the number of measurement sites having valid and long-term representative measurements of surface ozone in India are quite a few (Pulikesi et al., 2006; Debaje and Kakade, 2008), but the data are scattered and the coverage and time resolution is not sufficient and not very systematic which is essential for a correct and reliable assessment of regional distribution of AOT40. Regional chemical transport models have the potential to fill in this gap. There are few recent modeling studies aimed at describing gridded distribution of monthly AOT40 values over the Indian region (Mittal et al., 2007; Engardt, 2008). The former study was performed using a global emission inventory and the latter one uses an Asian emission inventory (Streets et al., 2003). However, recently a new gridded emission inventory of ozone precursors over the Indian geographical region have been prepared based on micro-level of all activity data (Dalvi et al., 2006; Beig and Brasseur, 2006) which are claimed to be better and found to be quite different than used by earlier modelers. Recently, Roy et al. (2008) have used this new emission inventory in the regional 3-D chemistry transport model for Indian region but the impact of cumulative ozone exposure (AOT40) over the entire Indian region has not been reported so far.

In view of the fact that sufficient measurements of AOT40 are very sparse over the Indian region, and there is a need for estimating the AOT40 over the complete Indian region, we made an attempt in this paper to simulate the ozone cumulative exposure indices and threshold exceedances using the regional chemistry-transport model (REMO-CTM) with new Indian emission inventory for the year 2003 as base case. We have also assessed the simulated results by comparing it with the statistically robust observational data (Beig et al., 2008) available for the suburban site Pune.

METHOD

Model Details

The model REMO-CTM is a 3-D regional, off-line, eulerian and hydrostatic model. The model has been applied with the ECHAM-4 physics (Jacob et al., 2001; Jacob et al., 2007) in the current study. A terrain following hybrid pressure-sigma coordinate is used in the vertical direction with 20 levels of increasing thickness between surface to 10 hPa pressure level. The horizontal resolution is 0.5 degree on a spherical rotated grid. Further detail of this model is available in Langmann (2000) with references therein. In this study the meteorological parameters generated from the atmospheric component of the on-line model REMO were used to drive the off-line chemical-transport model REMO-CTM. After initializing REMO-CTM once, the model was run in climate mode for the year 2003 using ECMWF analyses at the lateral boundaries which have been updated every 6 h. The emissions for all the species were obtained from the Reanalysis of the Tropospheric chemical composition (RETRO) data which provides $0.5^{\circ}\times0.5^{\circ}$ emission data set (Olivier et al., 2003) and the recent high-resolution estimates of the new Indian national emission inventory (Roy et al., 2008 and reference therein).

AOT40

A number of methodologies have been developed to assess the damaging effects of ozone on vegetation. AOT40 is most widely used exposure plant response index set by the United Nations Economic Commission for Europe (UNECE), United States Environmental Protection Agency (USEPA) and World Meteorological Organization (WMO) (World Health Organization (WHO), 2000). All these measurements are based on the concentrations occurring during the sunlit hours of the vegetation seasons. The AOT40 values of 3000 ppb h accumulated over a 3 months growing season, and 10 000 ppb h over 6 months, correspond to the critical levels (5% yield loss) for the protection of agricultural crops and forest respectively. It is implied that the critical level for daily AOT40 values is around 33 ppb h (~3000/90) which should not be exceeded. The Plant Exceedance Days (PED) are defined as those days when the daily AOT40 value exceeds the critical limit (Beig et al., 2008). The cumulative ozone exposure index (AOT40) is calculated as the sum of differences between the hourly ozone concentrations and 40 ppb, for each daylight hour with global radiation greater than or equal to 50 Wm^{-2} during the growing season. In the current study, we have calculated AOT40 for every day by summing the differences between the simulated hourly ozone concentrations at surface (in ppb) and 40 ppb, for every hour when the concentration is greater than 40 ppb during the daylight hours between 07:00 a.m.–07:00 p.m. The global radiation over different parts of the Indian geographical region varies depending on the location, month and season of the year. It is difficult to demarcate the different areas of the Indian region based on the distribution of global radiation during the different time periods of the year. Although the global radiation of 50 Wm^{-2} can be observed one hour after sunrise and one hour before sunset (eventually reducing the length of daylight hours by approximately two hours), yet the sunrise and sunset timings vary even in a much shorter distance. Therefore, accounting data as per exact timing and keeping the time duration equal (from statistical consideration) is very complicated issue. We have also assessed monthly modeled AOT40 data at our chosen site Pune where the observed monthly modeled AOT40 data is available for the 07:00 a.m.–07:00 p.m. time window (which is the average sunset and sunrise time at pune). In view of this we have considered uniform time window between 07:00 a.m. and 07:00 p.m. for calculations of AOT40 vales at each grid point over the Indian domain. Since the hourly concentration of ozone in general does not rise immediately close to sunrise and is mostly in the waning phase close to sunset, this addition of one or two hours near the sunrise and sunset time may not significantly effect the AOT40 values over the Indian domain, although our projection may be slightly on the higher side. In our expert opinion, we feel that the uncertainty due to choice of timing interval in the results of AOT40 should not exceed more than +/−5%. The daily AOT40 values are added when AOT40 over a month period is calculated. In addition to daily AOT40 values, we have also calculated daily 8-h mean (for 11–18 h) and daily maximum ozone concentration from the simulated results. It is important to note that the critical levels for AOT40 as mentioned in the EU directive are defined as multi-year means so as to compensate for annual variations (Klumpp et al., 2006) but the results presented here are based on the values for a particular year 2003. It needs to be mentioned here that we have extracted the daily, monthly and 3-monthly AOT40 values from the modeled hourly concentrations of ozone over the Indian region and from the observed hourly ozone data at the receptor site Pune, with the application of codes and scripts following the above mentioned definition.

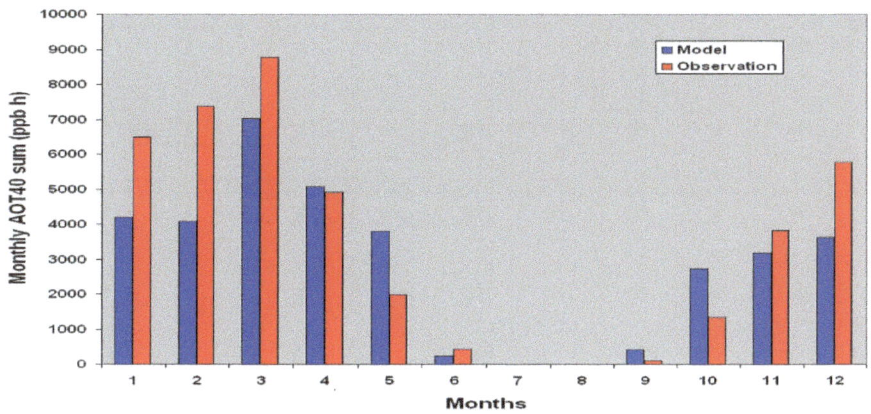

Figure 1. Comparison of model simulated monthly values of AOT40 with observational data for the year 2003 at Pune.

In the following section, we present the model calculated monthly and daily AOT40 values, daily 8-h average and the daily maximum ozone levels over the Indian region for the year 2003. We also calculated the model simulated monthly distribution of AOT40 values over the Indian region for the months of May 2003, July 2003 and October 2003. These three months have been chosen as the representative months for the pre-monsoon, monsoon and post-monsoon seasons over India. At the end we provide the modeled 3-monthly distribution of AOT40 over India for the months of November to January and February to April.

RESULTS AND DISCUSSION

Figure 1 shows the comparison between monthly AOT40 values calculated from the model and from the point observation at Pune, for the year 2003. It is found that observed monthly AOT40 values lie reasonably in the same range as the modeled values at Pune indicating that our model is capable of reproducing the monthly AOT40 values. It is interesting to note that the UNECE and WHO guideline for the critical limit of AOT40 (3000 ppb h for 3 months) is exceeded even in the individual months. At Pune, elevated monthly AOT40 values were found mostly between November and April both in observed and simulated results, which corresponds to the "Rabi" crop growing season in India. Rabi and Kharif are two types of crop seasons in India, based on the time of sowing and harvesting. The Rabi crop season starts from October/November and harvested in March/April. These are also called winter crops and some examples are Wheat, Barley, Mustard, Sesame, Peas etc. The highest level of monthly AOT40 is seen for the month of March'03 both in observed and model results. Model underestimates AOT40 values during late winter and pre-monsoon season at Pune, although the pattern of seasonal variation is similar between the observed and modeled AOT40 values. Over a 6-month period (November–April) the observed AOT40 cumulative ozone exposure exceeded up to 36 000 ppb h which is almost 3.6 times the critical level set for the protection of forest. While, modeled AOT40 cumulative ozone exposure exceeded up to 27 100 ppb h during this period, which is around 2.7 times higher that the critical level for the forest. Beig et al. (2008) reported that the elevated levels of surface ozone and excess of AOT40 values during this period is due to the long range transport of background ozone and its precursors at Pune from the northern plains of Indian sub-continent, which is affluent in ozone and precursor gases (Beig and Ali, 2006). The lowest monthly AOT40 values are generally found during monsoon months. This is due to the inflow of clean marine air mass towards the Indian region brought by the southwesterly winds prevailing during this period and lesser solar insolation (due to cloudy conditions) which leads to lower photochemical production of ozone during this time (Beig et al., 2008). A systematic underestimation in AOT40 (~35%) in the model results during the periods of highest ozone concentration (December to March) has been observed. The observational site is located at the northwestern parts of the Pune city (18.54 N, 73.81 E) which is a well industrialized semi urban city of India. Most of the industries are situated towards the north eastern and eastern side around 20–25 km away. Since the wind pattern is predominantly northeasterly during the said period, pollutants emitted from these industries can affect the ozone chemistry at the receptor site. Since the model is fed with monthly emissions on a 0.5×0.5 degree resolution, these local emission effects have not been incorporated as emission inputs to the model. This may have led to the observed systematic underestimation in AOT40 during the periods of highest ozone concentration (December to March). Therefore, there may be an underestimation or overestimation of AOT40 values depending on how much a particular site is affected by local emissions which has not been captured by the model due to coarser resolution. However, we wish to mention here that the ozone distribution over the IGP region, where hotspots are noticed, is largely controlled by the synoptic weather conditions and prevailing dynamics which are very well accounted in the model (Beig and Ali, 2006; Roy et al., 2008) as compared to local emissions and hence uncertainty on these hotspot regions are likely to be much less than Pune region.

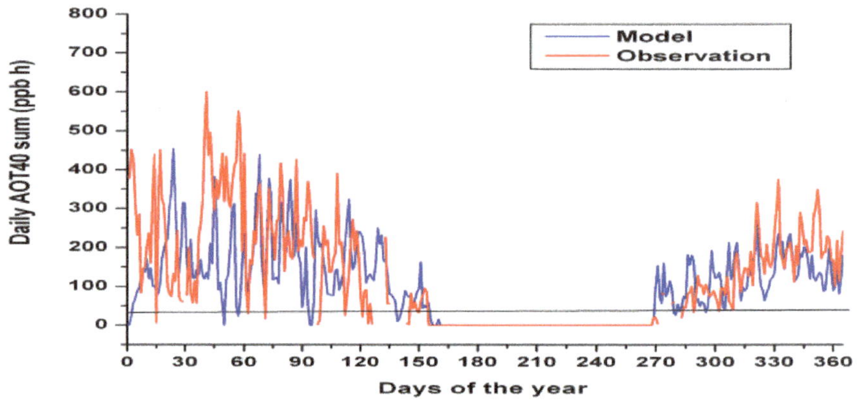

Figure 2a. Comparison of model calculated daily values of AOT40 with observed data for the year 2003 at Pune. Black horizontal grid line through 33 ppb h corresponds to the Plant Exceedance Days (PED).

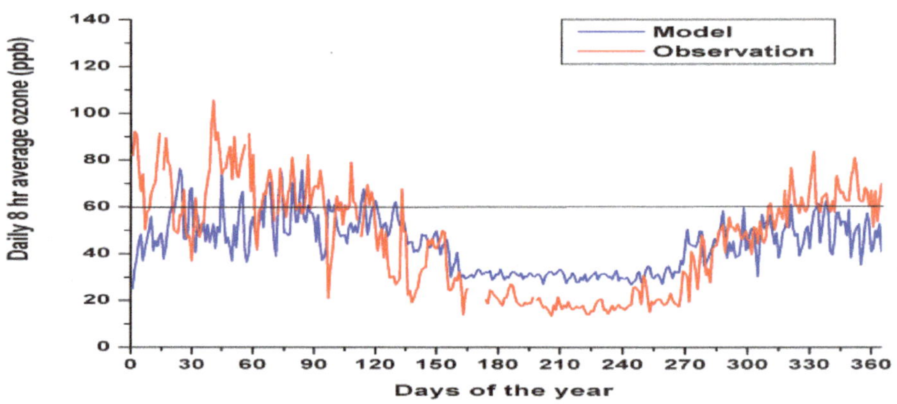

Figure 2b. Comparison between modeled and observed daily 8 h average of ozone concentration at Pune for the year 2003. Black horizontal grid line through 60 ppb corresponds to the health protection threshold (8 h>60 ppb).

Figure 2a, b and c compares the modeled and observed daily AOT40 values (PED), daily 8-h average and daily maximum ozone levels respectively at Pune for the year 2003. The modeled daily AOT40 values, daily 8-h average and daily maximum ozone concentration varies from 0 to 454 ppb h, 25 to 76 ppb and, 31 to 100 ppb respectively. Whereas, the corresponding observed values were found to vary approximately from 0 to 600 ppb h, 13 to 107 ppb and 15 to 122 ppb respectively. Daily AOT40 values were found to be almost zero during most of the time in the monsoon (June–September). Since July to September is mostly growing season for Kharif crops in India like mainly rice in Northern India, low values of AOT40 during this period indicates no threat to crop production. The daily maximum ozone levels and the daily 8-hour average ozone concentration were found to be comparatively lower during the same period. Reduced solar insolation due to frequent cloudy conditions along with the influx of fresh marine air from the oceanic regions towards the Indian land masses leads to lesser photochemical production of ozone during this period. Although the model qualitatively reproduces a similar phenomena, a reasonable quantitative agreement was not obtained in this regard possibly due to the inability of the regional model with 0.5×0.5 degree resolution, to capture any sudden, localized excess rainfall event which has happened during the monsoon months of the year 2003 as per the rainfall records (IMD-report, 2003) which is discussed in detail by us elsewhere (Roy et al., 2008). Although there are quite a few measurements of ozone reported from India, none of them have reported the AOT40 assessment. Such kind of calculation requires data in much more detailed form than reported by them and we do not have access to any of them in such a detailed manner and hence could perform the AOT40 calculations using observations only for Pune. In our complimentary

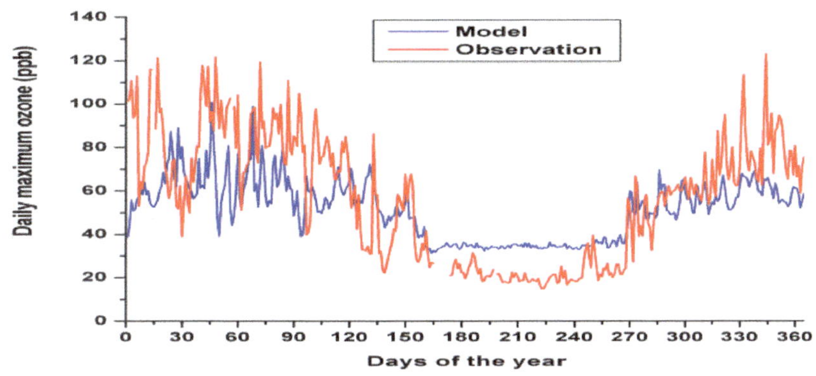

Figure 2c. Comparison between modeled and observed daily maximum ozone concentrations for the year 2003 at Pune.

JGR paper (Roy et al., 2008) we have shown that the model predicts the geographical distribution of chemical species including ozone concentration (from which AOT40 is derived) over India during the year 2003 to a reasonable extent when compared with observed data. The seasonal cycles of ozone at four receptor sites over the Indian region for the year 2003 has shown good qualitative agreement in comparison with observed data, although there were few quantitative disagreements due to the reasons which have been discussed in detail in (Roy et al., 2008). In addition, the meteorological as well as trace gas data obtained by satellite measurements are compared with model data for whole of the Indian geographical region. We have shown in Roy et al. (2008) that the model has been able to capture the gradient in ozone concentrations between the Indo-Gangetic plains and the remaining parts of India with higher values over the IGP area as has been reported earlier in several observational and modeling studies (Beig and Ali, 2006; Ali et al., 2004). So we are confident that the model performance at Pune with respect to AOT40 can be extrapolated to other sites with different meteorological environment and emission scenario like over the Indo-Gangetic plains where the emission hotspots have been identified, as model is able to capture the distinct features in ozone distribution as discussed above. However, uncertainties due to emission inputs can exist and it is difficult to estimate the uncertainty in the simulated (absolute) concentrations due to the uncertainties in emissions, but is likely to be within 2–5% (Beig and Brasseur, 2006).

Figure 3 shows the modeled AOT40 distribution at surface level over the Indian sub-continent for the months of May, July and October 2003. It is worth mentioning that even in a single month period (particularly during the month of May) large area of India shows AOT40 values above the 3000 ppb h (for 3 months), which is related to 5% decrease of the yield of many important crops in India. It is also observed that maximum AOT40 values are reached at different times at different regions across India. In all the three months discussed in this study, simulated AOT40 values are found to be substantially higher over the Indo-Gangetic (IG) plains than the other regions of India, which can have an adverse effect on vegetation in this region. However, during the southwest monsoon season comprising the months from June to September, comparatively reduced AOT40 values are observed over the IG region. The maximum level of AOT40 over the Indo-Gangetic plains of India is found to be ~14 000 ppb h during October and ~12 000 ppb h during May. While, during July the maximum level of AOT40 is found to be around 6000 ppb h over the IG-region. As can be seen from Fig. 3, the AOT40 values are higher than the critical level in some parts of western India during the month of October. The Kharif crop season lasts from June/July to September/October. These are also known as summer crops. Some examples of these crops are Millets, Paddy, Maize, Pulses, Groundnut, Red Chillies, Cotton, Soyabean, Sugarcane, Turmeric etc. Whereas over some parts of southern India, AOT40 values are higher than the critical level during the month of May. Comparatively low values of AOT40 are seen in the month of July 2003 over most of the geographical region of India except over the IG-region where maximum level of AOT40 is found to be 6500 ppb h. The reduced AOT40 values over much of the Indian region during July is mainly due to the south west monsoon circulation which helps in transporting pristine marine air towards the Indian land masses. Also lower solar radiation due to prevailing cloudy situations leads to lesser photochemical production of ozone. While, the higher AOT40 values over the IG-region as compared to the remaining parts of India in all the three months, can be attributed to the fact that this region is affluent in the ozone precursors due to high anthropogenic activity (Ghude et al., 2007; Ghude et al., 2008). The densely populated Indo-Gangetic plains have a substantial level of rural population as well as industries. This makes the IG-region prone to anthropogenic emissions. Included among the emission sources scattered over the entire IG region are coalbased thermal power plants, steel, sugar, and other small and medium industries (several of which use coal as fuel), fossil fuel extraction (coal mining, crude oil production, natural gas production, etc.) and open burning of litter and biofuels used for domestic cooking (Ghude et al., 2008; Beig et al., 2008; Sahu et al., 2008). All these sources lead to observed high ozone values over IG region. But the reasons for high concentration of ozone are both photochemical production as well as prevailing dynamics (Beig and Ali, 2006; Roy et al., 2008) since, this highly populated area (Ghude et al., 2008) is prone to confinement of tracers due to its favorable synoptic conditions and orographic features which favours the photochemical buildup of ozone (Beig et al., 2006; Kulkarni et al., 2009).

In Fig. 4. we provide the 3-monthly AOT40 values over the Indian region for the months from November to January and February to April. The AOT40 values reach as high as 50 000 ppb h during the months of November to January towards the eastern side of the Indo-Gangetic plains where the emission intensity of the precursor gases are quite high. In the months of February to April, the highest levels of AOT40 is seen to be around 40 000 ppb h in the gangetic West Bengal. In some parts of south west India where the precursor emissions are on the higher side, elevated levels of AOT40 of around 25 000 ppb h are found in both the 3- monthly periods. The AOT40 values for the 3-months period are seen to be comparatively higher over the Indo-Gangetic plains as compared to most of the remaining parts of India. This region is abundant in ozone precursors due to higher prevalence of anthropogenic activities. Also due to its suitable synoptic conditions and orographic features this area tends to confine tracers which enhances the photochemical buildup of ozone. In many parts of Central, Southeastern and extreme Northern India where lower levels of precursor emission prevails, the AOT40 values for 3-month periods are quite low. Overall the AOT40 values for both the 3-monthly periods from November to January and February to April show similar distribution pattern. We have compared our model calculated monthly and 3-monthly distribution of AOT40 values over the Indian region for the year 2003, with those obtained from two other studies (Mittal et. al., 2007; Engardt, 2008). The former study uses a regional model HANK, fed with a global emission inventory to predict the monthly AOT40 values (accumulated during daylight hours from 08:30 a.m.–04:30 p.m., IST) over the Indian region for few months from February 2000 to May 2000. The latter study uses a regional model MATCH with an Asian emission inventory for calculating the 3-monthly AOT40 distribution over India for the year 2000. However, the exact number of daylight hours for which the AOT40 has been calculated is not specifically mentioned in this study. Moreover the colour scales and colour legends along side the AOT40 plots in these studies seems to be poorly resolved. Considering the above mentioned facts, a detailed, point-to-point and sideby-side comparison was not possible in this regard. However, on a crude comparison, we find that the range of our model calculated 3-monthly AOT40 values over the Indian region is close to those predicted by the MATCH model but on the higher side as compared to the results obtained from the model HANK. The pattern of the geographical distribution of our modeled AOT40 values over India for 3-months period appears to be similar with those predicted over the Indian region by the MATCH model. This similarity is not so prominent when we compare our model calculated monthly AOT40 values over the Indian region with the ones obtained from the HANK model. A reason behind this may be that the cellular spatial structure of the distribution pattern is not distinctly visible when a continuous colour scale is used for the gridded plot as has been done for the AOT40 graphs obtained from HANK.

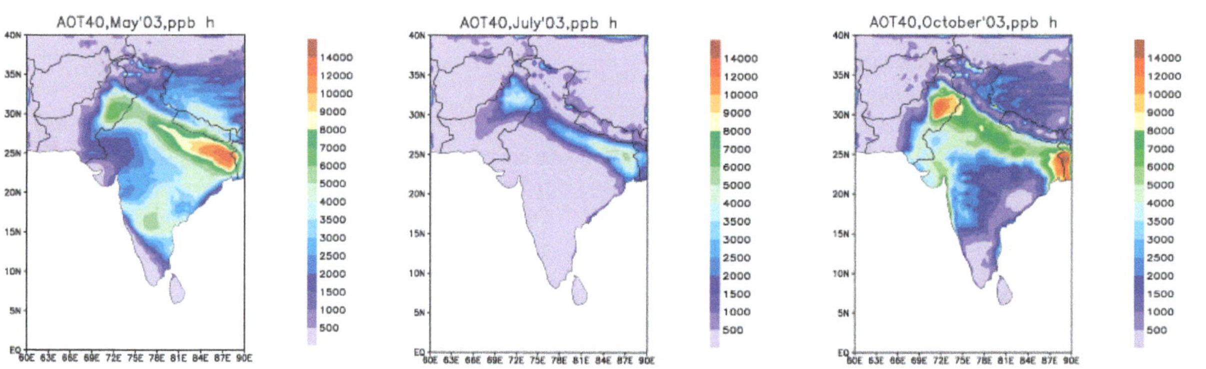

Figure 3. AOT40 values for the months of May 2003, July 2003 and October 2003 over the Indian region as simulated by REMO-CTM.

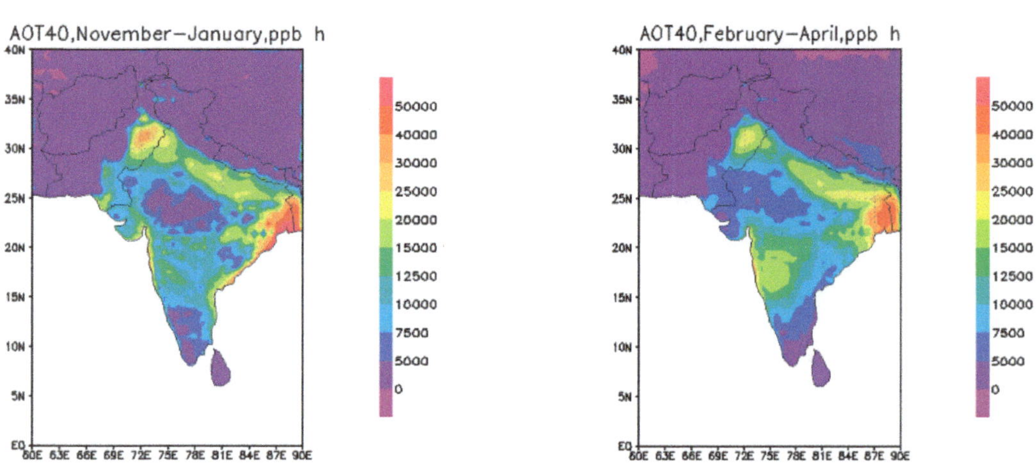

Figure 4. Modeled 3-monthly AOT40 values for the months of November to January and February to April over the Indian region.

SUMMARY AND CONCLUSIONS

In absence of sufficient observational information available about AOT40 over the Indian region, we use the regional chemistry-transport model to estimate cumulative exposure indices in the Indian region for the year 2003. We have also presented a comparison between modeled and measured monthly AOT40 values, daily 8-h average, daily maximum near surface ozone concentration and daily AOT40 vales at suburban station (Pune). Model is able to reproduce the observed monthly AOT40 values reasonably well. There is an underestimation of AOT40 in the model results during the periods of highest ozone concentration from December to March, since the model has been fed with input emissions on a 0.5×0.5 degree resolution and is unable to capture the effect of localized emission sources on a subgrid scale during the said period. The exposure-plant response index, is exceeded even in the individual months mostly between November and April at Pune. A substantial temporal and spatial variation in AOT40 values has been observed across the Indian region. Large areas of India show that the directive set by the UNECE and WHO for the critical limit of AOT40 is exceeded even in the individual months as well as in 3- months period. It is noticed that maximum AOT40 values are reached at different times at different regions across the Indian region. Comparatively low values of AOT40 are seen in monsoon months over most of the geographical region of India. These values are found to be substantially higher over the Indo-Gangetic plains than the other regions of India. Considering the fact that triple cropping is practiced in the IG-regions, the elevated AOT40 values can have an adverse effect on vegetation in this region. The results clearly indicate that presently the enhanced concentrations of ozone can have a potential impact on crop yield and forest over the large areas of India. Effective and appropriate regional/local emission control strategies needs to be developed to reduce the incessantly increasing ground level ozone pollution. Regional 3-dimensional photochemical air quality models capable of predicting the evolution of the chemical state of the atmosphere in response to natural or anthropogenic perturbations can play a significant role in defining such mitigation strategies.

Acknowledgements

We are thankful to Mrs.Daniela Jacob, MPIM, Germany and the director of IITM for their encouragement and support in this work.

REFERENCES

1. Ali, K., Momin, G. A., Safai, P. D., Chate, D. M., and Rao, P. S. P.: Surface ozone measurements over Himalayan region and Delhi, north India, Ind. J. Radio Space Phys., 33, 391–398, 2004.
2. Avol, E. L., Navidi, W. C., Rappaport, E. B., and Peters, J. M.: Acute effects of ambient ozone on asthmatic, wheezy, and healthy children, Res. Rep. 82(III), Health Eff. Inst., Boston, Mass., USA, 1–18, 1998.
3. Beig, G. and Brasseur, G. P.: Influence of anthropogenic emissions on tropospheric ozone and its precursors over the Indian tropical region during a Monsoon, Geophys. Res. Lett., 33, L07808, doi:10.1029/2005GL024949, 2006.
4. Beig, G. and Ali, K.: Behavior of boundary layer ozone and its precursors over a great alluvial plain of the world: Indo-Gangetic Plains, Geophys. Res. Lett., 33, L24813, doi:10.1029/2006GL028352, 2006.
5. Dalvi, M., Beig, G., Patil, U., Kaginalkar, A., Sharma, C., and Mitra, A. P.: A GIS based methodology for gridding large scale emission inventories: Application to carbon-monoxide emissions over Indian region, Atmos. Environ., 40, 2995, doi:10.1016/j.atmosenv.2006.01.013, 2006.
6. Fuhrer, J., Skarby, L., and Ashmore, M. R.: Critical levels for ozone ¨ effects on vegetation in Europe, Environ. Poll., 97(1–2), 91–106, 1997.
7. Ghude, S. D., Jain, S. L., Arya, B. C., Kulkarni, P. S., Kumar, A., and Ahmed, N.: Temporal and spatial variability of surface ozone at Delhi and Antarctica, Int. J. Climatol., 26, 2054–2062, 2006.
8. Ghude, S. D., Fadnavis, S., Beig, G., Polade, S. D., and Van der A, R. J.: Detection of surface emission hot spots, trends and seasonal cycle from satellite-retrieved NO2 over India, J. Geophys. Res., 113, D20305, doi:10.1029/2007JD009615, 2008.
9. Grunhage, L., J ¨ ager, H. -J., Haenel, H. -D., L ¨ opmeier, F. -J., and ¨ Hanewald, K.: The European critical levels for ozone: improving their usage, Environ. Poll., 105(2), 155–280, 1999.
10. IMD-report, Indian daily weather report, India Meteorological Department, New Delhi, India, 2003.
11. Jacob, D., Andrae, U., Elgered, G., Fortelius, C., Graham, L. P., Jackson, S. D., Karstens, U., Kopken, C., Lindau, R., Podzun, ¨ R., Rockel, B., Rubel, F., Sass, B. H., Smith, R. N. D., Van den Hurk, B. J. J. M., and Yang, X.: A comprehensive model intercomparison study investigating the water budget during the PIDCAP period, Meteorol. Atmos. Phys., 77(1–4), 19–43, doi:10.1007/s007030170015, 2001.
12. Jacob, D., Barring, L., Christensen, O. B., Christensen, J. H., Cas- ¨ tro, M. D., Dequ ´ e, M., Giorgi, F., Hagemann, S., Hirschi, M., ´ Jones, R., Kjellstrom, E., Lenderink, G., Rockel, B., S ¨ anchez, ´ E., Schar, C., Seneviratne, S. I., Somot, S., Ulden, A. V., and ¨ Van den Hurk, B.: An inter-comparison of regional climate models for Europe: Design of the experiments and model performance, Clim. Change, 81, 31–52, doi:10.1007/s10584-006-9213-4, 2007.
13. Keller, F., Bassin, S., Ammann, C., and Fuhrer, J.: High-resolution modelling of AOT40 and stomatal ozone uptake in wheat and grassland: A comparison between 2000 and the hot summer of 2003 in Switzerland, Environ. Pollut., 146(3), 671–677, 1999.
14. Klumpp, A., Ansel, W., Klumpp, G., Calatayud, V., Garrec, J. P., He, S., Penuelas, J., Ribas, A., Ro-Poulsen, H., Rasmussen, S., ˜ Sanz, M. J., and Vergne, P.: Ozone pollution and ozone biomonitoring in European cities. Part I: Ozone concentrations and cumulative exposure indices at urban and suburban sites, Atmos. Environ., 40, 7963–7974, 2006a.
15. Klumpp, A., Ansel, W., Klumpp, G., Vergne, P., Sifakis, N., Sanz, M.-J., Rasmussen, S., Ro-Poulsen, H., Ribas,

A., Penuelas, J., ˜ Kambezidis, H., Shang, H., Garrec, J. P., and Calatayud, V.: Ozone pollution and ozone biomonitoring in European cities. Part II: Ozone-induced plant injury and its relationship with descriptors of ozone pollution, Atmos. Environ., 40, 7437–7448, 2006b.

16. Langmann, B.: Numerical modelling of regional scale transport and photochemistry directly together with meteorological processes, Atmos. Environ., 34, 3585–3598, doi:10.1016/S1352- 2310(00)00114-X, 2000.

17. Mills, G., Buse, A., Gimeno, B., Bermejo, V., Holland, M., Emberson, L., and Pleijel, H.: A synthesis of AOT40-based response functions and critical levels of ozone for agricultural and horticultural crops, Atmos. Environ., 41, 2630– 2643, 2007.

18. Olivier, J., Peters, J., Granier, C., Petron, G., Muller, J.-F., and Wal- ¨ lens, S.: Present and future surface emissions of atmospheric compounds, POET Rep. 2, Eur. Union, Brussels, Belgium, 2003.

19. Pulikesia, M., Baskaralingama, P., Rayudub, V. N., Elangoc, D., Ramamurthia V., and Sivanesan, S.: Surface ozone measurements at urban coastal site Chennai, in India, J. Haz. Mat., 137(3), 1554– 1559, 2006.

20. Roy, S., Beig, G., and Jacob, D.: Seasonal distribution of ozone and its precursors over the tropical Indian region using regional chemistry-transport model, J. Geophys. Res., 113, D21307, doi:10.1029/2007JD009712, 2008.

21. Sahu, S. K., Beig, G., and Sharma, C.: Decadal growth of black carbon emissions in India, Geophys. Res. Lett., 35, L02807, doi:10.1029/2007GL032333, 2008.

22. Streets, D. G., Bond, T. C., Carmichael, G. R., Fernandes, S. D., Fu, Q., He, D., Klimont, Z., Nelson, S. M., Tsai, N. Y., Wang, M. Q., Woo, J.-H., and Yarber, K. F.: An inventory of gaseous and primary aerosol emissions in Asia in the year 2000, J. Geophys. Res., 108(D21), 8809, doi:10.1029/2002JD003093, 2003.

23. Tuovinen, J.-P.: Assessing vegetation exposure to ozone: is it possible to estimate AOT40 by passive sampling, Environ. Pollut., 119(2), 203–214, 2002.

24. Wahid, A.: Air pollution impacts on vegetation in Pakistan, in: Air pollution impacts on crops and forests: a global assessment, edited by: Emberson, L. D., Ashmore, M. R., and Murray, F., Imperial College Press, London, UK, 189–213, 2003.

25. World Health Organization (WHO): Air Quality Guidelines for Europe, Reg. Publ. Eur. Ser., WHO Reg. Off. Eur., Copenhagen, 2nd ed., 91, 288 pp., 2000.

CHAPTER 4 RESPONSE AND ADAPTATION OF PHOTOSYNTHESIS, RESPIRATION, AND ANTIOXIDANT SYSTEMS TO ELEVATED CO_2 WITH ENVIRONMENTAL STRESS IN PLANTS

Zhenzhu Xu, Yanling Jiang and Guangsheng Zhou

Keywords: Abiotic stress; antioxidant system; drought; elevated CO2; global warming; photosynthesis; respiration

INTRODUCTION

The major components of climate change include elevated atmospheric carbon dioxide concentrations (elevated CO_2), warming, and altered precipitation patterns, as well as their interactions within and with other environmental factors (IPCC, 2013). Based on updated information, with increases in global atmospheric CO_2 concentrations of 43% from the pre-industrial level of 280 μmol mol^{-1} in 1750 to the present level of 400 μmol mol^{-1} (an annual increase of 1.35%), the global CO_2 concentration has increased by about 1.55 ppm CO_2 per year over the past 55 years. It continues to be elevated at an unprecedented pace of ~1.0 μmol mol^{-1} per year, as a result of the further increase in the cumulative emissions of CO_2 to the atmosphere during the 21st century (400 μmol mol^{-1} in 2011 vs. 936 μmol mol^{-1} in 2100; IPCC, 2013; NASA, 2014). Meanwhile, the global mean surface temperature is expected to increase by 2.6–4.8°C by the end of the 21st century (2081–2100), relative to the 1986–2005 level under RCP8.5, based on a more undisciplined management scenario with higher greenhouse gas emissions (IPCC, 2013). The climate changes, such as elevated CO_2, rising temperature, and altered precipitation, have resulted in drastic impacts on the natural ecosystems, such as in vegetation function, sustainable food production, and crop yields (Lobell et al., 2011; Peñuelas et al., 2013; Ruiz-Vera et al., 2013; Xu et al., 2013a, 2014; Lavania et al., 2015), leading to more profound impacts when the climate changes are combined with other environmental constraints, such as air pollution, nutrition limitation, and their interactions (Gillespie et al., 2012; Peñuelas et al., 2012; Xu et al., 2013a,b; Wang et al., 2015).

Herein, we focus on the critical biological processes of plants with regard to climate change, including (mainly) photosynthesis, respiration, the antioxidant system, and the related metabolic activities. Photosynthesis and respiration are two fundamental physiological processes of plants, because the former involves initial carbon fixation, light energy transfer, and oxygen release, and the latter works on carbon efflux, energy production, and the relevant substrate metabolisms, such as those providing the carbon skeleton. They play a critical role in balancing the carbon budget and maintaining the carbon

Citation: Xu Z, Jiang Y and Zhou G (2015) Response and adaptation of photosynthesis, respiration, and antioxidant systems to elevated CO2 with environmental stress in plants Sci. 6:701. doi: 10.3389/fpls.2015.00701.

sink in terrestrial ecosystems, as well as in the response and feedback to climate change (Melillo et al., 1993; Prentice et al., 2001; Sage, 2004; Long et al., 2006; Atkin et al., 2010; Atkin, 2015). The excessive accumulation of reactive oxygen species (ROS) often occurs in plants grown under abiotic stress, while an enzymatic and non-enzymatic antioxidant defense system may work to protect plants against oxidative stress-induced damage, which can be affected by climate change (such as elevated CO_2, drought, and heat waves; Pérez-López et al., 2009; Gill and Tuteja, 2010; Xu et al., 2014; Zinta et al., 2014; Way et al., 2015).

Rising CO_2 has affected almost all crucial biological processes, including photosynthesis, respiration, and antioxidant systems, as well as other key secondary metabolisms in plants (Poorter et al., 1997; Long et al., 2004; Matros et al., 2006; Peñuelas et al., 2013; Singh and Agrawal, 2015). All other effects of elevated CO_2 on individual plants and ecosystems may be partly derived from these fundamental biological responses (Long et al., 2004; Ainsworth and Rogers, 2007; Peñuelas et al., 2013; Zinta et al., 2014). Genetic variations relative to the biological processes' traits might also be impacted by elevated CO_2, closely linking to these responses in various spatiotemporal aspects, from molecular, biochemical, and physiological, through individual levels and ecosystems, up to the entire Earth's life system, interacting with multiple environmental factors (both biotic and abiotic) as well as human-driven disturbances at different temporal scales (Long et al., 2004; Teng et al., 2009; Peñuelas et al., 2012, 2013; Jagadish et al., 2014; Zinta et al., 2014).

As stated above, plant responses to climate change have become a hot topic in botanical research across various scales in the recent decades. Many reports have reviewed the biological responses to CO_2 enrichment, and their interactions with environmental change, including photosynthesis and stomatal behavior (e.g., Long et al., 2004; Ainsworth and Long, 2005; Ainsworth and Rogers, 2007). Our earlier review by Xu et al. (2013a) examined plant growth, carbon and nitrogen (N) allocations, gas exchange responses to elevated CO_2 with drought and high temperature. Although this review discussed the changes in growth and photosynthesis, and water use efficiency (WUE) in higher plants exposed to CO_2 enrichment with abiotic variables, the various underlying mechanisms of the critical biological processes that are affected, modulated, and controlled by elevated CO_2 with other abiotic environmental variables were not fully covered, particular at the molecular, organelle, cell, biochemical, physiological, organ, individual, and ecosystem scales. Actually, no systematic synthesis of these has been well reviewed, thus far. Therefore, in this review, based on correcting and synthesizing any new progress of the relevant research concerning plant biology and climatic change, we attempted to systematically summarize the considerable study results that have reported the responses of photosynthesis, respiration, and the antioxidant systems, as well as the key substrate metabolisms to elevated CO_2 with other environmental variables. Particularly, we reviewed the underlying mechanisms and the response pathways, as well as their interrelationships. Finally, the future perspectives for this study related to the possible implications are briefly presented and discussed. Thus, the present review may be of current interest in terms of its interdisciplinary and systematic synthesis, providing comprehensive information on the important historical and new experimental results, relative theoretical analysis, underlying mechanisms, and potential applications to promote further research.

RESPONSES OF CRITICAL BIOLOGICAL PROCESSES TO ELEVATED CO_2

Photosynthetic Response to Elevated CO_2 Concentrations

Response Magnitude

The responses of photosynthesis to elevated CO_2 concentrations have been reviewed in many reports [e.g., Drake et al., 1997 most for enclosure results; Long et al., 2004; Nowak et al., 2004;Ainsworth and Long, 2005; Ainsworth and Rogers, 2007 for free-air CO_2 enrichment (FACE)]. The stimulation of the light-saturated photosynthetic CO_2 assimilation rate (A_{sat}) is a general response to CO_2 enrichment, with an average of 31% in the FACE experiments (Ainsworth and Rogers, 2007), and 23–58% in the potted plant experiments from earlier reports (Ryle et al., 1992; Drake et al., 1997). The magnitude of the stimulation by CO_2 enrichment varies with the different plant functional types (PFTs), with a maximum for trees and C_3 grasses; moderate for shrubs, C_3 and C_4crops, and legumes; and minimum for C_4 grass (even with a negative response; Drake et al., 1997;Ainsworth and Long, 2005; Ainsworth and Rogers, 2007). Therefore, there is greater variation in the stimulation by elevated CO_2, depending on the plant species, PFTs, and their surroundings, specifically environmental conditions like nutrition and water resource availability. For instance, elevated CO_2 leads to an increase in the A_{sat} of *Arabidopsis thaliana* leaves by 82%, since the N availability is ample (Markelz et al., 2014). However, a recent study of soybean plants indicated that elevated CO_2 did not produce significant effects on midday net photosynthetic rate (A_{net}), either in the FACE or open-top chamber (OTC) studies (Bunce, 2014), suggesting that the A_{net} at high photosynthetic photon flux density (PPFD) might be limited by a low ribulose 1, 5-bisphosphate carboxylase/oxygenase (Rubisco) carboxylation capacity (Bunce, 2014). Actually, other abiotic and biotic factors such as high temperature (e.g., Ruiz-Vera et al., 2013), drought (Xu et al., 2014), N deficit (Markelz et al., 2014), genetic variation (Ainsworth et al., 2004), and leaf senescence (Liu et al., 2014) may also diminish the photosynthetic response to elevated CO_2.

The same results appeared in *Lolium perenne* and *Medicago lupulina* plants in controlled chambers (Farfan-Vignolo and Asard, 2012). Moreover, C_4 plants may have no response to elevated CO_2, because their CO_2 concentrating mechanism (CCM) may concentrate the CO_2 12–20 times at the site of Rubisco, which is relatively higher than in C_3 species (von Caemmerer and Furbank, 2003; Ainsworth and Rogers, 2007). Case studies confirmed this theoretical conclusion under well-watered conditions in either enclosure (e.g., Xu et al., 2014) or FACE experiments (e.g., Leakey et al., 2006; Markelz et al., 2011). However, under a water deficit, the stimulation of the C_4 A_{sat} by elevated CO_2 still appears, because the drought-induced impairment of C_4 photosynthesis might be ameliorated by elevated CO_2 (Markelz et al., 2011; Meng et al., 2014; Xu et al., 2014). Moreover, C_4plants can avoid photorespiration to promote CO_2 fixation with higher light use efficiency (von Caemmerer and Furbank, 2003; Long et al., 2006). On the other hand, the down-regulation of the photosynthesis capacity is also more profound in C_3 species than in C_4 species (Morgan et al., 2001;Duarte et al., 2014), due in part to the N dilution, possibly because C_3 plants need to invest more N from the leaf into Rubisco, relative to the C_4 species, so that the former may easily undergo more severe N dilution under CO_2 enrichment (Sage et al., 1987; Yin, 2002; Luo et al., 2004; Sage, 2004), with no CCM (von Caemmerer and Furbank, 2003).

In addition to N limitation, photosynthetic acclimation under higher CO_2 levels may result from high stomatal and internal resistances, higher starch levels, and diluted chlorophyll concentrations (Delucia et al., 1985; Teng et al., 2009). Under elevated CO_2, carbohydrate accumulations, such as starch size and number of chloroplasts (Teng et al., 2006, 2009), can be enhanced, partially due to the carbon substrate increase. However, the excessive carbohydrate accumulation may cause feedback inhibition or physical damage at the chloroplast level, reducing the photosynthetic capacity (Delucia et al., 1985; Aranjuelo et al., 2011). More importantly, the Rubisco response, excessive sugar feedback, and the related gene expression may, together; play crucial roles in plants' photosynthetic acclimation under higher CO_2 concentrations, particularly for long-term CO_2 enrichment under a nitrogen availability deficit (see details below).

Molecular Mechanisms: Role of Rubisco

The stimulation of photosynthesis in C_3 species by short-term elevated CO_2 has been well established, and confirmed under almost all experimental conditions, particularly with FACE (e.g.,Long et al., 2004; Ainsworth and Rogers, 2007; Duarte et al., 2014). However, with long-term exposure to elevated CO_2 or other limitations, photosynthetic acclimation or the down-regulation of the photosynthetic capacity may occur, depending on the species, plant developmental stage, and environmental conditions (Moore et al., 1999; Urban et al., 2012; Sanz-Sáez et al., 2013).

Rubisco has been identified as a controlling rate enzyme for carbon fixation (Eichelmann et al., 2009). Here, we succinctly summarize the five major mechanisms that might explain the response to elevated CO_2, involving Rubisco: (1) under current CO_2 concentration levels, although the value of the Rubisco Michaelis–Menten constant (K_m) for CO_2 is close to the current intercellular CO_2concentration (C_i) (c. 190 µmol mol^{-1}) at the site of carboxylation (von Caemmerer and Evans, 1991;Ainsworth and Rogers, 2007). CO_2, as a substrate of photosynthesis, does not have to reach saturation; therefore, the rising CO_2 can lead to an immediate increase in the Rubisco carboxylation velocity, due to an increase in the carbon substrate availability. (2) The Rubisco catalyzing function has two intrinsic side features: carboxylation and oxygenation. The carboxylation rate is ~2.2 fold greater than the oxygenation rate at 25°C in C_3 plants; that is, about one-third of the ribulose-1,5-bisphosphate (RuBP) may be consumed in the oxygenation reaction (Ainsworth and Rogers, 2007). Thus, elevated CO_2, as a competing substrate, can competitively inhibit the oxygenation of RuBP (light-dependent photorespiration) through the down-regulation of Rubisco's affinity for O_2, while competitively promoting the carboxylation of RuBP via the up-regulation of Rubisco's affinity for CO_2 (Bowes, 1991; Long, 1991; Ainsworth and Rogers, 2007;Kane et al., 2013; Moroney et al., 2013). Consequently, this leads to the stimulation of photosynthesis, which may be compromised by heat and drought due to the enhancement of Rubisco's affinity for O_2 (Wingler et al., 1999; Tingey et al., 2003; Carmo-Silva et al., 2008;Moroney et al., 2013) (Figure 1). On the other hand, (3) with continually increasing CO_2, the ATP products may not meet enough of the demand for RuBP regeneration, and a reduction in Rubisco's activation state may occur, usually accompanied by a decrease in the capacity for RuBP regeneration, as well as in the RuBP pool, as indicated by a decline in the ATP:ADP ratio in the chloroplast (Eichelmann et al., 2009; Watanabe et al., 2014). (4) A reduction in the Rubisco content via N dilution, particularly under long-term elevated CO_2, may finally contribute to the reduction of carboxylation at the Rubisco active site. In addition, the nitrogen use efficiency (NUE) might be increased due to the optimization of the resource use (Moore et al., 1999; Luo et al., 2004;Fukayama et al., 2012; Urban et al., 2012; Palmroth et al., 2013; Sanz-Sáez et al., 2013). Because the leaf N of C_3 species can be more invested in Rubisco (more than 25% vs. 10–15% of the leaf N in C_3 and C_4 plants, respectively), the former may be affected more profoundly by N dilution.

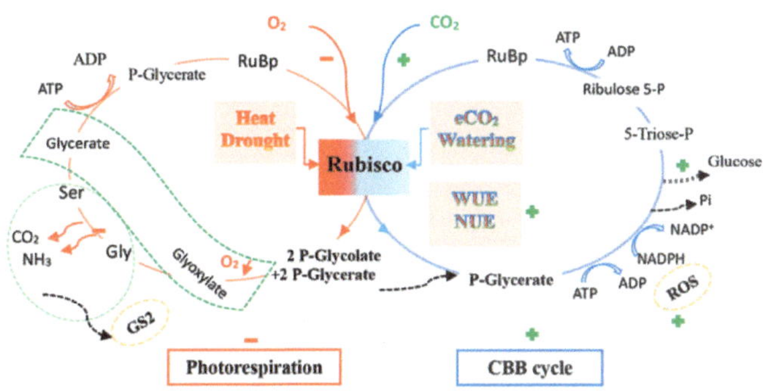

Figure 1. A diagrammatic outline of the Calvin–Benson–Basshan (CBB) cycle and photorespiration pathway in plants in response to elevated CO_2 with abiotic factors. Rubisco has two sites of carboxylation and oxygenation. Elevated CO_2(eCO_2) may promote carboxylation, but repress oxygenation under ample environmental conditions, such as well-watering, whereas extreme abiotic stress, such as heat and drought, may repress carboxylation but promote oxygenation (Wingler et al., 1999; Tingey et al., 2003). An energy consumption trade-off between the key cycles may occur, possibly modified by the CO_2 level, in which photorespiration may be promoted to quench reactive oxygen species (ROS), related to glutamine synthetase (GS2) to recycle ammonia, diminishing photo-oxidation and photo-inhibition (dotted orange line ellipse; Kozaki and Takeba, 1996; Watanabe et al., 2014). A low Gly:Ser ratio provides evidence that photorespiration is repressed in eCO_2 (Kebeish et al., 2007). Water use efficiency (WUE) and nitrogen use efficiency (NUE), despite the N dilution, should be enhanced by elevated CO_2, by decreasing stomatal conductance and investing relatively more N into the Rubisco protein (Palmroth et al., 2013). The photorespiration process is compartmentalized into the chloroplast (red line ellipse), peroxisome (dotted green line bent rectangle), and mitochondrion (dotted green line ellipse). The green plus and red minus signs denote the stimulation and suppression via rising CO_2, respectively (mainly referring to Kozaki and Takeba, 1996; Wingler et al., 1999; Tingey et al., 2003; Ainsworth and Rogers, 2007; Moroney et al., 2013; Xu et al., 2013a; Watanabe et al., 2014).

Finally, (5) Hexokinase (HXK), as a sensor of excessive photosynthate, may be involved in the downregulation of the

Rubisco content (Ainsworth and Rogers, 2007; Kirschbaum, 2011; see below). In summary, with respect to the Rubisco response, parts (1) and (2) above may explain the stimulation of photosynthesis by elevated CO_2, while the last three points may provide a mechanism for understanding the downregulation of the photosynthetic capacity under relatively long-term elevated CO_2 or other resource deficit conditions, such as a scarcity of N.

Sugar Feedback Inhibition of Photosynthesis

Under higher CO_2 concentrations, a prevailing explanation of the downregulation of photosynthesis may be ascribed to the sugar feedback inhibition hypothesis: certain reactive bioprocess activities within the Calvin–Benson–Basshan (CBB) cycle may be inhibited by elevated CO_2, due to the overload of the chemical reaction substrates. The hypothesis of the sugar feedback mechanism suggests that excessive photosynthate in chloroplasts under elevated CO_2 may trigger the sugar signal network (HXK acting as a flux sensor) to down-regulate the Rubisco contents through the gene expression processes, affecting the subunit of Rubisco (Drake et al., 1997; Stitt and Krapp, 1999; Long et al., 2004; Ainsworth and Rogers, 2007).

As noted above, HXK acting as a flux sensor in mesophyll cells may involve the down-regulation of the Rubisco content associated with genetic expressions under elevated CO_2; however, plants may prefer to reduce the Rubisco activity relative to the RuBP regeneration capacity (Ainsworth and Rogers, 2007). For instance, Aranjuelo et al. (2011) found a decline in wheat Rubisco and its activase protein content accompanying a photosynthetic down-regulation. In addition, the effects of the source-sink balance in response to CO_2 enrichment may play an important role in the regulation of the photosynthetic capacity.

Based on the "source-sink" hypothesis, some plants with a strong sink can encounter photosynthetic down-regulation, to some extent, under higher CO_2, which can generally be repressed by other limitations, such as intrinsic genetic constraints or the specific plant developmental stage (such as the flowering stage; Lewis et al., 2002; Ainsworth et al., 2004). Moreover, when enhanced carbohydrate availability exceeds the plants' ability to fully utilize carbohydrates, due to nutrient or inherent internal growth limitations, the feedback may lead to a lower level of photosynthesis (Kirschbaum, 2011), which may lead to an imbalance in the carbon sink:source ratio (Bryant et al., 1998; Aranjuelo et al., 2011). For instance, wheat plants exposed to high atmospheric CO_2 are incapable of excessive accumulation of leaf photoassimilate, due to the lack of an increase in the carbon sink strength (Aranjuelo et al., 2011).

Furthermore, the respiratory ATP may be consumed more under elevated CO_2 (Watanabe et al., 2014); for example, the rate of the carbohydrate/sugar export (i.e., the cost related to the carbohydrate export) is higher under elevated CO_2 than under normal CO_2 (Watanabe et al., 2014), which may cause a negative feedback effect on photosynthesis. This also highlights the close link between the photosynthetic and respiratory bioprocesses, between both the CBB and tricarboxylic acid (TCA) cycle, under climate change (Moroney et al., 2013; Watanabe et al., 2014).

Response of Respiration to Elevated CO_2

Photorespiration

Photorespiration enables the photosynthetic process to recycle the phosphoglycolate produced by the oxygenase reaction of Rubisco, consequently avoiding more carbon loss, with some protective regulation functions for plants, such as in the oxidative defense mechanism (Bowes et al., 1971;Husic et al., 1987; Kozaki and Takeba, 1996; Wingler et al., 1999; Carmo-Silva et al., 2008;Moroney et al., 2013). However, as reported in the review by Ainsworth and Rogers (2007), at room temperature (25°C), photorespiration can lead to a loss of 23–30% of the carbon fixed by photosynthesis with the rising temperature, whereas the CO_2 fixation may be increased by ~53% if only the carboxylation reaction occurs, without the oxygenation reaction (Monteith, 1977; Long et al., 2006).

Widely accepted results show that photorespiration can be restricted when C_3 plants are grown under high CO_2 concentrations (Bowes, 1991; Tingey et al., 2003; Long et al., 2004), because in C_3plants, the carboxylation capacity of Rubisco, with a low catalytic activity (operating below its $K_m$for CO_2), is easily promoted by high CO_2. Meanwhile, an increase in the CO_2 concentration, leading to a high CO_2:O_2 ratio, may reduce its oxygenation reaction capacity, inhibiting photorespiration (Bowes, 1991; Tingey et al., 2003; see above). For example, based on the earlier report by Sharkey (1988), the photorespiration rate should fall by ~50% when the CO_2 level is doubled. In *A. thaliana*plants grown under elevated CO_2, although the accumulations of several major amino acids (including glutamate, aspartate, asparagine, and alanine) were enhanced, a lower level of glycine (Gly), an intermediate of photorespiration, was observed in the plants, leading to a decline in the Gly:Ser ratio, indicating a lower photorespiration activity (Kebeish et al., 2007; Figure 1).

Enhanced photoperoxidation in chloroplasts can induce a destruction of the chlorophyll and a disassembly of the chloroplast membranes, leading to a decline in photosynthesis (Heath and Packer, 1968). Conversely, the constraints of photorespiration by elevated CO_2 may also reduce the H_2O_2 products, weakening oxidation stress, possibly protecting the photosynthetic apparatus (Watanabe et al., 2014; Zinta et al., 2014). Based on the fact that photorespiration has a protective function against photo-oxidation (Kozaki and Takeba, 1996; Zinta et al., 2014), possibly via the up-regulation of glutamine synthetase (GS2) to recycle ammonia, diminishing photo-oxidation and photo-inhibition (Kozaki and Takeba, 1996). This brings with it another dilemma: a decline in photorespiration under rising CO_2 levels may cancel the protective role, leading to a higher level of photo-oxidation than the higher rate of carboxylation stimulated by elevated CO_2 can maintain. In order to solve this dilemma, further research is required to cope with climate change, possibly by manipulating the modulated photorespiration bioprocess (Moroney et al., 2013).

Mitochondrial Respiration

Mitochondrial respiration involves the carbon balance in the whole plant, with 20–80% of the carbon fixed in photosynthesis being released again through the respiration process. The respiration of the leaves in both the light and dark can account for ~50% of the whole-plant respiratory CO_2 (Ayub et al., 2014). The response of dark leaf respiration (R_d) to elevated CO_2remains debatable, with a decrease in the major reports, while increasing or remaining stable in a number of experiments (e.g., Ryle et al., 1992; Curtis and Wang, 1998; Drake et al., 1999; Amthor, 2000; Gonzalez-Meler et al., 2004; Loreto et al., 2007; Ayub et al., 2014). For instance, there is a 15–18% range in the reduction of foliar respiration when plants are grown under a doubled CO_2concentration, relative to the ambient CO_2 level from one review (Drake et al., 1999; references). However, no significant response to the leaf R_d was observed in *L. perenne* plants exposed to high CO_2; although, the leaves grown in elevated CO_2 had a relative lower R_d (Ryle et al., 1992). A small response in the leaf respiration rate to a short-term CO_2 elevation (a 1.5% decrease) was obtained from the deciduous tree species used in an earlier experiment by Amthor (2000), with similar evidence found in soybean plants from a recent report by Ayub et al. (2014). Thus, for the plants grown under elevated CO_2, the R_d decrease response is *general*, not *universal*.

Correspondingly, the underlying mechanism has also been proposed in two contrasting hypotheses: elevated CO_2 may enhance the R_d due to the great increase in the respiratory substrates, such as sugar; whereas the N dilution induced by elevated CO_2 might reduce the demand on dark respiration to support the protein turnover, leading to a decline in the R_d (Thomas et al., 1993;Gonzalez-Meler et al., 2004; Fukayama et al., 2011; Markelz et al., 2014). A recent report showed that CO_2 enrichment can accelerate the accumulation of the relevant carbohydrates, such as sugar, starch, and respiratory glycolysis intermediates like hexose-P, phosphoglycerate (PGA), and phosphoenolpyruvate (PEP) in *A. thaliana* plants, which may enhance the respiration potential (Watanabe et al., 2014). Recent evidence has also indicated that the promotion to greater photo-assimilation availability at elevated CO_2 leads to a great transcriptional up-regulation of the genes, in association with the respiratory pathway (Leakey et al., 2009a; Fukayama et al., 2011; Markelz et al., 2014), supporting the first hypothesis. However, this may depend on the availability of the nutritional components, including nitrogen in the plants and/or the soil. For instance, based on a recent report by Markelz et al. (2014), widely and greatly adaptive responses of the expression of the respiratory genes were obtained when the plants were exposed to elevated CO_2. However, the transcriptional reprogramming with the stimulation of leaf respiration by elevated CO_2 can be suppressed by limited nitrogen availability (Markelz et al., 2014).

Response of Antioxidant System to Elevated CO_2

The ROS in plants, including superoxide radicals ($O_2^{\cdot-}$), hydrogen peroxide (H_2O_2), the hydroxyl radical ($OH^{\cdot}$), and the perhydroxy radical (HO_2), often accumulate when plants are subjected to abiotic stress, while the antioxidant defense system with enzymatic and non-enzymatic machinery may work to protect the plants against damage due to oxidative stress. This occurs particularly in the face of stressful environmental changes, such as adverse climatic changes like droughts and heat waves (Figure 2) (Schwanz and Polle, 1998; Pérez-López et al., 2009; Gill and Tuteja, 2010;Sekmen et al., 2014; Zinta et al., 2014). Generally, when plants become senesced, with some antioxidants increasing and others decreasing, the ROS may accumulate in a large amount, and the antioxidant system does not work well. This is often indicated by enhanced lipid peroxidation and decreased levels of antioxidant enzymes, such as superoxide dismutase (SOD) and catalase (CAT), leading to programmed cell death (PCD), particularly under severe abiotic stress (Dhindsa et al., 1981; Hodges and Forney, 2000; Gill and Tuteja, 2010; Duarte et al., 2013).

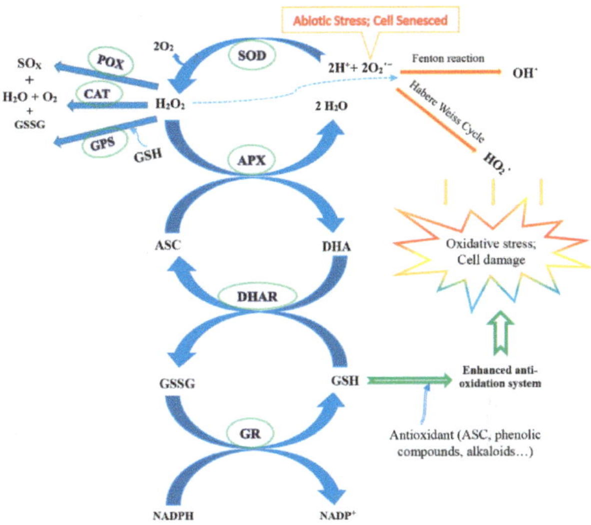

Figure 2. A diagrammatic outline of the antioxidant defense systems and the responses to elevated CO_2 with abiotic stress. Elevated CO_2 may alleviate the damage of oxidative stress from abiotic stress factors, such as heat, drought, and ozone, by ameliorating the antioxidant defense systems of non-enzymatic compounds, potentially including ascorbate (ASC), glutathione (GSH), phenolic compounds, and alkaloids, and the relevant enzymes, possibly including superoxide dismutase (SOD), ascorbate peroxidase (APX), dehydroascorbate reductase (DHAR), glutathione reductase (GR), peroxidase (POX), catalase (CAT), and glutathione peroxidase (GPX). ROSs, including superoxide radicals ($O_2^{\cdot-}$), hydrogen peroxide (H_2O_2), hydroxyl radicals ($OH^{\cdot}$), and perhydroxy radicals ($HO_2\cdot$), accumulate when plants undergo abiotic stress or are senesced by the Fenton reaction and/or the Habere Weiss mechanism (Hodges and Forney, 2000; Gill and Tuteja, 2010). Whether the rising CO_2 mitigates oxidative damage and the response magnitude, and which parts play major roles, depends on the plant species, crop varieties, developmental stage, abiotic factors, and their combinations (e.g., Hodges and Forney, 2000; Gill and Tuteja, 2010; Abd Elgawad and Asard, 2013; Kumari et al., 2013;Zinta et al., 2014). GSSG, oxidized glutathione; DHA, dehydroascorbate. This diagram is based mainly on the studies byGill and Tuteja (2010) and Zinta et al. (2014).

Elevated CO_2 may increase the levels of antioxidants, including polyphenols, ascorbate (ASC), alkaloids, and some antioxidant enzyme activities (such as CAT and SOD), with a significant enhancement in the antioxidant capacity, leading to declines in the ROS levels (Mishra and Agrawal, 2014; Zinta et al., 2014). For example, when the plants were exposed to elevated CO_2, increases in the ASC and phenol levels were obtained in *Beta vulgaris* (Kumari et al., 2013), and increases in the ASC, glutathione (GSH), and ASC/GSH, as well as in their redox status, were found in *L. perenne* and *M. lupulina* (Farfan-

Vignolo and Asard, 2012). Ascorbate synthesis can be triggered and enhanced by excessive carbohydrate production due to elevated CO_2 (Smirnoff and Wheeler, 2000; Zinta et al., 2014), which is closely linked to carbon metabolism (Smirnoff and Wheeler, 2000), and together improve the plant-antioxidant defense system. Moreover, with a delay in the onset of senescence and/or severe stress under elevated CO_2 conditions, it is suggested that the antioxidant profiles, such as the accumulation of antioxidant compounds and antioxidant enzyme activity, may show better performance in dealing with the biological process of senescence (Hodges and Forney, 2000). For instance, a reduction in the oxidative stress under elevated $CO_2$was found in *Zingiber officinale* (Ghasemzadeh et al., 2010), *Catharanthus roseus* (Singh and Agrawal, 2015), a temperate grassland shrub, *Caragana microphylla* (Xu et al., 2014), a bean,*Vigna radiate* (Mishra and Agrawal, 2014), and *A. thaliana* plants (Zinta et al., 2014).

However, with the elevated CO_2-alleviated oxidation stress evidence coming from a number of reports (e.g., Pérez-López et al., 2009; Xu et al., 2014), these results have not been confirmed in some species, such as in *Spinacia oleracea* leaves (Hodges and Forney, 2000). Farfan-Vignolo and Asard (2012) reported that CO_2 enrichment could exacerbate lipid peroxidation in *M. lupulina*, but not in *L. perenne* plants, with no rising-CO_2 responses in the ascorbate peroxidase (APX) and peroxidase (POX) in *M. lupulina*. No significant responses in the antioxidant enzyme activity, including APX, glutathione reductase (GR), POX, CAT, and SOD, were found when *B. vulgaris*plants were exposed to high levels of O_3 with elevated CO_2, except in the inhibition of APX (Kumari et al., 2013).

Instead, in *Quercus pubescens* and *Q. ilex* plants grown under elevated CO_2, down-regulation of the protective systems was observed (Schwanz and Polle, 1998). According to the findings of Schwanz and Polle (1998), although the GR in oak leaves remains stable, the activities of SOD, CAT, POX, and APX, as well as the sum of dehydroascorbate and ASC, were reduced in CO_2-elevated environments. Base on a recent report (Singh and Agrawal, 2015), the activities of the SOD, CAT, and APX declined, but the GR and POX were stimulated, finally leading to a significant reduction in the O·−2O2·−, H_2O_2, and malondialdehyde (MDA) contents in *C. roseus* plants grown under elevated CO_2. Recently, marked decreases in the ROS levels (O·−2O2·−, H_2O_2) and reductions in some antioxidant enzymes, such as CAT and SOD, were observed simultaneously in mung bean plants exposed to elevated CO_2, suggesting that a lower level of ROS might match the lower activity of antioxidant enzymes (Mishra and Agrawal, 2014).

Based on a report by AbdElgawad et al. (2015), who used C_3 grasses (*L. perenne, Poa pratensis*) and C_3 legumes (*M. lupulina, Lotus corniculatus*) as experimental materials, elevated CO_2 can reduce the H_2O_2 level, lipid peroxidation, and lipoxygenase (LOX) activities, while it decreased the SOD, CAT, glutathione peroxidase (GPX), and GR levels, but did not affect the ASC-GSH cycle (AbdElgawad et al., 2015). Thus, the predominant form of the enzymatic antioxidant defense may strongly depend on the species and the abiotic stress (Duarte et al., 2013; Singh and Agrawal, 2015).

The activities and gene transcription expression levels of ROS scavenging enzymes in *A. thaliana* at elevated CO_2 remained unchanged, particularly under well-watered conditions (Zinta et al., 2014). However, the excessive gene transcriptional response related to antioxidant metabolism due to O_3pollution was partly repressed by elevated CO_2 in soybean (*Glycine max*) plants under field conditions, again arguing the protective role of elevated CO_2 (Gillespie et al., 2012). Lipid peroxidation, indicated by MDA accumulation, would be lessened by CO_2 enrichment, especially under other severe abiotic stress conditions such as drought (Salazar-Parra et al., 2012; Xu et al., 2014; AbdElgawad et al., 2015), heat wave stress (Xu et al., 2014; Zinta et al., 2014; AbdElgawad et al., 2015), O_3 pollution (Yan et al., 2010; Kumari et al., 2013), and salinity (Pérez-López et al., 2009), implying that oxidative stress induced by severe environmental constraints may be mitigated, generally or at least partly, by CO_2 fertilization. It is again highlighted that the positive vs. negative roles of elevated CO_2 concentrations in antioxidant enzyme regulation under severe stressful abiotic environments may depend considerably on different species (Schwanz and Polle, 1998, 2001; Guo et al., 2006; Kumari et al., 2013; Xu et al., 2014; Zinta et al., 2014).

Moreover, based on a recent report using wheat plants, with increasing sugar levels via CO_2enrichment, sugar-derived reactive carbonyls (RCs; aggressive by-products of oxidative stress), including methylglyoxal (MG), were provoked by elevated CO_2, which can negate the functions of multiple proteins, and impair the biological membrane, suggesting that plant diabetes may be inducible (Takagi et al., 2014), supporting an earlier study by Schwanz and Polle (1998). Thus, whether and how much elevated CO_2 affects antioxidant systems in plant tissues depends on the plant species, crop variety, developmental stage, abiotic factors, and the combination of these (e.g.,Hodges and Forney, 2000; Gill and Tuteja, 2010; Mishra and Agrawal, 2014; Xu et al., 2014; Zinta et al., 2014). This is a debatable issue, requiring further research.

Response of Crucial Metabolites to Elevated CO_2

The metabolism changes of certain important metabolites, and the related genetic variations induced by elevated CO_2 have been found in a number of research reports. For instance, an accumulation of carbon compounds under elevated CO_2 occurs in wheat leaves accompanied by an up-regulation of phosphoglycerate mutase (PGAM) involving carbohydrate transport, but a down-regulation of the adenosine diphosphate glucose pyrophosphatase protein for synthesizing starch; thus affecting the carbon flux within the plants' tissues, and the balance between the carbon sink and source (Aranjuelo et al., 2011). Changes in the major chemical components induced by elevated CO_2 have also been reported in many studies. Generally, under elevated CO_2, there may be a decrease in the total N and organic N compounds, which define the elevated CO_2-induced dilution effectiveness. However, there is an increase in the total non-structural carbohydrates (TNC), including starch and sugar (e.g., glucose, fructose, sucrose; Lavola and Julkunen-Tiitto, 1994;Poorter et al., 1997; Luo et al., 2004; Markelz et al., 2014), with a mostly stable level in the total structural carbohydrates (cellulose plus hemicellulose), lignin, and lipids (review by Poorter et al., 1997; Markelz et al., 2014). Nitrogen assimilation may be enhanced by elevated CO_2 (Ribeiro et al., 2012), and a recent report indicated that elevated CO_2 may promote N assimilation and transamination-related enzyme activities, such as glutamate oxoglutarate aminotransferase (GOGAT) and glutamate oxalate transaminase (GOT), and lead to an increase in the phloem amino acid content in *M. truncatula* (Guo et al., 2013).

Elevated CO_2 can change not only the primary metabolic processes, but also the secondary metabolic composition in plant tissues (Lavola and Julkunen-Tiitto, 1994; Poorter et al., 1997;Matros et al., 2006; Lavola et al., 2013). Here, we mainly address the secondary metabolite responses, because there are fewer studies related to the key metabolites. Plant secondary metabolites often indicate that these compounds have no primary functions in the maintenance of life processes in plants; however, they are involved in the biological processes of plants dealing with environmental stress, with regard to adaptation and defense (Lavola and Julkunen-Tiitto, 1994;Ramakrishna and Ravishankar, 2011). Changes in the secondary metabolites with rising CO_2 have been reported in several relevant studies. For example, an alteration in the carbon allocation under elevated CO_2 has revised the carbon-nutrient balance (CNB) hypothesis (Bryant et al., 1983), increasing the C:N ratio in plant tissues (e.g., Poorter et al., 1997; Cotrufo et al., 1998; Xu et al., 2007), while increasing the levels of the C-based secondary compounds due to easier synthesis, in plants with excess carbon relative to the other nutrients (such as N; Lavola and Julkunen-Tiitto, 1994; Matros et al., 2006). However, a contradiction may arise when elevated CO_2-induced N dilution limits the carbohydrate reserves, leading instead to a reduction in some secondary substances (Lavola and Julkunen-Tiitto, 1994). However, with rising CO_2, a large accumulation of some secondary metabolites, including phenylpropanoids, tannins, triterpenoids, phenolic acids, and alkaloids, was observed, despite the effectiveness of the dilution (Lavola and Julkunen-Tiitto, 1994; Matros et al., 2006; Ghasemzadeh et al., 2010; Lavola et al., 2013). For example, in tobacco leaves there was a large accumulation of phenylpropanoids, including the major carbon-rich compound chlorogenic acid (CGA), and the scopolin and scopoletin coumarins (Matros et al., 2006). In the flavonoid response, although both the kaempferol and fisetin were increased by elevated CO_2 in ginger (*Zingiber officinale* Roscoe; Ghasemzadeh et al., 2010), whether there was a decrease or increase in birch plants grown in elevated CO_2 depended on the genetic type or environmental conditions (Lavola and Julkunen-Tiitto, 1994; Lavola et al., 2013). The glucosinolate accumulation was enhanced in *Brassica* plants exposed to elevated CO_2, possibly changing the feeding behavior of specialized herbivores (Klaiber et al., 2013). In other metabolites, including lignin, cell wall polysaccharides, and terpenes, no obvious response was found, depending on the compound composition, species, genotype, nutrient status (such as N availability), and other environmental factors (Poorter et al., 1997; Lindroth et al., 2001; Matros et al., 2006; Lavola et al., 2013; AbdElgawad et al., 2014; Singh and Agrawal, 2015). For example, the levels of the condensed tannins, most flavonols, and phenolic acids in birch plants can be stimulated by elevated CO_2 and elevated UVB, but this effect disappeared at high temperatures (Lavola et al., 2013).

Isoprene is a volatile hydrocarbon molecule, generally emitted by certain vegetation types, particularly tree species, protecting plants against damage from abiotic stress, and playing an important role in tropospheric chemistry and climate change due to its highly reactive molecular properties, especially in the formation processes of ozone and secondary organic aerosols (Sharkey and Singsaas, 1995; Claeys et al., 2004; Sun et al., 2012). However, it has been confirmed that isoprene may have an active function in protecting the photosynthetic apparatus against oxidative stress from abiotic stress (such as heat), by quenching the ROS via the promotion of oxidative defense machinery (Gill and Tuteja, 2010; Morfopoulos et al., 2014). In a number of related reports, elevated CO_2 has produced various effects on plant-derived isoprene emissions, including increases (Sharkey et al., 1991; Tognetti et al., 1998), remaining unchanged (Rosenstiel et al., 2003;Sun et al., 2012), and, most often, showing decreases (e.g., Wilkinson et al., 2009; Possell and Hewitt, 2011; Morfopoulos et al., 2014). The reason for the decreasing isoprene emission may be that the available reducing power captured by light may cause a large consumption, due to carbon fixation rather than isoprene synthesis, in CO_2 enrichment conditions, resulting in a reduction in isoprene emissions (Morfopoulos et al., 2014). A reduction in the isoprene emission capacity may be attributable to a decrease in both the isoprene synthase activity and pool size of dimethylallyldiphosphate (DMADP), an immediate isoprene precursor (Sun et al., 2012). Actually, DMADP synthesis is involved in the primary photosynthetic product of glyceraldehyde-3-phosphate (GAP), linked to a leaf's photosynthetic carbon metabolism (Loreto and Sharkey, 1993;Lichtenthaler, 1999; Sun et al., 2012; Trowbridge et al., 2012). Reduced ATP induced by elevated CO_2 may also diminish DMADP synthesis (Sun et al., 2012). Thus, the isoprene emission capacity may be determined by the status of the balance between the primary metabolites, such as sugar, and the secondary metabolites, such as isoprene (Loreto and Sharkey, 1993; Sun et al., 2012), again highlighting the importance of the primary-secondary metabolite balance (abbreviated by PSMB) with CO_2 enrichment. Based on the response model suggested by Morfopoulos et al. (2014), Figure3 succinctly describes a pathway involved in the downregulation of isoprene biosynthesis in response to elevated CO_2. Under elevated CO_2, more electron flux may be used in the CBB cycle for photosynthesis, whereas less electrons may flow into the photorespiration cycle, xanthophyll cycle, and the methylerythritol 4-phosphate (MEP, 5) pathway to synthesize isoprene, as well as other redox reactions, such as quenching ROS (e.g., GR reaction demands of NADPH; Gill and Tuteja, 2010). It is worth noting that the isoprene biosynthesis and emission, in and from plants, may be tightly associated with photosynthesis, photorespiration, the xanthophyll cycle, and oxidative defense systems in response to CO_2 enrichment, with abiotic environmental changes (Gill and Tuteja, 2010; Moroney et al., 2013; Morfopoulos et al., 2014; Figure 3).

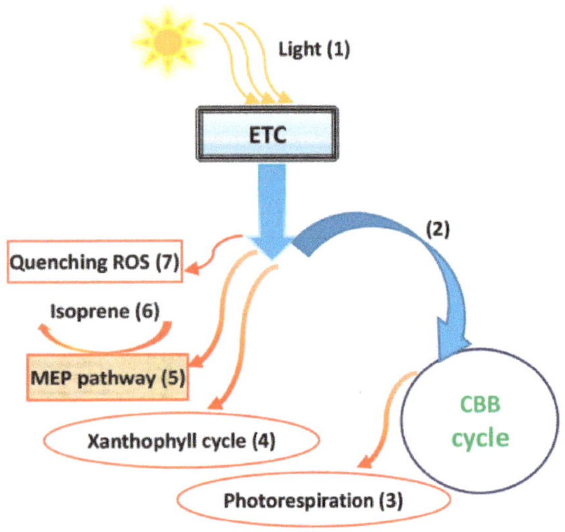

Figure 3. A diagrammatic representation of isoprene biosynthesis downregulation in response to elevated CO_2. Light energy from the sun (1) is transferred into the plant metabolic bioprocesses using an electron transport chain (ETC). Under elevated CO_2, more electron flux may be used for the CBB cycle for photosynthesis (2), while less electrons may flow into the photorespiration cycle (3), xanthophyll cycle (4), and the methylerythritol 4-phosphate pathway (MEP), (5) to synthesize isoprene (6), as well as other redox reactions, including the quenchers of ROS (7) (based mainly on Morfopoulos et al., 2014).

The role of hormone pathways in regulating the growth and metabolic responses to elevated CO_2 is not well known, despite there being a few reports (e.g., Li et al., 2011; Ribeiro et al., 2012; Zavala et al., 2013). Elevated CO_2 can promote an accumulation in the salicylic acid (SA, Zavala et al., 2013) and brassinosteroids (BR; Jiang et al., 2012), while reducing Jasmonates (JA) and ethylene concentrations (Zavala et al., 2013; Vaughan et al., 2014). Elevated CO_2-induced genes were associated with the metabolic processes of the BR regulator in plant tissues (Li et al., 2006), which can alleviate the heat-induced inhibition of photosynthesis, by increasing the carboxylation efficiency and enhancing the antioxidant systems in *Lycopersicon esculentum* (Ogweno et al., 2008). In one recent report, BRs were found to enhance the stimulation of plant growth and photosynthetic potential under elevated CO_2 (Jiang et al., 2012). An increase in the indole-3-acetic acid (IAA), isopentenyl-adenosine (iPA), and dihydrozeatin riboside (DHZR) was found, while a decrease in the ABA and unchangeable zeatin riboside (ZR) occurred in *Pinus tabulaeformis* plants exposed to elevated CO_2, which can encounter O_3 exposure effects to alleviate damage (Li et al., 2011). The iPA, DHZR, and ZR are recognized as the most commonly active cytokinins (CTKs) in plants. The results of the experiment by Ribeiro et al. (2012) indicated that elevated CO_2 may play a role similar to gibberellin (GA) in the integration of carbohydrate and nitrogen metabolisms underlying the optimal biomass determination. When the *Arabidopsis* plants exhibited the inhibition of growth via the GA biosynthesis inhibitor (low-GA regime), the activities of the enzymes involved in photosynthesis, including the CBB cycle enzymes [phosphoglycerate kinase (PGK) and transketolase (TK)], were enhanced by elevated CO_2, whereas the activities of the enzymes related to organic acid metabolism, such as the NAD-dependent malate dehydrogenase (MDH), were inhibited (Ribeiro et al., 2012). Moreover, nitrate reductase (NR) can be stimulated by elevated CO_2 (by 31%) in plants with a low-GA content, indicating that rising CO_2 may mediate inorganic N metabolism in association with GA (Ribeiro et al., 2012). This clearly indicates that elevated CO_2 can substitute for the relevant metabolic bioprocesses in the low-GA species, which may have a marked potential application for plants, particularly staple crops, to cope with future climate change in a high CO_2 concentration environment.

General Gene Expression Profile under Elevated CO_2

The genes expressed differently between ambient and elevated CO_2 might encode great changes in their metabolic functions (Li et al., 2006), including increases in the expression of a subset of genes encoding stress-related functions, and decreases in the expression of genes encoding chloroplast functions and other processes of photosynthesis (Moore et al., 1999; Miyazaki et al., 2004; Li et al., 2006). The decline in the gene expression may partly lead to so-called photosynthetic acclimation to long-term elevated CO_2, particularly under limited environmental conditions or in carbon sink limited species (Jifon and Wolfe, 2002; Ainsworth et al., 2004; Long et al., 2004; Fukayama et al., 2012; Xu et al., 2013a). For example, Fukayama et al. (2012) found the overexpression of Rubisco activase in rice leaves grown under elevated CO_2, possibly leading to a decrease in the photosynthetic capacity. However, the gene expression in response to CO_2 fumigation may depend on different developmental stages at the time of sampling, and different physiological conditions of the ecotypes of *A. thaliana* (Li et al., 2006). Moreover, limited N, a typical example of a nutrition resource deficit, may lower the stimulation of photosynthesis by elevated CO_2, due to excess photoassimilate availability, triggering sugar-signaling feedback. This reduces the expression of the photosynthetic genes, especially in Rubisco, leading to the allocation of photosynthetic N into sinks that are more necessary for relative biosynthesis (Moore et al., 1999; Leakey et al., 2009b; Markelz et al., 2014). Based on a report by Duanmu et al. (2009), the enhanced expression of the limited CO_2-induced gene *HLA3* may increase the HCO_3^- transport and photosynthetic C_i affinity, which may counter the down-regulation of the photosynthetic capacity under CO_2 enrichment, if the gene can be transported to higher plants (Price et al., 2011). This demonstrates the potential modified gene applications in the improvement of photosynthetic regulation traits in high CO_2 climates.

In a rice cultivar, gene expression for D1 protein (a protein of PSII gene) was down-regulated by 20% at heat stress under elevated CO_2, but this change did not occur in another cultivar, indicated that elevated CO_2 may enhance the damage of D1 protein, depending on genotypic variation (Gesch et al., 2003). Based on a recent study in poplar plants (Liu et al., 2014), only eight significantly changed key genes involved in crucial metabolisms in response to elevated CO_2 were identified by a qRT-PCR test. During wheat plant senescence, up-regulation of genes related to nitrogen remobilization, and down-regulation of genes related to carbon remobilization were observed under elevated CO_2, reflecting greater grain N-sink strength of developing grains (Buchner et al., 2015). Based on a microarray analysis, the *A. thaliana* photosynthetic gene expression can be most adversely affected by abiotic stress, such as heat and drought, where almost all of genes were down-regulated. However, the greatest down-regulation in gene expression can be diminished by elevated CO_2 (Zinta et al., 2014). From the genome-wide expression profiling of the mRNA in *A. thaliana*leaves (Zinta et al., 2014), 3643 differentially expressed genes appeared between plants exposed to climate extremes and ambient CO_2, whereas only 2841 genes were obtained when grown under elevated CO_2. Specifically, both the up-regulated and down-regulated genes were remarkably lower in plants exposed to elevated CO_2, than in ambient CO_2. For example, under stressful conditions such as heat and drought, the down-regulations of the genes involved in the light reactions (photosystem I and II, light-harvesting complex II), pigment synthesis, and the Calvin cycle can be dampened by elevated CO_2, being consistent with changes in photosynthetic rates. It is indicated that elevated CO_2 may repress the impact of climate extremes on gene expression in rosette leaves (Zinta et al., 2014). It is worth noting that we only presented a general description here, and a detailed list of the gene expression differences may be found in the report by Zinta et al. (2014).

Elevated CO_2 often can induce a marked decline in photorespiration (see above), suggesting that there may be an involvement of the expression of the genes related to photorespiration pathway including both transcripts and metabolite levels (Sharkey, 1988; Novitskaya et al., 2002; Foyer et al., 2009; Florian et al., 2014; Wang et al., 2014). A *BOUT DE SOUFFLE* (*BOU*) gene encoding a mitochondrial carrier may be involved in photorespiration in *Arabidopsis* because of the knockout mutant *bou-2* can arrest growth at ambient CO_2, but not at high CO_2 concentration, implying *BOU*gene linking glycine decarboxylase (GDC) activity, may regulate the response to CO_2 concentration changes (Eisenhut et al., 2013). Plants defective (*glyk1* mutants), the gene encoding glycerate kinase (GLYK), cannot grow in normal CO_2 level but fully recover at elevated CO_2, which the reasonable reason why the mutant requires a high CO_2 concentration is unknown (Timm and Bauwe, 2013). The transcript levels of some photorespiratory genes up-regulated such as plastid chaperonin proteins (CPN60B), and those down-regulated such as GDC under heat and drought stresses, were largely repressed under elevated CO_2, but that is not universal for all genes (Zinta et al., 2014). Furthermore, according to a study by Florian et al. (2014), the transcript levels of photorespiratory genes in *Arabidopsis* were almost unchanged at high CO_2 concentration except a decline in transcript levels of glycine decarboxylase H-protein (GDCH1) that functions in photorespiratory carbon recovery. Thus, whether and how the photorespiratory gene expression play a major role in responses to atmospheric CO_2 concentration changes are mostly unknown (Foyer et al., 2009; Timm and Bauwe, 2013; Florian et al., 2014), which needs to be tested further.

Because antioxidant defense systems would be enhanced by elevated CO_2, the gene expression levels of antioxidant enzymes may be also promoted accordingly (Gillespie et al., 2012; Mishra and Agrawal, 2014; Zinta et al., 2014). In *A. thaliana* plants, CO_2 enrichment can up-regulate the gene transcriptional expression of an antioxidant enzyme, dehydroascorbate reductase (DHAR), but down-regulate that of CAT, particularly under stressful environments. However, the gene expression changes in others such as APX, GR, GPX, POX, and SOD to elevated CO_2 were not significant (Zinta et al., 2014). Additionally, a high transcript abundance for the majority of the genes coding antioxidant recycling enzymes enhanced by high O_3 concentration was also not affected by elevated CO_2 (Gillespie et al., 2012). Elevated CO_2 did not modify the up-regulation of transcripts of oxidative-stress-related genes induced by herbivory or elevated O_3 in soybean plants (Casteel et al., 2008). Kontunen-Soppela et al. (2010) indicating that CO_2 enrichment cannot alleviate harmful effects from O_3 pollution based on a gene expression test in birch plants. Thus, the authors could not conclude that CO_2 enrichment can up-regulate the gene transcriptional expression levels of the antioxidant enzymes under stressful environment. Further studies are needed urgently to elucidate the molecular responses in the diverse antioxidant systems in responses to elevated CO_2 with the key environmental factors including drought, heat, and ozone (Gillespie et al., 2012; Zinta et al., 2014).

One recent research study described the results of a gene bioinformatics analysis of hardy winter wheat (*Triticum aestivum*), with different low temperature adaptive capacities in response to elevated CO_2 (Kane et al., 2013). The genes induced by elevated CO_2 was three times higher in the non-acclimated (NA) relative to cold-acclimated (CA) conditions (1,022 vs. 372). The greatest down-regulation of genes appeared in the plant defense responses in the NA plants. On the other hand, CA can reverse this down-regulation, due to the cold-induced genes involved in the plant's resistance to pathogenesis, and cellular and chloroplast protection (Kane et al., 2013), suggesting that cold-adapted hardy winter plants may be less affected by elevated CO_2. Conversely, the plants that are more sensitive to cold weather may be regulated both easily and drastically via CO_2enrichment. Of note is the down-regulative interaction of high CO_2 levels with low temperature adaptations, which requires further investigation.

Another recent microarray study describes the expression of the respiratory genes in *A. thaliana*plants exposed to elevated CO_2, with both limited and sufficient N availabilities (Markelz et al., 2014). This analysis showed that 4439 transcripts were significantly different between the ambient and elevated CO_2. Particularly, the transcriptional response of the genes related to protein synthesis was greatest during the day, due to elevated CO_2 induction. These genes included those related to the components of glycolysis, the TCA cycle, the mitochondrial electron transport chain (ETC), and the mitochondrial protein import complexes. The evidence of the up-regulation of the transcription of the genes, with relation to respiration under elevated CO_2 levels, has also been obtained from rice (Fukayama et al., 2011) and soybean plants (Leakey et al., 2009b). Furthermore, 1,708 transcripts differed significantly in abundance between the limited N and ample N availabilities, while 258 transcripts differed significantly due to the interactions of the CO_2 level and N availability, again indicating that the expression of the genes related to the key physiological bioprocesses in response to elevated CO_2 may be markedly affected by other environmental factors, such as N limitation (Markelz et al., 2014) and day length (Queval et al., 2012). It is worth pointing out that the systematicness and complexity of the underlying molecular mechanisms may coexist in the plant response to elevated CO_2, and its interaction with other multiple abiotic factors including nutrition condition. In addition to the relevant studies concerning the specific gene manipulation and genome-wide transcriptional analysis, with strong selective pressure due to the novel CO_2 level, the evolutionary adaption to an atmospheric CO_2 concentration change has been found in many reports of the stomatal developmental response (Gray et al., 2000; Ward and Kelly, 2004). Moreover, because the previous studies concerning the response to CO_2 enrichment are often limited to one generation of the plant life-cycle (Ward and Kelly, 2004), to further understand the genetic variations in the plants exposed to long-term elevated CO_2, Teng et al. (2009) found that the maternal genetic effects of elevated CO_2 cannot be retrieved in their offspring after undergoing 15 generations of *A. thaliana* grown in a long-term elevated CO_2 atmosphere, indicating the lack of genetic variation and specific adaptations for CO_2-enriched responsiveness (Teng et al., 2009). It is suggested that selective pressure from elevated CO_2 may be not enough to produce a genetic modification to adapt to new environmental changes. This issue should be investigated further, with long-term exposure to elevated CO_2.

Elevated CO_2 Interactions with Multiple Abiotic Stresses

There have been several review reports concerning the interactions between elevated CO_2 and other abiotic factors, such as temperature or drought, on plant growth and physiological processes (e.g.,Morison and Lawlor, 1999; Ainsworth and Rogers, 2007; Peñuelas et al., 2013; Ruiz-Vera et al., 2013). However, the underlying mechanism concerning the responses to CO_2 enrichment with multiple factors has rarely been systematically reviewed (Xu et al., 2013a; Jagadish et al., 2014;Way et al., 2015). Although the related descriptions have been presented in the appropriate places above, here, we present a succinct statement, particularly for the underlying mechanisms in physiological responses to elevated CO_2, in combination with several abiotic factors, such as drought and heat waves.

Water deficits and heat waves are considered to be the most critical stress factors with markedly potential effects on plant growth, crop yield, vegetation productivity, photosynthetic capacity, promotion of ROS accumulation (such as H_2O_2), and the oxidative enhancement of functional molecules, such as active proteins and DNA (e.g., Mittler, 2006; Xu and Zhou, 2006; Barnabás et al., 2008; Xu et al., 2009b, 2013a, 2014; Zinta et al., 2014). With the exception of a few reports (e.g., Coleman et al., 1991; Roden and Ball, 1996), most of the relevant studies have concluded that CO_2 fertilization may mitigate the adverse impacts of environmental stresses, such as heat, drought, O_3 pollution, and their combinations (Biswas et al., 2013; Xu et al., 2013a, 2014; Zinta et al., 2014). These aspects of the mitigation of CO_2 enrichment include relatively increased individual growth (e.g., Xu et al., 2014) and community production (Naudts et al., 2014), enhanced photosynthesis (Biswas et al., 2013; Xu et al., 2014; Zinta et al., 2014), elevated WUE and NUE (Palmroth et al., 2013), optimized chlorophyll fluorescence (Biswas et al., 2013; Xu et al., 2014;Zinta et al., 2014), up-regulated antioxidant defense metabolism via increased lipophilic antioxidants and membrane-protecting enzymes (Naudts et al., 2014; Xu et al., 2014), and decreased photorespiration with low H_2O_2 production (Foyer and Noctor, 2009; Munne-Bosch et al., 2013; Zinta et al., 2014).

Elevated CO_2 may help the leaf tissues of a dominant grass in Northern China to partly escape the negative effects of heat and drought stresses on plant growth, canopy structure, leaf development, photosynthetic potential, and antioxidant systems (Xu et al., 2014). For *A. thaliana* plants, the combination of the heat and drought-induced inhibition of photosynthesis was 62% under ambient CO_2, but the reduction in photosynthesis was only 40% with elevated CO_2. Furthermore, the protein carbonyl content, a marker of protein oxidation, increased significantly during a heat wave and drought, in which the effects were repressed by increased CO_2 (Zinta et al., 2014). The dramatic differences between the altered transcriptional expression of *A. thaliana* plants subjected to a combination of heat and drought stresses were demonstrated in the presence and absence of elevated CO_2, with less down-regulation of the genes involved in the light reactions (photosystem I and II, light-harvesting complex II), pigment synthesis, and the CBB under elevated CO_2. Additionally, there was less limitation to the photosynthetic parameters, such as A_{net}, maximum photochemical efficiency (F_v/F_m), g_s, and chlorophyll content (Zinta et al., 2014), possibly due to the effectiveness of the mitigation of the CO_2 enrichment. Moreover, following cancelation of the extreme heat and drought stresses, the plant growth and physiological activities related to the positive responses to growth may partly resume at high CO_2 concentrations, and the oxidative stress can be greatly alleviated, although they cannot reach the control levels (Xu et al., 2009a,2010; Xu and Zhou, 2011; Zinta et al., 2014). Again, this implies that CO_2 fertilization may alleviate the damage of extreme climatic events, such as snap heat waves and droughts, compromising part of the loss and accelerating recovery in case of the elimination of severe abiotic stress. However, a recent report indicated that high temperature, with no elevated CO_2, provokes the drought sensitivity of the leaf to gas exchange, while the latter did not affect the *Eucalyptus radiata*seedling response to drought, and cannot alleviate the negative effects of rising temperature on drought stress (Duan et al., 2014). From another point of view, drought, warming, air pollution, and, particularly, their combination may substantially negate the elevated-CO_2 stimulation in photosynthesis, plant growth, and productivity (Biswas et al., 2013; Ruiz-Vera et al., 2013; Xu et al., 2013a; Duan et al., 2014), which is worth noting.

A hot issue has arisen, in which the photosynthetic responses to elevated CO_2 and its combination with climatic change may differ completely between plant species within their photosynthetic pathways. Because of the C_4 specific photosynthetic pathway with a CO_2-concentrating pump (von Caemmerer and Furbank, 2003), they cannot benefit from elevated CO_2 relative to C_3 plants (Leakey et al., 2006; Morgan et al., 2011; Bütof et al., 2012; Xu et al., 2014). However, there is a practical and explicable positive response of growth and photosynthesis to elevated CO_2, with drought and heat stress in C_4 plants. (1) Although no obvious response to CO_2 enrichment occurs under ample water availability, great stimulation in the growth and photosynthetic capacity may be obtained under water deficits, due to the marked alleviations of drought stress via g_s reduction and WUE elevation, and oxidative stress mitigation under elevated CO_2 (e.g., Long et al., 2004;Ghasemzadeh et al., 2010). (2) The higher temperature might benefit the C_4 species that originate from, and currently grow under warming conditions, and because the photorespiration of C_3 plants increases with rising temperature, leading to a reduction in the A_{net}. The C_4 plants lack photorespiration pathways, with no effect on photosynthesis (Long et al., 2006; Long and Ort, 2010; Morgan et al., 2011). Actually, both hypotheses have been well tested in several reports (Morgan et al., 2001, 2011; Leakey et al., 2006; Xu et al., 2014). This positive response to a combination of CO_2 enrichment and warming, as well as water deficits, highlights the fact that C_4plants may have a great potential advantage in future climatic change. A higher CO_2 concentration with warming and drought suggests that C_4 plants may prosper in these vulnerable ecosystems in arid and semiarid regions in the future (Morgan et al., 2011; Lobell et al., 2013; Xu et al., 2014). However, elevated CO_2 induced an electron transfer rate (ETR) enhancement in one C_3 species,*Halimione portulacoides*, and one C_4 species, *Spartina maritime*, but with lower photosynthetic efficiency in the C_4 plants due to an increase in the dissipated energy flux, indicated by higher non-photochemical quenching (NPQ), suggesting that the abundance of C_3 species may increase in Mediterranean halophyte vegetation (Duarte et al., 2014). Thus, the future climatic change may induce a rapid shift in some terrestrial vegetation, because of the different responses between the species with the specific photosynthetic pathways, such as C_3 and C_4 plants, depending on the combination of multiple climatic factors.

CONCLUSION

We briefly summarize several key points. (1) Elevated CO_2 generally increases the A_{net}, in which the positive responses strongly depend on the plant functional groups and species, with the expected stimulation from rising CO_2, for almost all of the C_3 species, but only for C_4 plants under water deficit conditions (due to the CCM). The performance of Rubisco in fixing carbon is promoted by CO_2 enrichment, because of its dual character. However, a downregulation in the photosynthetic capacity may occur because of the decreased ATP:ADP ratio, diluted N, and excessive photosynthate accumulation under continually rising CO_2, particularly under N and/or carbon sink limitations. (2) An elevated CO_2-induced suppression of photorespiration has been tested using a lower Gly:Ser ratio as an indicator, while a general negative response in mitochondrial respiration varies, depending on the species. The balance between the increased respiratory substrate and diluted N may play a key role in the rising CO_2-induced response, with evidence from the expression up-regulation of the genes related to the respiratory pathway. (3) Plants may run an antioxidant defense system with both the enzymatic and non-enzymatic machinery protected from the damage of oxidative stress due to the generation of ROS under abiotic stresses (such as drought and heat), while elevated CO_2 may partly promote the accumulation of antioxidants like polyphenols and ascorbate, and enhance some antioxidant enzyme activities to diminish the oxidative stress from abiotic factors, alone or combination, depending on the genetic variations and plant developmental stage. (4) Elevated CO_2 leads to a lower N level and higher content of the total non-structural carbohydrates (TNC), including starch and sugars, while remaining mostly stable in the totals of the structural carbohydrates, lignin, and lipids. However, some secondary metabolites, such as phenylpropanoids, tannins, and phenolic acids, are enhanced by CO_2 enrichment. Isoprene emissions may be weakened by elevated CO_2, because biosynthesis may need to balance the ATP and NADPH with photosynthetic metabolism. (5) Elevated CO_2 might mitigate the adverse effects of abiotic stresses via relatively increased individual growth, enhanced photosynthesis, increased resource use efficiency, promoted antioxidant defense metabolism, and decreased photorespiration under multiple environmental stresses. In terms of the photosynthetic pathway, CO_2 enrichment did not affect C_4 plants under ample environmental conditions, but promoted it when exposed to drought, warming and their combination, predicting a great potential advantage in future climatic change scenarios for the C_4 species, particularly in arid and semiarid areas.

FUTURE PERSPECTIVES

Promotion of the Relevant Research

In the future, we may focus on several crucial research aspects: (1) to further elucidate the underlying mechanisms of the response to CO_2 enrichment in key biological processes, including photosynthesis, antioxidant machinery, and other related critical metabolic bioprocesses, such as hormone-involved regulation, as well as the relevant biochemical signal cascades; (2) to disentangle and compare the diverse responses from different species and PFTs to elevated CO_2 or its combination with other abiotic factors; (3) to integrate various spatial-temporal scales from molecular, cellular, biochemical, physiological, individual, ecosystem, and global vegetation levels, and from instantaneous to annual or longer time-scales to elucidate the underlying genetic mechanisms in association with key biological processes under the effects of global environmental factors, including elevated CO_2, warming, drought, and air pollution; (4) to strengthen the linkages to other relevant research subjects, including ecological, biogeoscience, environmental, climatic, and social-economic aspects, to find appropriate synthetic solutions to urgent practical issues like environmental contamination, ecosystem damage, and global warming impacts.

Potential Applications under Future Climate Change

Future climate change may impact key biological metabolic processes and their feedback. For example, environmental stresses may provoke the generation of ROS in chloroplasts, the site of photosynthesis, while future high CO_2 levels may alleviate the limitations of these stresses. We might also use biotechnological tools such as the protection function against ROS to deal with future climatic change. In addition, Rubisco properties may be improved by regulating the transgenic expression of Rubisco activase in crops such as rice, possibly enhancing the photosynthetic capacity under rising CO_2 (Fukayama et al., 2012), while the high CO_2-induced downregulation of the photosynthetic capacity might induce the modification of the photosynthetic pathway (Price et al., 2011). Furthermore, the modified genetic capacity for the high utilization of photosynthate to strengthen sink storage may make plants capable of sustaining increased photosynthesis when the plants are grown in elevated atmospheric CO_2, while additional thermo-tolerant transgenic crops may be required to cope simultaneously with climatic warming (Lavania et al., 2015). Finally, research should be conducted to strengthen the feasible applications from the relevant research results in response to CO_2 enrichment, and its combination with multiple environmental factors for ecological management, climate change mitigation, sustainable development, and related policy decisions, but not at the expense of environment.

Author Contributions

YJ is co-first author, ZX and YJ collected and analyzed the data, ZX, YJ, and GZ wrote the manuscript.

CONFLICT OF INTEREST STATEMENT

The authors declare that the research was conducted in the absence of any commercial or financial relationships that could be construed as a potential conflict of interest.

Acknowledgments

The study was funded by the National Natural Science Foundation of China (41330531, 31170456), and China Special Fund for Meteorological Research in the Public Interest (Major projects) (GYHY201506001-3).

REFERENCES

1. Abd Elgawad, H., and Asard, H. (2013). Elevated CO_2 attenuates oxidative stress caused by drought and elevated temperature in four C_3 plant species. *Biotechnologia* 94, 156–202.
2. AbdElgawad, H., Farfan-ignolo, E. R., de Vos, D., and Asard, H. (2015). Elevated CO_2 mitigates drought and temperature-induced oxidative stress differently in grasses and legumes. *Plant Sci.* 231, 1–10. doi: 10.1016/j.plantsci.2014.11.001
3. AbdElgawad, H., Peshev, D., Zinta, G., Van den Ende, W., Janssens, I. A., and Asard, H. (2014). Climate extreme effects on the chemical composition of temperate grassland species under ambient and elevated CO_2: a comparison of fructan and non-fructan accumulators. *PLoS ONE* 9:e92044. doi: 10.1371/journal.pone.0092044
4. Ainsworth, E. A., and Long, S. P. (2005). What have we learned from 15 years of free-air CO_2 enrichment (FACE)? A meta-analytic review of the responses of photosynthesis, canopy properties and plant production to rising CO_2. *New Phytol.* 165, 351–372. doi: 10.1111/j.1469-8137.2004.01224.x
5. Ainsworth, E. A., and Rogers, A. (2007). The response of photosynthesis and stomatal conductance to rising [CO_2]: mechanisms and environmental interactions. *Plant Cell Environ.* 30, 258–270. doi: 10.1111/j.1365-3040.2007.01641.x
6. Ainsworth, E. A., Rogers, A., Nelson, R., and Long, S. P. (2004). Testing the 'source-sink' hypothesis of down-regulation of photosynthesis in elevated [CO_2] in the field with single gene substitutions in *Glycine max*. *Agr. Forest Meteorol.* 122, 85–94. doi: 10.1016/j.agrformet.2003.09.002
7. Amthor, J. S. (2000). Direct effect of elevated CO_2 on nocturnal in situ leaf respiration in nine temperate deciduous tree species is small. *Tree Physiol.* 20, 139–144. doi: 10.1093/treephys/20.2.139
8. Aranjuelo, I., Cabrera-Bosquet, L., Morcuende, R., Avice, J. C., Nogués, S., Araus, J. L., et al. (2011). Does ear C sink strength contribute to overcoming photosynthetic acclimation of wheat plants exposed to elevated CO_2? *J. Exp. Bot.* 62, 3957–3969. doi: 10.1093/jxb/err095
9. Atkin, O. (2015). New Phytologist and the 'fate' of carbon in terrestrial ecosystems. *New Phytol.* 205, 1–3. doi: 10.1111/nph.13185
10. Atkin, O., Millar, H. A., and Turnbull, M. H. (2010). Plant respiration in a changing world. *New Phytol.* 187, 268–272. doi: 10.1111/j.1469-8137.2010.03343.x
11. Ayub, G., Zaragoza-Castells, J., Griffin, K. L., and Atkin, O. K. (2014). Leaf respiration in darkness and in the light under pre-industrial, current and elevated atmospheric CO_2 concentrations. *Plant Sci.* 226, 120–130. doi: 10.1016/j.plantsci.2014.05.001
12. Barnabás, B., Jäger, K., and Fehér, A. (2008). The effect of drought and heat stress on reproductive processes. *Plant Cell Environ.* 31, 11–38. doi: 10.1111/j.1365-3040.2007.01727.x
13. Biswas, D. K., Xu, H., Li, Y. G., Ma, B. L., and Jiang, G. M. (2013). Modification of photosynthesis and growth responses to elevated CO_2 by ozone in two cultivars of winter wheat with different years of release. *J. Exp. Bot.* 64, 1485–1496. doi: 10.1093/jxb/ert005
14. Bowes, G. (1991). Growth at elevated CO_2: photosynthetic responses mediated through Rubisco. *Plant Cell Environ.* 14, 795–806. doi: 10.1111/j.1365-3040.1991.tb01443.x
15. Bowes, G., Ogren, W. L., and Hageman, R. H. (1971). Phosphoglycolate production catalyzed by ribulose diphosphate carboxylase. *Biochem. Biophys. Res. Commun.* 45, 716–722. doi: 10.1016/0006-291X(71)90475-X
16. Bryant, J. P., Chapin, F. S. III, Reichardt, P., and Clausen, T. (1983). Carbon/nutrient balance of boreal plants in relation to vertebrate herbivory. *Oikos* 40, 357–368. doi: 10.2307/3544308
17. Bryant, J., Taylor, G., and Frehner, M. (1998). Photosynthetic acclimation to elevated CO_2 is modified by source: sink balance in three component species of chalk grassland swards grown in a free air carbon dioxide enrichment (FACE) experiment. *Plant Cell Environ.* 21, 159–168. doi: 10.1046/j.1365-3040.1998.00265.x
18. Buchner, P., Tausz, M., Ford, R., Leo, A., Fitzgerald, G. J., Hawkesford, M. J., et al. (2015). Expression patterns of C- and N-metabolism related genes in wheat are changed during senescence under elevated CO_2 in dry-land agriculture. *Plant Sci.* 236, 239–249. doi: 10.1016/j.plantsci.2015.04.006
19. Bunce, J. A. (2014). Limitations to soybean photosynthesis at elevated carbon dioxide in free-air enrichment and open top chamber systems. *Plant Sci.* 226, 131–135. doi: 10.1016/j.plantsci.2014.01.002
20. Bütof, A., von Riedmatten, L. R., Dormann, C. F.,

Scherer-Lorenzen, M., Welk, E., and Bruelheide, H. (2012). The responses of grassland plants to experimentally simulated climate change depend on land use and region. *Globe Change Biol.* 18, 127–137. doi: 10.1111/j.1365-2486.2011.02539.x

21. Carmo-Silva, A. E., Powers, S. J., Keys, A. J., Arrabaca, M. C., and Parry, M. A. J. (2008). Photorespiration in C_4 grasses remains slow under drought conditions. *Plant Cell Environ.* 31, 925–940. doi: 10.1111/j.1365-3040.2008.01805.x

22. Casteel, C. L., O'Neill, B. F., Zavala, J. A., Bilgin, D. D., Berenbaum, M. R., and Delucia, E. H. (2008). Transcriptional profiling reveals elevated CO_2 and elevated O_3 alter resistance of soybean *(Glycine max)* to Japanese beetles (*Popillia japonica*). *Plant Cell Environ.* 31, 419–434. doi: 10.1111/j.1365-3040.2008.01782.x

23. Claeys, M., Graham, B., Vas, G., Wang, W., Vermeylen, R., Pashynska, V., et al. (2004). Formation of secondary organic aerosols through photooxidation of isoprene. *Science* 303, 1173–1176. doi: 10.1126/science.1092805

24. Coleman, J. S., Rochefort, L., Bazzaz, F. A., and Woodward, F. I. (1991). Atmospheric CO_2, plant nitrogen status and the susceptibility of plants to an acute increase in temperature. *Plant Cell Environ.* 14, 667–674. doi: 10.1111/j.1365-3040.1991.tb01539.x

25. Cotrufo, M. F., Ineson, P., and Scott, A. (1998). Elevated CO_2 reduces the nitrogen concentration of plant tissues. *Global Change Biol.* 4, 43–54. doi: 10.1046/j.1365-2486.1998.00101.x

26. Curtis, P. S., and Wang, X. (1998). A meta-analysis of elevated CO_2 effects on woody plant mass, form, and physiology.*Oecologia* 113, 299–313. doi: 10.1007/s004420050381

27. Delucia, E. H., Sasek, T. W., and Strain, B. R. (1985). Photosynthetic inhibition after long-term exposure to elevated levels of atmospheric carbon dioxide. *Photosyn. Res.* 7, 175–184. doi: 10.1007/BF00037008

28. Dhindsa, R. S., Plumb-Dhindsa, P. A., and Thorpe, T. A. (1981). Leaf senescence: correlated with increased levels of membrane permeability and lipid peroxidation, and decreased levels of superoxide dismutase and catalase. *J. Exp. Bot.* 32, 93–101. doi: 10.1093/jxb/32.1.93

29. Drake, B. G., Berry, J., Bunce, J., Dijkstra, P., Farrar, J., and Gifford, R. M. (1999). Does elevated atmospheric CO_2concentration inhibit mitochondrial respiration in green plants? *Plant Cell Environ.* 22, 649–657. doi: 10.1046/j.1365-3040.1999.00438.x

30. Drake, B. G., Gonzalez-Meler, M. A., and Long, S. P. (1997). More efficient plants: a consequence of rising atmospheric CO_2?*Annu. Rev. Plant Physiol. Plant Mol. Biol.* 48, 609–639. doi: 10.1146/annurev.arplant.48.1.609

31. Duan, H. L., Duursma, R. A., Huang, G. M., Smith, R. A., Choat, B., O'Grady, A. P., et al. (2014). Elevated [CO_2] does not ameliorate the negative effects of elevated temperature on drought-induced mortality in *Eucalyptus radiata* seedlings. *Plant Cell Environ.* 37, 1598–1613. doi: 10.1111/pce.12260

32. Duanmu, D., Miller, A. R., Horken, K. M., Weeks, D. P., and Spalding, M. H. (2009). Knockdown of limiting-CO_2–induced gene HLA3 decreases HCO3- transport and photosynthetic Ci affinity in *Chlamydomonas reinhardtii*. *Proc. Nati. Acad. Sci. U.S.A.*106, 5990–5995. doi: 10.1073/pnas.0812885106

33. Duarte, B., Santos, D., Marques, J. C., and Caçador, I. (2013). Ecophysiological adaptations of two halophytes to salt stress: photosynthesis, PS II photochemistry and anti-oxidant feedback – Implications for resilience in climate change. *Plant Physiol. Biochem.* 67, 178–188. doi: 10.1016/j.plaphy.2013.03.004

34. Duarte, B., Santos, D., Silva, H., Marques, J. C., and Caçador, I. (2014). Photochemical and biophysical feedbacks of C_3 and C_4Mediterranean halophytes to atmospheric CO_2 enrichment confirmed by their stable isotope signatures. *Plant Physiol. Biochem.* 80, 10–22. doi: 10.1016/j.plaphy.2014.03.016

35. Eichelmann, H., Talts, E., Oja, V., Padu, E., and Laisk, A. (2009). Rubisco in planta kcat is regulated in balance with photosynthetic electron transport. *J. Exp. Bot.* 60, 4077–4088. doi: 10.1093/jxb/erp242

36. Eisenhut, M., Planchais, S., Cabassa, C., Guivarćh, A., Justin, A.-M., Taconnat, L., et al. (2013). *Arabidopsis* a bout de souffle is a putative mitochondrial transporter involved in photorespiratory metabolism and is required for meristem growth at ambient CO_2 levels. *Plant J.* 73, 836–849. doi: 10.1111/tpj.12082

37. Farfan-Vignolo, E. R., and Asard, H. (2012). Effect of elevated CO_2 and temperature on the oxidative stress response to drought in *Lolium perenne* L. and *Medicago sativa* L. *Plant Physiol. Biochem.* 59, 55–62. doi: 10.1016/j.plaphy.2012.06.014

38. Florian, A., Timm, S., Nikoloski, Z., Tohge, T., Bauwe, H., Araújo, W. L., et al. (2014). Analysis of metabolic alterations in*Arabidopsis* following changes in the carbon dioxide and oxygen partial pressures. *J. Integr. Plant Biol.* 56, 941–959. doi: 10.1111/jipb.12237

39. Foyer, C. H., Bloom, A. J., Queval, G., and Noctor, G. (2009). Photorespiratory metabolism: genes, mutants, energetics, and redox signaling. *Annu. Rev. Plant Biol.* 60, 455–484. doi: 10.1146/annurev.arplant.043008.091948

40. Foyer, C. H., and Noctor, G. (2009). Redox regulation in photosynthetic organisms: signaling, acclimation, and practical implications. *Antioxid. Redox Signal.* 11, 861–905. doi: 10.1089/ars.2008.2177

41. Fukayama, H., Sugino, M., Fukuda, T., Masumoto, C., Taniguchi, Y., and Okada, M, et al. (2011). Gene expression profiling of rice grown in free air CO_2 enrichment (FACE) and elevated soil temperature. *Field Crops Res.* 121, 195–199. doi: 10.1016/j.fcr.2010.11.018

42. Fukayama, H., Ueguchi, C., Nishikawa, K., Katoh, N., Ishikawa, C., Masumoto, C., et al. (2012). Overexpression of Rubisco activase decreases the photosynthetic CO_2 assimilation rate by reducing Rubisco content in rice leaves. *Plant Cell Physiol.* 53, 976–986. doi: 10.1093/pcp/pcs042

43. Gesch, R. W., Kang, I.-H., Gallo-Meagher, M., Vu, J. C. V., Boote, K. J., Allen, L. H., et al. (2003). Rubisco

expression in rice leaves is related to genotypic variation of photosynthesis under elevated growth CO_2 and temperature. *Plant Cell Environ.* 26, 1941–1950. doi: 10.1046/j.1365-3040.2003.01110.x

44. Ghasemzadeh, A., Jaafar, H. Z. E., and Rahmat, A. (2010). Elevated carbon dioxide increases contents of flavonoids and phenolic compounds, and antioxidant activities in Malaysian young ginger *(Zingiber officinale* roscoe.) varieties. *Molecules* 15, 7907–7922. doi: 10.3390/molecules15117907
45. Gill, S. S., and Tuteja, N. (2010). Reactive oxygen species and antioxidant machinery in abiotic stress tolerance in crop plants.*Plant Physiol. Biochem.* 48, 909–930. doi: 10.1016/j.plaphy.2010.08.016
46. Gillespie, K. M., Xu, F., Richter, K. T., McGrath, J. M., Markelz, R. C., Ort, D. R., et al. (2012). Greater antioxidant and respiratory metabolism in field – grown soybean exposed to elevated O_3 under both ambient and elevated CO_2. *Plant Cell Environ.* 35, 169–184. doi: 10.1111/j.1365-3040.2011.02427.x
47. Gonzalez-Meler, M. A., Taneva, L., and Trueman, R. J. (2004). Plant respiration and elevated atmospheric CO_2 concentration: cellular responses and global significance. *Ann. Bot.* 94, 647–656. doi: 10.1093/aob/mch189
48. Gray, J. E., Holroyd, G. H., van der Lee, F. M., Bahrami, A. R., Sijmons, P. C., Woodward, F. I., et al. (2000). The HIC signalling pathway links CO_2 perception to stomatal development. *Nature* 408, 713–716. doi: 10.1038/35047071
49. Guo, H., Sun, Y., Li, Y., Tong, B., Herris, M., Zhu-Salzman, K., et al. (2013). Pea aphid promotes amino acid metabolism both in*Medicago truncatula* and bacteriocytes to favor aphid population growth under elevated CO_2. *Global Change Biol.* 19, 3210–3223. doi: 10.1111/gcb.12260
50. Guo, Z., Ou, W., Lu, S., and Zhong, Q. (2006). Differential responses of antioxidative system to chilling and drought in four rice cultivars differing in sensitivity. *Plant Physiol. Biochem.* 44, 828–836. doi: 10.1016/j.plaphy.2006.10.024
51. Heath, R. L., and Packer, L. (1968). Photoperoxidation in isolated chloroplasts: I. Kinetics and stoichiometry of fatty acid peroxidation. *Arch. Biochem. Biophys.* 125, 189–198. doi: 10.1016/0003-9861(68)90654-1
52. Hodges, D. M., and Forney, C. F. (2000). The effects of ethylene, depressed oxygen and elevated carbon dioxide on antioxidant profiles of senescing spinach leaves. *J. Exp. Bot.* 51, 645–655. doi: 10.1093/jexbot/51.344.645
53. Husic, D. W., Husic, H. D., Tolbert, N. E., and Black, Jr. C. C. (1987). The oxidative photosynthetic carbon cycle or C2 cycle. *Crit. Rev. Plant Sci.* 5, 45–100. doi: 10.1080/07352688709382234
54. IPCC. (2013). "Summary for Policymakers," in *Climate Change 2013: The Physical Science Basis. Contribution of Working Group I to the Fifth Assessment Report of the Intergovernmental Panel on Climate Change*, eds T. F. Stocker, D. Qin, G. K. Plattner, M. Tignor, S. K. Allen, J. Boschung, et al. (New York, NY: Cambridge University Press).
55. Jagadish, K. S. V., Kadam, N. N., Xiao, G., Melgar, R. J., Bahuguna, R. N., Quinones, C., et al. (2014). Agronomic and physiological responses to high temperature, drought, and elevated CO_2 interactions in cereals. *Adv. Agron.* 127, 111–156. doi: 10.1016/B978-0-12-800131-8.00003-0
56. Jiang, Y. P., Cheng, F., Zhou, Y. H., Xia, X. J., Shi, K., and Yu, J. Q. (2012). Interactive effects of CO_2 enrichment and brassinosteroid on CO_2 assimilation and photosynthetic electron transport in *Cucumis sativus*. *Environ. Exp. Bot.* 75, 98–106. doi: 10.1016/j.envexpbot.2011.09.002
57. Jifon, J., and Wolfe, D. (2002). Photosynthetic acclimation to elevated CO_2 in *Phaseolus vulgaris* L. is altered by growth response to nitrogen supply. *Global Change Biol.* 8, 101–1027. doi: 10.1046/j.1365-2486.2002.00531. x
58. Kane, K., Dahal, K. P., Badawi, M. A., Houde, M., Hüner, N. P., and Sarhan, F. (2013). Long-term growth under elevated CO_2suppresses biotic stress genes in non-acclimated, but not cold-acclimated winter wheat. *Plant Cell Physiol.* 54, 1751–1768. doi: 10.1093/pcp/pct116
59. Kebeish, R., Niessen, M., Thiruveedhi, K., Bari, R., Hirsch, H. J., Rosenkranz, R., et al. (2007). Chloroplastic photorespiratory bypass increases photosynthesis and biomass production in *Arabidopsis thaliana*. *Nat. Biotechnol.* 25, 593–599. doi: 10.1038/nbt1299
60. Kirschbaum, M. U. (2011). Does enhanced photosynthesis enhance growth? Lessons learned from CO_2 enrichment studies. *Plant Physiol.* 155, 117–124. doi: 10.1104/pp.110.166819
61. Klaiber, J., Dorn, S., and Najar-Rodriguez, A. J. (2013). Acclimation to elevated CO_2 increases constitutive glucosinolate levels of *Brassica* plants and affects the performance of specialized herbivores from contrasting feeding guilds. *J. Chem. Ecol.* 39, 653–665. doi: 10.1007/s10886-013-0282-3
62. Kontunen-Soppela, S., Parviainen, J., Ruhanen, H., Brosché, M., Keinänen, M., and Thakur, R. C. (2010). Gene expression responses of paper birch (*Betula papyrifera*) to elevated CO_2 and O_3 during leaf maturation and senescence. *Environ. Poll.* 158, 959–968. doi: 10.1016/j.envpol.2009.10.008
63. Kozaki, A., and Takeba, G. (1996). Photorespiration protects C_3 plants from photooxidation. *Nature* 384, 557–560. doi: 10.1038/ 384557a0
64. Kumari, S., Agrawal, M., and Tiwari, S. (2013). Impact of elevated CO_2 and elevated O_3 on *Beta vulgaris* L.: pigments, metabolites, antioxidants, growth and yield. *Environ. Poll.* 174, 279–288. doi: 10.1016/j.envpol.2012.11.021
65. Lavania, D., Dhingra, A., Siddiqui, M. H., Al-Whaibi, M. H., and Grover, A. (2015). Current status of the production of high temperature tolerant transgenic crops for cultivation in warmer climates. *Plant Physiol. Biochem.* 86, 100–108. doi: 10.1016/j.plaphy.2014.11.019
66. Lavola, A., and Julkunen-Tiitto, R. (1994). The effect of elevated carbon dioxide and fertilization on

primary and secondary metabolites in birch, *Betula pendula* (Roth). *Oecologia* 99, 315–321. doi: 10.1007/BF00627744

67. Lavola, A., Nybakken, L., Rousi, M., Pusenius, J., Petrelius, M., Kellomaki, S., et al. (2013). Combination treatment of elevated UVB radiation, CO_2 and temperature has little effect on silver birch (*Betula pendula*) growth and phytochemistry. *Physiol. Plant.* 149, 499–514. doi: 10.1111/ppl.12051
68. Leakey, A. D. B., Ainsworth, E. A., Bernacchi, C. J., Rogers, A., Long, S. P., and Ort, D. R. (2009a). Elevated CO_2 effects on plant carbon, nitrogen, and water relations: six important lessons from FACE. *J. Exp. Bot.* 60, 2859–2876. doi: 10.1093/jxb/erp096
69. Leakey, A. D. B., Xu, F., Gillespie, K. M., McGrath, J. M., Ainsworth, E. A., and Ort, D. R. (2009b). Genomic basis for stimulated respiration by plants growing under elevated carbon dioxide. *Proc. Natl. Acad. Sci. U.S.A.* 106, 3597–3602. doi: 10.1073/pnas.0810955106
70. Leakey, A. D. B., Uribelarrea, M., Ainsworth, E. A., Naidu, S. L., Rogers, A., Ort, D. R., et al. (2006). Photosynthesis, productivity and yield of maize are not affected by open-air elevation of CO_2 concentration in the absence of drought. *Plant Physiol.* 140, 779–790. doi: 10.1104/pp.105.073957
71. Lewis, J. D., Wang, X. Z., Griffin, K. L., and Tissue, D. T. (2002). Effects of age and ontogeny on photosynthetic responses of a determinate annual plant to elevated CO_2 concentration. *Plant Cell Environ.* 25, 359–368. doi: 10.1046/j.0016-8025.2001.00815.x
72. Li, P., Sioson, A., Mane, S. P., Ulanov, A., Grothaus, G., Heath, L. S., et al. (2006). Response diversity of *Arabidopsis thaliana*ecotypes in elevated [CO_2] in the field. *Plant Mol. Biol.* 62, 593–609. doi: 10.1007/s11103-006-9041-y
73. Li, X. M., Zhang, L. H., Ma, L. J., and Li, Y. Y. (2011). Elevated carbon dioxide and/or ozone concentrations induce hormonal changes in *Pinus tabulaeformis*. *J. Chem. Ecol.* 37, 779–784. doi: 10.1007/s10886-011-9975-7
74. Lichtenthaler, H. K. (1999). The 1-deoxy-D-xylulose-5-phosphate pathway of isoprenoid biosynthesis in plants. *Annu. Rev. Plant Physiol. Plant Mol. Biol.* 50, 47–65. doi: 10.1146/annurev.arplant.50.1.47
75. Lindroth, R. L., Roth, S., and Nordheim, E. V. (2001). Genotypic variation in response of quaking aspen (*Populus tremuloides*) to atmospheric CO_2 enrichment. *Oecologia* 126, 371–379. doi: 10.1007/s004420000521
76. Liu, J., Zhang, J., He, C., and Duan, A. (2014). Genes responsive to elevated CO_2 concentrations in triploid white poplar and integrated gene network analysis. *PLoS ONE* 9:e98300. doi: 10.1371/journal.pone.0098300
77. Lobell, D. B., Hammer, G. L., McLean, G., Messina, C., Roberts, M. J., and Schlenker, W. (2013). The critical role of extreme heat for maize production in the United States. *Nat. Clim. Change* 3, 497–501. doi: 10.1038/nclimate1832
78. Lobell, D. B., Schlenker, W., and Costa-Roberts, J. (2011). Climate trends and global crop production since 1980. *Science* 333, 616–620. doi: 10.1126/science.1204531
79. Long, S. P. (1991). Modification of the response of photosynthetic productivity to rising temperature by atmospheric CO_2concentrations: has its importance been underestimated? *Plant Cell Environ.* 14, 729–739. doi: 10.1111/j.1365-3040.1991.tb01439.x
80. Long, S. P., Ainsworth, E. A., Rogers, A., and Ort, D. R. (2004). Rising atmospheric carbon dioxide: plants FACE the future.*Annu. Rev. Plant Biol.* 55, 591–628. doi: 10.1146/annurev.arplant.55.031903.141610
81. Long, S. P., and Ort, D. R. (2010). More than taking the heat: crops and global change. *Curr. Opin. Plant Biol.* 13, 241–248. doi: 10.1016/j.pbi.2010.04.008
82. Long, S. P., Zhu, X.-G., Naidu, S., and Ort, D. R. (2006). Can improvement in photosynthesis increase crop yields? *Plant Cell Environ.* 29, 315–330. doi: 10.1111/j.1365-3040.2005.01493.x
83. Loreto, F., Centritto, M., Barta, C., Calfapietra, C., Fares, S., and Monson, R. K. (2007). The relationship between isoprene emission rate and dark respiration rate in white poplar (*Populus alba* L.) leaves. *Plant Cell Environ.* 30, 662–669. doi: 10.1111/j.1365-3040.2007.01648.x
84. Loreto, F., and Sharkey, T. D. (1993). On the relationship between isoprene emission and photosynthetic metabolites under different environmental conditions. *Planta* 189, 420–424. doi: 10.1007/BF00194440
85. Luo, Y., Su, B., Currie, W. S., Dukes, J. S., Finzi, A., and Hartwig, U. et al. (2004). Progressive nitrogen limitation of ecosystem responses to rising atmospheric carbon dioxide. *Bioscience* 54, 731–739. doi: 10.1641/0006-3568(2004)054
86. Markelz, R. C., Lai, L. X., Vosseler, L. N., and Leakey, A. D. (2014). Transcriptional reprogramming and stimulation of leaf respiration by elevated CO_2 concentration is diminished, but not eliminated, under limiting nitrogen supply. *Plant Cell Environ.*37, 886–898. doi: 10.1111/pce.12205
87. Markelz, R. C., Strellner, R. S., and Leakey, A. D. (2011). Impairment of C_4 photosynthesis by drought is exacerbated by limiting nitrogen and ameliorated by elevated [CO_2] in maize. *J. Exp. Bot.* 62, 3235–3246. doi: 10.1093/jxb/err056
88. Matros, A., Amme, S., Kettig, B., Buck-Sorlin, G. H., Sonnewald, U., and Mock, H. P. (2006). Growth at elevated CO_2concentrations leads to modified profiles of secondary metabolites in tobacco cv. SamsunNN and to increased resistance against infection with potato virus Y. *Plant Cell Environ.* 29, 126–137. doi: 10.1111/j.1365-3040.2005.01406.x
89. Melillo, J. M., McGuire, A. D., Kicklighter, D. W., Moore, B. III, Vorosmarty, J., and Schloss, A. L. (1993). Global climate change and terrestrial net primary production. *Nature* 363, 234–240. doi: 10.1038/363234a0
90. Meng, F., Zhang, J., Yao, F., and Hao, C. (2014). Interactive effects of elevated CO_2 concentration and irrigation on photosynthetic parameters and yield of maize in northeast China. *PLoS ONE* 9:e98318. doi:

10.1371/journal.pone.0098318

91. Mishra, A. K., and Agrawal, S. B. (2014). Cultivar specific response of CO_2 fertilization on two tropical mung bean (*Vigna radiata*L.) cultivars: ROS generation, antioxidant status, physiology, growth, yield and seed quality. *J. Agron. Crop Sci.* 20, 273–289. doi: 10.1111/jac.12057

92. Mittler, R. (2006). Abiotic stress, the field environment and stress combination. *Trends Plant Sci.* 11, 15–19. doi: 10.1016/j.tplants.2005.11.002

93. Miyazaki, S., Fredricksen, M., Hollis, K. C., Poroyko, V., Shepley, D., and Galbraith, D. W., et al. (2004). Transcript expression profiles of *Arabidopsis thaliana* grown under controlled conditions and open-air elevated concentrations of CO_2 and of O_3. *Field Crops Res.* 90, 47–59. doi: 10.1016/j.fcr.2004.07.010

94. Monteith, J. L. (1977). Climate and the efficiency of crop production in Britain. *Philos. Trans. R. Soc. B Biol.* 281, 277–294. doi: 10.1098/rstb.1977.0140

95. Moore, B. D., Cheng, S.-H., Sims, D., and Seemann, J. R. (1999). The biochemical and molecular basis for photosynthetic acclimation to elevated atmospheric CO_2. *Plant Cell Environ.* 22, 567–582. doi: 10.1046/j.1365-3040.1999.00432.x

96. Morfopoulos, C., Sperlich, D., Peñuelas, J., Filella, I., Llusià, J., Medlyn, B. E., et al. (2014). A model of plant isoprene emission based on available reducing power captures responses to atmospheric CO_2. *New Phytol.* 203, 125–139. doi: 10.1111/nph.12770

97. Morgan, J. A., Lecain, D. R., Mosier, A. R., and Milchunas, D. G. (2001). Elevated CO_2 enhances water relations and productivity and affects gas exchange in C_3 and C_4 grasses of the Colorado shortgrass steppe. *Global Change Biol.* 7, 451–466. doi: 10.1046/j.1365-2486.2001.00415.x

98. Morgan, J. A., LeCain, D. R., Pendall, E., Blumenthal, D. M., Kimball, B. A., and Carrillo, Y. et al. (2011). C_4 grasses prosper as carbon dioxide eliminates desiccation in warmed semi-arid grassland. *Nature* 476, 202–205. doi: 10.1038/nature10274

99. Morison, J. I. L., and Lawlor, D. W. (1999). Interactions between increasing CO_2 concentration and temperature on plant growth. *Plant Cell Environ.* 22, 659–682. doi: 10.1046/j.1365-3040.1999.00443.x

100. Moroney, J. V., Jungnick, N., DiMario, R. J., and Longstreth, D. J. (2013). Photorespiration and carbon concentrating mechanisms, two adaptations to high O_2, low CO_2 conditions. *Photosyn. Res.* 117, 121–131. doi: 10.1007/s11120-013-9865-7

101. Munne-Bosch, S., Queval, G., and Foyer, C. H. (2013). The impact of global change factors on redox signaling underpinning stress tolerance. *Plant Physiol.* 161, 5–19. doi: 10.1104/pp.112.205690

102. NASA. (2014). *Global Climate Change: Vital Signs of the Planet.* Available at: http://climate.nasa.gov/400ppmquotes[accessed December 4, 2014].

103. Naudts, K., Van Den Berge, J., Farfan, E., Rose, P., AbdElgawad, H., Ceulemans, R., et al. (2014). Future climate alleviates stress impact on grassland productivity through altered antioxidant capacity. *Environ. Exp. Bot.* 99, 150–158. doi: 10.1016/j.envexpbot.2013.11.003

104. Novitskaya, L., Trevanion, S. J., Driscoll, S., Foyer, C. H., and Noctor, G. (2002). How does photorespiration modulate leaf amino acid contents—A dual approach through modelling and metabolite analysis. *Plant Cell Environ.* 25, 821–835. doi: 10.1046/j.1365-3040.2002.00866.x

105. Nowak, R. S., Ellsworth, D. S., and Smith, S. D. (2004). Functional responses of plants to elevated atmospheric CO_2 - do photosynthetic and productivity data from FACE experiments support early predictions? *New Phytol.* 162, 253–280. doi: 10.1111/j.1469-8137.2004.01033.x

106. Ogweno, J. O., Song, X. S., Shi, K., Hu, W. H., Mao, W. H., Zhou, Y. H., et al. (2008). Brassinosteroids alleviate heat-induced inhibition of photosynthesis by increasing carboxylation efficiency and enhancing antioxidant systems in *Lycopersicon esculentum*. *J. Plant Growth Regul.* 27, 49–57. doi: 10.1007/s00344-007-9030-7

107. Palmroth, S., Katul, G. G., Maier, C. A., Ward, E., Manzoni, S., and Vico, G. (2013). On the complementary relationship between marginal nitrogen and water-use efficiencies among *Pinus taeda* leaves grown under ambient and CO_2-enriched environments.*Ann. Bot.* 111, 467–477. doi: 10.1093/aob/mcs268

108. Peñuelas, J., Sardans, J., Estiarte, M., Ogaya, R., Carnicer, J., Coll, M., et al. (2013). Evidence of current impact of climate change on life: a walk from genes to the biosphere. *Global Change Biol.* 19, 2303–2338. doi: 10.1111/gcb.12143

109. Peñuelas, J., Sardans, J., Rivas-Ubach, A., and Janssess, I. (2012). The human-induced imbalances between C, N and P in Earth's life system. *Global Change Biol.* 18, 3–6. doi: 10.1111/j.1365-2486.2011.02568.x

110. Pérez-López, U., Robredo, A., Lacuesta, M., Sgherri, C., Muñoz-Rueda, A., Navari-Izzo, F., et al. (2009). The oxidative stress caused by salinity in two barley cultivars is mitigated by elevated CO_2. *Physiol. Plant.* 135, 29–42. doi: 10.1111/j.1399-3054.2008.01174.x

111. Poorter, H., Berkel, V., Baxter, R., den Hertog, J., Dijkstra, P., Gifford, R. M., et al. (1997). The effect of elevated carbon dioxide on the chemical composition and construction costs of leaves of 27 C_3 species. *Plant Cell Environ.* 20, 472–482. doi: 10.1046/j.1365-3040.1997.d01-84.x

112. Possell, M., and Hewitt, C. N. (2011). Isoprene emissions from plants are mediated by atmospheric CO_2 concentrations. *Global Change Biol.* 17, 1595–1610. doi: 10.1111/j.1365-2486.2010.02306.x

113. Prentice, I. C., Farquhar, G. D., Fasham, M., Goulden, M. L., Heimann, M., Jaramillo, V. J., et al. (2001). "The carbon cycle and atmospheric carbon dioxide," in *Climate Change 2001: The Scientific Basis. Contribution of Working Group I to the Third Assessment Report of the Intergovernmental Panel on Climate Change*, eds J. Houghton, Y. Ding, and D. Griggs (New York, NY: Cambridge University Press), 185–237.

114. Price, G. D., Badger, M. R., and von Caemmerer, S.

(2011). The prospect of using cyanobacterial bicarbonate transporters to improve leaf photosynthesis in C_3 crop plants. *Plant Physiol.* 155, 20–26. doi: 10.1104/pp.110.164681

115. Queval, G., Neukermans, J., Vanderauwera, S., Van Breusegem, F., and Noctor, G. (2012). Day length is a key regulator of transcriptomic responses to both CO_2 and H_2O_2 in *Arabidopsis*. *Plant Cell Environ.* 35, 374–387. doi: 10.1111/j.1365-3040.2011.02368.x
116. Ramakrishna, A., and Ravishankar, G. A. (2011). Influence of abiotic stress signals on secondary metabolites in plants. *Plant Signal. Behav.* 6, 1720–1731. doi: 10.4161/psb.6.11.17613
117. Ribeiro, D. M., Araujo, W. L., Fernie, A. R., Schippers, J. H. M., and Mueller-Roeber, B. (2012). Action of Gibberellins on growth and metabolism of *Arabidopsis* plants associated with high concentration of carbon dioxide. *Plant Physiol.* 160, 1781–1794. doi: 10.1104/pp.112.204842
118. Roden, J. S., and Ball, M. C. (1996). Growth and photosynthesis of two eucalypt species during high temperature stress under ambient and elevated CO_2. *Global Change Biol.* 2, 115–128. doi: 10.1111/j.1365-2486.1996.tb00056.x
119. Rosenstiel, T. N., Potosnak, M. J., Griffin, K. L., Fall, R., and Monson, R. K. (2003). Increased CO_2 uncouples growth from isoprene emission in an agroforestry ecosystem. *Nature* 421, 256–259. doi: 10.1038/nature01312
120. Ruiz-Vera, U. M., Siebers, M., Gray, S. B., Drag, D. W., Rosenthal, D. M., Kimball, B. A., et al. (2013). Global warming can negate the expected CO_2 stimulation in photosynthesis and productivity for soybean grown in the Midwestern United States. *Plant Physiol.* 162, 410–423. doi: 10.1104/pp.112.211938
121. Ryle, G. J. A., Powell, C. E., and Tewson, V. (1992). Effect of elevated CO_2 on the photosynthesis, respiration and growth of perennial ryegrass. *J. Exp. Bot.* 43, 811–818. doi: 10.1093/jxb/43.6.811
122. Sage, R. F. (2004). The evolution of C_4photosynthesis. *New Phytol.* 161, 341–370. doi: 10.1111/j.1469-8137.2004.00974.x
123. Sage, R. F., Pearcy, W. R., and Seemann, J. R. (1987). The nitrogen use efficiency of C_3 and C_4 plants. III Leaf nitrogen effects on the activity of carboxylating enzymes in *Chenopodium album* (L.) and *Amaranthus retroflexus* (L.). *Plant Physiol.* 85, 355–359. doi: 10.1104/pp.85.2.355
124. Salazar-Parra, C., Aguirreolea, J., Sánchez-Díaz, M., JoséIrigoyen, J., and Morales, F. (2012). Climate change (elevated CO_2, elevated temperature and moderate drought) triggers the antioxidant enzymes response of grapevine cv. Tempranillo, avoiding oxidative damage. *Physiol. Plant.* 144, 99–110. doi: 10.1111/j.1399-3054.2011.01524.x
125. Sanz-Sáez,Á., Erice, G., Aranjuelo, I., Aroca, R., Ruíz-Lozano, J. M., Aguirreolea, J., et al. (2013). Photosynthetic and molecular markers of CO_2-mediated photosynthetic downregulation in nodulated alfalfa. *J. Integr. Plant Biol.* 55, 721–734. doi: 10.1111/jipb.12047
126. Schwanz, P., and Polle, A. (1998). Antioxidative systems, pigment and protein contents in leaves of adult Mediterranean oak species (*Quercus pubescens* and Q. ilex) with lifetime exposure to elevated CO_2. *New Phytol.* 140, 411–423. doi: 10.1111/j.1469-8137.1998.00290.x
127. Schwanz, P., and Polle, A. (2001). Differential stress responses of antioxidative systems to drought in pendunculate oak (*Quercus robur*) and maritime pine (*Pinus pinaster*) grown under high CO_2 concentrations. *J. Exp. Bot.* 52, 133–143. doi: 10.1093/jexbot/52.354.133
128. Sekmen, A. H., Ozgur, R., Uzilday, B., and Turkan, I. (2014). Reactive oxygen species scavenging capacities of cotton (*Gossypium hirsutum*) cultivars under combined drought and heat induced oxidative stress. *Environ. Exp. Bot.* 99, 141–149. doi: 10.1016/j.envexpbot.2013.11.010
129. Sharkey, T. D. (1988). Estimating the rate of photorespiration in leaves. *Physiol. Plant.* 73, 147–152. doi: 10.1111/j.1399-3054.1988.tb09205.x
130. Sharkey, T. D., Loreto, F., and Delwiche, C. F. (1991). High carbon dioxide and sun/shade effects on isoprene emission from oak and aspen tree leaves. *Plant Cell Environ.* 14, 333–338. doi: 10.1111/j.1365-3040.1991.tb01509.x
131. Sharkey, T. D., and Singsaas, E. L. (1995). Why plants emit isoprene. *Nature* 374, 769. doi: 10.1038/374769a0
132. Singh, A., and Agrawal, M. (2015). Effects of ambient and elevated CO_2 on growth, chlorophyll fluorescence, photosynthetic pigments, antioxidants, and secondary metabolites of *Catharanthus roseus* (L.) G Don. grown under three different soil N levels. *Environ. Sci. Poll. Res. Int.* 22, 3936–3946. doi: 10.1007/s11356-014-3661-6
133. Smirnoff, N., and Wheeler, G. L. (2000). Ascorbic acid in plants: biosynthesis and function. *Crit. Rev. Plant Sci.* 19, 267–290. doi: 10.1016/S0735-2689(00)80005-2
134. Stitt, M., and Krapp, A. (1999). The interaction between elevated carbon dioxide and nitrogen nutrition: the physiological and molecular background. *Plant Cell Environ.* 22, 583–621. doi: 10.1046/j.1365-3040.1999.00386.x
135. Sun, Z., Niinemets,Ü., Hüve, K., Noe, S. M., Rasulov, B., Copolovici, L., et al. (2012). Enhanced isoprene emission capacity and altered light responsiveness in aspen grown under elevated atmospheric CO_2 concentration. *Global Change Biol.* 18, 3423–3440. doi: 10.1111/j.1365-2486.2012.02789.x
136. Takagi, D., Inoue, H., Odawara, M., Shimakawa, G., and Miyake, C. (2014). The Calvin cycle inevitably produces sugar-derived reactive carbonyl methylglyoxal during photosynthesis: a potential cause of plant diabetes. *Plant Cell Physiol.* 55, 333–340. doi: 10.1093/pcp/pcu007
137. Teng, N., Jin, B., Wang, Q., Hao, H., Ceulemans, R., Kuang, T., et al. (2009). No detectable maternal effects of elevated CO_2 on*Arabidopsis thaliana* over 15 generations. *PLoS ONE* 4:e6035. doi: 10.1371/journal.pone.0006035

138. Teng, N., Wang, J., Chen, T., Wu, X., Wang, Y., and Lin, J. (2006). Elevated CO_2 induces physiological, biochemical and structural changes in leaves of *Arabidopsis thaliana*. *New Phytol.* 172, 92–103. doi: 10.1111/j.1469-8137.2006.01818.x

139. Thomas, R. B., Reid, C. D., Ybema, R., and Strain, B. R. (1993). Growth and maintenance components of leaf respiration of cotton grown in elevated carbon dioxide partial pressure. *Plant Cell Environ.* 16, 539–546. doi: 10.1111/j.1365-3040.1993.tb00901.x

140. Timm, S., and Bauwe, H. (2013). The variety of photorespiratory phenotypes—employing the current status for future research directions on photorespiration. *Plant Biol.* 15, 737–747. doi: 10.1111/j.1438-8677.2012.00691.x

141. Tingey, D. T., McKane, R. B., Olszyk, D. M., Johnson, M. G., Rygiewicz, P. T., and Lee, E. H. (2003). Elevated CO_2 and temperature alter nitrogen allocation in Douglas-fir. *Global Change Biol.* 9, 1038–1050. doi: 10.1046/j.1365-2486.2003.00646.x

142. Tognetti, R., Johnson, J. D., Michelozzi, M., and Raschi, A. (1998). Response of foliar metabolism in mature trees of *Quercus pubescens* and *Quercus ilex* to long-term elevated CO_2. *Environ. Exp. Bot.* 39, 233–245. doi: 10.1016/S0098-8472(98)00013-6

143. Trowbridge, A. M., Asensio, D., Eller, A. S., Way, D. A., Wilkinson, M. J., Schnitzler, J. P., et al. (2012). Contribution of various carbon sources toward isoprene biosynthesis in poplar leaves mediated by altered atmospheric CO_2 concentrations. *PLoS ONE*7:e32387. doi: 10.1371/journal.pone.0032387

144. Urban, O., Hrstka, M., Zitová, M., Holišová, P., Šprtová, M., Klem, K., et al. (2012). Effect of season, needle age and elevated CO_2 concentration on photosynthesis and Rubisco acclimation in *Picea abies*. *Plant Physiol. Biochem.* 58, 135–141. doi: 10.1016/j.plaphy.2012.06.023

145. Vaughan, M. M., Huffaker, A., Schmelz, E. A., Dafoe, N. J., Christensen, S., Sims, J., et al. (2014). Effects of elevated [CO_2] on maize defence against mycotoxigenic *Fusarium verticillioides*. *Plant Cell Environ.* 37, 2691–2706. doi: 10.1111/pce.12337

146. von Caemmerer, S., and Evans, J. R. (1991). Determination of the average partial pressure of CO_2 in chloroplasts from leaves of several C_3 plants. *Aust. J. Plant Physiol.* 18, 287–305. doi: 10.1071/PP9910287

147. von Caemmerer, S., and Furbank, R. T. (2003). The C_4 pathway: an efficient CO_2 pump. *Photosyn. Res.* 77, 191–207. doi: 10.1023/A:1025830019591

148. Wang, J., Cheung, M., Rasooli, L., Amirsadeghi, S., and Vanlerberghe, G. C. (2014). Plant respiration in a high CO_2 world: how will alternative oxidase respond to future atmospheric and climatic conditions? *Can. J. Plant Sci.* 94, 1091–1101. doi: 10.4141/cjps2013-176

149. Wang, X., Taub, D. R., and Jablonski, L. M. (2015). Reproductive allocation in plants as affected by elevated carbon dioxide and other environmental changes: a synthesis using meta-analysis and graphical vector analysis. *Oecologia* 177, 1075–1087. doi: 10.1007/s00442-014-3191-4

150. Ward, J. K., and Kelly, J. K. (2004). Scaling up evolutionary responses to elevated CO_2: lessons from *Arabidopsis*. *Ecol. Lett.* 7, 427–440. doi: 10.1111/j.1461-0248.2004.00589.x

151. Watanabe, C. K., Sato, S., Yanagisawa, S., Uesono, Y., Terashima, I., and Noguchi, K. (2014). Effects of elevated CO_2 on levels of primary metabolites and transcripts of genes encoding respiratory enzymes and their diurnal patterns in *Arabidopsis thaliana*: possible relationships with respiratory rates. *Plant Cell Physiol.* 55, 341–357. doi: 10.1093/pcp/pct185

152. Way, D. A., Oren, R., and Kroner, Y. (2015). The space-time continuum: the effects of elevated CO_2 and temperature on trees and the importance of scaling. *Plant Cell Environ.* 38, 991–1007. doi: 10.1111/pce.12527

153. Wilkinson, M. J., Monson, R. K., Trahan, N., Lee, S., Brown, E., and Jackson, R. B, et al. (2009). Leaf isoprene emission rate as a function of atmospheric CO_2 concentration. *Global Change Biol.* 15, 1189–1200. doi: 10.1111/j.1365-2486.2008. 01803.x

154. Wingler, A., Quick, W. P., Bungard, R. A., Bailey, K. J., Lea, P. J., and Leegood, R. C. (1999). The role of photorespiration during drought stress: an analysis utilizing barley mutants with reduced activities of photorespiratory enzymes. *Plant Cell Environ.* 22, 361–373. doi: 10.1046/j.1365-3040.1999.00410.x

155. Xu, Z. Z., Shimizu, H., Ito, S., Yagasaki, Y., Zou, C. J., Zhou, G. S., et al. (2014). Effects of elevated CO_2, warming and precipitation change on plant growth, photosynthesis and peroxidation in dominant species from North China grassland. *Planta*239, 421–435. doi: 10.1007/s00425-013-1987-9

156. Xu, Z. Z., Shimizu, H., Yagasaki, Y., Ito, S., Zheng, Y. R., and Zhou, G. S. (2013a). Interactive effects of elevated CO_2, drought, and warming on plants. *J. Plant Growth Regul.* 32, 692–707. doi: 10.1007/s00344-013-9337-5

157. Xu, Z. Z., Yu, Z. W., and Zhao, J. Y. (2013b). Theory and application for the promotion of wheat production in china: past, present and future. *J. Sci. Food Agrc.* 93, 2339–2350. doi: 10.1002/jsfa.6098

158. Xu, Z. Z., and Zhou, G. S. (2006). Combined effects of water stress and high temperature on photosynthesis, nitrogen metabolism and lipid peroxidation of a perennial grass *Leymus chinensis*. *Planta* 224, 1080–1090. doi: 10.1007/s00425-006-0281-5

159. Xu, Z. Z., and Zhou, G. S. (2011). Responses of photosynthetic capacity to soil moisture gradient in perennial rhizome grass and perennial bunchgrass. *BMC Plant Biol.* 11:21. doi: 10.1186/1471-2229-11-21

160. Xu, Z. Z., Zhou, G. S., and Shimizu, H. (2009a). Are plant growth and photosynthesis limited by pre-drought following rewatering in grass? *J. Exp. Bot.* 60, 3737–3749. doi: 10.1093/jxb/erp216

161. Xu, Z. Z., Zhou, G. S., and Shimizu, H. (2009b). Effects of soil drought with nocturnal warming on leaf stomatal traits and mesophyll cell ultrastructure of a perennial grass. *Crop Sci.* 49, 1843–1851. doi: 10.2135/cropsci2008.12.0725

162. Xu, Z. Z., Zhou, G. S., and Shimizu, H. (2010). Plant responses to drought and rewatering. *Plant Signal. Behav.* 5, 649–654. doi: 10.4161/psb.5.6.11398

163. Xu, Z. Z., Zhou, G. S., and Wang, Y. H. (2007). Combined effects of elevated CO_2 and soil drought on carbon and nitrogen allocation of the desert shrub *Caragana intermedia*. *Plant Soil* 301, 87–97. doi: 10.1007/s11104-007-9424-0

164. Yan, K., Chen, W., Zhang, G., Xu, S., Liu, Z., He, X., et al. (2010). Elevated CO_2 ameliorated oxidative stress induced by elevated O_3 in *Quercus mongolica*. *Acta Physiol. Plant.* 32, 375–385. doi: 10.1007/s11738-009-0415-z

165. Yin, X. W. (2002). Responses of leaf nitrogen concentration and specific leaf area to atmospheric CO_2 enrichment: a retrospective synthesis across 62 species. *Global Change Biol.* 8, 631–642. doi: 10.1046/j.1365-2486.2002.00497.x

166. Zavala, J. A., Nabity, P. D., and Delucia, E. H. (2013). An emerging understanding of mechanisms governing insect herbivory under elevated CO_2. *Annu. Rev. Entomol.* 58, 79–97. doi: 10.1146/annurev-ento-120811-153544

167. Zinta, G., AbdElgawad, H., Domagalska, M. A., Vergauwen, L., Knapen, D., Nijs, I., et al. (2014). Physiological, biochemical, and genome-wide transcriptional analysis reveals that elevated CO_2 mitigates the impact of combined heat wave and drought stress in *Arabidopsis thaliana* at multiple organizational levels. *Global Change Biol.* 20, 3670–3685. doi: 10.1111/gcb.12626

CHAPTER 5

PLANT RESPONSES TO STRESSES: ROLE OF ASCORBATE PEROXIDASE IN THE ANTIOXIDANT PROTECTION

Andréia Caverzan, Gisele Passaia, Silvia Barcellos Rosa, Carolina Werner Ribeiro, Fernanda Lazzarotto, Márcia Margis-Pinheiro

Keywords: Ascorbate peroxidase, antioxidant system, reactive oxygen species, abiotic stress, mutant plants.

PLANT RESPONSES TO STRESSES

The exposure of plants to unfavorable environmental conditions increases the production of reactive oxygen species (ROS) such as, singlet oxygen ($^1O_2$), superoxide ($O_2^{\bullet-}$), hydrogen peroxide (H_2O_2), and hydroxyl radical ($OH^\bullet$). The ROS detoxification process in plants is essential for the protection of plant cells and their organelles against the toxic effect of these species (Apel and Hirt, 2004; Mittler, 2002). The differences in subcellular localization and biochemical properties of antioxidant enzymes and the distinct responses in gene expression, in addition to the presence of non-enzymatic mechanisms, result in a versatile and flexible antioxidant system able to control the optimum ROS levels (Vranova *et al.*, 2002). The ROS detoxification systems include enzymatic and non-enzymatic antioxidant com-ponents (Scandalios, 2005). Ascorbate (AsA) and glutathione (GSH), non-enzymatic antioxidants are crucial for plant defense against oxidative stress, playing a key role as antioxidant buffers (Foyer and Noctor, 2005; Mittler, 2002). Other non-enzymatic antioxidants involved include flavonoids, phenolic compounds, alkaloids, tocopherol and carotenoids (Gratão *et al.*, 2005).

Enzymatic antioxidants comprise superoxide dismutase (SOD), ascorbate peroxidase (APX), catalase (CAT), glutathione peroxidase (GPX) and peroxiredoxin (PrxR). These enzymes are present in practically all subcellular compartments. Usually, an organelle has more than one enzyme able to scavenge a single ROS (Mittler, 2002; Mittler *et al.*, 2004; Scandalios, 2005). The main hydrogen peroxide-detoxification system in plant chloroplasts is the ascor-bate-glutathione cycle, in which APX is a key enzyme (Asada, 1992). APX utilizes AsA as specific electron donor to reduce H_2O_2 to water. The importance of APX and ascorbate-glutathione cycle is not restricted to chloroplasts; it also plays a role in ROS scavenging in cytosol, mitochondria and peroxisomes (Asada, 1992, 1999; Mittler *et al.*, 2004; Noctor and Foyer, 1998; Shigeoka *et al.*, 2002). The ROS-scavenging enzymes in plants have been widely studied and the results have demonstrated that, in response to environmental stress, APX activity generally increases along with other enzymes activities, such as CAT, SOD, and GSH reductase (Shigeoka *et*

Citation: Andréia Caverzan; Gisele Passaia; Silvia Barcellos Rosa; Carolina Werner Ribeiro; Fernanda Lazzarotto; Márcia Margis-Pinheiro, Plant responses to stresses: role of ascorbate peroxidase in the antioxidant protection, http://dx.doi.org/10.1590/S1415-47572012000600016.

al., 2002). Over the past ten years substantial efforts have been made to understand plant antioxidant system mechanisms (Figure1). The increasing number of publications addressing CAT, SOD, GPX and APX enzymes in plants are examples of this tendency, especially APX, which had the number of articles doubled from 158 published in 2000 to 368 in 2010 (ISI Web of Knowledge database). Publications related more specifically to ROS in plants increased 18 times in the same period. The data presented in this study confirm those -reported by Azevedo and Azevedo (2006) in which the number of publications addressing antioxidant mechanisms increased after 2000. This shows the relevance of studying these enzymes to further understand the biological processes dealing with oxidative stress responses in plants. The focus of this review is to discuss the main findings related to the APX enzyme at molecular and physiological levels, in different plant species. *APX* gene modulation in response to abiotic stress conditions, especially temperature, high light, drought, salinity and heavy metals will also be reviewed.

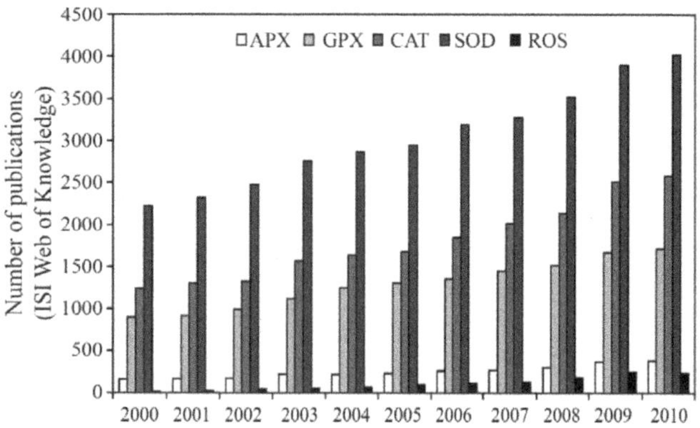

Figure 1 - Number of publication evolution addressing antioxidant enzymes in plants in the last ten years. CAT (catalase), SOD (Superoxide dismutase), GPX (Gluthatione peroxidase) and APX (Ascorbate peroxidase) enzymes, Reactive Oxygen Species (ROS).

ASCORBATE PEROXIDASE IN PLANTS

Ascorbate peroxidase (APX) (EC 1.11.1.11) belongs to the class I heme-peroxidases that is found in higher plants, chlorophytes (Takeda *et al.*, 1998, 2000), red algae (Sano *et al.*, 2001), and members of the protist kingdom (Shigeoka *et al.*, 1980; Wilkinson *et al.*, 2002). APX and other peroxidase sequences from all kingdoms of life are stored in the database Peroxibase (Oliva *et al.*, 2009), which also provides a series of bioinformatics tools useful for analyzing the peroxidases stored sequences.

Genomic and cDNA *APX* sequences were obtained from a great variety of plant species, showing that APX are widely distributed in the vegetal kingdom. These enzymes are encoded by small gene families in these organisms (Passardi *et al.*, 2007). The different isoforms are classified according to their subcellular localization. Soluble iso-forms are found in cytosol (cAPX), mitochondria (mitAPX) and chloroplast stroma (sAPX), while membrane-bound isoforms are found in microbody (including peroxisome and glyoxisome) (mAPX) and chloroplast thyla-koids (tAPX). The presence of organelle-specific targeting peptides and transmembrane domains found in the N- and C- terminal protein regions determine the final subcellular localization of the isoenzyme (Shigeoka *et al.*, 2002; Teixeira *et al.*, 2004, 2006). Plant chloroplastic APX (chlAPX) isoenzymes encoding genes are divided into two groups. The first group comprises single genes encoding two isoenzymes through a post-transcriptional alternative splicing regulation. This group includes genes from spinach (*S. oleracea*), tobacco (*N. tabacum*), pumpkin (*Cucurbita sp*) and ice plant (*M. crystallium*). In the second group, individual genes codify different isoenzymes which are individually regulated. This group includes genes from *Arabidopsis,* rice, and tomato. The mechanism of alternative splicing in chlAPX has been studied in spinach (Ishikawa and Shigeoka, 2008) and the results showed that alternative splicing is fundamental for controlling the expression of stromal (sAPX) and thylakoid (tAPX) isoenzymes. This regulation occurs in a tissue-dependent manner.

Ascorbate peroxidases have been partially characterized in some plant species. In spinach, the APX family is formed by genes encoding one cytosolic and two chloro-plastic (sAPX and tAPX membrane) isoenzymes, one targeted to microbody membrane and an unknown putative cytosol-soluble isoenzyme (Ishikawa *et al.*, 1995, 1996, 1998). In cowpea, four cDNAs were isolated and characterized, corresponding to putative cytosolic, peroxissomal and chloroplastic (thylakoid and stromal) APX isoforms (D'Arcy-Lameta *et al.*, 2006). Six loci encoding *APX* were identified in *Eucalyptus grandis* and their subcellular localizations were indicated by prediction programs. Among the six isoforms, three were putatively identified as cytosolic, one as a putative peroxisomal protein and two predicted to be associated with chloroplasts (Teixeira *et al.*, 2005). In tomato, seven members were identified, three cytosolic, two peroxisomal, and two chloroplastic (Najami *et al.*, 2008). In the model plant *Arabidopsis thaliana*, the presence of nine *APX* genes was described, two choroplastic, one thylakoid-bound and one member whose product is targeted to both chloroplast stroma and mitochondria (Chew*et al.*, 2003); the intracellular localization of an additional member is yet unknown. In addition, three cytosolic and three microsomal proteins were also described (Mittler *et al.*, 2004; Narendra *et al.*, 2006; Panchuk *et al.*, 2002). In another important model plant, rice, the *APX* gene family comprises eight members, *viz.* two cytosolic, two peroxi-somal, two chloroplastic (stromal and thylakoid-bound) and two mitochondrial ones (Teixeira *et al.*, 2004, 2006). Recently, a new protein has also been identified as functionally associated with APX in rice, the APX-R (Ascor-bate peroxidase-related) (Lazzarotto *et al.*, 2011). Detailed analyses of evolution and structure of *APX-R* genes indicate that these genes correspond to a new class of heme-peroxidases (Lazzarotto *et al.*, 2011).

APX isoenzymes are labile in the absence of AsA. Thus, high level of endogenous AsA is essential to effectively maintain the antioxidant system that protects plants from oxidative damage (Asada, 1992; Shigeoka *et al.*, 2002). Under special conditions in which the concentration of AsA is lower than 20 ∝M, the APX activity is quickly lost, this making the chlAPX the least stable isoform. Both cAPX and mAPX have half-inactivation times of around one hour or more, while that for mitAPX and chlAPX is less than 30 seconds (Chen and Asada, 1989; Miyake *et al.*, 1993; Ishikawa *et al.*, 1998; Yoshimura *et al.*, 1998; Leo-nardis *et al.*, 2000).

APX ENZYME RESPONSES UNDER ABIOTIC STRESS

The expression of *APX* encoding genes is modulated by various environmental stimuli, such as drought and salt stress, high light, high and low temperatures, pathogen attacks, H_2O_2 and abscisic acid (Zhang *et al.*, 1997; Yoshi-mura *et al.*, 2000; Agrawal et al., 2003; Fryer *et al.*, 2003; Menezes-Benavente *et al.*, 2004; Teixeira *et al.*, 2006; Rosa *et al.*, 2010; Bonifacio *et al.*, 2011). Furthermore, the transcriptional expression of *APX* genes is tissue and developmental stage dependent (Agrawal *et al.*, 2003; Teixeira *et al.*, 2006).

Salt Stress

Plants are greatly affected by salinity, which causes alteration in nutrient uptake, accumulation of toxic ions, osmotic stress, and oxidative stress (Verslues *et al.*, 2006). Consequently, salinity results in molecular damage, growth arrest, and even cell death (Wang *et al.*, 2008). Salt stress induces the production of ROS, and the response of *APX* genes to this condition is tissue and developmental stage regulated. When the response of major antioxidant enzymes transcripts was analyzed for different developmental stages in salt stressed rice, *cAPX* was up-regulated in 11-day-old seedlings, while in 6-week-old plants salt had no significant effect on this gene (Menezes-Benavente *et al.*, 2004). In addition to *APX* expression alteration, discrimination on *CAT* transcript accumulation was also noticed in the basal region of rice leaves under salinity (Yamane *et al.*, 2010). Concerning the APX rice isoforms, induction was observed for the *OsAPX1*, *OsAPX4*, *OsAPX6* and *OsAPX7* genes, whereas cytosolic *OsAPX2* gene expression was not altered by salinity (Yamane *et al.*, 2010).

Teixeira *et al.* (2006) reported that three rice *APX* genes, *OsAPX2*, *OsAPX7*, and *OsAPX8*, showed altered transcript levels in response to NaCl treatment. The expression of *OsAPX2* and *OsAPX7* was increased, whereas *OsAPX8* transcript accumulation was strongly suppressed in plants undergoing salt stress (Teixeira *et al.*, 2006). The transcript level of *OsAPX8* was slightly decreased by salinity in the basal region of rice leaves (Yamane *et al*, 2010). On the other hand, *OsAPX8* expression in rice roots was enhanced by all NaCl concentrations tested (150, 200 and 300 mM), and *OsAPX7* expression was down-regulated by 300 mM NaCl. This discrepancy in regulation for the *OsAPX* genes might be due to differences in cultivars, organs, plant age and growth conditions (Hong *et al.*, 2007). An increase in rice cytosolic *APX2* gene transcript levels after treatment with salt has previously been shown by our group (Mene-zes-Benavente *et al.*, 2004; Teixeira *et al.*, 2006). In accordance, transgenic *Arabidopsis* plants over-expressing cyto-solic *OsAPXb* (*OsAPX2*) showed higher tolerance to NaCl than those over-expressing cytosolic *OsAPXa* (*OsAPX1*) (Lu *et al.*, 2007). A similar increment in salt stress tolerance was also observed in transgenic tobacco over-expressing the *Arabidopsis cAPX* gene (Badawi *et al.*, 2004) and also in tobacco plants over-expressing a *Solanum lycopersicum* thylakoid-bound ascorbate pero-xi-dase gene (*StAPX*) (Sun *et al.*, 2010b). Transgenic tobacco plants that simultaneously expressed CuZnSOD, APX, and DHAR in chloroplasts presented increased protection against salt induced injury (Lee *et al.*, 2007b). Transgenic tobacco BY-2 cells with 50 and 75% lower cAPX activity showed higher intracellular content of ROS. On the other hand, the tobacco cells showed a potential enhancement in tolerance to heat and salt stress, perhaps by induction of stress-related gene expression. However, no substantial differences were observed in the activity levels of the other antioxidant enzymes (Ishikawa *et al.*, 2005).

In barley, the transcript level of peroxisomal *APX* gene (*HvAPX1*) increased significantly under salt stress (Shi *et al.*, 2001). However, *Arabidopsis apx3* knockout mutants exposed to normal or stressful conditions did not present disturbed growth or development. In these plants, other antioxidant enzymes possibly compensate the lack of the peroxisomal isoform (Narendra *et al.*, 2006). In contrast, the overexpression of a *Populus* peroxisomal *APX* (*PpAPX*) gene in transgenic tobacco improved salt tolerance at the vegetative stage and plants were more resistant to oxidative damage induced by methyl viologen (MV) and, in addition, the plants had longer roots (Li *et al.*, 2009). Lin and Pu (2010) studied changes in enzymes involved in ROS scavenging in sweet potato plants tolerant and sensitive to salinity. After exposure to salinity (450 mM NaCl), APX activity increased in plants at 24 and 48 h, and this response was higher in a salt-stress tolerant genotype than in the salt sensitive ones. The expression of *cAPX*, *mAPX* and *chlAPX* in response to salinity was tissue specific and dependent on stress duration (Lin and Pu, 2010). Taken together, these studies put in evidence that salt stress causes disturbances in antioxidant gene expression by producing alterations in the transcriptional pattern in several plant species, and that the expression of distinct APX isoform may result in redox homeostasis regulation in each cellular compartment.

Temperature Stress

Extreme temperatures affect the growth, yield and quality of plant production. ROS levels tend to increase if plants are exposed to stressful conditions such as low or high temperatures (Mittler *et al.*, 2004; Scandalios, 2005). In potato tubers, the transient accumulation of *cAPX* mRNA after storage at low-temperature was greater than after high-temperature storage, showing that *APX* expression was induced in response to low temperature (Kawa-kami *et al.*, 2002). Likewise, the two rice *cAPX* (*OsAPX1* and *OsAPX2*) genes were induced after rice plants were exposed to low temperatures. Furthermore, *OsAPX3*, *OsAPX4, OsAPX6* and *OsAPX7* were also significantly induced, while *OsAPX8* were repressed after 24 h under low temperature (unpublished data). The sweet potato *cAPX* gene was highly induced in leaves after exposure to high temperature (Park *et al.*, 2004). In cucumber plants submitted to heat treatment, the activities of cAPX, sAPX and mAPX increased after an initial slight decline during the course of the experiment. The expression of *sAPX* followed a similar pattern (Song *et al.*, 2005). In response to cold, the expression of a peroxisomal *APX* gene increased slightly in *Arabidopsis* (Zhang *et al.*, 1997).

These results were corroborated in *Arabidopsis* by overexpressing a putative peroxisomal membrane-bound *APX* from barley, resulting in an increased tolerance to higher temperature treatment (Shi *et al.*, 2001). Furthermore, the overexpression of chloroplastic *tAPX* in tobacco plants improved the tolerance to chilling stress combined with high light intensity (Yabuta *et al.*, 2002). On the other hand, *Arabidopsis* plants lacking *tAPX* had enhanced tolerance to heat stress (Miller *et al.*, 2007). Recently, Sato *et al.* (2011) showed that transgenic rice plants overexpressing a cytosolic *APX1* gene (*OsAPXa*) which exhibited higher APX activity in spikelets than in wild type (WT) plants, sustained higher levels of APX activity under cold stress, resulting in enhanced cold tolerance at the booting stage.

Plants with enhanced tolerance to multiple environmental stresses were obtained through induced expression of*CuZnSod* and *APX* genes. *Sod* and *APX* genes were expressed in chloroplasts of potato plants under the control of an oxidative stress inducible promoter - *SWPA2*. These plants showed enhanced tolerance to MV and when exposed to 42 ºC for 20 h, the photosynthetic activity of these transgenic plants decreased by only 6%, whereas in non-transformed (NT) plants it decreased by 29% (Tang *et al.*, 2006). Sweet potato plants expressing both*CuZnSod* and *APX* in chloroplasts through the inducible promoter also showed higher tolerance to MV-mediated oxidative stress and chilling stress (Lim *et al.*, 2007). The tolerance to high and low temperature stresses was studied in tobacco plants overexpressing a tomato *tAPX* gene. The overexpression of chloroplastic *APX* played a significant role in H_2O_2 detoxification and in minimizing photooxidative damage during temperature stress. The transgenic plants showed a higher photochemical efficiency of photosystem II when compared to WT plants under cold and heat stresses (Sun *et al.*, 2010a). These results put in evidence that the manipulation of the antioxidative mechanism in chloroplasts may be applied in the development of plants with increased tolerance to multiple environmental stresses.

High light Stress

Plants exposed to excessive light can suffer photo-inhibition, serious damage to the photosynthetic apparatus, and degradation of photosynthetic proteins (Demmig-Adams and Adams III, 1992). Light stress can also lead to ROS accumulation and antioxidant enzymes activation (Mit-tler, 2002). The responses of APX isoenzymes to photo-oxidative stress were studied in spinach leaves during high light stress. cAPX activity and transcripts increased during high light stress, however protein levels were not altered. The activities of chlAPX isoforms showed a gradual decrease, while the other isoenzymes showed no significant variation in transcript and protein levels, as well as activities (Yoshimura *et al.*, 2000). In wheat, a mutant line showing decreased tAPX activity presented reduced photo-synthetic activity and biomass accumulation when growing under high-light intensity, suggesting that tAPX is essential for photosynthesis (Danna *et al.*, 2003). Single mutants of *Arabidopsis* lacking tAPX or sAPX presented higher levels of H_2O_2 and oxidized proteins than WT plants when exposed to high light and MV stresses. The strongest effect of photooxidative stress was observed in plants lacking tAPX, these showing increased H_2O_2 accumulation and oxidized proteins (Maruta *et al.*, 2010).

Double mutants deficient in two *APX* genes, thyla-koid-bound and a cytosolic one (*tylapx/apx1*), resulted in different signals in *Arabidopsis* plants, such as late flowering, low protein oxidation during light stress and enhanced accumulation of anthocyanins (Miller *et al.*, 2007). Mutants lacking a functional copy of tAPX, sAPX or both, were characterized in *Arabidopsis* under photooxidative stress during germination. The stress led to chloroplast bleaching in *sapx* single-mutant and *tapx/sapx* double-mutant plants, while the greening process of WT and *tapx* plants was partially impaired (Kangasjarvi *et al.*, 2008). When mature leaves of *tapx/sapx* double mutants were submitted to short-term photooxidative stress induced by high light or MV treatment, the plants showed susceptibility (Kangasjarvi *et al.*, 2008). These results indicate that the APXs isoenzymes are indispensable under environmental stresses in different species, especially under light stress conditions.

In *Arabidopsis* leaves, high light treatment induced the expression of cytosolic *APX2*, which has its expression restricted to bundle sheath cells of the vascular tissue (Fryer *et al.*, 2003). In *Arabidopsis,* APX1 knockout plants showed suppressed growth and development, altered stomatal responses and induction of heat shock proteins during light stress. The inactivation of cytosolic APX resulted in the alteration of several transcripts involved in different functions. In transgenic APX1 plants kept under optimal conditions, the transcripts encoding APX enzymes were not elevated. However, during light stress, certain enzymes were induced in knockout-APX1 plants (Pnueli *et al.*, 2003). In another study (Davletova *et al.*, 2005) with APX1-deficient *Arabidopsis* plants the observation was that the entire chloroplastic H_2O_2-scavenging system collapsed, H_2O_2 levels increased and protein oxidation -occurred in leaves subjected to a moderate light stress, suggesting that the absence of cytosolic APX1 resulted not only in the accumulation of H_2O_2 but also in damage to specific proteins in leaf cells. On the other hand, rice plants double silenced for cytosolic APXs up-regulated other peroxi-dases, making these transgenic plants able to survive under stress, such as salt, heat, high light and MV, similar to NT plants. The antioxidative compensatory mechanism exhibited by the silenced plants was associated with increased expression of *Gpx* genes. The transcript levels of *OsCatA* and *OsCatB* and the activities of CAT and guaiacol peroxidase (GPOD; type III peroxidases) were also up-regulated. In contrast, none of the other isoforms of*OsAPX* were up-regulated under normal growth conditions. These results suggested that signaling mechanisms triggered in rice could be distinct from those proposed for *Arabidopsis* (Bonifacio *et al.*, 2011).

Drought Stress

Drought stress in plants leads to severe effects such as reduction in vegetative growth and cell division. As a consequence of drought stress several changes occur inside the cell, including changes in gene expression levels, synthesis of molecular chaperones, and activation of enzymes involved in the production and removal of ROS (Mahajan and Tuteja, 2005). In two cowpea (*Vigna unguiculata*) cultivars, one drought-tolerant and the other drought-sensi-tive, APX activity was 60% higher in tolerant plants cultivated under control conditions. In response to drought stress, a higher increase in transcript levels of cytosolic and peroxisomal *APX* genes was observed in the sensitive culti-var (D'Arcy-Lameta *et al.*, 2006). Chloroplastic *APX* genes expression was stimulated earlier in the tolerant cultivar when submitted to drought stress. These data suggest the capacity of these enzymes to efficiently detoxify ROS at their production site (D'Arcy-Lameta *et al.*,

2006). Relative *APX*transcript levels showed distinct changes in two genotypes of wheat exposed to mild water deficit. Cyto-solic*APX1* expression levels increased in both genotypes, while cytosolic *APX2* was up-regulated only in the drought-tolerant genotype. The transcript level of thylakoid *APX* increased in the drought-tolerant genotype, while stromal *APX2* showed higher expression levels in the drought-sensitive cultivar (Secenji *et al.*, 2010).

APX gene expression patterns in rice were studied after 15 days of drought stress. In marked contrast with the experiments with wheat, thylakoid *APX* (*OsAPX8*) expression was down-regulated in this condition, while the*OsAPX1*, *OsAPX2*, *OsAPX5*, *OsAPX6* and *OsAPX7* genes were up-regulated. The peroxisomal *OsAPX3* gene was not affected, while *OsAPX4* was slightly but significantly down-regulated by this treatment (Rosa *et al.*, 2010). This discrepancy could be due to distinct responses of *APX* genes in different species and to different magnitudes of stress. In *Arabidopsis*, APX1 protein and mRNA accumulated during combination of heat and drought stress. A cytosolic *APX1*-deficient mutant accumulated more H_2O_2 and was more sensitive to stress combination than WT plants when exposed to heat and drought stress combined. In contrast, plants deficient in thylakoid APX were not more sensitive to this stress combination than *APX1*-deficient mutant or WT plants. The cytosolic *APX1* gene may thus play a key role in the acclimation of plants to combined stress such drought and heat (Koussevitzky *et al.*, 2008). Indeed, when the overexpression of cytosolic *APX* was studied in tobacco chloroplasts, its overexpression protected the plant from several oxidative stresses, including drought and polyethylene glycol-induced stress (Badawi *et al.*, 2004). Plants overexpressing other antioxidant enzymes in different species showed increased tolerance to various stresses, including drought resistance. The overexpression of a *Populus* peroxisomal ascorbate peroxidase (*PpAPX*) gene in transgenic tobacco improved drought resistance in these plants (Li *et al.*, 2009). The overexpression of tomato (*Solanum lycopersicum*) thylakoid-bound *APX* (*StAPX*) gene in tobacco plants enhanced the tolerance of these plants to salt and osmotic stress (Sun *et al.*, 2010b).

Heavy Metals

The contamination of soils with heavy metals is a serious environmental problem that limits crop production. Exposure at higher concentrations of heavy metals can increase the production of ROS and change antioxidant response (Gratão *et al.*, 2005). Exposure of pea (*Pisum sativum* L.) plants to cadmium changed enzymatic and non-enzymatic antioxidant defenses, however, APX activity or accumulation of its transcripts were not significantly different (Romero-Puertas *et al.*, 2007). However, it was observed that in coffee cells, the activity of APX was increased at the lower cadmium concentration. On the other hand, APX activity was not detectable in cells submitted to the higher cadmium concentration after 24 h of treatment (Gomes-Junior *et al.*, 2006b).

An increase in APX activity was also observed in response to other heavy metals such as aluminum (Sharma and Dubey, 2007). In rice, the transcript levels of all *OsAPX* genes, except *OsAPX6*, were significantly increased after eight hours of 20 ppm aluminum exposure (Rosa *et al.*, 2010). In pea plants, *cAPX* expression increased in the shoots under aluminum treatment, but APX activity presented a significant decline at 10 ∝M aluminum in roots and shoots after 24 and 48 h of stress; at 50 ∝M aluminum treatment, however, APX activity did not show any significant changes (Panda and Matsumoto, 2010). Transgenic rice plants double silenced for *APX1* and *APX2*(APX1/2s plants) exhibited normal development and enhanced tolerance to a toxic concentration of aluminum (Rosa *et al.*, 2010). In bean plants, the expression of cytosolic APX was induced both at mRNA and protein levels in leaves of de-rooted plants in response to iron overload. Likewise, transgenic tobacco plants with suppressed cytosolic APX levels were more sensitive to iron application than WT plants (Pekker *et al.*, 2002). In coffee cells treated with nickel showed a rapid increase in APX activity, although the activity trends were slightly different between the two nickel concentrations tested (0.05 mM and 0.5 mM) (Gomes-Junior *et al.*, 2006a). Transgenic tall fescue plants expressing the CuZnSOD and APX genes in chloroplasts were submitted to copper, cadmium or arsenic treatment. Of the metals tested, copper and cadmium increased SOD and APX activities in control and transgenic plants, with a higher increase observed in transgenic plants. In contrast, in leaves exposed to arsenic, both enzymes exhibited less activity when compared to other treatments and no significant differences were observed between control and transgenic plants (Lee *et al.*, 2007a). These results emphasize the important role of APX and other antioxidant enzymes in H_2O_2 scavenging under toxic metals levels in the soil.

CONCLUSIONS

Ascorbate peroxidase is a key enzyme regulating ROS levels acting in different subcellular compartments (Figure 2). The expression of *APX* encoding genes is differentially modulated by several abiotic stresses in different plant species. All the data collected so far firmly indicate that APX isoforms play important and direct roles as protective elements against adverse environmental conditions. The diverse effects of knockdown or knockout of different *APX* genes on the plant growth, physiology and antioxidant metabolism indicate that APX may also regulate redox signaling pathways involved in plant development. These results emphasize the importance and complexity of the interactions of APX with other antioxidants in fine tuning the vegetal antioxidant metabolism.

ACKNOWLEDGMENTS

The authors are grateful to Conselho Nacional de Desenvolvimento Científico e Tecnológico (CNPq) and Coordenação de Aperfeiçoamento de Pessoal de Nível Superior (CAPES) for financial support.

REFERENCES

1. Agrawal GK, Jwa NS, Iwahashi H and Rakwal R (2003) Importance of ascorbate peroxidases OsAPX1 and OsAPX2 in the rice pathogen response pathways and growth and reproduction revealed by their transcriptional profiling. Gene 322:93-103.
2. Apel K and Hirt H (2004) Reactive oxygen species: Metabolism, oxidative stress, and signal transduction.

Annu Rev Plant Biol 55:373-399.

3. Asada K (1992) Ascorbate peroxidase - A hydrogen peroxide-sca-venging enzyme in plants. Physiol Plantarum 85:235-241.
4. Asada K (1999) The water cycle in chloroplasts: Scavenging of active oxygens and dissipation of excess photons. Annu Rev Plant Physiol Plant Mol Biol 50:601-639.
5. Azevedo JA and Azevedo RA (2006) Heavy metals and oxidative stress: Where do we go from here? Commun Biometry Crop Sci 1:135-138.
6. Badawi GH, Kawano N, Yamauchi Y, Shimada E, Sasaki R, Kubo A and Tanaka K (2004) Over-expression of ascorbate peroxidase in tobacco chloroplasts enhances the tolerance to salt stress and water deficit. Physiol Plant 121:231-238.
7. Bonifacio A, Martins MO, Ribeiro CW, Fontenele AV, Carvalho FE, Margis-Pinheiro M and Silveira JA (2011) Role of peroxidases in the compensation of cytosolic ascorbate pero-xidase knockdown in rice plants under abiotic stress. Plant Cell Environ 34:1705-1722.
8. Chen GX and Asada K (1989) Ascorbate peroxidase in tea leaves: Occurrence of two isozymes and the differences in their enzymatic and molecular properties. Plant Cell Physiol 30:897-998.
9. Chew O, Whelan J and Millar AH (2003) Molecular definition of the ascorbate-glutathione cycle in Arabidopsis mitochondria reveals dual targeting of antioxidant defenses in plants. J Biol Chem 278:46869-46877.
10. D'Arcy-Lameta A, Ferrari-Iliou R, Contour-Ansel D, Pham-Thi AT and Zuily-Fodil Y (2006) Isolation and characterization of four ascorbate peroxidase cDNAs responsive to water deficit in cowpea leaves. Ann Bot 97:133-140.
11. Danna CH, Bartoli CG, Sacco F, Ingala LR, Santa-Maria GE, Guiamet JJ and Ugalde RA (2003) Thylakoid-bound ascor-bate peroxidase mutant exhibits impaired electron transport and photosynthetic activity. Plant Physiol 132:2116-2125.
12. Davletova S, Rizhsky L, Liang H, Shengqiang Z, Oliver DJ, Coutu J, Shulaev V, Schlauch K and Mittler R (2005) Cyto-solic ascorbate peroxidase 1 is a central component of the reactive oxygen gene network of Arabidopsis. Plant Cell 17:268-281.
13. Demmig-Adams B and Adams WW III (1992) Photoprotection and other responses of plants to high light stress. Annu Rev Plant Physiol and Plant Mol Biol 43:599-626.
14. Foyer CH and Noctor G (2005) Redox homeostasis and antioxidant signaling: A metabolic interface between stress perception and physiological responses. Plant Cell 17:1866-1875.
15. Fryer MJ, Ball L, Oxborough K, Karpinski S, Mullineaux PM and Baker NR (2003) Control of Ascorbate Peroxidase 2 expression by hydrogen peroxide and leaf water status during excess light stress reveals a functional organisation of Arabidopsis leaves. Plant J 33:691-705.
16. Gomes-Junior RA, Moldes CA, Delite FS, Gratão PL, Mazzafera P, Lea PJ and Azevedo RA (2006a) Nickel elicits a fast antioxidant response in *Coffea arabica* cells. Plant Physiol Biochem 44:420-429.
17. Gomes-Junior RA, Moldes CA, Delite FS, Pompeu GB, Gratão PL, Mazzafera P, Lea PJ and Azevedo RA (2006b) Antioxidant metabolism of coffee cell suspension cultures in response to cadmium. Chemosphere 65:1330-1337.
18. Gratão PL, Polle A, Lea PJ and Azevedo RA (2005) Making the life of heavy metal-stressed plants a little easier. Funct Plant Biol 32:481-494.
19. Hong CY, Hsu YT, Tsai YC and Kao CH (2007) Expression of ASCORBATE PEROXIDASE 8 in roots of rice (*Oryza sativa* L.) seedlings in response to NaCl. J Exp Bot 58:3273-3283.
20. Ishikawa T and Shigeoka S (2008) Recent advances in ascorbate biosynthesis and the physiological significance of ascorbate peroxidase in photosynthesizing organisms. Biosci Bio-tech-nol Biochem 72:1143-1154.
21. Ishikawa T, Sakai K, Takeda T and Shigeoka S (1995) Cloning and expression of cDNA encoding a new type of ascorbate peroxidase from spinach. FEBS Lett 367:28-32.
22. Ishikawa T, Sakai K, Yoshimura K, Takeda T and Shigeoka S (1996) cDNAs encoding spinach stromal and thylakoid-bound ascorbate peroxidase, differing in the presence or absence of their 3'-coding regions. FEBS Lett 384:289-293.
23. Ishikawa T, Yoshimura K, Sakai K, Tamoi M, Takeda T and Shigeoka S (1998) Molecular characterization and physiological role of a glyoxysome-bound ascorbate peroxidase from spinach. Plant Cell Physiol 39:23-34.
24. Ishikawa T, Morimoto Y, Madhusudhan R, Sawa Y, Shibata H, Yabuta Y, Nishizawa A and Shigeoka S (2005) Acclimation to diverse environmental stresses caused by a suppression of cytosolic ascorbate peroxidase in tobacco BY-2 cells. Plant Cell Physiol 46:1264-1271.
25. Kangasjarvi S, Lepisto A, Hannikainen K, Piippo M, Luomala EM, Aro EM and Rintamaki E (2008) Diverse roles for chloroplast stromal and thylakoid-bound ascorbate peroxi-dases in plant stress responses. Biochem J 412:275-285.
26. Kawakami S, Matsumoto Y, Matsunaga A, Mayama S and Mizu-no M (2002) Molecular cloning of ascorbate peroxidase in potato tubers and its response during storage at low temperature. Plant Sci 163:829-836.
27. Koussevitzky S, Suzuki N, Huntington S, Armijo L, Sha W, Cortes D, Shulaev V and Mittler R (2008) Ascorbate peroxi-dase 1 plays a key role in the response of *Arabidopsis thaliana* to stress combination. J Biol Chem 283:34197-34203.
28. Lazzarotto F, Teixeira FK, Rosa SB, Dunand C, Fernandes CL, de Vasconcelos Fontenele A, Silveira JA, Verli H, Margis R and Margis-Pinheiro M (2011) Ascorbate peroxidase-related (APx-R) is a new heme-containing protein functionally associated with ascorbate peroxidase but evolutionarily divergent. New Phytol 191:234-250.

29. Lee SH, Ahsan N, Lee KW, Kim DH, Lee DG, Kwak SS, Kwon SY, Kim TH and Lee BH (2007a) Simultaneous overex-pression of both CuZn superoxide dismutase and ascorbate peroxidase in transgenic tall fescue plants confers increased tolerance to a wide range of abiotic stresses. J Plant Physiol 164:1626-1638.

30. Lee YP, Kim SH, Bang JW, Lee HS, Kwak SS and Kwon SY (2007b) Enhanced tolerance to oxidative stress in transgenic tobacco plants expressing three antioxidant enzymes in chloroplasts. Plant Cell Rep 26:591-598.

31. Leonardis S, Dipierro N and Dipierro S (2000) Purification and characterization of an ascorbate peroxidase from potato tuber mitochondria. Plant Physiol Biochem 38:773-779.

32. Li Y-J, Hai R-L, Du X-H, Jiang X-N and Lu H (2009) Over-expression of a Populus peroxisomal ascorbate peroxidase (PpAPX) gene in tobacco plants enhances stress tolerance. Plant Breed 128:404-410.

33. Lim S, Kim Y-H, Kim S-H, Kwon S-Y, Lee H-S, Kim J-S, Cho K-Y, Paek K-Y and Kwak S-S (2007) Enhanced tolerance of transgenic sweetpotato plants that express both CuZnSOD and APX in chloroplasts to methyl viologen-mediated oxidative stress and chilling. Mol Breed 19:227-239.

34. Lin KH and Pu SF (2010) Tissue- and genotype-specific ascorbate peroxidase expression in sweet potato in response to salt stress. Biol Plantarum 54:664-670.

35. Lu Z, Liu D and Liu S (2007) Two rice cytosolic ascorbate peroxidases differentially improve salt tolerance in transgenic Arabidopsis. Plant Cell Rep 26:1909-1917.

36. Mahajan S and Tuteja N (2005) Cold, salinity and drought stres-ses: An overview. Arch Biochem Biophys 444:139-158.

37. Maruta T, Tanouchi A, Tamoi M, Yabuta Y, Yoshimura K, Ishikawa T and Shigeoka S (2010) Arabidopsis chloro-plastic ascorbate peroxidase isoenzymes play a dual role in photoprotection and gene regulation under photooxidative stress. Plant Cell Physiol 51:190-200.

38. Menezes-Benavente L, Teixeira FK, Kamei CLA and Margis-Pinheiro M (2004) Salt stress induces altered expression of genes encoding antioxidant enzymes in seedlings of a Brazilian indica rice (*Oryza sativa* L.) Plant Sci 166:323-331.

39. Miller G, Suzuki N, Rizhsky L, Hegie A, Koussevitzky S and Mittler R (2007) Double mutants deficient in cytosolic and thylakoid ascorbate peroxidase reveal a complex mode of interaction between reactive oxygen species, plant development, and response to abiotic stresses. Plant Physiol 144:1777-1785.

40. Mittler R (2002) Oxidative stress, antioxidants and stress tolerance. Trends Plant Sci 7:405-410.

41. Mittler R, Vanderauwera S, Gollery M and Van Breusegem F (2004) Reactive oxygen gene network of plants. Trends Plant Sci 9:490-498.

42. Miyake C, Cao WH and Asada K (1993) Purification and molecular properties of thylakoid-bound ascorbate peroxidase in spinhach chloroplasts. Plant Cell Physiol 34:881-889.

43. Najami N, Janda T, Barriah W, Kayam G, Tal M, Guy M and Volokita M (2008) Ascorbate peroxidase gene family in tomato: Its identification and characterization. Mol Genet Geno-mics 279:171-182.

44. Narendra S, Venkataramani S, Shen G, Wang J, Pasapula V, Lin Y, Kornyeyev D, Holaday AS and Zhang H (2006) The Arabidopsis ascorbate peroxidase 3 is a peroxisomal membrane-bound antioxidant enzyme and is dispensable for Arabi-dopsis growth and development. J Exp Bot 57:3033-3042.

45. Noctor G and Foyer CH (1998) Ascorbate glutathione: Keeping active oxygen under control. Annu Rev Plant Physiol Plant Mol Biol 49:249-279.

46. Oliva M, Theiler G, Zamocky M, Koua D, Margis-Pinheiro M, Passardi F and Dunand C (2009) PeroxiBase: A powerful tool to collect and analyse peroxidase sequences from Viridiplantae. J Exp Bot 60:453-459.

47. Panchuk II, Volkov RA and Schöffl F (2002) Heat stress- and heat shock transcription factor-dependent expression and activity of ascorbate peroxidase in Arabidopsis. Plant Physiol 129:838-853.

48. Panda SK and Matsumoto H (2010) Changes in antioxidant gene expression and induction of oxidative stress in pea (*Pisum sativum* L.) under Al stress. Biometals 23:753-762.

49. Park SY, Ryu SH, Jang IC, Kwon SY, Kim JG and Kwak SS (2004) Molecular cloning of a cytosolic ascorbate peroxi-dase cDNA from cell cultures of sweet potato and its expression in response to stress. Mol Genet Genomics 271:339-346.

50. Passardi F, Bakalovic N, Teixeira FK, Margis-Pinheiro M, Penel C and Dunand C (2007) Prokaryotic origins of the non-animal peroxidase superfamily and organelle-mediated trans-mission to eukaryotes. Genomics 89:567-579.

51. Pekker I, Tel-Or E and Mittler R (2002) Reactive oxygen intermediates and glutathione regulate the expression of cytosolic ascorbate peroxidase during iron-mediated oxidative stress in bean. Plant Mol Biol 49:429-438.

52. Pnueli L, Liang H, Rozenberg M and Mittler R (2003) Growth suppression, altered stomatal responses, and augmented induction of heat shock proteins in cytosolic ascorbate pero-xidase (Apx1)-deficient Arabidopsis plants. Plant J 34:187-203.

53. Romero-Puertas MC, Corpas FJ, Rodríguez-Serrano M, Gómez M, Río LAd and Sandalio LM (2007) Differential expression and regulation of antioxidative enzymes by cadmium in pea plants. J Plant Physiol 164:1346-1357.

54. Rosa SB, Caverzan A, Teixeira FK, Lazzarotto F, Silveira JA, Ferreira-Silva SL, Abreu-Neto J, Margis R and Margis-Pinheiro M (2010) Cytosolic APx knockdown indicates an ambiguous redox responses in rice. Phytochemistry 71:548-558.

55. Sano S, Ueda M, Kitajima S, Takeda T, Shigeoka S, Kurano N, Miyachi S, Miyake C and Yokota A (2001) Characterization of ascorbate peroxidases from unicellular red alga Galdieria partita. Plant Cell Physiol 42:433-440.

56. Sato Y, Masuta Y, Saito K, Murayama S and Ozawa K (2011) Enhanced chilling tolerance at the booting stage in rice by transgenic overexpression of the ascorbate peroxidase gene, OsAPXa. Plant Cell Rep 30:299-406.

57. Scandalios JG (2005) Oxidative stress: Molecular perception and transduction of signals triggering antioxidant gene defenses. Braz J Med Biol Res 38:995-1014.

58. Secenji M, Hideg E, Bebes A and Gyorgyey J (2010) Transcrip-tional differences in gene families of the ascorbate-gluta-thione cycle in wheat during mild water deficit. Plant Cell Rep 29:37-50.

59. Sharma P and Dubey RS (2007) Involvement of oxidative stress and role of antioxidative defense system in growing rice seedlings exposed to toxic concentrations of aluminum. Plant Cell Rep 26:2027-2038.

60. Shi WM, Muramoto Y, Ueda A and Takabe T (2001) Cloning of peroxisomal ascorbate peroxidase gene from barley and enhanced thermotolerance by overexpressing in *Arabidopsis thaliana*. Gene 273:23-27.

61. Shigeoka S, Ishikawa T, Tamoi M, Miyagawa Y, Takeda T, Yabuta Y and Yoshimura K (2002) Regulation and function of ascorbate peroxidase isoenzymes. J Exp Bot 53:1305-1319.

62. Shigeoka S, Nakano Y and Kitaoka S (1980) Metabolism of hydrogen peroxide in *Euglena gracilis* Z by L-ascorbic acid peroxidase. Biochem J 186:377-380.

63. Song XS, Hu WH, Mao WH, Ogweno JO, Zhou YH and Yu JQ (2005) Response of ascorbate peroxidase isoenzymes and ascorbate regeneration system to abiotic stresses in *Cucumis sativus* L. Plant Physiol Biochem 43:1082-1088.

64. Sun W-H, Duan M, Li F, Shu D-F, Yang S and Meng Q-W (2010a) Overexpression of tomato tAPX gene in tobacco improves tolerance to high or low temperature stress. Biol Plantarum 54:614-620.

65. Sun WH, Duan M, Shu DF, Yang S and Meng QW (2010b) Over-expression of StAPX in tobacco improves seed germination and increases early seedling tolerance to salinity and osmotic stresses. Plant Cell Rep 29:917-926.

66. Takeda T, Yoshimura K, Ishikawa T and Shigeoka S (1998) Purification and characterization of ascorbate peroxidase in *Chlorella vulgaris*. Biochimie 80:295-301.

67. Takeda T, Yoshimura K, Yoshii M, Kanahoshi H, Miyasaka H and Shigeoka S (2000) Molecular characterization and physiological role of ascorbate peroxidase from halotolerant Chlamy-domonas sp. W80 strain. Arch Biochem Biophys 376:82-90.

68. Tang L, Kwon SY, Kim SH, Kim JS, Choi JS, Cho KY, Sung CK, Kwak SS and Lee HS (2006) Enhanced tolerance of transgenic potato plants expressing both superoxide dismutase and ascorbate peroxidase in chloroplasts against oxidative stress and high temperature. Plant Cell Rep 25:1380-1386.

69. Teixeira FK, Menezes-Benavente L, Margis R and Margis-Pinheiro M (2004) Analysis of the molecular evolutionary history of the ascorbate peroxidase gene family: Inferences from the rice genome. J Mol Evol 59:761-770.

70. Teixeira FK, Menezes-Benavente L, Galvão VC and Margis-Pinheiro M (2005) Multigene families encode the major enzymes of antioxidant metabolism in Eucalyptus grandis L. Genet Mol Biol 28:529-538.

71. Teixeira FK, Menezes-Benavente L, Galvão VC, Margis R and Margis-Pinheiro M (2006) Rice ascorbate peroxidase gene family encodes functionally diverse isoforms localized in different subcellular compartments. Planta 224:300-314.

72. Verslues PE, Agarwal M, Katiyar-Agarwal S, Zhu J and Zhu JK (2006) Methods and concepts in quantifying resistance to drought, salt and freezing, abiotic stresses that affect plant water status. Plant J 45:523-539.

73. Vranova E, Inze D and Van Breusegem F (2002) Signal transduc-tion during oxidative stress. J Exp Bot 53:1227-1236.

74. Wang R, Chen S, Zhou X, Shen X, Deng L, Zhu H, Shao J, Shi Y, Dai S, Fritz E, Huttermann A and Polle A (2008) Ionic homeostasis and reactive oxygen species control in leaves and xylem sap of two poplars subjected to NaCl stress. Tree Physiol 28:947-957.

75. Wilkinson SR, Obado SO, Mauricio IL and Kelly JM (2002) T*rypanosoma cruzi* expresses a plant-like ascorbate-de-pendent hemoperoxidase localized to the endoplasmic reticulum. Proc Natl Acad Sci USA 99:13453-13458.

76. Yabuta Y, Motoki T, Yoshimura K, Takeda T, Ishikawa T and Shigeoka S (2002) Thylakoid membrane-bound ascorbate peroxidase is a limiting factor of antioxidative systems under photo-oxidative stress. Plant J 32:915-925.

77. Yamane K, Mitsuya S, Taniguchi M and Miyake H (2010) Transcription profiles of genes encoding catalase an ascorbate peroxidase in rice leaf tissues under salinity. Plant Prod Sci 13:164-168.

78. Yoshimura K, Ishikawa T, Nakamura Y, Tamoi M, Takeda T, Tada T, Nishimura K and Shigeoka S (1998) Comparative study on recombinant chloroplastic and cytosolic ascorbate peroxidase isozymes of spinach. Arch Biochem Biophys 353:55-63.

79. Yoshimura K, Yabuta Y, Ishikawa T and Shigeoka S (2000) Expression of spinach ascorbate peroxidase isoenzymes in response to oxidative stresses. Plant Physiol 123:223-234.

80. Zhang H, Wang J, Nickel U, Allen RD and Goodman HM (1997) Cloning and expression of an Arabidopsis gene encoding a putative peroxisomal ascorbate peroxidase. Plant Mol Biol 34:967-971.

CHAPTER 6

HEAVY METAL POLLUTED SOILS: EFFECT ON PLANTS AND BIOREMEDIATION METHODS

G. U. Chibuike and S. C. Obiora

INTRODUCTION

Although heavy metals are naturally present in the soil, geologic and anthropogenic activities increase the concentration of these elements to amounts that are harmful to both plants and animals. Some of these activities include mining and smelting of metals, burning of fossil fuels, use of fertilizers and pesticides in agriculture, production of batteries and other metal products in industries, sewage sludge, and municipal waste disposal [1–3].

Growth reduction as a result of changes in physiological and biochemical processes in plants growing on heavy metal polluted soils has been recorded [4–6]. Continued decline in plant growth reduces yield which eventually leads to food insecurity. Therefore, the remediation of heavy metal polluted soils cannot be overemphasized.

Various methods of remediating metal polluted soils exist; they range from physical and chemical methods to biological methods. Most physical and chemical methods (such as encapsulation, solidification, stabilization, electrokinetics, vitrification, vapour extraction, and soil washing and flushing) are expensive and do not make the soil suitable for plant growth [7]. Biological approach (bioremediation) on the other hand encourages the establishment/reestablishment of plants on polluted soils. It is an environmentally friendly approach because it is achieved via natural processes. Bioremediation is also an economical remediation technique compared with other remediation techniques. This paper discusses the nature and properties of soils polluted with heavy metals. Plant growth and performance on these soils were examined. Biological approaches employed for the remediation of heavy metal polluted soils were equally highlighted.

HEAVY METAL POLLUTED SOILS

Heavy metals are elements that exhibit metallic properties such as ductility, malleability, conductivity, cation stability, and ligand specificity. They are characterized by relatively high density and high relative atomic weight with an atomic number greater than 20 [2]. Some heavy metals such as Co, Cu, Fe, Mn, Mo, Ni, V, and Zn are required in minute quantities by organisms. However, excessive amounts of these elements can become harmful to organisms. Other heavy metals such as Pb, Cd, Hg, and As (a metalloid but generally referred to as a

Citation: G. U. Chibuike and S. C. Obiora, "Heavy Metal Polluted Soils: Effect on Plants and Bioremediation Methods vol. 2014, Article ID 752708, 12 pages, 2014. doi:10.1155/2014/752708.

heavy metal) do not have any beneficial effect on organisms and are thus regarded as the "main threats" since they are very harmful to both plants and animals.

Metals exist either as separate entities or in combination with other soil components. These components may include exchangeable ions sorbed on the surfaces of inorganic solids, nonexchangeable ions and insoluble inorganic metal compounds such as carbonates and phosphates, soluble metal compound or free metal ions in the soil solution, metal complex of organic materials, and metals attached to silicate minerals [7]. Metals bound to silicate minerals represent the background soil metal concentration and they do not cause contamination/pollution problems compared with metals that exist as separate entities or those present in high concentration in the other 4 components [8].

Soil properties affect metal availability in diverse ways. Harter [9] reported that soil pH is the major factor affecting metal availability in soil. Availability of Cd and Zn to the roots of Thlaspi caerulescens decreased with increases in soil pH [10]. Organic matter and hydrous ferric oxide have been shown to decrease heavy metal availability through immobilization of these metals [11]. Significant positive correlations have also been recorded between heavy metals and some soil physical properties such as moisture content and water holding capacity [12].

Other factors that affect the metal availability in soil include the density and type of charge in soil colloids, the degree of complexation with ligands, and the soil's relative surface area [7, 13]. The large interface and specific surface areas provided by soil colloids help in controlling the concentration of heavy metals in natural soils. In addition, soluble concentrations of metals in polluted soils may be reduced by soil particles with high specific surface area, though this may be metal specific [7]. For instance, Mcbride and Martínez [14] reported that addition of amendment consisting of hydroxides with high reactive surface area decreased the solubility of As, Cd, Cu, Mo, and Pb while the solubility of Ni and Zn was not changed. Soil aeration, microbial activity, and mineral composition have also been shown to influence heavy metal availability in soils [15].

Conversely, heavy metals may modify soil properties especially soil biological properties [16]. Monitoring changes in soil microbiological and biochemical properties after contamination can be used to evaluate the intensity of soil pollution because these methods are more sensitive and results can be obtained at a faster rate compared with monitoring soil physical and chemical properties [17]. Heavy metals affect the number, diversity, and activities of soil microorganisms. The toxicity of these metals on microorganisms depends on a number of factors such as soil temperature, pH, clay minerals, organic matter, inorganic anions and cations, and chemical forms of the metal [16, 18, 19].

There are discrepancies in studies comparing the effect of heavy metals on soil biological properties. While some researchers have recorded negative effect of heavy metals on soil biological properties [16, 17, 20], others have reported no relationship between high heavy metal concentrations and some soil (micro)biological properties [21]. Some of the inconsistencies may arise because some of these studies were conducted under laboratory conditions using artificially contaminated soils while others were carried out using soils from areas that are actually polluted in the field. Regardless of the origin of the soils used in these experiments, the fact that the effect of heavy metals on soil biological properties needs to be studied in more detail in order to fully understand the effect of these metals on the soil ecosystem remains. Further, it is advisable to use a wide range of methods (such as microbial biomass, C and N mineralization, respiration, and enzymatic activities) when studying effect of metals on soil biological properties rather than focusing on a single method since results obtained from use of different methods would be more comprehensive and conclusive.

The presence of one heavy metal may affect the availability of another in the soil and hence plant. In other words, antagonistic and synergistic behaviours exist among heavy metals. Salgare and Acharekar [22] reported that the inhibitory effect of Mn on the total amount of mineralized C was antagonized by the presence of Cd. Similarly, Cu and Zn as well as Ni and Cd have been reported to compete for the same membrane carriers in plants [23]. In contrast, Cu was reported to increase the toxicity of Zn in spring barley [24]. This implies that the interrelationship between heavy metals is quite complex; thus more research is needed in this area. Different species of the same metal may also interact with one another. Abedin et al. [25] reported that the presence of arsenite strongly suppressed the uptake of arsenate by rice plants growing on a polluted soil.

EFFECT OF HEAVY METAL POLLUTED SOIL ON PLANT GROWTH

The heavy metals that are available for plant uptake are those that are present as soluble components in the soil solution or those that are easily solubilized by root exudates [26]. Although plants require certain heavy metals for their growth and upkeep, excessive amounts of these metals can become toxic to plants. The ability of plants to accumulate essential metals equally enables them to acquire other nonessential metals [27]. As metals cannot be broken down, when concentrations within the plant exceed optimal levels, they adversely affect the plant both directly and indirectly.

Some of the direct toxic effects caused by high metal concentration include inhibition of cytoplasmic enzymes and damage to cell structures due to oxidative stress [28, 29]. An example of indirect toxic effect is the replacement of essential nutrients at cation exchange sites of plants [30]. Further, the negative influence heavy metals have on the growth and activities of soil microorganisms may also indirectly affect the growth of plants. For instance, a reduction in the number of beneficial soil microorganisms due to high metal concentration may lead to decrease in organic matter decomposition leading to a decline in soil nutrients. Enzyme activities useful for plant metabolism may also be hampered due to heavy metal interference with activities of soil microorganisms. These toxic effects (both direct and indirect) lead to a decline in plant growth which sometimes results in the death of plant [31].

The effect of heavy metal toxicity on the growth of plants varies according to the particular heavy metal involved in the process. Table 1 shows a summary of the toxic effects of specific metals on growth, biochemistry, and physiology of various plants. For metals such as Pb, Cd, Hg, and As which do not play any beneficial role in plant growth, adverse effects have been recorded at very low concentrations of these metals in the growth medium. Kibra [32] recorded significant reduction in height of rice plants growing on a soil contaminated with 1 mgHg/kg. Reduced tiller and panicle formation also occurred at this concentration of Hg in the soil. For Cd, reduction in shoot and root growth in wheat plants occurred when Cd in the soil solution was as low as 5 mg/L [33]. Most of the reduction in growth parameters of plants growing on polluted soils can be attributed to reduced photosynthetic activities, plant mineral nutrition, and reduced activity of some enzymes [34].

Table 1: Effect of heavy metal toxicity on plants.

Heavy metal	Plant	Toxic effect on plant	Reference
As	Rice (*Oryza sativa*)	Reduction in seed germination; decrease in seedling height; reduced leaf area and dry matter production	[35, 36]
	Tomato (*Lycopersicon esculentum*)	Reduced fruit yield; decrease in leaf fresh weight	[37]
	Canola (*Brassica napus*)	Stunted growth; chlorosis; wilting	[38]
Cd	Wheat (*Triticum* sp.)	Reduction in seed germination; decrease in plant nutrient content; reduced shoot and root length	[33, 39]
	Garlic (*Allium sativum*)	Reduced shoot growth; Cd accumulation	[40]
	Maize (*Zea mays*)	Reduced shoot growth; inhibition of root growth	[41]
Co	Tomato (*Lycopersicon esculentum*)	Reduction in plant nutrient content	[42]
	Mung bean (*Vigna radiata*)	Reduction in antioxidant enzyme activities; decrease in plant sugar, starch, amino acids, and protein content	[43]
	Radish (*Raphanus sativus*)	Reduction in shoot length, root length, and total leaf area; decrease in chlorophyll content; reduction in plant nutrient content and antioxidant enzyme activity; decrease in plant sugar, amino acid, and protein content	[44]
Cr	Wheat (*Triticum* sp.)	Reduced shoot and root growth	[45, 46]
	Tomato (*Lycopersicon esculentum*)	Decrease in plant nutrient acquisition	[47, 48]
	Onion (*Allium cepa*)	Inhibition of germination process; reduction of plant biomass	[49]
Cu	Bean (*Phaseolus vulgaris*)	Accumulation of Cu in plant roots; root malformation and reduction	[50]
	Black bindweed (*Polygonum convolvulus*)	Plant mortality; reduced biomass and seed production	[51]
	Rhodes grass (*Chloris gayana*)	Root growth reduction	[52]
Hg	Rice (*Oryza sativa*)	Decrease in plant height; reduced tiller and panicle formation; yield reduction; bioaccumulation in shoot and root of seedlings	[32, 53]
	Tomato (*Lycopersicon esculentum*)	Reduction in germination percentage; reduced plant height; reduction in flowering and fruit weight; chlorosis	[54]
Mn	Broad bean (*Vicia faba*)	Mn accumulation shoot and root; reduction in shoot and root length; chlorosis	[55]
	Spearmint (*Mentha spicata*)	Decrease in chlorophyll a and carotenoid content; accumulation of Mn in plant roots	[56]
	Pea (*Pisum sativum*)	Reduction in chlorophylls a and b content; reduction in relative growth rate; reduced photosynthetic O_2 evolution activity and photosystem II activity	[57]
	Tomato (*Lycopersicon esculentum*)	Slower plant growth; decrease in chlorophyll concentration	[58]
Ni	Pigeon pea (*Cajanus cajan*)	Decrease in chlorophyll content and stomatal conductance; decreased enzyme activity which affected Calvin cycle and CO_2 fixation	[59]
	Rye grass (*Lolium perenne*)	Reduction in plant nutrient acquisition; decrease in shoot yield; chlorosis	[60]
	Wheat (*Triticum* sp.)	Reduction in plant nutrient acquisition	[61, 62]
	Rice (*Oryza sativa*)	Inhibition of root growth	[63]

For other metals which are beneficial to plants, "small" concentrations of these metals in the soil could actually improve plant growth and development. However, at higher concentrations of these metals, reductions in plant growth have been recorded. For instance, Jayakumar et al. [42] reported that, at 50 mgCo/kg, there was an increase in nutrient content of tomato plants compared with the control. Conversely, at 100 mgCo/kg to 250 mgCo/kg, reductions in plant nutrient content were recorded. Similarly, increase in plant growth, nutrient content, biochemical content, and antioxidant enzyme activities (catalase) was observed in radish and mung bean at 50 mgCo/kg soil concentration while reductions were recorded at 100 mgCo/kg to 250 mgCo/kg soil concentration [43, 44]. Improvements in growth and physiology of cluster beans have also been reported at Zn concentration of 25 mg/L of the soil solution. On the other hand, growth reduction and adverse effect on the plant's physiology started when the soil solution contained 50 mgZn/L [67].

It is worth mentioning that, in most real life situations (such as disposal of sewage sludge and metal mining wastes) where soil may be polluted with more than one heavy metal, both antagonistic and synergistic relationships between heavy metals may affect plant metal toxicity. Nicholls and Mal [70] reported that the combination of Pb and Cu at both high concentration (1000 mg/kg each) and low concentration (500 mg/kg) resulted in a rapid and complete death of the leaves and stem of Lythrum salicaria. The authors reported that there was no synergistic interaction between these heavy metals probably because the concentrations used in the experiment were too high for interactive relationship to be observed between the metals. Another study [71] examined the effect of 6 heavy metals (Cd, Cr, Co, Mn, and Pb) on the growth of maize. The result showed that the presence of these metals in soil reduced the growth and protein content of maize. The toxicity of these metals occurred in the following order: Cd > Co > Hg > Mn > Pb > Cr. It was also observed in this study that the combined effect of 2 or more heavy metals was only as harmful as the effect of the most toxic heavy metal. The researcher attributed this result to the antagonistic relationship which exists between heavy metals. It is important to note that certain plants are able to tolerate high concentration of heavy metals in their environment. Baker [72] reported that these plants are able to tolerate these metals via 3 mechanisms, namely, (i) exclusion: restriction of metal transport and maintenance of a constant metal concentration in the shoot over a wide range of soil concentrations; (ii) inclusion: metal concentrations in the shoot reflecting those in the soil solution through a linear relationship; and (iii) bioaccumulation: accumulation of metals in the shoot and roots of plants at both low and high soil concentrations.

BIOREMEDIATION OF HEAVY METAL POLLUTED SOILS

Bioremediation is the use of organisms (microorganisms and/or plants) for the treatment of polluted soils. It is a widely accepted method of soil remediation because it is perceived to occur via natural processes. It is equally a cost effective method of soil remediation. Blaylock et al. [73] reported 50% to 65% saving when bioremediation was used for the treatment of 1 acre of Pb polluted soil compared with the case when a conventional method (excavation and landfill) was used for the same purpose. Although bioremediation is a nondisruptive method of soil remediation, it is usually time consuming and its use for the treatment of heavy metal polluted soils is sometimes affected by the climatic and geological conditions of the site to be remediated [74].

Heavy metals cannot be degraded during bioremediation but can only be transformed from one organic complex or oxidation state to another. Due to a change in their oxidation state, heavy metals can be transformed to become either less toxic, easily volatilized, more water soluble (and thus can be removed through leaching), less water soluble (which allows them to precipitate and become easily removed from the environment) or less bioavailable [75, 76].

Bioremediation of heavy metals can be achieved via the use of microorganisms, plants, or the combination of both organisms.

Using Microbes for Remediation of Heavy Metal Polluted Soils

Several microorganisms especially bacteria (Bacillus subtilis, Pseudomonas putida, and Enterobacter cloacae) have been successfully used for the reduction of Cr (VI) to the less toxic Cr (III) [77–80]. B. subtilis has also been reported to reduce nonmetallic elements. For instance, Garbisu et al. [81] recorded that B. subtilis reduced the selenite to the less toxic elemental Se. Further, B. cereus and B. thuringiensis have been shown to increase extraction of Cd and Zn from Cd-rich soil and soil polluted with effluent from metal industry [82]. It is assumed that the production of siderophore (Fe complexing molecules) by bacteria may have facilitated the extraction of these metals from the soil; this is because heavy metals have been reported to simulate the production of siderophore and this consequently affects their bioavailability [83]. For instance, siderophore production by Azotobacter vinelandii was increased in the presence of Zn (II) [84]. Hence, heavy metals influence the activities of siderophore-producing bacteria which in turn increases mobility and extraction of these metals in soil.

Bioremediation can also occur indirectly via bioprecipitation by sulphate reducing bacteria (Desulfovibrio desulfuricans) which converts sulphate to hydrogen sulphate which subsequently reacts with heavy metals such as Cd and Zn to form insoluble forms of these metal sulphides [85].

Most of the above microbe assisted remediation is carried out ex situ. However, a very important in situmicrobe assisted remediation is the microbial reduction of soluble mercuric ions Hg (II) to volatile metallic mercury and Hg (0) carried out by mercury resistant bacteria [86]. The reduced Hg (0) can easily volatilize out of the environment and subsequently be diluted in the atmosphere [87].

Genetic engineering can be adopted in microbe assisted remediation of heavy metal polluted soils. For instance, Valls et al. [88] reported that genetically engineered Ralstonia eutropha can be used to sequester metals (such as Cd) in polluted soils. This is made possible by the introduction of metallothionein (cysteine rich metal binding protein) from mouse on the cell surface on this organism. Although the sequestered metals remain in the soil, they are made less bioavailable and hence less harmful. The controversies surrounding genetically modified organisms [89] and the fact that the heavy metal remains in the soil are major limitations to this approach to bioremediation.

Making the soil favourable for soil microbes is one strategy employed in bioremediation of polluted soils. This process known as biostimulation involves the addition of nutrients in the form of manure or other organic amendments which serve as C source for microorganisms present in the soil. The added nutrients increase the growth and activities of microorganisms involved in the remediation process and thus this increases the efficiency of bioremediation.

Although biostimulation is usually employed for the biodegradation of organic pollutants [90], it can equally be used for the remediation of heavy metal polluted soils. Since heavy metals cannot be biodegraded, biostimulation can indirectly enhance remediation of heavy metal polluted soil through alteration of soil pH. It is well known that the addition of organic materials reduces the pH of the soil [91]; this subsequently increases the solubility and hence bioavailability of heavy metals which can then be easily extracted from the soil [92].

Biochar is one organic material that is currently being exploited for its potential in the management of heavy metal polluted soils. Namgay et al. [93] recorded a reduction in the availability of heavy metals when the polluted soil was amended with biochar; this in turn reduced plant absorption of the metals. The ability of biochar to increase soil pH unlike most other organic amendments [94] may have increased sorption of these metals, thus reducing their bioavailability for plant uptake. It is important to note that, since the characteristics of biochar vary widely depending on its method of production and the feedstock used in its production, the effect different biochar amendments will have on the availability of heavy metals in soil will also differ. Further, more research is needed in order to understand the effect of biochar on soil microorganisms and how the interaction between biochar and soil microbes influences remediation of heavy metal polluted soils because such studies are rare in literature.

Using Plants for Remediation of Heavy Metal Polluted Soils

Phytoremediation is an aspect of bioremediation that uses plants for the treatment of polluted soils. It is suitable when the pollutants cover a wide area and when they are within the root zone of the plant [76]. Phytoremediation of heavy metal polluted soils can be achieved via different mechanisms. These mechanisms include phytoextraction, phytostabilization, and phytovolatilization.

Phytoextraction

This is the most common form of phytoremediation. It involves accumulation of heavy metals in the roots and shoots of phytoremediation plants. These plants are later harvested and incinerated. Plants used for phytoextraction usually possess the following characteristics: rapid growth rate, high biomass, extensive root system, and ability to tolerate high amounts of heavy metals. This ability to tolerate high concentration of heavy metals by these plants may lead to metal accumulation in the harvestable part; this may be problematic through contamination of the food chain [7].

There are two approaches to phytoextraction depending on the characteristics of the plants involved in the process. The first approach involves the use of natural hyperaccumulators, that is, plants with very high metal-accumulating ability, while the second approach involves the use of high biomass plants whose ability to accumulate metals is induced by the use of chelates, that is, soil amendments with metal mobilizing capacity [95].

Hyperaccumulators accumulate 10 to 500 times more metals than ordinary plant [96]; hence they are very suitable for phytoremediation. An important characteristic which makes hyperaccumulation possible is the tolerance of these plants to increasing concentrations of these metals (hypertolerance). This could be a result of exclusion of these metals from the plants or by compartmentalization of these metal ions; that is, the metals are retained in the vacuolar compartments or cell walls and thus do not have access to cellular sites where vital functions such as respiration and cell division take place [76, 96].

Generally, a plant can be called a hyperaccumulator if it meets the following criteria: (i) the concentration of metal in the shoot must be higher than 0.1% for Al, As, Co, Cr, Cu, Ni, and Se, higher than 0.01% for Cd, and higher than 1.0% for Zn [97]; (ii) the ratio of shoot to root concentration must be consistently higher than 1 [98]; this indicates the capability to transport metals from roots to shoot and the existence of hypertolerance ability [7]; (iii) the ratio of shoot to root concentration must be higher than 1; this indicates the degree of plant metal uptake [7, 98]. Reeves and Baker [99] reported some examples of plants which have the ability to accumulate large amounts of heavy metals and hence can be used in remediation studies. Some of these plants include Haumaniastrum robertii (Co hyperaccumulator); Aeollanthus subacaulis (Cu hyperaccumulator);Maytenus bureaviana (Mn hyperaccumulator); Minuartia verna and Agrostis tenuis (Pb hyperaccumulators);Dichapetalum gelonioides, Thlaspi tatrense, and Thlaspi caerulescens (Zn hyperaccumulators); Psycotria vanhermanni and Streptanthus polygaloides (Ni hyperaccumulators); Lecythis ollaria (Se hyperaccumulator). Pteris vittata is an example of a hyperaccumulator that can be used for the remediation of soils polluted with As [100]. Some plants have the ability to accumulate more than one metal. For instance, Yang et al. [101] observed that the Zn hyperaccumulator, Sedum alfredii, can equally hyperaccumulate Cd.

The possibility of contaminating the food chain through the use of hyperaccumulators is a major limitation in phytoextraction. However, many species of the Brassicaceae family which are known to be hyperaccumulators of heavy metals contain high amounts of thiocyanates which make them unpalatable to animals; thus this reduces the availability of these metals in the food chain [102].

Most hyperaccumulators are generally slow growers with low plant biomass; this reduces the efficiency of the remediation process [103]. Thus, in order to increase the efficiency of phytoextraction, plants with high growth rate as well as high biomass (e.g., maize, sorghum, and alfalfa) are sometimes used together with metal chelating substances for soil remediation exercise. It is important to note that some hyperaccumulators such as certain species within the Brassica genus (Brassica napus, Brassica juncea, and Brassica rapa) are fast growers with high biomass [104].

In most cases, plants absorb metals that are readily available in the soil solution. Although some metals are present in soluble forms for plant uptake, others occur as insoluble precipitate and are thus unavailable for plant uptake. Addition of chelating substances prevents precipitation and metal sorption via the formation of metal chelate complexes; this subsequently increases the bioavailability of these metals [7]. Further, the addition of chelates to the soil can transport more metals into the soil solution through the dissolution of precipitated compounds and desorption of sorbed species [13]. Certain chelates are also able to translocate heavy metal into the shoots of plants [73].

Marques et al. [7] documented examples of synthetic chelates which have successfully been used to extract heavy metals from polluted soils. Some of these chelates include EDTA (ethylenediaminetetraacetic acid), EDDS (SS-ethylenediamine disuccinic acid), CDTA (trans-1,2-diaminocyclohexane-N,N,N′,N′-tetraacetic acid), EDDHA (ethylenediamine-di-o-hydroxyphenylacetic acid), DTPA (diethylenetriaminepentaacetic acid), and HEDTA (N-hydroxyethylenediaminetriacetic acid). EDTA is a synthetic chelate that is widely used not only because it is the least expensive compared with other synthetic chelates [105] but also because it has a high ability to successfully improve plant metal uptake [106–108]. Organic chelates such as citric acid and malic acid can also be used to improve phytoextraction of heavy metals from polluted soils [109].

One major disadvantage of using chelates in phytoextraction is the possible contamination of groundwater via leaching of these heavy metals [110]. This is because of the increased availability of heavy metals in the soil solution when these chelates are used. In addition, when chelates (especially synthetic chelates) are used in high concentrations, they can become toxic to plants and soil microbes [106]. In general, solubility/availability of heavy metals for plant uptake and suitability of a site for phytoextraction are additional factors that should be considered (in addition to suitability of plants) before using phytoextraction for soil remediation [26].

Phytostabilization

Phytostabilization involves using plants to immobilize metals, thus reducing their bioavailability via erosion and leaching. It is mostly used when phytoextraction is not desirable or even possible [98]. Marques et al. [7] argued that this form of phytoremediation is best applied when the soil is so heavily polluted so that using plants for metal extraction would take a long time to be achieved and thus would not be adequate. Jadia and Fulekar [111] on the other hand showed that the growth of plants (used for phytostabilization) was adversely affected when the concentration of heavy metal in the soil was high.

Phytostabilization of heavy metals takes place as a result of precipitation, sorption, metal valence reduction, or complexation [29]. The efficiency of phytostabilization depends on the plant and soil amendment used. Plants help in stabilizing the soil through their root systems; thus, they prevent erosion. Plant root systems equally prevent leaching via reduction of water percolation through the soil. In addition, plants prevent man's direct contact with pollutants and they equally provide surfaces for metal precipitation and sorption [112].

Based on the above factors, it is important that appropriate plants are selected for phytostabilization of heavy metals. Plants used for phytostabilization should have the following characteristics: dense rooting system, ability to tolerate soil conditions, ease of establishment and maintenance under field conditions, rapid growth to provide adequate ground coverage, and longevity and ability to self-propagate.

Soil amendments used in phytostabilization help to inactivate heavy metals; thus, they prevent plant metal uptake and reduce biological activity [7]. Organic materials are mostly used as soil amendments in phytostabilization. Marques et al. [113] showed that Zn percolation through the soil reduced by 80% after application of manure or compost to polluted soils on which Solanum nigrum was grown.

Other amendments that can be used for phytostabilization include phosphates, lime, biosolids, and litter [114]. The best soil amendments are those that are easy to handle, safe to workers who apply them, easy to produce, and inexpensive and most importantly are not toxic to plants [113]. Most of the times, organic amendments are used because of their low cost and the other benefits they provide such as provision of nutrients for plant growth and improvement of soil physical properties [7].

In general, phytostabilization is very useful when rapid immobilization of heavy metals is needed to prevent groundwater pollution. However, because the pollutants remain in the soil, constant monitoring of the environment is required and this may become a problem.

Phytovolatilization

In this form of phytoremediation, plants are used to take up pollutants from the soil; these pollutants are transformed into volatile forms and are subsequently transpired into the atmosphere [115]. Phytovolatilization is mostly used for the remediation of soils polluted with Hg. The toxic form of Hg (mercuric ion) is transformed into the less toxic form (elemental Hg). The problem with this process is that the new product formed, that is, elemental Hg, may be redeposited into lakes and rivers after being recycled by precipitation; this in turn repeats the process of methyl-Hg production by anaerobic bacteria [115].

Raskin and Ensley [116] reported the absence of plant species with Hg hyperaccumulating properties. Therefore, genetic engineered plants are mostly used in phytovolatilization. Examples of transgenic plants which have been used for phytovolatilization of Hg polluted soils are Nicotiana tabacum, Arabidopsis thaliana,and Liriodendron tulipifera [117, 118]. These plants are usually genetically modified to include gene for mercuric reductase, that is, merA. Organomercurial lyase (merB) is another bacterial gene used for the detoxification of methyl-Hg. Both merA and merB can be inserted into plants used to detoxify methyl-Hg to elemental Hg [119]. Use of plants modified with merA and merB is not acceptable from a regulatory perspective [119]. However, plants altered with merB are more acceptable because the gene prevents the introduction of methyl-Hg into the food chain [120].

Phytovolatilization can also be employed for the remediation of soils polluted with Se [7]. This involves the assimilation of inorganic Se into organic selenoamino acids (selenocysteine and selenomethionine). Selenomethionine is further biomethylated to dimethylselenide which is lost in the atmosphere via volatilization [121]. Plants which have successfully been used for phytovolatilization of soils polluted with Se are Brassica juncea and Brassica napus [122].

Combining Plants and Microbes for the Remediation of Heavy Metal Polluted Soils

The combined use of both microorganisms and plants for the remediation of polluted soils results in a faster and more efficient clean-up of the polluted site [123]. Mycorrhizal fungi have been used in several remediation studies involving heavy metals and the results obtained show that mycorrhizae employ different mechanisms for the remediation of heavy metal polluted soils. For instance, while some studies have shown enhanced phytoextraction through the accumulation of heavy metals in plants [124–126], others reported enhanced phytostabilization through metal immobilization and a reduced metal concentration in plants [127, 128].

In general, the benefits derived from mycorrhizal associations—which range from increased nutrient and water acquisition to the provision of a stable soil for plant growth and increase in plant resistance to diseases [129–131]—are believed to aid the survival of plants growing in polluted soils and thus help in the vegetation/revegetation of remediated soils [132]. It is important to note that mycorrhiza does not always assist in the remediation of heavy metal polluted soils [133, 134] and this may be attributed to the species of mycorrhizal fungi and the concentration of heavy metals [7, 132]. Studies have also shown that activities of mycorrhizal fungi may be inhibited by heavy metals [135, 136]. In addition, Weissenhorn and Leyval [137] reported that certain species of mycorrhizal fungi (arbuscular mycorrhizal fungi) can be more sensitive to pollutants compared to plants.

Other microorganisms apart from mycorrhizal fungi have also been used in conjunction with plants for the remediation of heavy metal polluted soils. Most of these microbes are the plant growth-promoting rhizobacteria (PGPR) that are usually found in the rhizosphere. These PGPR stimulate plant growth via several mechanisms such as production of phytohormones and supply of nutrients [138], production of siderophores and other chelating agents [139], specific enzyme activity and N fixation [140], and reduction in ethylene production which encourages root growth [141].

In general, PGPR have been used in phytoremediation studies to reduce plant stress associated with heavy metal polluted soils [142]. Enhanced accumulation of heavy metals such as Cd and Ni by hyperaccumulators (Brassica juncea and Brassica napus) has been observed when the plants were inoculated with Bacillus sp. [143,144]. On the other hand, Madhaiyan et al. [145] reported increased plant growth due to a reduction in the accumulation of Cd and Ni in the shoot and root tissues of tomato plant when it was inoculated with Methylobacterium oryzae and Burkholderia spp. Thus, this indicates that the mechanisms employed by PGPR in the phytoremediation of heavy metal polluted soils may be dependent on the species of PGRP and plant involved in the process. Although studies involving both the use of mycorrhizal fungi and PGPR are uncommon, Vivas et al. [146] reported that PGPR (Brevibacillus sp.) increased mycorrhizal efficiency which in turn decreased metal accumulation and increased the growth of white clover growing on a heavy metal (Zn) polluted soil.

CONCLUSION

Plants growing on heavy metal polluted soils show a reduction in growth due to changes in their physiological and biochemical activities. This is especially true when the heavy metal involved does not play any beneficial role towards the growth and development of plants. Bioremediation can be effectively used for the treatment of heavy metal polluted soil. It is most appropriate when the remediated site is used for crop production because it is a nondisruptive method of soil remediation. Using plants for bioremediation (phytoremediation) is a more common approach to bioremediation of heavy metal compared with the use of microorganisms. Plants employ different mechanisms in the remediation of heavy metal polluted soils. Phytoextraction is the most common method of phytoremediation used for treatment of heavy metal polluted soils. It ensures the complete removal of the pollutant. Combining both plants and microorganisms in bioremediation increases the efficiency of this method of remediation. Both mycorrhizal fungi and other PGPR have been successfully incorporated in various phytoremediation programmes. The success of the combined use of these organisms depends on the species of microbe and plants involved and to some extent on the concentration of the heavy metal in soil.

Conflict of Interests

The authors declare that there is no conflict of interests regarding the publication of this paper.

REFERENCES

1. B. J. Alloway, Heavy Metal in Soils, John Wiley & Sons, New York, NY, USA, 1990.
2. I. Raskin, P. B. A. N. Kumar, S. Dushenkov, and D. E. Salt, “Bioconcentration of heavy metals by plants,”Current Opinion in Biotechnology, vol. 5, no. 3, pp. 285–290, 1994.
3. Z. Shen, X. Li, C. Wang, H. Chen, and H. Chua, “Lead phytoextraction from contaminated soil with high-biomass plant species,” Journal of Environmental Quality, vol. 31, no. 6, pp. 1893–1900, 2002.
4. J. Chatterjee and C. Chatterjee, “Phytotoxicity of cobalt, chromium and copper in cauliflower,”Environmental Pollution, vol. 109, no. 1, pp. 69–74, 2000. · ·
5. I. Öncel, Y. Keleş, and A. S. Üstün, “Interactive effects of temperature and heavy metal stress on the growth and some biochemical compounds in wheat seedlings,” Environmental Pollution, vol. 107, no. 3, pp. 315–320, 2000.
6. S. Oancea, N. Foca, and A. Airinei, “Effects of heavy metals on plant growth and photosynthetic activity,”Analele Ştiinţifice ale Universităţii “AL. I. CUZA1 IAŞI, Tomul I, s. Biofizică, Fizică medicală şi Fizica mediului, pp. 107–110, 2005.
7. A. P. G. C. Marques, A. O. S. S. Rangel, and P. M. L. Castro, “Remediation of heavy metal contaminated soils: phytoremediation as a potentially promising clean-up technology,” Critical Reviews in Environmental Science and Technology, vol. 39, no. 8, pp. 622–654, 2009.
8. L. Ramos, L. M. Hernandez, and M. J. Gonzalez, “Sequential fractionation of copper, lead, cadmium and zinc in soils from or near Donana National Park,” Journal of Environmental Quality, vol. 23, no. 1, pp. 50–57, 1994. ·
9. R. D. Harter, “Effect of soil pH on adsorption of lead, copper, zinc, and nickel,” Soil Science Society of America Journal, vol. 47, no. 1, pp. 47–51, 1983.
10. A. S. Wang, J. S. Angle, R. L. Chaney, T. A. Delorme, and R. D. Reeves, “Soil pH effects on uptake of Cd and Zn by Thlaspi caerulescens,” Plant and Soil, vol. 281, no. 1-2, pp. 325–337, 2006. · ·
11. L. Yi, Y. Hong, D. Wang, and Y. Zhu, “Determination of free heavy metal ion concentrations in soils around a cadmium rich zinc deposit,” Geochemical Journal, vol. 41, no. 4, pp. 235–240, 2007.
12. M. S. Rakesh Sharma and N. S. Raju, “Correlation of heavy metal contamination with soil properties of industrial areas of Mysore, Karnataka, India by cluster analysis,” International Research Journal of Environment Sciences, vol. 2, no. 10, pp. 22–27, 2013.
13. W. A. Norvell, “Comparison of chelating agents as extractants for metals in diverse soil materials,” Soil Science Society of America Journal, vol. 48, no. 6, pp. 1285–1292, 1984.
14. M. B. Mcbride and C. E. Martínez, “Copper phytotoxicity in a contaminated soil: remediation tests with adsorptive materials,” Environmental Science and Technology, vol. 34, no. 20, pp. 4386–4391, 2000.
15. M. L. Magnuson, C. A. Kelty, and K. C. Kelty, “Trace metal loading on water-borne soil and dust particles characterized through the use of Split-flow thin-cell

fractionation," Analytical Chemistry, vol. 73, no. 14, pp. 3492–3496, 2001.

16. M. Friedlová, "The influence of heavy metals on soil biological and chemical properties," Soil and Water Research, vol. 5, no. 1, pp. 21–27, 2010. ·

17. P. Nannipieri, L. Badalucco, L. Landi, and G. Pietramellara, "Measurement in assessing the risk of chemicals to the soil ecosystem," in Ecotoxicology: Responses, Biomarkers and Risk Assessment, J. T. Zelikoff, Ed., pp. 507–534, OECD Workshop, SOS Publ., Fair Haven, NY, USA, 1997.

18. E. Baath, "Effects of heavy metals in soil on microbial processes and populations (a review)," Water, Air, & Soil Pollution, vol. 47, no. 3-4, pp. 335–379, 1989.

19. K. E. Giller, E. Witter, and S. P. Mcgrath, "Toxicity of heavy metals to microorganisms and microbial processes in agricultural soils," Soil Biology and Biochemistry, vol. 30, no. 10-11, pp. 1389–1414, 1998. · ·

20. M. Šmejkalova, O. Mikanova, and L. Borůvka, "Effects of heavy metal concentrations on biological activity of soils microorganisms," Plant, Soil and Environment, vol. 49, pp. 321–326, 2003.

21. S. Castaldi, F. A. Rutigliano, and A. Virzo de Santo, "Suitability of soil microbial parameters as indicators of heavy metal pollution," Water, Air, & Soil Pollution, vol. 158, no. 1, pp. 21–35, 2004. · ·

22. S. A. Salgare and C. Acharekar, "Effect of industrial pollution on growth and content of certain weeds,"Journal for Nature Conservation, vol. 4, pp. 1–6, 1992.

23. D. T. Clarkson and U. Luttge, "Mineral nutrition: divalent cations, transport and compartmentation,"Progress in Botany, vol. 51, pp. 93–112, 1989.

24. Y. Luo and D. L. Rimmer, "Zinc-copper interaction affecting plant growth on a metal-contaminated soil,"Environmental Pollution, vol. 88, no. 1, pp. 79–83, 1995. · ·

25. M. J. Abedin, J. Feldmann, and A. A. Meharg, "Uptake kinetics of arsenic species in rice plants," Plant Physiology, vol. 128, no. 3, pp. 1120–1128, 2002.

26. M. J. Blaylock and J. W. Huang, "Phytoextraction of metals," in Phytoremediation of Toxic Metals: Using Plants to Clean up the Environment, I. Raskin and B. D. Ensley, Eds., pp. 53–70, Wiley, New York, NY, USA, 2000.

27. R. Djingova and I. Kuleff, "Instrumental techniques for trace analysis," in Trace Elements: Their Distribution and Effects in the Environment, J. P. Vernet, Ed., Elsevier, London, UK, 2000.

28. F. Assche and H. Clijsters, "Effects of metals on enzyme activity in plants," Plant, Cell and Environment, vol. 24, pp. 1–15, 1990.

29. C. D. Jadia and M. H. Fulekar, "Phytoremediation of heavy metals: recent techniques," African Journal of Biotechnology, vol. 8, no. 6, pp. 921–928, 2009. ·

30. L. Taiz and E. Zeiger, Plant Physiology, Sinauer Associates, Sunderland, Mass, USA, 2002.

31. A. Schaller and T. Diez, "Plant specific aspects of heavy metal uptake and comparison with quality standards for food and forage crops," in Der Einfluß von festen Abfällen auf Böden, Pflanzen, D. Sauerbeck and S. Lübben, Eds., pp. 92–125, KFA, Jülich, Germany, 1991, (German).

32. M. G. Kibra, "Effects of mercury on some growth parameters of rice (Oryza sativa L.)," Soil & Environment, vol. 27, no. 1, pp. 23–28, 2008.

33. I. Ahmad, M. J. Akhtar, Z. A. Zahir, and A. Jamil, "Effect of cadmium on seed germination and seedling growth of four wheat (Triticum aestivum L.) cultivars," Pakistan Journal of Botany, vol. 44, no. 5, pp. 1569–1574, 2012. ·

34. A. Kabata-Pendias, Trace Elements in Soils and Plants, CRC Press, Boca Raton, Fla, USA, 3rd edition, 2001.

35. A. R. Marin, S. R. Pezeshki, P. H. Masscheleyn, and H. S. Choi, "Effect of dimethylarsinic acid (DMAA) on growth, tissue arsenic and photosynthesis of rice plants," Journal of Plant Nutrition, vol. 16, no. 5, pp. 865–880, 1993. ·

36. M. J. Abedin, J. Cotter-Howells, and A. A. Meharg, "Arsenic uptake and accumulation in rice (Oryza sativa L.) irrigated with contaminated water," Plant and Soil, vol. 240, no. 2, pp. 311–319, 2002.

37. A. C. Barrachina, F. B. Carbonell, and J. M. Beneyto, "Arsenic uptake, distribution, and accumulation in tomato plants: effect of arsenite on plant growth and yield," Journal of Plant Nutrition, vol. 18, no. 6, pp. 1237–1250, 1995.

38. M. S. Cox, P. F. Bell, and J. L. Kovar, "Differential tolerance of canola to arsenic when grown hydroponically or in soil," Journal of Plant Nutrition, vol. 19, no. 12, pp. 1599–1610, 1996.

39. M. S. Yourtchi and H. R. Bayat, "Effect of cadmium toxicity on growth, cadmium accumulation and macronutrient content of durum wheat (Dena CV.)," International Journal of Agriculture and Crop Sciences, vol. 6, no. 15, pp. 1099–1103, 2013.

40. W. Jiang, D. Liu, and W. Hou, "Hyperaccumulation of cadm ium by roots, bulbs and shoots of garlic,"Bioresource Technology, vol. 76, no. 1, pp. 9–13, 2001. ·

41. M. Wang, J. Zou, X. Duan, W. Jiang, and D. Liu, "Cadmium accumulation and its effects on metal uptake in maize (Zea mays L.)," Bioresource Technology, vol. 98, no. 1, pp. 82–88, 2007.

42. K. Jayakumar, M. Rajesh, L. Baskaran, and P. Vijayarengan, "Changes in nutritional metabolism of tomato (Lycopersicon esculantum Mill.) plants exposed to increasing concentration of cobalt chloride,"International Journal of Food Nutrition and Safety, vol. 4, no. 2, pp. 62–69, 2013.

43. K. Jayakumar, C. A. Jaleel, and M. M. Azooz, "Phytochemical changes in green gram (Vigna radiata) under cobalt stress," Global Journal of Molecular Sciences, vol. 3, no. 2, pp. 46–49, 2008.

44. K. Jayakumar, C. A. Jaleel, and P. Vijayarengan, "Changes in growth, biochemical constituents, and antioxidant potentials in radish (Raphanus sativus L.) under cobalt stress," Turkish Journal of Biology, vol. 31, no. 3, pp. 127–136, 2007. ·

45. D. C. Sharma and C. P. Sharma, "Chromium uptake and its effects on growth and biological yield of wheat," Cereal Research Communications, vol. 21, no. 4, pp. 317–322, 1993. ·

46. S. K. Panda and H. K. Patra, "Nitrate and ammonium ions effect on the chromium toxicity in developing wheat seedlings," Proceedings of the National Academy of Sciences, India, vol. 70, pp. 75–80, 2000.

47. R. Moral, J. Navarro Pedreno, I. Gomez, and J. Mataix, "Effects of chromium on the nutrient element content and morphology of tomato," Journal of Plant Nutrition, vol. 18, no. 4, pp. 815–822, 1995.

48. R. Moral, I. Gomez, J. N. Pedreno, and J. Mataix, "Absorption of Cr and effects on micronutrient content in tomato plant (Lycopersicum esculentum M.)," Agrochimica, vol. 40, no. 2-3, pp. 132–138, 1996. ·

49. N. Nematshahi, M. Lahouti, and A. Ganjeali, "Accumulation of chromium and its effect on growth of (Allium cepa cv. Hybrid)," European Journal of Experimental Biology, vol. 2, no. 4, pp. 969–974, 2012.

50. C. M. Cook, A. Kostidou, E. Vardaka, and T. Lanaras, "Effects of copper on the growth, photosynthesis and nutrient concentrations of Phaseolus plants," Photosynthetica, vol. 34, no. 2, pp. 179–193, 1997.

51. C. Kjær and N. Elmegaard, "Effects of copper sulfate on black bindweed (Polygonum convolvulus L.),"Ecotoxicology and Environmental Safety, vol. 33, no. 2, pp. 110–117, 1996.

52. A. R. Sheldon and N. W. Menzies, "The effect of copper toxicity on the growth and root morphology of Rhodes grass (Chloris gayana Knuth.) in resin buffered solution culture," Plant and Soil, vol. 278, no. 1-2, pp. 341–349, 2005.

53. X. Du, Y.-G. Zhu, W.-J. Liu, and X.-S. Zhao, "Uptake of mercury (Hg) by seedlings of rice (Oryza sativaL.) grown in solution culture and interactions with arsenate uptake," Environmental and Experimental Botany, vol. 54, no. 1, pp. 1–7, 2005.

54. C. H. C. Shekar, D. Sammaiah, T. Shasthree, and K. J. Reddy, "Effect of mercury on tomato growth and yield attributes," International Journal of Pharma and Bio Sciences, vol. 2, no. 2, pp. B358–B364, 2011. ·

55. S. K. Arya and B. K. Roy, "Manganese induced changes in growth, chlorophyll content and antioxidants activity in seedlings of broad bean (Vicia faba L.)," Journal of Environmental Biology, vol. 32, no. 6, pp. 707–711, 2011. ·

56. Z. Asrar, R. A. Khavari-Nejad, and H. Heidari, "Excess manganese effects on pigments of Mentha spicataat flowering stage," Archives of Agronomy and Soil Science, vol. 51, no. 1, pp. 101–107, 2005.

57. S. Doncheva, K. Georgieva, V. Vassileva, Z. Stoyanova, N. Popov, and G. Ignatov, "Effects of succinate on manganese toxicity in pea plants," Journal of Plant Nutrition, vol. 28, no. 1, pp. 47–62, 2005.

58. M. Shenker, O. E. Plessner, and E. Tel-Or, "Manganese nutrition effects on tomato growth, chlorophyll concentration, and superoxide dismutase activity," Journal of Plant Physiology, vol. 161, no. 2, pp. 197–202, 2004.

59. I. S. Sheoran, H. R. Singal, and R. Singh, "Effect of cadmium and nickel on photosynthesis and the enzymes of the photosynthetic carbon reduction cycle in pigeonpea (Cajanus cajan L.)," Photosynthesis Research, vol. 23, no. 3, pp. 345–351, 1990.

60. B. Y. Khalid and J. Tinsley, "Some effects of nickel toxicity on rye grass," Plant and Soil, vol. 55, no. 1, pp. 139–144, 1980.

61. T. Pandolfini, R. Gabbrielli, and C. Comparini, "Nickel toxicity and peroxidase activity in seedlings ofTriticum aestivum L.," Plant, Cell and Environment, vol. 15, no. 6, pp. 719–725, 1992. · ·

62. V. S. Barsukova and O. I. Gamzikova, "Effects of nickel surplus on the element content in wheat varieties contrasting in Ni resistance," Agrokhimiya, vol. 1, pp. 80–85, 1999.

63. Y.-C. Lin and C.-H. Kao, "Nickel toxicity of rice seedlings: Cell wall peroxidase, lignin, and $NiSO_4$-inhibited root growth," Crop, Environment Bioinformatics, vol. 2, pp. 131–136, 2005.

64. A. Hussain, N. Abbas, F. Arshad et al., "Effects of diverse doses of lead (Pb) on different growth attributes of Zea mays L.," Agricultural Sciences, vol. 4, no. 5, pp. 262–265, 2013.

65. M. Kabir, M. Z. Iqbal, and M. Shafiq, "Effects of lead on seedling growth of Thespesia populnea L.,"Advances in Environmental Biology, vol. 3, no. 2, pp. 184–190, 2009. ·

66. M. Moustakas, T. Lanaras, L. Symeonidis, and S. Karataglis, "Growth and some photosynthetic characteristics of field grown Avena sativa under copper and lead stress," Photosynthetica, vol. 30, no. 3, pp. 389–396, 1994. ·

67. R. Manivasagaperumal, S. Balamurugan, G. Thiyagarajan, and J. Sekar, "Effect of zinc on germination, seedling growth and biochemical content of cluster bean (Cyamopsis tetragonoloba (L.) Taub)," Current Botany, vol. 2, no. 5, pp. 11–15, 2011.

68. S. Doncheva, Z. Stoynova, and V. Velikova, "Influence of succinate on zinc toxicity of pea plants," Journal of Plant Nutrition, vol. 24, no. 6, pp. 789–804, 2001.

69. M. Bonnet, O. Camares, and P. Veisseire, "Effects of zinc and influence of Acremonium lolii on growth parameters, chlorophyll a fluorescence and antioxidant enzyme activities of ryegrass (Lolium perenne L. cv Apollo)," Journal of Experimental Botany, vol. 51, no. 346, pp. 945–953, 2000.

70. A. M. Nicholls and T. K. Mal, "Effects of lead and copper exposure on growth of an invasive weed,Lythrum salicaria L. (Purple Loosestrife)," Ohio Journal of Science, vol. 103, no. 5, pp. 129–133, 2003. ·

71. A. Ghani, "Toxic effects of heavy metals on plant growth and metal accumulation in maize (Zea maysL.)," Iranian Journal of Toxicology, vol. 3, no. 3, pp. 325–334, 2010.

72. A. J. M. Baker, "Accumulators and excluders strategies in the response of plants to heavy metals," Journal of

Plant Nutrition, vol. 3, pp. 643–654, 1981.

73. M. J. Blaylock, D. E. Salt, S. Dushenkov et al., "Enhanced accumulation of Pb in Indian mustard by soil-applied chelating agents," Environmental Science and Technology, vol. 31, no. 3, pp. 860–865, 1997.

74. M. E. V. Schmoger, M. Oven, and E. Grill, "Detoxification of arsenic by phytochelatins in plants," Plant Physiology, vol. 122, no. 3, pp. 793–801, 2000.

75. C. Garbisu and I. Alkorta, "Bioremediation: principles and future," Journal of Clean Technology, Environmental Toxicology and Occupational Medicine, vol. 6, no. 4, pp. 351–366, 1997. ·

76. C. Garbisu and I. Alkorta, "Basic concepts on heavy metal soil bioremediation," The European Journal of Mineral Processing and Environmental Protection, vol. 3, no. 1, pp. 58–66, 2003.

77. P. Wang, T. Mori, K. Komori, M. Sasatsu, K. Toda, and H. Ohtake, "Isolation and characterization of anEnterobacter cloacae strain that reduces hexavalent chromium under anaerobic conditions," Applied and Environmental Microbiology, vol. 55, no. 7, pp. 1665–1669, 1989. ·

78. Y. Ishibashi, C. Cervantes, and S. Silver, "Chromium reduction in Pseudomonas putida," Applied and Environmental Microbiology, vol. 56, no. 7, pp. 2268–2270, 1990. ·

79. C. Garbisu, M. J. Llama, and J. L. Serra, "Effect of heavy metals on chromate reduction by Bacillus subtilis," Journal of General and Applied Microbiology, vol. 43, no. 6, pp. 369–371, 1997. · ·

80. C. Garbisu, I. Alkorta, M. J. Llama, and J. L. Serra, "Aerobic chromate reduction by Bacillus subtilis,"Biodegradation, vol. 9, no. 2, pp. 133–141, 1998.

81. C. Garbisu, S. González, W.-H. Yang et al., "Physiological mechanisms regulating the conversion of selenite to elemental selenium by Bacillus subtilis," BioFactors, vol. 5, no. 1, pp. 29–37, 1995. ·

82. R. Ajaz Haja Mohideena, V. Thirumalai Arasuc, K. R. Narayananb, and M. I. Zahir Hussaind, "Bioremediation of heavy metal contaminated soil by the exigobacterium and accumulation of Cd, Ni, Zn and Cu from soil environment," International Journal of Biological Technology, vol. 1, no. 2, pp. 94–101, 2010.

83. D. van der Lelie, P. Corbisier, L. Diels et al., "The role of bacte ria in the phytoremediation of heavy metals," in Phytoremediation of Contaminated Soil and Water, N. Terry and E. Banuelos, Eds., pp. 265–281, G Lewis, Boca Raton, Fla, USA, 1999.

84. M. Huyer and W. J. Page, "Zn^{2+} increases siderophore production in Azotobacter vinelandii," Applied and Environmental Microbiology, vol. 54, no. 11, pp. 2625–2631, 1988. ·

85. C. White, A. K. Sharman, and G. M. Gadd, "An integrated microbial process for the bioremediation of soil contaminated with toxic metals," Nature Biotechnology, vol. 16, no. 6, pp. 572–575, 1998.

86. J. L. Hobman and N. L. Brown, "bacterial mercury-resistance genes," Metal ions in biological systems, vol. 34, pp. 527–568, 1997. ·

87. D. R. Lovley and J. R. Lloyd, "Microbes with a mettle for bioremediation," Nature Biotechnology, vol. 18, no. 6, pp. 600–601, 2000.

88. M. Valls, S. Atrian, V. de Lorenzo, and L. A. Fernández, "Engineering a mouse metallothionein on the cell surface of Ralstonia eutropha CH34 for immobilization of heavy metals in soil," Nature Biotechnology, vol. 18, no. 6, pp. 661–665, 2000.

89. M. Urgun-Demirtas, B. Stark, and K. Pagilla, "Use of genetically engineered microorganisms (GEMs) for the bioremediation of contaminants," Critical Reviews in Biotechnology, vol. 26, no. 3, pp. 145–164, 2006. · ·

90. O. P. Abioye, "Biological remediation of hydrocarbon and heavy metals contaminated soil," in Soil Contamination, S. Pascucci, Ed., InTech, Vienna, Austria, 2011. ·

91. A. McCauley, C. Jones, and J. Jacobsen, "Soil pH and organic matter," in Nutrient Management Module, vol. 8, Montana State University Extension, Bozeman, Mont, USA, 2009.

92. A. Karaca, "Effect of organic wastes on the extractability of cadmium, copper, nickel, and zinc in soil,"Geoderma, vol. 122, no. 2–4, pp. 297–303, 2004.

93. T. Namgay, B. Singh, and B. P. Singh, "Influence of biochar application to soil on the availability of As, Cd, Cu, Pb, and Zn to maize (Zea mays L.)," Soil Research, vol. 48, no. 6-7, pp. 638–647, 2010.

94. J. M. Novak, W. J. Busscher, D. L. Laird, M. Ahmedna, D. W. Watts, and M. A. S. Niandou, "Impact of biochar amendment on fertility of a southeastern coastal plain soil," Soil Science, vol. 174, no. 2, pp. 105–112, 2009.

95. D. E. Salt, R. D. Smith, and I. Raskin, "Phytoremediation," Annual Review of Plant Biology, vol. 49, pp. 643–668, 1998. ·

96. R. L. Chaney, M. Malik, Y. M. Li et al., "Phytoremediation of soil metals," Current Opinion in Biotechnology, vol. 8, no. 3, pp. 279–284, 1997.

97. A. J. M. Baker and R. R. Brooks, "Terrestrial higher plants which hyperaccumulate metallic elements: a review of their distribution, ecology and phytochemistry," Biorecovery, vol. 1, pp. 81–126, 1989.

98. S. P. McGrath and F. Zhao, "Phytoextraction of metals and metalloids from contaminated soils," Current Opinion in Biotechnology, vol. 14, no. 3, pp. 277–282, 2003. · ·

99. R. D. Reeves and A. J. M. Baker, "Metal-accumulating plants," in Phytoremediation of Toxic Metals: Using Plants to Clean Up the Environment, I. Raskin and B. D. Ensley, Eds., pp. 193–229, Wiley, New York, NY, USA, 2000.

100. L. Q. Ma, K. M. Komar, C. Tu, W. Zhang, Y. Cai, and E. D. Kenelley, "A fern that hyperaccumulates arsenic—a hardy, versatile, fast-growing plant helps to remove arsenic from contaminated soils," Nature, vol. 409, p. 579, 2001.

101. X. E. Yang, X. X. Long, H. B. Ye, Z. L. He, D. V. Calvert, and P. J. Stoffella, "Cadmium tolerance and

hyperaccumulation in a new Zn-hyperaccumulating plant species (Sedum alfredii Hance)," Plant and Soil, vol. 259, no. 1-2, pp. 181–189, 2004.

102. F. Navari-Izzo and M. F. Quartacci, "Phytoremediation of metals," Minerva Biotecnologica, vol. 13, no. 2, pp. 73–83, 2001. ·

103. L. Van Ginneken, E. Meers, R. Guisson et al., "Phytoremediation for heavy metal-contaminated soils combined with bioenergy production," Journal of Environmental Engineering and Landscape Management, vol. 15, no. 4, pp. 227–236, 2007. ·

104. S. D. Ebbs and L. V. Kochian, "Toxicity of zinc and copper to Brassica species: implications for phytoremediation," Journal of Environmental Quality, vol. 26, no. 3, pp. 776–781, 1997. ·

105. R. L. Chaney, S. L. Brown, L. Yin-Ming et al., "Progress in risk assessment for soil metals, and in-situ remediation and phytoextraction of metals from hazardous contaminated soils," in Proceedings of the US EPA's Conference Phytoremediation: State of the Science Conference, Boston, Mass, USA, 2000.

106. Y. Chen, X. Li, and Z. Shen, "Leaching and uptake of heavy metals by ten different species of plants during an EDTA-assisted phytoextraction process," Chemosphere, vol. 57, no. 3, pp. 187–196, 2004.

107. H. Lai and Z. Chen, "The EDTA effect on phytoextraction of single and combined metals-contaminated soils using rainbow pink (Dianthus chinensis)," Chemosphere, vol. 60, no. 8, pp. 1062–1071, 2005.

108. S. C. Wu, K. C. Cheung, Y. M. Luo, and M. H. Wong, "Effects of inoculation of plant growth-promoting rhizobacteria on metal uptake by Brassica juncea," Environmental Pollution, vol. 140, no. 1, pp. 124–135, 2006.

109. K. K. Chiu, Z. H. Ye, and M. H. Wong, "Growth of Vetiveria zizanioides and Phragmities australis on Pb/Zn and Cu mine tailings amended with manure compost and sewage sludge: a greenhouse study,"Bioresource Technology, vol. 97, no. 1, pp. 158–170, 2006. · ·

110. E. Lombi, F. J. Zhao, S. J. Dunham, and S. P. McGrath, "Phytoremediation of heavy metal-contaminated soils: Natural hyperaccumulation versus chemically enhanced phytoextraction," Journal of Environmental Quality, vol. 30, no. 6, pp. 1919–1926, 2001.

111. C. D. Jadia and M. H. Fulekar, "Phytotoxicity and remediation of heavy metals by fibrous root grass (sorghum)," Journal of Applied Biosciences, vol. 10, pp. 491–499, 2008.

112. V. Laperche, S. J. Traina, P. Gaddam, and T. J. Logan, "Effect of apatite amendments on plant uptake of lead from contaminated sail," Environmental Science and Technology, vol. 30, no. 10, pp. 1540–1552, 1997.

113. A. P. G. C. Marques, R. S. Oliveira, A. O. S. S. Rangel, and P. M. L. Castro, "Application of manure and compost to contaminated soils and its effect on zinc accumulation by Solanum nigrum inoculated with arbuscular mycorrhizal fungi," Environmental Pollution, vol. 151, no. 3, pp. 608–620, 2008.

114. D. C. Adriano, W. W. Wenzel, J. Vangronsveld, and N. S. Bolan, "Role of assisted natural remediation in environmental cleanup," Geoderma, vol. 122, no. 2–4, pp. 121–142, 2004.

115. United States Environmental Protection Agency, Electrokinetic and Phytoremediation In Situ Treatment of Metal-Contaminated Soil: State-of-the-Practice, EPA/542/R-00/XXX, Environmental Protection Agency, Office of Solid Waste and Emergency Response Technology Innovation Office, Washington, DC, USA, 2000.

116. I. Raskin and B. D. Ensley, Phytoremediation of Toxic Metals: Using Plants to Clean Up the Environment, John Wiley & Sons, New York, NY, USA, 2000.

117. C. L. Rugh, J. F. Senecoff, R. B. Meagher, and S. A. Merkle, "Development of transgenic yellow poplar for mercury phytoremediation," Nature Biotechnology, vol. 16, no. 10, pp. 925–928, 1998. · ·

118. R. B. Meagher, C. L. Rugh, M. K. Kandasamy, G. Gragson, and N. J. Wang, "Engineered phytoremediation of mercury pollution in soil and water using bacterial genes," in Phytoremediation of Contaminated Soil and Water, N. Terry and G. Bañuelos, Eds., pp. 201–219, Lewis Publishers, Boca Raton, Fla, USA, 2000.

119. United States Environmental Protection Agency (USEPA), "Introduction to phytoremediation," EPA600/R-99/107, U.S. Environmental Protection Agency, Office of Research and Development, Cincinnati, Ohio, USA, 2000.

120. R. B. Meagher, "Phytoremediation: An Affordable, Friendly Technology to Restore Marginal Lands in the Twenty-First Century," 1998, http://www.lsc.psu.edu/nas/Panelists/Meagher%20comment.html.

121. N. Terry, A. M. Zayed, M. P. de Souza, and A. S. Tarun, "Selenium in higher plants," Annual Review of Plant Biology, vol. 51, pp. 401–432, 2000. ·

122. G. S. Bañuelos, H. A. Ajwa, B. Mackey et al., "Evaluation of different plant species used for phytoremediation of high soil selenium," Journal of Environmental Quality, vol. 26, no. 3, pp. 639–646, 1997. ·

123. N. Weyens, D. van der Lelie, S. Taghavi, L. Newman, and J. Vangronsveld, "Exploiting plant-microbe partnerships to improve biomass production and remediation," Trends in Biotechnology, vol. 27, no. 10, pp. 591–598, 2009.

124. E. J. Joner and C. Leyval, "Time-course of heavy metal uptake in maize and clover as affected by root density and different mycorrhizal inoculation regimes," Biology and Fertility of Soils, vol. 33, no. 5, pp. 351–357, 2001.

125. A. Jamal, N. Ayub, M. Usman, and A. G. Khan, "Arbuscular mycorrhizal fungi enhance zinc and nickel uptake from contaminated soil by soybean and lentil," International Journal of Phytoremediation, vol. 4, no. 3, pp. 205–221, 2002.

126. A. P. G. C. Marques, R. S. Oliveira, A. O. S. S. Rangel, and P. M. L. Castro, "Zinc accumulation inSolanum nigrum is enhanced by different arbuscular mycorrhizal fungi," Chemosphere, vol. 65, no. 7, pp. 1256–1263, 2006.

127. A. Heggo, J. S. Angle, and R. L. Chaney, "Effects of vesicular-arbuscular mycorrhizal fungi on heavy metal uptake by soybeans," Soil Biology & Biochemistry, vol. 22, no. 6, pp. 865–869, 1990.

128. M. Janoušková, D. Pavlíková, and M. Vosátka, "Potential contribution of arbuscular mycorrhiza to cadmium immobilisation in soil," Chemosphere, vol. 65, no. 11, pp. 1959–1965, 2006.

129. L. A. Harrier and C. A. Watson, "The potential role of arbuscular mycorrhizal (AM) fungi in the bioprotection of plants against soil-borne pathogens in organic and/or other sustainable farming systems," Pest Management Science, vol. 60, no. 2, pp. 149–157, 2004.

130. I. M. Cardoso and T. W. Kuyper, "Mycorrhizas and tropical soil fertility," Agriculture, Ecosystems and Environment, vol. 116, no. 1-2, pp. 72–84, 2006.

131. S. F. Wright, V. S. Green, and M. A. Cavigelli, "Glomalin in aggregate size classes from three different farming systems," Soil & Tillage Research, vol. 94, no. 2, pp. 546–549, 2007.

132. G. U. Chibuike, "Use of mycorrhiza in soil remediation: a review," Scientific Research and Essays, vol. 8, no. 35, pp. 1679–1687, 2013.

133. G. Díaz, C. Azcón-Aguilar, and M. Honrubia, "Influence of arbuscular mycorrhizae on heavy metal (Zn and Pb) uptake and growth of Lygeum spartum and Anthyllis cytisoides," Plant and Soil, vol. 180, no. 2, pp. 241–249, 1996.

134. E. J. Joner and C. Leyval, "Uptake of ^{109}Cd by roots and hyphae of a Glomus mosseae/Trifolium subterraneum mycorrhiza from soil amended with high and low concentrations of cadmium," New Phytologist, vol. 135, no. 2, pp. 353–360, 1997.

135. C. C. Chao and Y. P. Wang, "Effects of heavy-metals on the infection of vesicular arbuscular mycorrhizae and the growth of maize," Journal of the Agricultural Association of China, vol. 152, pp. 34–45, 1990.

136. C. Del Val, J. M. Barea, and C. Azcón-Aguilar, "Diversity of arbuscular mycorrhizal fungus populations in heavy-metal- contaminated soils," Applied and Environmental Microbiology, vol. 65, no. 2, pp. 718–723, 1999.

137. I. Weissenhorn and C. Leyval, "Spore germination of arbuscular mycorrhizal fungi in soils differing in heavy metal content and other parameters," European Journal of Soil Biology, vol. 32, no. 4, pp. 165–172, 1996.

138. B. R. Glick, D. M. Karaturovic, and P. C. Newell, "A novel procedure for rapid isolation of plant growth promoting pseudomonads," Canadian Journal of Microbiology, vol. 41, no. 6, pp. 533–536, 1995.

139. A. A. Kamnev and D. van der Lelie, "Chemical and biological parameters as tools to evaluate and improve heavy metal phytoremediation," Bioscience Reports, vol. 20, no. 4, pp. 239–258, 2000.

140. A. G. Khan, "Role of soil microbes in the rhizospheres of plants growing on trace metal contaminated soils in phytoremediation," Journal of Trace Elements in Medicine and Biology, vol. 18, no. 4, pp. 355–364, 2005.

141. B. R. Glick, D. M. Penrose, and J. Li, "A model for the lowering of plant ethylene concentrations by plant growth-promoting bacteria," Journal of Theoretical Biology, vol. 190, no. 1, pp. 63–68, 1998.

142. M. L. E. Reed and B. R. Glick, "Growth of canola (Brassica napus) in the presence of plant growth-promoting bacteria and either copper or polycyclic aromatic hydrocarbons," Canadian Journal of Microbiology, vol. 51, no. 12, pp. 1061–1069, 2005.

143. X. Sheng and J. Xia, "Improvement of rape (Brassica napus) plant growth and cadmium uptake by cadmium-resistant bacteria," Chemosphere, vol. 64, no. 6, pp. 1036–1042, 2006.

144. S. Zaidi, S. Usmani, B. R. Singh, and J. Musarrat, "Significance of Bacillus subtilis strain SJ-101 as a bioinoculant for concurrent plant growth promotion and nickel accumulation in Brassica juncea,"Chemosphere, vol. 64, no. 6, pp. 991–997, 2006.

145. M. Madhaiyan, S. Poonguzhali, and S. A. Torgmin, "Metal tolerating methylotrophic bacteria reduces nickel and cadmium toxicity and promotes plant growth of tomato (Lycopersicon esculentum L.),"Chemosphere, vol. 69, no. 2, pp. 220–228, 2007.

146. A. Vivas, B. Biró, J. M. Ruíz-Lozano, J. M. Barea, and R. Azcón, "Two bacterial strains isolated from a Zn-polluted soil enhance plant growth and mycorrhizal efficiency under Zn-toxicity," Chemosphere, vol. 62, no. 9, pp. 1523–1533, 2006.

CHAPTER 7

CELL WALL METABOLISM IN RESPONSE TO ABIOTIC STRESS

Hyacinthe Le Gall, Florian Philippe, Jean-Marc Domon, Françoise Gillet, Jérôme Pelloux and Catherine Rayon

Keywords: Plant cell wall; abiotic stress; water deficit; temperature; salt; flooding; heavy metals; light; air pollutants

INTRODUCTION

During their entire life cycle, plants are exposed to various environmental stresses. Two major categories can be distinguished: abiotic stress, which encompasses a variety of unfavorable environmental conditions, such as drought, submergence, salinity, heavy metal contamination and nutrient deficiency, and biotic stress caused by infectious living organisms, such as bacteria, viruses, fungi, or nematodes. Both types of stress negatively affect the productivity and survival of plants [1]. Abiotic stress reduces agricultural yield, so that novel crop genotypes adapted to environmental stress need to be developed [2,3,4]. Restriction of plant growth by retarding cell extension is often the earliest visible effect of stress. Plants have developed a range of morphological, physiological and biochemical mechanisms that enable them to avoid and/or tolerate stress factors and survive [5]. Reactive oxygen species (ROS) homeostasis is changed in response to stress; they are regarded as molecules causing damage to cells at high concentration as well as ubiquitous signaling molecules at low concentration, thus participating in recognizing and responding to stress factors [6]. Transduction of the signal modifies gene and protein expression levels leading to physiological responses. ROS are produced in most plant subcellular locations including the plasma membrane, mitochondria, nucleus, chloroplast and cell wall [7].

The plant cell wall determines cell size and shape through the mechanical control of cell expansion, which governs tissue and organ morphology. It can be considered a protective barrier and is a complex structure composed of cellulose microfibrils and non-cellulosic neutral polysaccharides embedded in a physiologically active pectin matrix, cross-linked with structural proteins and, depending on the tissue/organ, with lignin [8,9]. Plant cell walls are divided into a primary wall, which is initiated during cell division and deposited during cell elongation, and a secondary wall, which is deposited within the primary wall of specific cell types with specialized functions. In parallel, depending on the presence/absence of some particular polysaccharides, type I and type II cell walls can be distinguished. Type I primary walls are a framework of cellulose microfibrils, mostly cross-linked with xyloglucans (XG) and embedded in a complex matrix

Citation: Hyacinthe Le Gall, Florian Philippe, Jean-Marc Domon, Françoise Gillet, Jérôme Pelloux and Catherine Rayon, Cell Wall Metabolism in Response to Abiotic Stress doi:10.3390/plants4010112.

of pectic polysaccharides [8,9]. Unlike type I, type II primary cell walls contain small amounts of pectic polysaccharides and xyloglucans, whereas glucuronoarabinoxylans (GAX) and mixed-linkage (1→3),(1→4)-β-D-glucan (β-glucan) are the main matrix polymers that interlace cellulose microfibrils [10]. Furthermore, ferulic and *p*-coumaric acid arabinosyl esters can cross-link GAX in type II primary walls, as well as in the secondary wall, to stiffen the matrix [11,12]. Cell wall-related proteins are believed to play a central role in modulating cell wall extensibility, which mediates cell enlargement and expansion. These proteins include xyloglucan endo-β-transglucosylases/hydrolases (XET/XTHs; GH16), endo-1,4-β-D-glucanase (EGase, GH9), and expansins (EXPA, EXPB, EXPLA, EXPLB; GH45) [13,14]. Other cell wall-modifying enzymes play a major role in controlling cell wall plasticity/rheology. These include pectin-modifying enzymes such as pectin methylesterase (PME; CE8), polygalacturonase (PG; GH28), pectin/pectate lyase-like (PLL; PL1) and pectin acetylesterase (PAE; CE13) [15].

The massive deposition of cellulose and hemicelluloses inside primary walls gives secondary walls their characteristic thickness [16]. During secondary cell wall formation, monolignols, precursors of lignin, are secreted into the cell wall space and randomly cross-linked through oxidative polymerization [17]. The cross-linking is dependent on the availability of ROS generated by laccases and cell wall peroxidases (PRX). This process reinforces the strength and rigidity of the cell walls and can be a key component of the plant response to environmental factors [17,18,19]. Cell wall architecture is important in plant resistance to abiotic stress and essential in stress sensing and signal transduction [20]. A summary of state of the art of recent reviews is presented in Table 1.

Table 1. List of references of recent reviews on the effects of abiotic stress on cell wall metabolism. References to the papers are given in brackets.

Reference	Stress	Cell Wall	Data Studied
[21]	C, Fl, L, Wd	Primary cell wall	Transcriptomic and proteomic
[22]	Wd	Primary and secondary cell wall	Physiological, proteomic and cell wall composition
[23]	Ap, HM, L, Wd	Secondary cell wall	Transcriptomic, proteomic and metabolomic
[24]	Wd, S, C, Ap, HM, L	Secondary cell wall	Transcriptomic, proteomic and cell wall composition
[25]	S	Primary cell wall	Transcriptomic and proteomic
[26]	S	Primary and secondary cell wall	Proteomic
[27]	HM	Primary cell wall	Transcriptomic and proteomic
[28]	HM	Primary and secondary cell wall	Transcriptomic and proteomic

Ap: Air pollutants; C: Cold; Fl: Flooding; HM: Heavy Metal; L: Light; S: Salt stress; Wd: Water deficit.

The present review summarizes the most recent insights into the effects of eight abiotic stress factors (drought, flooding, heat, cold, salinity, metals, light irradiance, and air pollutants) on primary and secondary cell wall metabolism. Although callose synthesis and deposition have been observed in response to described abiotic stress factors, this polymer is usually not present in the walls of most cell types and is thus beyond the scope of the present review (For reviews: [29,30]). Although the cuticle composed of cutin and waxes, interacts with cell wall, this review will not extend on the effect of abiotic stress on plant cuticle, and will only focus on cell wall components involved in cell wall architecture [31].

WATER DEFICIT

Plant growth requires concerted water uptake and irreversible cell wall expansion to enlarge cells [32]. Drought or dehydration results in water deficit stress. Water deficit is defined as an imbalance between soil water availability and evaporative demand [33]. It is one of the major environmental stresses that reduces plant growth and productivity [34]. It is generally agreed that drought stress generates physiological changes in higher plants, including loss of turgor, osmotic adjustment, and reduced leaf water potential (for reviews: [35,36]). Turgor pressure is a critical factor in regulating cell growth. It is the physical force needed to drive cell enlargement, which notably depends upon the extensibility of the cell wall [37,38]. A low turgor pressure caused by water stress leads to a reduction or cessation of growth by decreasing cell extensibility and cell expansion [39]. Plants exposed to water deficit display morphological changes that are the result of plant cell wall modifications. In Black spruce (*Picea mariana*) saplings, drought stress induced a decrease in cell wall thickness, reflecting a lower carbon allocation to cell wall formation and preventing the adaptation of the hydraulic system to drought [40]. Cell expansion can only occur when turgor pressure is greater than the cell wall yield threshold [33,35]. In winter triticale, water deficit induced leaf rolling, which was correlated with a higher level of cell wall-bound phenolics [41]. This adaptive mechanism can limit water loss by restricting the leaf transpiration surface [42,43].

In response to stress, an increase in cell wall elasticity (CWE) can contribute to the maintenance of cell turgor or symplast volume [44,45,46]. CWE can be correlated with plant drought tolerance [46,47], as shown by a study performed on six cultivars of common bean, which highlighted that the cultivars most resistant to drought showed a strong decrease in their elasticity modulus (ε) in association with a higher CWE; this ultimately enabled an osmotic adjustment [47]. In contrast, the highly drought-sensitive cultivars did not show any significant change in their ε and CWE. These variations in ε and CWE in response to water stress presumably reflect differences in wall structure. However, these changes could depend on the species considered, as in *Ziziphus jujuba* and smooth cordgrass (*Spartina alterniflora*) a large increase in ε was observed, thus indicating that rigid cell walls in leaves may be necessary to maintain cell/tissue integrity during rehydration following a period of stress [37,48,49].

A number of studies report that cellulose biosynthesis can be altered in response to water deficit, as shown by the decreased level of cellulose content in several different species including Arabidopsis, tobacco suspension cells, grape leaves and wheat roots [50,51,52,53]. However, in other studies, an increased level of UDP-Glc in the expression of *SuSy* (sucrose synthase) and UDP-glucose pyrophosphorylase (*UGPase*) encoding genes was observed in cotton under drought stress, suggesting a potentially higher cellulose biosynthesis [54]. Increased cellulose synthesis could be a means by which cell wall integrity and cell turgor pressure are maintained, thus allowing continuous cell growth under low water potential [55]. Xyloglycan biosynthesis seems to follow a similar trend in the elongation zone of rice roots under water deficit [56]. In that study, *XTH* and xylose isomerase genes encoding two xyloglucan-modifying enzymes were up-regulated in the early stages following stress application, indicating a role of xyloglucan in the maintenance of root growth [56]. Similarly, the overexpression of an *XTH* gene from pepper in transgenic plants confirmed the role of XTH in enabling better drought tolerance [57,58]. This appears to be a common feature of distinct species, as transgenic Arabidopsis plants overexpressing a hot pepper (*Capsicum annuum*, cv. Pukang) *XTH* (*CaXTH3*) exhibited abnormal leaf morphology resulting in a severely wrinkled leaf shape, an increase in the number of small-sized cells in the leaf mesophyll cells and, ultimately, an increased tolerance to severe water deficit. This suggests that CaXTH3 may be involved in cell wall remodeling to strengthen the wall layers, and hence it could participate in the protection of mesophyll cells against water deficit [57]. In another study performed on transgenic tomato plants overexpressing *CaXTH3*, similar results were shown, highlighting the key role of XTH in water-deficit tolerance [58]. This can be further illustrated by a proteomic analysis in maize, which showed that XTH proteins were differentially regulated upon drought stress exposure [59]. In tomato, a drought responsive transcription factor ASR1 (Abscisic Stress Ripening) was identified as interacting with *Solyc08g082650.2.1*, which was annotated as a cellulose synthase-like (*CSL-G*) involved in hemicellulose biosynthesis [55]. Expansin, another cell wall protein involved in cell wall extension, appears to be strongly regulated by water stress deficiency [60]. In maize roots, expansin activity increased by 4-fold (apical region) and by 2-fold (basal region) under low water potential, which could be related to the higher expression of expansin encoding genes in the root apical zone, corresponding to maintaining elongation [61,62]. However, surprisingly, in the basal region of maize roots, the cell wall underwent a stiffening reaction rather than extension in response to expansin, thus indicating that loss of susceptibility to expansin was associated with growth arrest in this area [61,63]. In soybean, the expression of *GmEXPB2*, a gene that encodes a β-expansin (*EXPB*), was induced by mild water deficiency [64]. This gene, which is intrinsically involved in the response of the root system architecture to some abiotic stresses, including water deficiency, could be an interesting target for improving crop production under these conditions. In a similar way, the overexpression of the wheat β-expansin *EXPB23* under the control of a stress inducible promoter (RD29) in transgenic tobacco plant conferred a higher tolerance to drought stress [65]. The potential key role of expansin in drought tolerance can be further illustrated by studies showing that, in rose (*Rosa hybrida*, cv. Samantha), the virus-induced gene silencing of *RhNAC2*, a transcription factor, or *RhEXPA4*, decreased dehydration tolerance during cell expansion of rose petals [66]. The authors suggested that *RhEXPA4* expression might be regulated by *RhNAC2*. Overexpression of *RhEXPA4* in Arabidopsis transgenic plants conferred a strong phenotype with shorter stems, curly leaves, compact inflorescences and strong drought tolerance, indicating the key role of EXP in response to water deficit in the plant [67]. In wheat coleoptiles, PEG treatment, which can be used to modify osmotic potential and thus induce plant water deficit, limited the increase in the amounts of middle- and low-molecular-weight polysaccharides. In contrast, no effect was shown on high-molecular-weight polysaccharides, such as (1→3),(1→4)-β-D-glucan and arabinoxylan, despite growth arrest [68]. These results demonstrated that the cell wall maintained its ability to extensibility.

In parallel with observations on cellulose and hemicellulose, pectins have been shown to play a key role in modulating cell wall structure in response to drought stress. This fits well with the recognized importance of pectin in controlling plant growth and development (for reviews: [38,69]). In particular, the large amount of homogalacturonan (HG), the most abundant pectic polymer in type I cell walls, combined with a precise regulation of its degree of methylesterification, is likely to be a key element in the control of the stiffness and hydration status of the pectic matrix during drought stress. In addition, it was shown that the amount of side chains of RGI and/or RGII determined the hydration status of the cell wall matrix [70]. The comparison of two wheat cultivars, one tolerant and one sensitive to water stress deficiency, shed new light on the role of this polymer. In the tolerant cultivar, the amount of side chains was indeed increased during water stress, with consequent effects on the viscosity status of the cell wall [52,70]. This latter parameter could be an indicator of drought tolerance [52,70]. In some plants, known as resurrection plants, which can recover completely from a fully-dehydrated state, side chains of pectin are highly enriched in arabinose-rich polymers, including pectin-arabinan, arabinogalactan proteins and arabinoxylans [22,71,72]. The presence of arabinan-rich pectin would prevent water loss during desiccation. In parallel, pectin-degrading enzymes, including PG, can be down-regulated by water stress with consequences on cell wall integrity and cell expansion [53,73]. Alternatively, PG could be involved in controlling cellular water relationships [53,74]. In another study on date palm fruit, the level of methylesterification and *O*-acetylation of xyloglucan was reduced up to 75% and 38% respectively 232 days after pollination, indicating cell wall remodeling of fruits in a water deficit environment [71]. The overexpression of *OsBURP16*, which encodes a putative precursor of PG1β, one of the subunits that regulates PG activity in rice transgenic plants, led to an increase in PG activity and a decrease in pectin content [75,76]. Thus, an increase in the sensitivity of these transgenic rice plants to water stress deficit was shown. In these plants, the intercellular spaces of leaves were larger, suggesting that cell adhesion could be reduced. HG modifications are likely to be an important determinant of stress tolerance as transgenic Arabidopsis, overexpressing a pepper PME inhibitor protein (CaPMEI1), exhibited enhanced tolerance to drought stress during post-germination growth [77].

The secondary cell wall can be strengthened by the incorporation of lignin. Lignification is a complex process involving several different phenolic substrates and enzymes (for a review: [78]). It can occur prematurely to avoid cell wall damage when plants are exposed to a long water stress deficit [23]. Two recent reviews highlighted changes in lignin content and structure following biotic and abiotic stress ([23,24]). Phenylalanine ammonia-lyase (PAL) is a key intermediate at the crossroads of phenolics and lignin synthetic pathways. PAL, which catalyzes the deamination of phenylalanine to trans-cinnamate, was shown to be regulated upon stress exposure. In rice, a proteomic approach showed an increased level of PAL following drought stress [73]. In Grey-maranta (*Ctenanthe setosa*) exposed to long-term drought stress, PAL activity was increased up to 16-fold, with consequent effects on lignin content, ionically wall-bound peroxidase and polyphenol oxidase activities and leaf rolling [79]. The caffeoyl-CoA 3-*O*-methyl-transferase (*CCoAOMT*) encoding gene, which converts caffeoyl-CoA into feruloyl-CoA and thus plays a crucial role in the synthesis of monolignols, was up-regulated in roots of wild water melon subjected to drought [80]. The increase in cinnamoyl-CoA reductase (*CCR*) transcripts in maize roots was associated with a progressive inhibition of wall extensibility and root growth, and may facilitate root acclimation to drying environments [81]. In Eucalyptus, drought stress caused disruption of lignin deposition in leaves and increased the S/G unit ratio [82]. In roots of rice seedlings, the increase in peroxidase activity was correlated with cell wall stiffening, which could be involved in the regulation of root growth reduction caused by water deficit [83]. In white clover leaf, a low water potential enhanced the activation of guaiacol peroxidase and coniferyl alcohol peroxidase, which was closely correlated with an increase in lignin content [84]. Similarly, in the leaf elongation zone of darnel ryegrass (*Lolium temulentum* L.), a grass plant, an increase in cell wall peroxidase activity was associated with a decrease in cell expansion during drought stress, due to phenolic cross-linkage formation between cell wall components [85]. However, the association of free phenolic compounds with the cell wall can be an indicator of plant resistance to drought stress [41,81,86]. The introduction of phenolic compounds linked to the cell wall matrix could generate a stiffening of the cell wall, which might alter cell wall elongation after rehydration. The wall is thus more compact, tight, and less permeable to water. This could allow the plant to maintain leaf or root turgor under a low water potential due to the hydrophobic stabilizing properties of these molecules, thus preventing water loss to the apoplast [41,81,86]. In contrast to free phenolics, cross-linked phenolic compounds and lignin could trigger wall rigidification and growth arrest in the later stage of drought stress, which could lead to a loss of productivity, especially for crops [56]. A Diagram summarizing the plant cell wall response to water deficit is presented inFigure 1.

Although much progress has been made over recent years in understanding the consequences of water deficit on cell wall structure and dynamics, much remains to be investigated before targeting key cell wall polymers that could be useful in engineering drought-resistant crops.

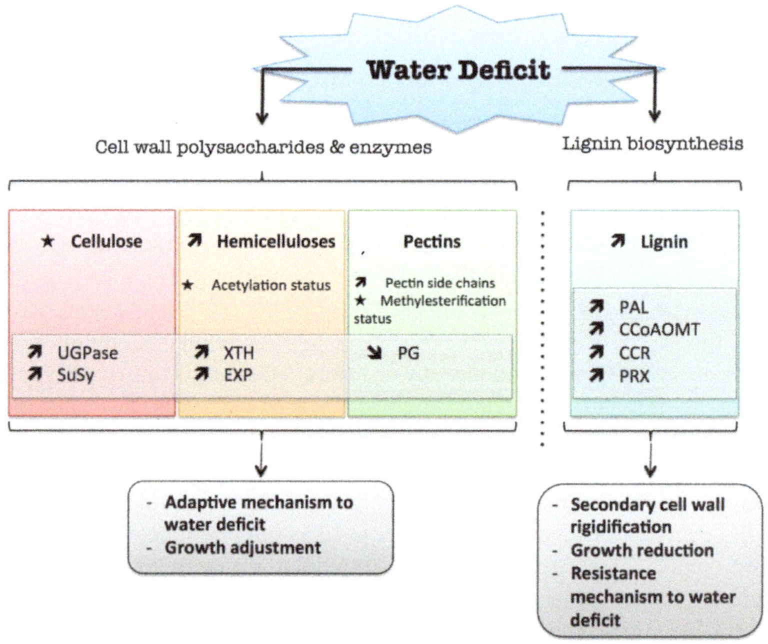

Figure 1. Diagram summarizing the plant cell wall response to water deficit. The schematic presentation is deduced from the results of different studies reported in the review. Arrow (↗) means increased abundance and arrow (↘) means decreased abundance of the molecules. Star (★) means contrasting data on the gene/protein or the molecule studied according to the literature cited in the review. Cell wall (CW); xyloglucan endo-β-transglucosylases/hydrolases (XET/XTH); expansin (EXP); sucrose synthase (SuSy); UDP-glucose pyrophosphorylase (UGPase); polygalacturonase (PG); pectin methylesterase (PME); phenylalanine ammonia-lyase (PAL); caffeoyl-CoA 3-*O*-methyl-transferase (CCoAOMT); cinnamoyl-CoA reductase (CCR); cell wall peroxidases (PRX).

FLOODING

The occurrence of floods, which is often associated with climate change, has increased over recent decades. Flooding stress in plants results in a lower oxygen level in the root zone due to this molecule's low diffusion rate in water [87]. In soybean seedlings and wheat roots exposed to flooding stress, various deleterious effects have been observed, such as suppression or reduction of hypocotyl and root elongation, and suppression of lateral root development. These effects are dependent upon the flooding intensity (water volume and duration) and the extent of recovery periods [88,89,90]. In Arabidopsis root, flooding affected the rapid hyponastic movement of the leaf, which was related to differential growth between the abaxial and adaxial leaf parts [91]. The hyponastic movement could be a locally restricted expansion of the cells in the proximal, abaxial region while adaxial cells remained largely unaffected. *EXP* and *XTH*, whose expression is modulated by SHYG (a NAC transcription factor that regulates 1-aminocyclopropane-1-carboxylic acid oxidase, a key enzyme in ethylene biosynthesis under waterlogging conditions) could contribute to cell wall remodeling, enabling differential growth [91]. Flooding alters cell wall polysaccharides. FTIR analysis performed on maize seedlings exposed to flooding showed a degradation of cell wall polysaccharides and a decrease in pectin content [92]. A similar trend was observed in azuki bean seedling (*Vigna angularis*), where a decrease in cell wall sugar content was observed, correlated with cell wall thinning [93]. Furthermore, the xyloglucan content had a lower molecular mass and the apoplastic pH increased, which might inhibit the activity of expansin [93]. Cell wall polysaccharides play an essential role in maintaining the seedling establishment of an Amazonian plant, *Himatanthus sucuuba*, in floodplains. In the flood zone, the seed endosperm of this tree is richer in mannan while species grown in non-flooded areas contain galactomannan and arabinogalactan, indicating that the presence of galactose on mannan in addition of pectin makes the seed tissues more water-permeable. Thus, the presence of pure mannan slowed down water uptake in seeds adapted to environments with plenty of water [94]. Proteomic analysis performed on soybean seedlings revealed that proteins involved in cell wall metabolism were highly regulated in response to flooding stress [90]. For instance, polygalacturonase inhibitor 2-like protein, expansin-like B1-like and rhamnogalacturonate lyase B-like proteins were flooding-responsive proteins common to all the organs studied (root tip, root without tip and hypocotyl). Polygalacturonase inhibitor 2-like protein and EXPLB1 protein levels were greatly increased, while the abundance of other cell wall-related proteins, including cellulose synthase-interactive protein 1 and CAD, decreased, which was dependent on the severity of the flooding. In wheat roots under flooding stress, similar effects to those observed with soybean have been reported. Most down-regulated proteins, such as β-glucanase, β-glucosidase, and β-galactosidase, are related to cell wall metabolism and structural modification [88]. It has been suggested that these down-regulated proteins might be correlated with the inhibition of cell wall elongation under flooding stress, which significantly suppressed the growth of wheat plants.

These alterations in protein abundance are regulated by transcriptional changes since RT-qPCR analysis showed that flooding treatment resulted in the down-regulation of genes encoding cellulose synthase in soybean roots; this could have consequences on cellulose synthesis [95]. Similar trends were observed in woody plants such as gray poplar where the expression of genes encoding cellulose synthases, required for secondary cell wall formation, was strongly down-regulated (up to 44-fold) in roots, while those involved in primary cell wall biosynthesis were less drastically affected (up 4.4-fold). These effects were dependent upon the period of flooding exposure. The primary wall would therefore be a key target for mediating cellular integrity in response to flooding [96]. In addition, other genes encoding proteins involved in cell wall degradation, including cellulase, β-(1,4) glucanase, pectate lyase and polygalacturonase, were down-regulated suggesting the reduced enzyme activity of these proteins could alter the cell wall structure with consequent effects on root growth, leading to growth arrest and suppression of lateral root formation [95]. Another transcriptome analysis performed on soybean seedlings, showed differential mRNA abundance of PGIP and expansin encoding genes. The discrepancy between the inhibition of cell elongation of soybean seedling under flooding and the relative increase in the abundance of EXPLB1 could indicate that flooding recruits proteins with distinct functions in the cell wall. To a certain extent, PGIP, expansin and rhamnoglacturonate lyase B could be flooding stress indicator proteins, which could trigger defense systems [90].

Flooding alters lignification as shown by the decrease in lignin deposition, assessed by phloroglucinol staining, being suppressed in roots and hypocotyls of soybean seedlings under flooding stress [97]. In parallel, proteomic analysis showed a down-regulation of an extracellular superoxide dismutase (Cu-Zn), which might play a role in the regulation of lignification under flooding. Germin-like protein with oxalate oxidase or superoxide dismutase activity produces H_2O_2, which could be used in cross-linking reactions within the cell wall. Four germin-like protein precursors were down-regulated during flooding, thus indicating a reduction in the linking of polysaccharides and proteins within the cell wall, as well as in lignification [97]. In addition, in poplar, the expression of genes encoding cell wall degradation enzymes, such as genes encoding lignin biosynthesis enzymes, including PAL, trans-cinnamate 4-hydroxylase (C4H), 4 coumarate CoA-ligase (4CL), ferulate 5-hydroxylase (F5H), COMT, and CCoAOMT, were repressed indicating the alteration of lignification biosynthesis under flooding.

These data are mostly derived from genomic or proteomic approaches and provide valuable knowledge about the role of the plant cell wall in response to oxygen deprivation (Figure 2). However, very few data are currently available at the biochemical level on cell wall polymers, cell wall enzyme activities and signaling processes that regulate this response.

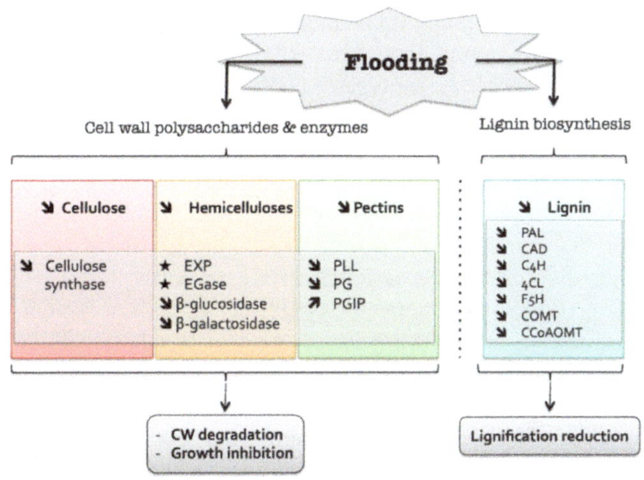

Figure 2. Overview of the plant cell wall response to flooding. The schematic presentation is deduced from the results of different studies reported in the review. Arrow (↗) means increased abundance and arrow (↘) means decreased abundance of the molecules. Star (★) means contrasting data on the gene/protein or the molecule studied according to the literature cited in the review. Cell wall (CW); expansin (EXP); endoglucanase (EGase); polygalacturonase (PG); polygalacturonase inhibitor protein (PGIP); pectin/pectate lyase-like (PLL); phenylalanine ammonia-lyase (PAL); hydroxycinnamyl alcohol dehydrogenase (CAD); trans-cinnamate 4-hydroxylase (C4H); 4 coumarate CoA-ligase (4CL); ferulate 5-hydroxylase (F5H); caffeate *O*-methyltransferase (COMT); caffeoyl-CoA 3-*O*-methyl-transferase (CCoAOMT).

HEAT

In the context of global warming, an increase in temperature is likely to be a key abiotic stress to which plants will have to adjust to survive [98]. Heat stress is often considered an increase in temperature beyond a threshold level for a period of time that induces irreversible damage to plant growth (*i.e.*, 10–15 °C above the basal temperature). The upper threshold temperature depends on the plant species and genotypes within the species (for a review: [98]). Heat stress is an important agricultural problem in many areas, which can lead to a drastic reduction in economic yield, especially for cereals [99,100]. The application of transitory or constantly high temperatures causes a range of morpho-anatomical, physiological and biochemical changes in plants, particularly observed in Arabidopsis, maize and wheat [101,102,103,104]. These affect plant growth and development but in a different way, depending on the developmental stage at which plants are exposed [99]. Various studies have demonstrated that roots are more sensitive to heat stress, suggesting that high soil temperature is more detrimental than high air temperature for whole-plant growth [105]. To maintain growth and productivity, specific tolerance mechanisms must be implemented by plants [106]. Although the cell wall is not the structure primarily altered under heat shock, some studies have revealed that it could be affected at various levels. Cell wall composition and sugar metabolism were compared between maize genotypes during corn expansion, as a target for genetic improvement during heat stress [107]. In sweet corn plants, sugar concentrations in both hemicellulose and cellulose fractions of cobs decreased under stress treatment. The hemicellulose fraction in the shank from the first ear also decreased. In contrast, in dent corn plants, there was no reduction in sugar content, except in the cellulose fraction. Overall, this suggested that synthesis of cell wall components was affected by heat stress and differed between genotypes. High temperature treatment enhanced growth of vegetative plant parts but reduced husk expansion, by suppressing cob extensibility through impairing hemicellulose and cellulose synthesis following a reduction in photosynthesis supply. Therefore, the reduction in grain yield after high temperature treatment was due to effects on sink rather than on source activities [107]. In coffee leaves, heat treatment (37 °C) generated changes in cell wall polymers, as illustrated by a ~40% increased level in leaf hemicelluloses and a 50% decreased level in pectin, with contrasting effects depending on the sugars considered [108]. For instance, in the pectic fraction, arabinose and galactose contents were increased by 40% and 44% respectively, whereas uronic acid and rhamnose decreased by 60% and 40% respectively. An increase in arabinose and galactose contents could be related to a larger amount of type II arabinogalactan, which could contribute to the rigidification of the cell wall by either oxidative cross-linking or water retention. In contrast, the decrease in rhamnose and uronic contents probably reflected changes in the structural organization of cell wall pectin. In the non-cellulosic fraction, the increased xylose content (~30%) was associated with a decrease in arabinose (~33%), thus indicating a lower substitution of arabinose in the xylose backbone of xylan under heat stress. Furthermore, the analysis of leaf apoplastic proteomes of *Coffea arabica* plants reflected these changes in cell wall metabolism in a changing environment related to the greenhouse effect [109]. In contrast, in wheat grain, which displays a primary type II cell wall, arabinoxylan content increased under heat stress [110]. Overall, this shows that the plant response might be dependent upon its cell wall type. Cell-wall related gene expression can be altered in response to heat stress as well. A transcriptomic study performed on Chinese cabbage (*Brassica rapa* L.) showed that some cell wall-related genes could play a role in the acquisition of thermotolerance [111]. Indeed, several genes, encoding proteins from the XTH family, were up-regulated following heat treatment. Other genes, encoding cell wall enzymes or proteins as diverse as arabinogalactan protein, β-glucosidase, cellulose synthase, expansin, extensin, glycosyl transferase, pectin esterase and xylosidase, were up-regulated up to 2–3 fold following their first exposure to permissive high temperature conditions (37 °C). This corresponds to acclimation to a non-lethal temperature to acquire thermotolerance, *i.e.*, the ability for the plant to survive an otherwise lethal high temperature. These results support the idea that thermotolerance acquisition partially requires cell wall-remodeling enzymes, which could be involved in the biosynthesis or *in muro* modification of major cell wall polymers [111]. The authors suggested that the functional significance of XTH during this period could correspond to establishing secondary cell wall deposition, after cell elongation has ceased. Following high temperature exposure, these proteins may have a role in wall strengthening, in addition to that classically observed for cell expansion, which might help plants to adapt to high temperature. XTH could be involved in the reinforcement of connections between the primary and secondary cell walls, thus increasing XG polymerization. In another study, performed on grapevine during fruit ripening, an increased level of *XET* transcripts was measured following heat stress [112]. This could be related to the adaptation of berry volume to temperature and the need for more flexible cell walls. As observed for water stress, the fine-tuning of the cellulose-hemicellulose network appears to be a key component of heat tolerance. In *Agrostis* grass species, expansin is one of the major genes that is up-regulated in response to heat stress at 40 °C [113]. Cell walls with increased levels of expansin may therefore be loosened, relaxed, and become more elastic, which could help maintain cellular functions during heat stress. The overexpression of a grass expansin gene, *PpEXP1* from *Poa pratensis*, in tobacco plants led to a better heat tolerance (35 °C), with less structural damage to cells. This confirms that expansin plays an important role in plant adaptation to heat stress [114]. However, it appears specific to species and/or expansin isoforms, as in *Brassica napus*seedlings subjected to heat stress, where the expression of *EXPA5* was down-regulated up to 10-fold compared to the control [115]. Similar results were observed in poplar (*Populus simonii)*, where thirteen genes encoding *EXP* were down-regulated [116]. Other cell wall-remodeling enzymes are regulated under heat stress at gene expression level, such as a PME gene,*PME35*, which was reduced nearly 10-fold in *Brassica napus* [115].

In addition to its effects on the primary cell wall, heat stress can mediate changes in secondary cell wall metabolism. Synthesis of lignin can be altered during the application of elevated temperature. For instance, in *Coffea* leaves, heat stress increased G and S monolignol contents while a decrease in H monolignol was observed, indicating a structural change in lignin [108]. A proteomic analysis on rice seedlings, growing under various high temperatures, highlighted a differential expression of proteins involved in lignin biosynthesis, including PRX and PAL, with consequent effects on the inhibition of lignification [117]. In contrast to these results, a gene encoding a laccase was up-regulated in Chinese cabbage under heat treatment [113]. This enzyme is suspected of being involved in lignin biosynthesis, based on its ability to oxidize lignin precursors. This result is in accordance with observations in strawberry where elevated temperature possibly regulated the lignin biosynthetic pathway through increased PRX activity [118]. Heat causes changes in cell wall metabolism and this is an important physiological mechanism of heat tolerance (Figure 3). However, much more work is needed to gain a clearer understanding of the cell wall at the physiological, genetic, and biochemical levels of plant heat tolerance.

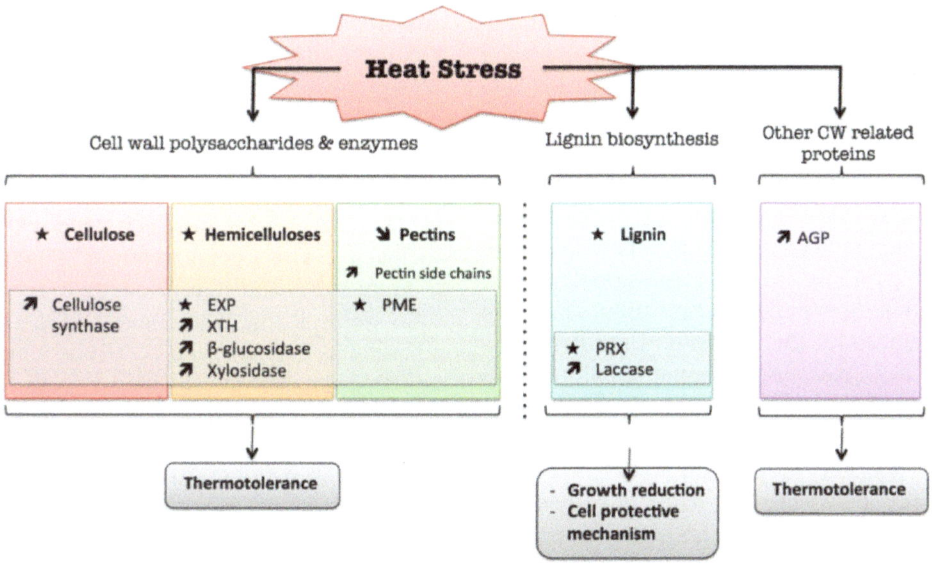

Figure 3. Diagram summarizing the plant cell wall response to heat. The schematic presentation is deduced from the results of different studies reported in the review. Arrow (↗) means increased abundance and arrow (↘) means decreased abundance of the molecules. Star (★) means contrasting data on the gene/protein or the molecule studied according to the literature cited in the review. Cell wall (CW); xyloglucan endo-β-transglucosylases/hydrolases (XET/XTH); expansin (EXP); pectin methylesterase (PME); cell wall peroxidases (PRX); arabinogalactan protein (AGP).

COLD

Cold stress, which includes both chilling and freezing injuries, modifies gene expression and plant metabolism with consequent effects on many biological functions [119,120,121,122]. The aerial parts of the plant are affected, including leaf expansion, wilting, chlorosis, and necrosis [123,124]. Root growth can also be altered under cold stress [125]. During drastic cold exposure, freezing can induce extracellular ice formation, which leads to cell dehydration and may result in cell collapse [126]. This could be related to the mechanical constraint generated by the enlargement of ice crystals on the cell wall and the plasma membrane, ultimately leading to cell rupture [123,127]. In relation to global warming, the seasons are often altered with changes in temperature and occurrence of freezing stress that can appear without any previous chilling period [128]. A sudden shift to extreme temperature can induce severe physiological damage, which is more important for native plants of a warm habitat cultivated in a temperate climate, most of which are of agronomic interest (maize, soybean, tomato,*etc.*). Such stress can cause a marked decrease in production and significant economic loss [124]. The challenge is thus to improve crops to winter hardiness [129]. In the past twenty years, major efforts have been directed towards understanding the plant response to chilling (0–15 °C) and frost (<0 °C) temperatures [121,127,130,131,132,133]. Temperate plants are often chilling-tolerant. They have also developed mechanisms to increase their ability to withstand freezing temperatures following a period of low but non-freezing temperatures [134,135,136,137]. This process, called cold acclimation, is a multigenic and quantitative trait that is associated with complex physiological and biochemical changes [119,137] (for a review: [138]). Aside from other factors such as membrane integrity and protoplast properties, plant resistance to cold stress often depends on tissue morphology and cell wall mechanical properties [139,140,141,142]. It has been proposed that cell wall rigidity may be an important factor in cell resistance to freeze-induced dehydration [143].

Cold acclimation induces changes in cell wall polysaccharide composition and in the activities of cell wall-modifying enzymes. In a suspension culture of grape cells or in apple, cold acclimation increased cell wall strength and decreased the pore-size of the cell wall [143]. Pectins appear to be a key element of the plant response to cold stress, as shown by a number of studies on various species. In oilseed rape leaves, the cold response was associated with an increase in pectin content [142,144]. This was also observed in pea epicotyls during cold acclimation (2 °C), with a higher level of arabinosyl residues, and in bromeliad (*Nidularium minutum*), a tropical plant, where pectin content increased in plants grown at 10 °C and 15 °C compared to those grown at 25 °C [145,146]. In a frost-tolerant pea genotype, cold acclimation (10 °C day/2 °C night) was accompanied by an increase in homogalacturonan (HG), xylogalacturonan and highly branched rhamnogalacturonan I with branched and unbranched (1→5)-α-arabinans and (1→4)-β-galactans. HG was more esterified in the frost-tolerant pea genotype. The increased cold tolerance in the frost genotype might be related to an increased synthesis of arabinan and galactan side chains and HG methylesterification status, which may act as a gelling component, or regulators of pore size as well as a cold protectant following frost exposure [147]. Changes in the activities of pectin-modifying enzymes have been shown for several species in variable growth conditions in response to cold. For instance, increased PME activity was observed in cold-acclimated oilseed rape plants, which correlated with an increased rigidity of the cell wall in the leaf. This rigidity could be related to a demethylesterification of pectin, which generates free carboxyl groups, and Ca^{2+} ion-mediated cross-linking of these carboxyl groups to create a stiff pectate gel [142]. In contrast, in two chicory root varieties subjected to low temperature in field experiment, PME activity was differentially affected [148]. PME activity was higher during summer and then decreased consistently in both varieties (Nausica and Arancha). However, the decrease in PME activity was stronger in the Nausica variety, which is more cold-sensitive, compared to the Arancha variety where PME activity almost recovered by the end of the cold season [148]. Unlike in oilseed rape, PME activity was not increased in a frost-tolerant pea genotype [147]. During pollen germination, the expression levels of many genes responsible for pollen germination and tube growth were not influenced by cold stress, in particular PME, which were strongly expressed under normal conditions and had steady-state levels of expression in cold-treated samples [149]. This discrepancy could indicate distinct genotype-, species- or tissue-dependent mechanisms of temperature control of PME activity. PG is also regulated in response to cold stress. In young rice leaves of *Ubi::OsBURP16* transgenic plants, which overexpressed a beta-subunit of PG, reduced cell adhesion caused by pectin degradation, which affected cell wall integrity, was associated with a decreased tolerance to cold stress [76]. Cold treatment does not only affect pectin-remodeling enzymes but also pectin-associated receptors or WAKs (wall-associated kinases). WAKs are required for cell elongation and development and were shown to play a role in the plant/pathogen interaction [38,150,151,152,153]. An up-regulation of *WAK* genes was observed in rice seedling and a cold-tolerant maize genotype under cold treatment [154]. WAKs could bind either oligogalacturonides or pectin in the cell wall in response to cold stress, indicating a potential role of the cell wall as a cold stress sensor [154].

This modification in pectin structure is often accompanied by changes in neutral sugar composition in vegetative tissues [126,145]. An increase in hemicellulose content could enhance cell wall stiffening and prevent cell collapse caused by dehydration [144,155]. A reduction in xylose and glucose contents in non-cellulosic components was observed in cold-acclimated winter oilseed rape leaves [144]. In contrast, in the winter/dormant stems of poplar, xylose and glucose contents increased and the cellulose content increased by almost 3-fold [156]. Unlike a frost-tolerant pea genotype, a frost-sensitive pea cultivar accumulated substantial amounts of (1→4)-β-xylans and glucuronoxylans, but not pectins, indicating a premature end to cell growth and an initiation of vascularization [147]. Some alterations in the type II cell wall have been shown upon cold exposure. In Miscanthus plants grown at the juvenile stage, an increase in (1→3),(1→4)-β-D-glucan content was observed during cold acclimation (12 °C day/12 °C night) while in wheat roots cultivated over a low temperature acclimation (2 °C) period, hemicellulose synthesis was modified [155,157]. Hemicellulose synthesis increased during the first 24 h following application of stress. In contrast, in the first hour of low temperature exposure, a decrease in the proportion of β-glucans and branched glucuronoarabinoxylans occurred while an increase in the activities of β-fucosidase and β-glucosidase was observed. The activities of glycosidases could account for the transient accumulation of GXAG (oligosaccharin) in winter wheat tissues during the first hours of cold acclimation at 2 °C [158]. Such an oligosaccharin could be produced as a result of the increased hemicellulose turnover and could be physiologically active oligosaccharides that could stimulate the acquisition of freezing tolerance in a winter variety of *Triticum aestivum* L. [155]. Moreover, the authors hypothesized that GXAG could increase cell receptivity to ABA signaling [155,158].

The observed changes in cell wall composition correlate with a strong up-regulation of genes encoding proteins involved in cell wall modification [155]. During cold acclimation, *EXP* and *XTH* gene expression was down-regulated in sweet potato while *EXP* genes were up-regulated in Arabidopsis and rice, showing a contrasting regulation of these genes at low temperatures [125,159,160]. In a distinct experiment, extensin and *XTH* encoding genes were down-regulated in Arabidopsis under cold response [161]. The functional role of XTH was shown in Arabidopsis transgenic plants, overexpressing*AtXTH21*, which had an improved freezing tolerance [162]. Similarly, in rice, a xyloglucan endotransglucosylase (*OsXET9*) was highly induced under cold stress, especially in seedlings [163]. Transcriptomic data in cotton showed an up-regulation of genes encoding a cellulose synthase, *PME3*, and *MUR4*, a UDP-D-xylose 4-epimerase, an enzyme that catalyzes UDP-xylose-4-epimerase to UDP-L-arabinose, in response to cold [164]. In potato plants, the expression of xylose isomerase, which catalyzes the conversion of D-xylulose into D-xylose, a constituent of non-cellulosic polysaccharides, was increased [165]. Overall, cold stress had a drastic effect on the regulation of genes encoding cellulose synthase, as illustrated by the results obtained on cotton, in which cellulose synthases were down-regulated by 2-fold upon cold treatment [164]. In poplar during the winter dormancy period, cellulose synthesis was increased in stems compared to summer stems [156,166], but most cellulose synthase genes were surprisingly down-regulated, which might suggest an unknown control mechanism of cellulose deposition in the winter cell wall of stems [156]. In contrast, in a rice cold-tolerant genotype, cellulose synthase genes were up-regulated by 5 to 8-fold after 48h of cold exposure [154]. The effects of cold stress can also target other cell wall compounds such as arabinogalactan proteins (AGPs), which are located in plant cell walls and belong to a super family of highly glycosylated hydroxyproline-rich glycoproteins. In cotton, the expression of a non-classic AGP which contains a predicted signal peptide, a short AGP domain of seven amino acids, a His-stretch, a Pro-rich domain and a PAC (PRP-AGP containing Cys) domain, was found to be up-regulated in response to cold stress [167]. Transgenic Arabidopsis plants overexpressing *GhAGP31* confirmed that GhAGP31 improved the freezing tolerance of Arabidopsis seedlings [167].

Lignification and changes in lignin content can occur during growth at low temperature as well [23]. Lignin synthesis can be enhanced during cold acclimation and participate in cell wall modification by strengthening it, thus preventing freezing damage and cell collapse. A large increase in CAD (hydroxycinnamyl alcohol dehydrogenase) activity under cold stress was observed in frost-sensitive or cold-tolerant Miscanthus genotypes, but it was largest in the frost-tolerant clone *M. sinensis*August Feder [157]. Furthermore, PAL enzyme activity was higher in the frost-tolerant genotype during cold acclimation [157]. In *Brassica napus*, PAL activity, as well as the rate of accumulation of different phenolic compounds in leaves, depended on the range of low temperature applied to the plant [168,169,170]. The accumulation of ferulic acid observed in these leaves could increase cell wall rigidity and cell resistance to mechanical stress, thus explaining the growth delay by limited cell expansion [23,170,171]. These data suggest that the phenolic pathway activated by cold stress participates in protecting against freezing damage [23,157]. Lignin biosynthesis genes are affected in the cold response. It was shown that*CAD* gene expression was induced in sweet potato or Arabidopsis in response to cold [161,172]. Other genes involved in lignin biosynthesis, including *PAL*, *CCR*, *COMT* (caffeate *O*-methyltransferase), and *CCoAOMT*, were shown to be up-regulated in pea and *Phaseolus vulgaris* in response to cold [161,164,173,174]. This is concomitant with the increase in cell wall strengthening in response to low temperature stress.

All these data show that the cell wall structure changes in response to cold (Figure 4). Although it is likely that the cell wall could modulate different aspects of the cold response, an investigation of the cell wall as a primary effector is still required.

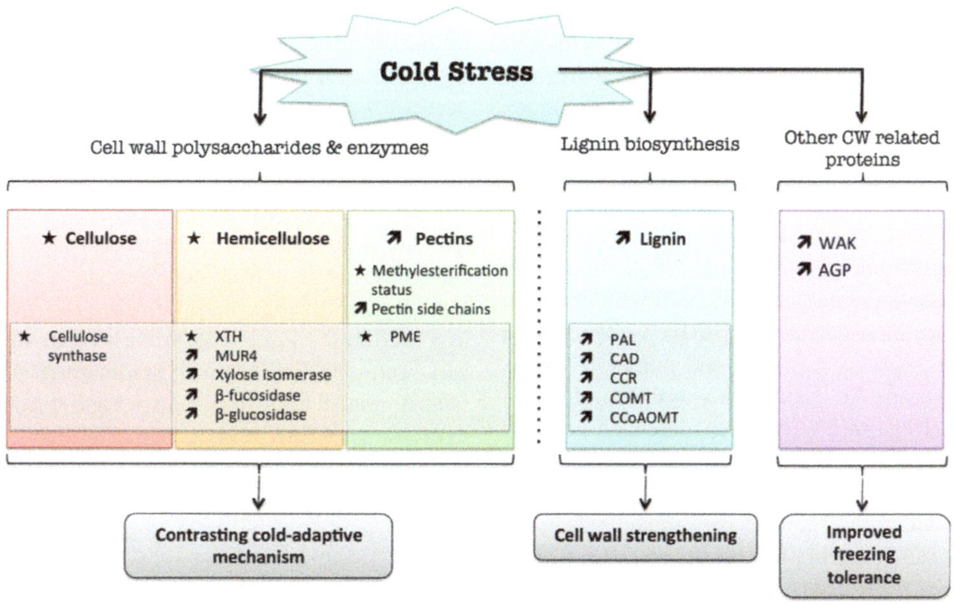

Figure 4. Overview of the plant cell wall response to cold. The schematic presentation is deduced from the results of different studies reported in the review. Arrow (↗) means increased abundance and arrow (↘) means decreased abundance of the molecules. Star (★) means contrasting data on the gene/protein or the molecule studied according to the literature cited in the review. Cell wall (CW); xyloglucan endo-β-transglucosylases/hydrolases (XET/XTH); pectin methylesterase (PME); UDP-D-xylose 4-epimerase (MUR4); phenylalanine ammonia-lyase (PAL); hydroxycinnamyl alcohol dehydrogenase (CAD); caffeate *O*-methyltransferase (COMT); caffeoyl-CoA 3-*O*-methyl-transferase (CCoAOMT); cinnamoyl-CoA reductase (CCR); cell wall peroxidases (PRX); arabinogalactan protein (AGP); wall-associated kinase (WAK).

SALT

Another major abiotic stress is due to soil salinity, which can affect plant productivity. More than 6% of the world's total land and approximately 20% of irrigated land is affected by this abiotic stress [175]. Increasing salinization of arable land is expected to have drastic effects, with up to 30% land loss within the next 15 years [106]. Salt stress has two origins: the deposition of oceanic salts by wind and rain and the erosion of rocks that release soluble salts [175,176]. The major effect of salt stress is a decrease in available soil water due to the reduction in osmotic potential of the soil solution leading to water deficit-related plant growth impairment [177,178,179]. Plants display shoot growth inhibition during the first phase of salt stress, partly due to a loss of cell wall extensibility. In maize, reduced cell wall extensibility was related to the formation and accumulation of diferulic acid cross-links in the cell [180]. Changes in cell wall structure appear to be a common feature of the response to salt stress. For instance, in coffee plants grown with 150 mM NaCl during 25 days, the palisade and spongy parenchyma exhibited thinner cells, which might be related to decreasing water content in leaf cells [181]. Such physiological response was reported in salt-sensitive plants, called glycophytes. This group includes a number of cultivated plants such as rice, soybean and maize [180,182,183]. The resistance of glycophytes to lethal salinity can be improved by short exposure to low levels of salt stress, a process known as salt acclimation [177,184]. This acclimation mechanism has been described and shown to involve changes in cell wall-related proteins [185,186]. Candidates that could participate in salt acclimation are receptor-like protein kinases (RLKs), primary cell wall "sensors", such as WAKs [187]. Although in abiotic stress the function of this protein is not so well defined, induction of *WAKL4* (WAK-like kinase 4) gene expression was observed in Arabidopsis roots under high salt concentration (750 mM NaCl) [188]. These data suggest that the main salt stress perception-to-signaling event could occur at the cell wall-plasma membrane interface. Following salt stress perception, the plant response was shown to involve the regulation of a number of primary cell wall protein-encoding genes, through the recruitment of specific transcription factors. For example, *AtMYB41* or the rice R2R3-type MYB transcription factor *OsMPS* (MULTIPASS) were activated by salt stress (100 mM NaCl) and shown to regulate the biosynthesis of cell wall-related genes [189,190]. *OsMPS* is transiently expressed in vegetative and reproductive tissues and negatively regulates the expression of expansin and endoglucanase genes under salt stress. The hormone-dependent regulation of *OsMPS* expression allowed root growth adaptation to changing environmental conditions while *AtMYB41* was shown to down-regulate the expression of *EXP* or *XTH* encoding genes and up-regulate the expression of *HRGP* when Arabidopsis plants were exposed to 200 mM NaCl [191]. There appear to be distinct responses when species and/or salt-stress conditions are compared. For instance, *RhEXPA4* expression in rose was increased by salt while transgenic Arabidopsis plants overexpressing *RhEXPA4* displayed a compact phenotype (shorter stems, curly leaves, compact inflorescence) and were salt-tolerant [67]. Similar results were observed with the overexpression of wheat expansin, *TaEXPB23*, in transgenic tobacco plants [192]. However, Arabidopsis transgenic plants overexpressing *AtEXP3* or *AtEXP-B1* were salt-sensitive following exposure from 250 mM to 300 mM NaCl treatment [193]. In physic nut roots (*Jatropha curcas*), *EXP* genes were both up-regulated and down-regulated while *XET* genes were up-regulated during different periods of salt exposure at 100 mM NaCl [194]. In parallel, other structural cell wall proteins, including AGP and proline-rich protein (PRP), were affected by salt [25,195]. The expression of rice and poplar (*Populus euphratica*) genes encoding AGP and PRP was up-regulated in response to salt [196,197,198,199]. The glycine-rich proteins (GRP), another class of cell wall structural proteins, appear to be involved in salt tolerance. Genes encoding these proteins were induced under salt exposure, with consequent effects on seedling growth and seed germination [200]. MsGRP, a GRP from *Medicago sativa*, probably participates in iron or water transport to maintain cellular homeostasis [200]. *SOS6*, which encodes a cellulose synthase-like protein *AtCSLD5*, plays a role in the regulation by decreasing ROS content under salt stress. The Arabidopsis *sos6* mutant displayed a decreased level of arabinose, rhamnose and galacturonic acid sugar contents under salt stress suggesting that SOS6 is required to maintain cell wall integrity in these conditions [201]. In the primary cell wall of glycophytes, other classes of proteins and polysaccharides are modified upon salt stress treatment. These include RGP (reversibly glycosylated polypeptide), such as RGP1 in pea roots (*Pisum sativum*), or glycosyl hydrolase 1 (GH1 family), GRP in Arabidopsis root, which are up-regulated under salt treatment [200,202,203,204]. In potato, an increased expression of a UDP-glucose-4-epimerase (UGE), which interconverts UDP-glucose and UDP-galactose, was observed under various abiotic stresses such as high salinity [165]. Moreover, in red-osier dogwood seedlings, a salt-adapted plant, the hemicellulose composition and the lignification status were not altered but an increase in the rhamnose content in the pectic fraction was observed, accompanied by a decrease in plant growth [205]. In addition, cellulose and hemicellulose displayed a steady-state level while a slight decrease in uronic acid was observed in root of a salt-tolerant soybean genotype in response to salt stress. In the root elongation zone of the salt-sensitive soybean genotype, a significant decrease in the cell wall total sugar content was shown, with a more severe decrease (~70%–80%) in the abundance of pectin, compared to that observed in the salt-tolerant one which only displayed a 30% decrease [206]. The higher abundance of pectin in the salt-tolerant cultivar could explain its salt tolerance. A similar study performed on three maize hybrids with contrasting salt tolerance showed a decrease in cellulose concentration in all genotypes and an accumulation of non-methylated uronic acid, which was more pronounced in the salt-sensitive maize genotype [207]. This accumulation was delayed in the salt-tolerant maize genotype, particularly in the youngest shoots, which favored their elongation due to the smaller increase in non-methylated uronic acid [207]. Unlike maize, highly methylated pectin, as shown by immunolabeling with JIM7, increased within the cell wall of Aspen petiole cells [208]. This was associated with an increase in the elasticity modulus, indicating a decrease in cell wall plasticity to maintain the turgor pressure necessary for plant growth. According to these authors, the increase in highly methylated pectin could prevent water molecules from binding to pectin leading to a less hydrated cell wall, and thus favoring water dehydration in the plant during salt stress.

Salt stress has been shown to impact secondary cell wall formation and structure, as revealed by an altered lignin biosynthesis. Salt treatment increased root lignification and the number of lignified vessels [182,209]. Many enzymes involved in the lignin biosynthesis pathway were altered upon salt treatment. For instance, in tomato, *SlPAL5* gene expression was increased after treatment with 200 mM of NaCl while the expression of PAL was decreased in rice [210,211]. Proteomic studies have also revealed that COMT and CCoAOMT are salt-responsive proteins [26]. This was confirmed at the activity level as in rice roots, COMT enzyme activity was decreased in salt stress conditions [211]. However, the CCoAOMT enzyme was salt-induced in the roots of Arabidopsis and rice [212,213]. The action of these enzymes might help reduce the bypass water flow that allows Na^+ ions to enter rice roots *via* an apoplastic pathway [214].

The response of halophyte plants to salt stress appears to differ from that of glycophytes. Indeed, in these plants, cell homeostasis and osmotic adjustment are maintained, with consequences on the synthesis of proteins and cell wall pectins [215]. For instance, in *Sonneratia alba* (a halophyte plant), a higher calcium concentration was measured suggesting that the plant could sequester and efficiently use Ca^{2+} within cell walls to increase their rigidity through pectin egg-box formation [216]. Like in glycophytes, WAK protein could be a salt-stress sensor. In favor of this hypothesis, the results obtained in *Nitraria sphaerocarpa* showed that the expression of *WAK4* was increased after long exposure to salt [217]. Salt also has consequences on proteins that are involved in the biosynthesis or remodeling of cell wall polymers. In roots of *Salicornia europaea*, genes encoding cell wall proteins of the primary cell wall, including UDP-L-rhamnose synthase and cellulose synthases, were down-regulated while other genes, encoding pectin methylesterase inhibitor proteins, were increased under saline conditions [218].

In conclusion, the cell wall plays an important role in salt tolerance, especially in the detection of salt stress. However, the cell wall modifications are different in glycophytes and halophytes. In glycophytes, there is an increase in primary and secondary cell walls whereas in halophytes only the secondary cell wall structure is altered (Figure 5).

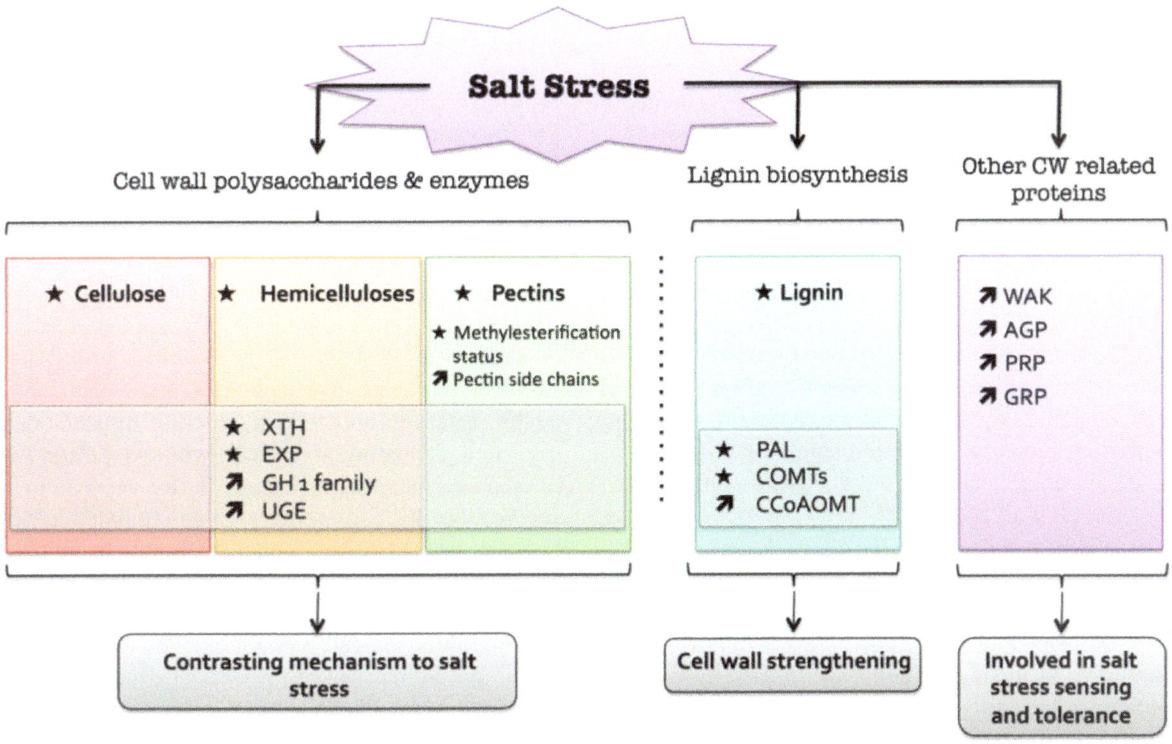

Figure 5. Diagram summarizing the plant cell wall response to salt. The schematic presentation is deduced from the results of different studies reported in the review. Arrow (↗) means increased abundance and arrow (↘) means decreased abundance of the molecules. Star (★) means contrasting data on the gene/protein or the molecule studied according to the literature cited in the review. Cell wall (CW); xyloglucan endo-β-transglucosylases/hydrolases (XET/XTH); glycosyl hydrolase family (GH); reversibly glycosylated polypeptide (RGP); phenylalanine ammonia-lyase (PAL); caffeate *O*-methyltransferase (COMT); caffeoyl-CoA 3-*O*-methyl-transferase (CCoAOMT); arabinogalactan protein (AGP); wall-associated kinase (WAK); proline-rich protein (PRP); glycine-rich protein (GRP).

HEAVY METALS

Heavy metal (HM) pollution has become an increasingly serious environmental problem over recent decades due to its toxicity, leading to many detrimental effects on living organisms, from plants to animals [219,220]. For instance, heavy metal exposure was shown to reduce root growth and to enhance formation of lateral roots [221,222,223]. When considering plant mineral nutrition, metals are often categorized into those that are essential for many physiological processes, e.g., copper (Cu), zinc (Zn) and iron (Fe), and non-essential ones, such as lead (Pb), cadmium (Cd), aluminum (Al) and mercury (Hg). An excess, as well as a deficiency, of the former can cause harmful effects in plant cells, while the latter may be toxic even at relatively low concentrations [224]. It is also possible to classify them as redox-active metals (e.g., Cu and Fe), which catalyze the formation of hydroxyl radicals by directly participating in the Haber-Weiss reaction, or metals without a redox capacity (e.g., Cd, Pb, Hg, Zn) which inactivate the cellular antioxidant pool and disrupt the metabolic balance, thus enhancing the load of ROS [225,226]. This affects the distribution of calcium (Ca^{2+}), Ca^{2+}-binding protein activity, signaling processes and MAPK pathways [227,228].

The cell wall represents a physical barrier against the entry of HM into the symplastic compartment. One strategy of plant cells for coping with HM is to remove them from the protoplast by sequestration into extra-cytoplasmic compartments such as the cell wall [27,28]. This protects the most sensitive sites within the protoplast from HM toxicity [27,229,230]. Plant cell wall polysaccharides play a crucial role in HM binding and accumulation in the cell wall, although other compounds, such as proteins, amino acids and phenolics, can also take part in this process [27,231]. One of the advantages of precipitating metals within the cell wall is their strong metabolic inactivation [232].

Pectin, together with hemicellulose, plays a crucial role in binding HM [233,234]. It was shown that the presence of pectin containing boron cross-linked rhamnogalacturonan II in the cell wall promoted the formation of an Al-complex, preventing damage and the irreversible inhibition of growth [235,236]. The presence of exogenous silicon increased cell wall affinity for Zn in maize roots and rice seedlings, thereby reducing its uptake and translocation [237,238]. The degree of methylesterification of pectins appears to contribute to HM tolerance [239]. Some amounts of heavy metal bound to low-methylesterified pectins in the cell wall can enter the protoplast together with this compound by endocytosis, in particular in meristematic dividing cells [229,240]. A striking example of metal accumulation in the cell wall was observed in the common reed (*Phragmites australis)*, which is considered to be a plant with a high detoxification and phytoremediation potential. Zn concentrations followed a gradient with the sequence: intercellular space >cell wall > vacuole > cytoplasm, indicating that most Zn was accumulated within the apoplast [232]. The ability to bind divalent metal cations depends on the number of functional groups, such as –COOH, –OH, and –SH present in the cell wall [241]. The main pectic domain responsible for binding divalent and trivalent metal cations is HG. The HG fraction, particularly those with a low degree of methylesterification, recognized by JIM 5 antibodies containing some amounts of free carboxyl groups, can bind divalent and trivalent metal ions [242]. It is known that low methylesterified HG can interact with calcium and form an egg-box structure in the cell wall. Calcium ions can be replaced by trivalent Al^{3+}, and also by divalent HM cations, including Pb^{2+}, Cu^{2+}, Cd^{2+} and Zn^{2+} [243]. For instance, under short-term Al supply, Al accumulated primarily in the root apoplast where Al^{3+} strongly binds to the negatively-charged binding sites provided by non-methylesterified pectin in the cell wall [239]. In addition, in flax hypocotyls, Cd^{2+} caused reorganization in the distribution of pectin epitopes related to PME activities [244]. Enhanced JIM 5 and 2F4 labeling at the junctions of the inner tissues indicated that the presence of blockwise de-esterified HGA might oppose cell separation under Cd stress [245,246]. This was accompanied by an increase in PME and peroxidase enzyme activities, indicating a reinforcement of cell junctions leading to better tissue cohesion and Cd sequestration [246]. An increase in cell wall polysaccharides confers on the plant the capacity to store metal in the cell wall. In the common bean root apex, the reduction in pectin in the cell wall following PEG treatment is likely to change cell wall structure, thus affecting its porosity and inhibiting Al accumulation [247]. In another study, Al induced pectin biosynthesis in the roots of an Al-tolerant rye cultivar, but reduced it in the sensitive one [248]. This was associated with a decreased level in root PME gene expression in the tolerant rye genotype, although Al treatment increased PME activity in both cultivars.

The degree of methylesterification and acetylation of pectin decreased its affinity for HM. In a copper-tolerant plant, *Silene paradoxa*, a reduction in the root pectin content, together with an increase in its methylesterification status, was observed, in contrast to the copper-sensitive cultivar, indicating a Cu-tolerant/excluder phenotype [249]. In *Medicago truncatula*, the silencing of two Al-induced genes, pectin acetylesterase and annexin, in hairy roots slightly increased the sensitivity of the plant to Al, suggesting that these genes play a role in Al resistance [250].

In addition to pectin-remodeling enzymes, other cell wall enzymes are regulated following HM exposure. For instance, in Arabidopsis, XET enzyme activity was inhibited under exposure to Al, which was mostly bound to hemicellulose [251]. This was associated with root growth inhibition. Interestingly, transgenic poplar overexpressing *XET* had a higher Cd tolerance by reducing Cd uptake and accumulation in roots. This was accompanied by a higher xyloglucan degradation activity, leading to a reduction in XG content in the root cell wall [252].

The structure of the cell wall has an effect on the translocation of HM within xylem. In rice plant, mutation in a cellulose synthase gene (*CESA9*) modified the structure of the secondary cell wall, which further reduced translocation of Cd within xylem and its accumulation inside the plant [253]. In *Elsholtzia splendens*, a hyperaccumulator of Cu, excess Cu altered the composition and distribution of cell wall sugars in roots [254]. Cu enhanced cellulose, hemicellulose and pectin contents with a particular increase in galacturonic acids and KDO levels during HM exposure. In *Sedum alfredii Hance*, an HM-hyperaccumulator, transcriptome analysis showed that cell wall biosynthesis genes encoding *CesA* and *XTH* were up-regulated [255]. In addition to the above-mentioned study, several transcriptome analyses have been performed on plants exposed to HM, revealing that cell wall-related genes are altered. A large-scale rice transcriptome analysis under arsenate (As(V)) stress showed changes in gene expression including 40 cell wall-related genes among 637 transcripts [222]. Of these, 13 were up-regulated and 27 down-regulated. Down-regulated genes were cellulose synthase like A (*CslA*), xyloglucan galactosyltransferases (*XGT*), xyloglucan xylosyltransferases, galactomannan galactosyltransferases (*XXT*), glycosyl transferase (GT8, pectin synthesis), xyloglucan endotransglycosylases/hydrolases (*XTH*), β-galactosidases (βGAL), glycoside hydrolases 9 (GH9) and polygalacturonases (GH28). The down-regulation of these genes could limit cell expansion, as root growth is inhibited under arsenic stress. In *Medicago truncatula*, cell wall-modifying genes, including pectin esterase and pectin esterase precursors, were up-regulated following treatment with Al [250]. Furthermore, microarray analysis performed on rice exposed to metal stress showed that WAK proteins were up-regulated in rice subjected to As(V) stress, indicating that WAK may be involved in heavy metal stress tolerance [222]. The presence of heavy metals in the cell wall does not always result in cell wall stiffening and inhibition of cell elongation. For example, the presence of Cu and Fe can lead to polysaccharide scission, promoting cell elongation. Pectin and xyloglucan can be broken down by hydroxyl radical (·OH) production generated by hydrogen peroxide in the presence of ascorbate and a transient metal ion [256]. Thus, (·OH) free radicals produced in the cell wall under metal stress could act as a site-specific oxidant, targeted to play a role in the cell wall loosening process supporting cell expansion [257].

Lignin biosynthesis is induced under HM exposure. Cu had a positive effect on the biosynthesis of lignin and increased activities of several enzymes, including PRX and laccases, which carry out the polymerization of monolignol precursors of lignin [258]. In addition, in *Panax ginseng* exposed to a higher level of Cu, PAL and CAD enzyme activities were induced. This was correlated with an accumulation of lignin, indicating a protective response to higher levels of Cu [259]. Furthermore, transcriptome analysis of rice seedling revealed that the expression of 6 lignin biosynthesis genes was increased in rice roots under As(V) exposure. A previous study indicated that lignification might be an important step in root growth reduction in plants exposed to heavy metal stress. Therefore, As(V) may have a significant effect on the inhibition of root elongation by regulating cell wall organization. The ROS production induced by As(V) exposure may contribute to cell wall rigidity and inhibition of rice root growth. A similar trend has been observed for *Medicago truncatula*exposed to Al stress, where genes encoding *4CL* and *COMT* were up-regulated in response to Al stress in root tips [250]. It was proposed that the extent of root growth inhibition in response to Al was closely correlated with the extent of lignin deposition. It is conceivable that the Al-induced increase in activity of peroxidases triggered lignin production resulting in cell wall stiffening and reduced root growth in *M. truncatula* [250]. In a Cd-hyperaccumulator plant, *Sedum alfredii Hance*, genes encoding laccase and *C4H* were not induced but expressed at a higher basal level compared with the non-hyperaccumulating genotype [255]. Taken together, these data show a key role for cell wall composition particularly that of pectins, in the tolerance of plants to heavy metals, and could open new perspectives to engineer HM-resistant genotypes (Figure 6).

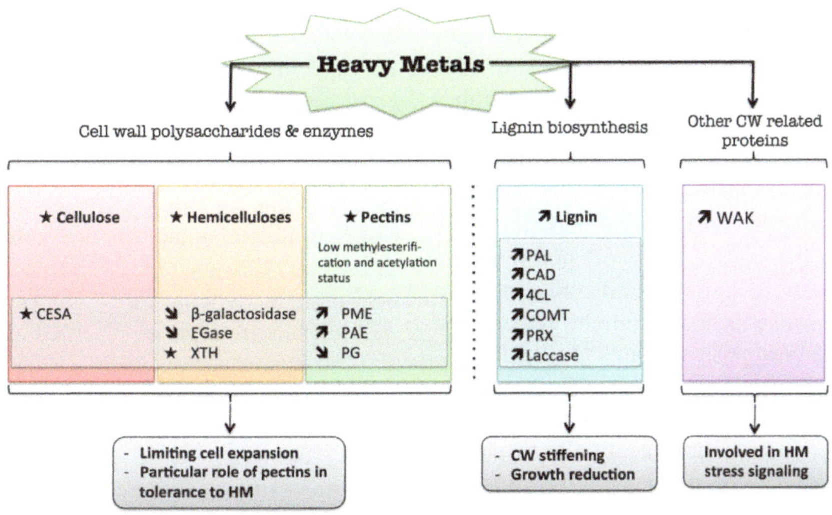

Figure 6. Diagram summarizing the plant cell wall response to heavy metals. The schematic presentation is deduced from the results of different studies reported in the review. Arrow (↗) means increased abundance and arrow (↘) means decreased abundance of the molecules. Star (★) means contrasting data on the gene/protein or the molecule studied according to the literature cited in the review. Cell wall (CW); xyloglucan endo-β-transglucosylases/hydrolases (XET/XTH); endoglucanase (EGase); pectin methylesterase (PME); pectin acetylesterase (PAE); polygalacturonase (PG); phenylalanine ammonia-lyase (PAL); hydroxycinnamyl alcohol dehydrogenase (CAD); 4 coumarate CoA-ligase (4CL); caffeate *O*-methyltransferase (COMT); cell wall peroxidases (PRX).

LIGHT

Although light, which contains Ultra Violet (UV), visible and far-red wavelengths, is essential for plants to ensure normal growth, its excess can cause cellular damage. Plants have therefore developed means of protection, notably through changes in the cell wall.

Shade Avoidance

Plant-to-plant competition is an example of environment sensing and response. A shade avoidance system (SAS) is present in several plant species upon the detection of shade signals from neighboring plants, notably within dense populations [260]. Following sensing through the absorption of light by photoreceptors, it enables morphological responses and differential growth to reach the upper canopy [261]. A reduction in blue and red light (red to far-red ratio) is an important factor that is sensed by the plant photoreceptors that trigger stem and petiole elongation, primarily through cellular expansion [260]. This elongation is a consequence of cell wall loosening mediated by cell wall-modifying enzymes such as expansins and XTHs [262].

A microarray analysis performed on Arabidopsis seedlings identified two *EXP* genes (N37536 and R29778) whose expression was modified by changes in light quality [263]. Similarly, in *Stellaria longipes*, five expansin genes, among the 12 expansin sequences identified from the internodes, were up-regulated relative to the controls in response to changes in light regimes [264]. The key role of the fine-tuning of the cellulose-hemicellulose network in response to light can be further illustrated by the regulation of *XTH*. In Arabidopsis, the expression of five *XTH* genes (*XTH9*, *-15*, *-16*, *-17*, and *-19*) out of 33 was increased by low red to far-red light conditions. In parallel, the expression of *XTH16*, *-17*, *-19* and *-22* was significantly increased in response to green shade which mimics white light conditions in a dense canopy (combined reductions of blue, R/FR, and total light intensity). The functional role of XTH in the shade response was shown by characterizing knockout mutants for two of these *XTH* genes (*XTH15* and *XTH17*), which revealed reduced or absent shade avoidance responses to these light signals. This ultimately demonstrated that cell wall loosening is a key element in the plant response to shade avoidance [265].

Light Irradiance

As observed in pea and lettuce, light is involved in the regulation of cell wall sugar content in plants, which depends on the tissue or organ considered. [266,267]. In etiolated pea seedling, stem elongation is inhibited by about 75% following both blue and red light exposure as compared to dark controls. The fast inhibitory response was mediated by blue light while the long one was induced by red light. The degree of growth rate inhibition was increased by longer exposure to red light [268]. That inhibition was due to changes of viscoelastic properties of the cell wall.

During Bartlett pear (*Pyrus communis* L.) ripening, cell wall catabolism occurred in both sun- and shade-grown fruits, but pectin solubilization was clearly delayed in sun-exposed fruit. This was associated with a decreased removal of RG I-arabinan side chains rather than reduced depolymerization of the entire polymer. After ripening, sun-exposed pears remained firmer, and there was not a great impact on pectin molecular weight. At harvest, however, a higher proportion of water-solubilized uronic acids and alkali-solubilized neutral sugars and a larger mean molecular size of tightly bound glucans were found in sun-exposed pears [269]. Light radiation altered genes encoding proteins involved in cell wall modifications [155]. The overexpression of a blue light photoreceptor, *BnCRY1*, in *Brassica napus* reduced seedling elongation. In parallel, the expression of cell wall related genes including *XTH* and *PME* showed a 2.5- and 4-fold decreased level respectively in that transgenic line [270]. The lignin biosynthetic pathway is regulated by the photoperiod and light quality [24]. Changes in light conditions have been shown to be associated with an increase in lignin content in a number of different developmental processes and species. *Ebenus cretica* L. seedlings grown in the light displayed 2.5 times more lignin than those grown in the dark [271]. The observed alteration in leaf morphology in response to changing light conditions could thus be the consequence of modifications of the structural components of lignin. Similar results were observed in coffee leaves exposed to sunlight, where the lignin content in leaves has been positively linked to radiation, with an increase in the bulk elasticity modulus that could be explained by the higher content of lignin when compared with shade leaves [24]. Light intensity has also been shown to exert an effect on the activity of enzymes involved in the lignin biosynthetic pathway. Analyses of orchid plants exposed to different light intensities, after transfer from *in vitro* to soil, revealed that PAL and CAD enzyme activities were induced in leaves [272]. In Arabidopsis, 7000 genes exhibited altered expression when plants were subjected to high light intensity. Among them, 110 showed an increased expression, and several of them, including *4CL*, are involved in the lignin biosynthetic pathway [273].

UV Light

In recent years, there has been an increasing interest in the biological effects of UV light (230–400 nm) on plants. UV-B radiation (280–315 nm) reduced leaf elongation in Antarctic grass (*Deschampsia antarctica*) and in lettuce, when compared with plants exposed to reduced UV-B. This could be partly explained by the accumulation of insoluble hydroxycinnamic acids (*p*-coumaric, caffeic and ferulic acids) cross-linking the cell wall polysaccharide network and cell wall PRX thus limiting the expansion of epidermal cells and leaf elongation [267,274]. Among the effects of UV-B, cell wall elasticity was reduced in grapevine cells [275]. In cell wall isolates from pea leaves or on citrus fruit pectin, UV-A (315–400 nm) or UV-B produced hydroxyl radical (·OH) in presence of hydrogen peroxide. Pectin is able to transform (·OH) to superoxide ($O_2^{\cdot-}$), which may have an intracellular impact, since it can pass cellular membrane [276]. (·OH) released could contribute to methane production from galacturonic acid methyl ester groups, and also, at a lower level, from acetate groups, [276,277]. This indicates that pectin might be involved in the UV radiation signaling pathway as a source of superoxide [276,278]. In tomato pericarp, UV-C irradiation (230–280 nm) exposure during postharvest retarded cell wall disassembly. This was notably related to the down-regulation of transcriptional expression or through the inhibition of the activities of many cell wall-degrading enzymes, including PME, PG and cellulase, during ripening [279,280].

Changes in cell wall structure and composition under UV light could produce mechanical barriers and protective substances such as secondary compounds including lignin [281]. In cucumber, cotyledons exposed to UV-B radiation lignin accumulated in cell walls of the parenchyma of tracheid elements and in the intercellular space [282,283]. Phenylpropanoids, such as hydroxycinnamic acids and their derivatives (sinapate esters), as well as flavonoids, might play complementary roles in UV-B protection. In addition, in grapevine leaves, the expression of genes of the lignin biosynthetic pathway, such as *PAL*, *C4H*, *4CL* and *CCoAOMT*, were all up-regulated in response to UV-B radiation while genes associated with cell wall loosening were down-regulated [284,285].

These rather scarce data show that light intensity and quality influence cell wall architecture, particularly lignin, which could attenuate radiation and thus protect plant tissues and organs. However, the mechanism of cell wall regulation by light remains to be investigated. An overview of the plant cell wall response to light is presented in Figure 7.

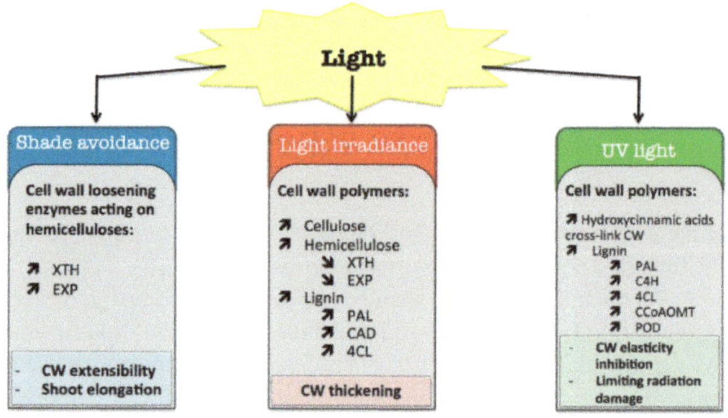

Figure 7. Diagram summarizing the plant cell wall response to light. The schematic presentation is deduced from the results of different studies reported in the review. Arrow (↗) means increased abundance and arrow (↘) means decreased abundance of the molecules. Star (★) means contrasting data on the gene/protein or the molecule studied according to the literature cited in the review. Cell wall (CW); xyloglucan endo-β-transglucosylases/hydrolases (XET/XTH); expansin (EXP); phenylalanine ammonia-lyase (PAL); hydroxycinnamyl alcohol dehydrogenase (CAD); 4 coumarate CoA-ligase (4CL); caffeoyl-CoA 3-*O*-methyl-transferase (CCoAOMT); trans-cinnamate 4-hydroxylase (C4H); cell wall peroxidases (PRX).

AIR POLLUTANTS

According to the US environment protection Agency (EPA) the main air pollutants are ozone, particle, carbon monoxide, nitrogen oxide, sulfur dioxide and lead. Among these factors the main pollutants affecting ecosystems are carbon dioxide (CO_2) and ozone (O_3) [286]. CO_2 is the primary greenhouse gas emitted through human activities (U.S. environment protection Agency, EPA) and is the substrate for photosynthesis. Its increased concentration in the atmosphere can stimulate plant growth photosynthesis and the accumulation of plant biomass in the short-term. But, elevated atmospheric CO_2 is accompanied by a decrease in oxygen levels which thus can reduce photorespiration, a major source of hydrogen peroxide. This may then reduce abiotic stress response signaling pathway [287]. The tropospheric ozone concentrations (O_3) are more than double compared to the levels measured over a century ago [288]. (O_3) are high enough to cause damage to crops and thus decrease crop yields [289]. Unlike CO_2, O_3 is rapidly converted to ROS. It then interferes with basic physiological functions by accelerating foliar damage, senescence and reducing photosynthesis [290]. This chapter only reports data on the effects of ozone exposure and elevated atmospheric CO_2 on cell wall metabolism since these two main air pollutants for which a number of data is available.

Ozone (O_3)

Ozone (O_3) is considered the most important consequence of air pollution and is harmful to plants, animals and humans [291]. The cell wall has been shown to be an early target of ozone [292]. The plant response to ozone is thus dependent on the species, the age of the plant, the season, the time of day and, ultimately, ozone concentration [293,294]. For instance, tobacco was more sensitive to O_3 than deciduous trees while juvenile Manna ash (*Fraxinus ornus* L.) trees were more sensitive than mature trees [295]. O_3 induces mainly leaf and root injuries, which can lead to biomass reduction, as observed in tobacco, poplar and birch [291,295,296]. In Manna ash (*Fraxinus ornus* L.), one of the main structural impacts of O_3 was punctures within the cell wall, as observed on leaf mesophyll cells [295]. The development of symptoms was accompanied by the thickening of Manna ash (*Fraxinus ornus* L.) palisade mesophyll cell walls, which appears to be a common feature in several species [295,297]. Cell wall thickening is accompanied by an accumulation of cellulose layers in woody plant species including, *Fagus sylvatica*, *Acer pseudoplatanus*, which reduces the inner space of the cells [297,298]. Cell wall thickening facilitates detoxification processes and confers greater mechanical resistance to cell collapse [297,299]. In Manna ash, pectins and other cell wall polysaccharides are likely to be key components of this cell wall thickening, as shown by the cytochemical cell wall composition of stomata, which displayed more cellulose and lignin in the outer cell wall layer and more pectin in the inner cell wall layer [293,295]. This might contribute to maintaining the mechanical properties of these cells [295]. In ragweed (*Ambrosia artemisiifolia* L.) pollen, FTIR analysis showed a decrease in phenolic compounds and an increase in pectin components (rhamnogalacturonic acid, arabinan) under elevated ozone [300]. Notably, a decrease in acetylester groups, indicating a de-esterification of pectin in response to ozone, was shown. Transcriptome analysis further revealed that the expression of pectin methylesterase and pectate lyase encoding genes was strongly induced in ragweed pollen in response to ozone. Although Amb a1, an allergen found in *Ambrosia artemisiifolia* pollen, which is a pectate lyase, was highly expressed in response to ozone, no significant difference was measured at the protein level, as tested by an ELISA experiment [300]. Cell wall sugar content increased under ozone exposure in strawberry plant, which was related to increased pectin content (100% more) after a short treatment (7 days) rather than long exposure (2 months). These changes were primarily due to water-soluble pectin suggesting that either pectin degradation was decreased or *de novo* synthesis increased. In parallel, long-term exposure reduced non-soluble pectin by 60% and 80% in more and less O_3-tolerant strawberry cultivars, respectively. This appeared to be related to changes in the amount of low methoxy pectin content, which increased in the tolerant genotype and decreased in the sensitive one.

O_3 can be used to delay fruit ripening during cold storage, due to its ability to inhibit ethylene oxidizing capacity. In kiwifruit, O_3 treatment at 0 °C reduced cell wall disassembly of the fruits during ripening at room temperature [301]. In that study, the swelling capacity of the cell wall associated with the solubilization of pectin was reduced when fruit was treated with O_3. The treatment induced lower PG and endo-1,4-β-glucanase activities, enzymes involved in pectin and hemicellulose degradation, compared with the control environment. In contrast to the above-mentioned findings, in O_3-sensitive aspen, leaves were thinner compared to the tolerant genotype, concomitant with a reduction in cell walls. The average thinner cell walls in mesophyll cells in the sensitive aspen clone could decrease cell protection from O_3 present in the apoplast [302]. At the transcriptional level, in an O_3-sensitive rice cultivar, genes encoding cell wall degradation enzymes, including cellulase and β-galactosidase, were up-regulated. Other genes encoding cell wall-related proteins, such as cell wall invertase inhibitor or glycine-rich protein, were similarly up-regulated while the expression of genes encoding proteins involved in cell wall biosynthesis, such as *XET* and *COBRA*, was down-regulated [303]. In contrast, in the O_3-tolerant rice genotype, a different transcriptional regulation is likely to occur, with an increased expression of genes encoding some expansin and xylanase inhibitors.

O_3 also has an impact on wood formation. In poplar, stems were more affected than leaves, with a reduction in stem diameter and the anatomy of the tension wood [296]. O_3 reduced cambial growth and xylem differentiation, which was associated with a reduction in cellulose and lignin biosynthesis [296]. This was accompanied by changes in the expression of specific genes, and by a decrease in activity of enzymes involved in cellulose (SuSy, UGPase) and lignin formation (PAL, CAD) [296]. In poplar exposed to ozone, changes in lignin structure occurred, with enrichment in condensed lignin by a higher proportion of H-units near the foliar necrotic area [304]. CAD activity was rapidly and strongly increased (10-fold) whatever the foliar developmental stage, while PAL enzyme activity was increased in old and mid-aged leaves [304]. In non-woody plants, at the transcriptome level, *PAL* mRNA, a key enzyme involved in the phenylpropanoid pathway including lignin biosynthesis, was rapidly and transiently induced reaching a 3-fold higher level compared to control Arabidopsis or parsley plants [305,306]. In contrast, O_3 was also found to modify cell wall components by depolymerization of lignin, which released small phenolic compounds that could be used as elicitor molecules [307].

Cell wall peroxidases might regulate the level of ROS by scavenging H_2O_2 to polymerize lignin and cross-link cell wall proteins and polysaccharides [308]. In sunflower leaves exposed to ozone, cell wall PRX was induced thus leading to cell wall stiffening and contributing to slowing down O_3 penetration into the cell [308,309]. Furthermore, PRX displayed higher activity with syringyl compared with guaiacyl molecules. In addition, PRX activity in strawberry leaves decreased in the less tolerant cultivar while it increased in the ozone-tolerant genotype under short term ozone exposure, indicating a significant contribution of cell wall PRX to the detoxification of ozone in the apoplast [310] These data, which are mostly derived from cytochemical analysis and enzyme activity measurements, implies a contribution of cell wall proteins to apoplastic ozone detoxification. In addition, some genomic approaches provided valuable knowledge about the role of the plant cell wall in response to ozone (Figure 8). However, very few data are currently available at the biochemical level on cell wall polymers, cell wall enzyme activities and signaling processes that regulate this response.

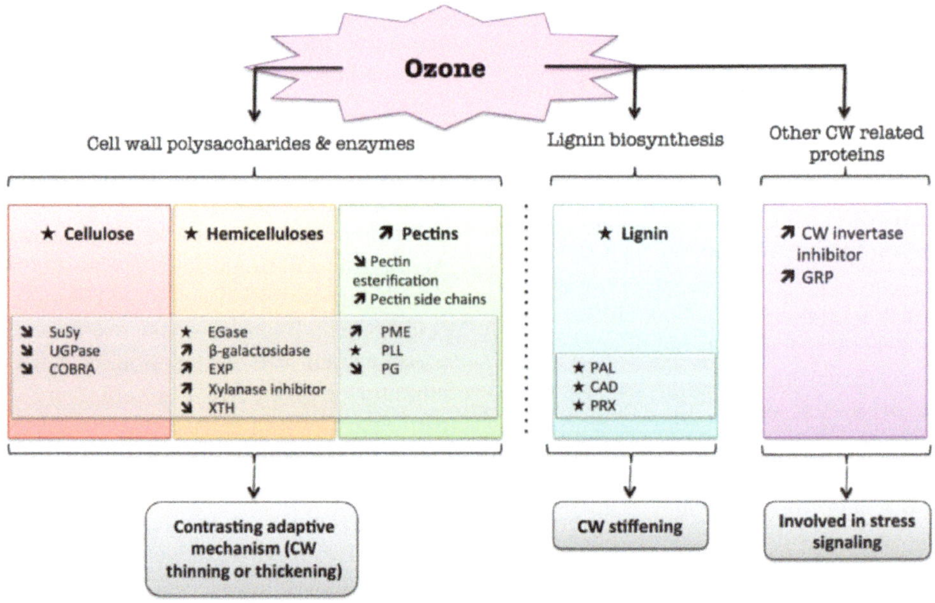

Figure 8. Diagram summarizing the plant cell wall response to ozone. The schematic presentation is deduced from the results of different studies reported in the review. Arrow (↗) means increased abundance and arrow (↘) means decreased abundance of the molecules. Star (★) means contrasting data on the gene/protein or the molecule studied according to the literature cited in the review. Cell wall (CW); xyloglucan endo-β-transglucosylases/hydrolases (XET/XTH); expansin (EXP); sucrose synthase (SuSy); UDP-glucose pyrophosphorylase (UGPase); endoglucanase (EGase); pectin methylesterase (PME); pectin/pectate lyase-like (PLL); polygalacturonase (PG); phenylalanine ammonia-lyase (PAL); hydroxycinnamyl alcohol dehydrogenase (CAD); glycine-rich protein (GRP); cell wall peroxidases (PRX).

CO_2

The atmospheric CO_2 level has increased by 30% over the last 250 years and is forecast to rise by 1%–2% per year [311]. This has an impact on plant development as C3 crops show a significant increase in photosynthesis and growth in response to rising CO_2 while C4 plants, including maize, Miscanthus, and sugarcane differ in their response [312,313].

In woody plants, including poplar, growing under elevated CO_2, trees have a larger leaf size, increased stem and branch growth and root biomass [314,315,316,317,318,319,320]. CO_2 responsiveness was found to be dependent on genotype. The unresponsive aspen clone grew slower than the responsive one, suggesting different strategies used in carbon partitioning [317]. A significant decrease in cell wall thickness was observed in the CO_2-responsive clone [317]. During the late season, the CO_2 unresponsive clone displayed a cell wall thickening in the leaves, which is likely to be related to an increased level of UDP-sugar pyrophosphorylase transcripts. In contrast, a long term exposure (11 years) to an elevated CO_2, decreased cell wall thickness in different aspen genotypes and in sweetgum (*Liquidambar styraciflua* L.), suggesting that a short term exposure may not provide a realistic view of a long term exposure [319,320]. In addition, no significant changes in the cell wall sugar content were observed [320]. In herbaceous plants, including Arabidopsis, grown under elevated CO_2 conditions, a rise in whole plant biomass (+20%) was observed before flowering. UDP-Glc dehydrogenase, a key sugar in the nucleotide interconversion pathway, necessary for cell wall polysaccharide biosynthesis, was increased in these young plants. However, although the CO_2 concentration increased the dry weight of the cell wall, it had no significant effect on cell wall composition as observed in sweetgum [320,321]. In Arabidopsis, an increase in plasmodesmata formation in the midrib of the first leaves exposed to elevated CO_2 was observed. This might be related to a higher sucrose translocation in the sieve elements [322]. In barley, under ambient CO_2, cell wall elasticity was decreased compared to elevated CO_2 levels, which could be considered as a positive mechanism for maintaining cell turgor and preventing cell wall rupture [323]. In addition, high CO_2 was shown to have an effect on post-harvest biochemical and textural properties of white asparagus spears, based on changes in cell wall components [324]. For instance, during post-harvest storage, high CO_2 levels inhibited the excessive accumulation of secondary wall components, including cellulose and lignin, at room temperature (20 °C), thus decreasing the toughening of asparagus spears. Pectin content was less influenced by elevated CO_2 compared to secondary cell wall components. In contrast, at lower temperature (10 °C) and elevated CO_2 atmosphere, no significant changes in cell wall composition and distribution were observed compared with cold applied on its own. Thickening of cell walls and lignification could be delayed by elevated CO_2 at room temperature. This could explain why elevated CO_2 partially inhibited PAL activity, as observed in green asparagus grown in an oxygen modified atmosphere [325]. In arrowhead, shoot elongation was stimulated by elevated CO_2. This growth phenotype was associated with higher transcript levels of genes encoding *EXP* and *XTH* genes, both of which are known to control cell wall loosening during cell growth [326]. MicroRNAs which are important regulators of gene expression are altered under long-term elevated CO_2 in Arabidopsis. For instance, miRPAL2 which regulates, cellulose synthase genes including *AtCesA10*, pectin biosynthesis gene including galacturonosyltransferase-like 9 or glucosyltransferase family 8 (GT8) is up-regulated, indicating an alteration of plant cell wall biosynthesis under elevated CO_2 [327].

Another, transcriptome analysis performed on old aspen trees after 12 years of exposure to elevated CO_2, showed a significantly up-regulatation of genes encoding enzymes involved in cell wall loosening and expansion (*XTH*, *EXP*). This indicates a radial growth expansion in response to elevated CO_2 [318]. In addition, other genes involved in cellulose biosynthesis, non-cellulosic and pectin biosynthesis had increased transcript levels. The overall average expression of these genes was 33% higher in the vascular cambium zone (VCZ), which reflects a more intensive cell wall biosynthesis there. Pectin-remodeling enzymes, including pectate lyase, pectin acetylesterase and pectin methylesterase, were highly up-regulated in response to elevated CO_2, and a *PME* orthologous to *AtPME3* was up-regulated by at least 25% in the VCZ. Similar patterns of expression were shown for a putative pectin acetylesterase gene and a number of NDP-sugar epimerases (UDP-glucuronate 4-epimerase, UDP-glucose 4-epimerase, and UDP-arabinose 4-epimerase (MUR4)). Overall, this indicates a significant increase in pectin and hemicellulose biosynthesis in vascular meristematic cells in response to growth under elevated CO_2. In contrast, genes involved in lignin biosynthesis were mostly down-regulated in the VCZ (24%) and in leaves (11.4%) indicating that down-regulation of lignin biosynthesis is important for maintaining cell growth in this zone [318]. In another atudy, performed on Aspen leaves of both unresponsive and CO_2 responsive genotypes revealed a distinct response between them during both early and late growing seasons. During the early season, the expression of genes encoding cell wall glycosyltransferase (GT43) was increased in both clones in elevated CO_2 conditions. In contrast, some other genes encoding cell wall-remodeling enzymes were differentially regulated. For instance, genes encoding enzymes involved in secondary cell wall metabolism, including cell wall *PRX*, were up-regulated in the CO_2-unresponsive clone. Furthermore, a strong up-regulation of NAC protein (for NAM, no apical meristem), a transcription factor known to regulate secondary cell wall synthesis genes in fibers, was observed in the sensitive clone leaves. This would suggest that NAC could control the carbon deposition in the leaf secondary cell wall thickening of the sensitive clone in response to elevated CO_2 [317]. In contrast, in the genotype that displayed a high responsiveness to elevated CO_2, genes encoding enzymes involved in lignin biosynthesis, including *CCoAOMT* and laccase, were down-regulated in leaves during the early growing season. No significant changes were observed in the late growing season in that genotype. A similar trend was observed for fascilin-like arabinogalactan protein and pectin methylesterase genes. The responsive clone is likely to have a strategy that maintains the ability to grow while the sensitive one devotes more energy to passive defense, *i.e.*, lignin biosynthesis and cell wall thickening.

All these data confirm that atmospheric pollution has an impact on plant cell wall metabolism (Figure 8 and Figure 9). However, these data are scarce and studies need to be developed to supply plants adapted to such environments in the future.

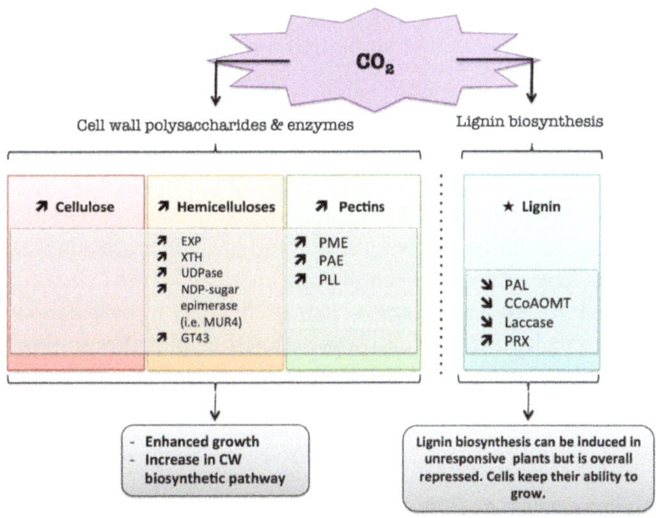

Figure 9. Diagram summarizing the plant cell wall response to elevated CO_2. The schematic presentation is deduced from the results of different studies reported in the review. Arrow (↗) means increased abundance and arrow (↘) means decreased abundance of the molecules. Star (★) means contrasting data on the gene/protein or the molecule studied according to the literature cited in the review. Cell wall (CW); xyloglucan endo-β-transglucosylases/hydrolases (XET/XTH); expansin (EXP); UDP-glucose pyrophosphorylase (UGPase); pectin methylesterase (PME); pectin acetylesterase (PAE); pectin/pectate lyase-like (PLL); glycosyltransferase family 43 (GT43); phenylalanine ammonia-lyase (PAL); caffeoyl-CoA 3-*O*-methyl-transferase (CCoAOMT); cell wall peroxidases (PRX).

CONCLUSIONS

The present review outlined the impact of various abiotic stresses on the plant cell wall. Overall, cell wall architecture is affected by abiotic stress, but the extent of the available data differs depending on the stress considered. Although changes in the cell wall in response to drought or cold have been quite well documented, data concerning some other stresses, including flooding and air pollutants, are scarcer. Furthermore, the data available in the literature are more related to transcriptome and proteome analysis than to the detailed chemical composition of the cell wall in response to stress. Stress exposure, depending on the species and stress intensity, appears to result in both plant cell wall loosening and tightening. One possible strategy is thus to improve the viscoelastic properties of the primary wall by increasing levels of cell wall remodeling and biosynthesis enzymes, and by modulating other wall loosening agents, including pectin, thus contributing to increasing the hydration status of the plant and maintaining turgor pressure for growth (Figure 10A). This could enable a relative growth status to be maintained. Another option is to modulate the viscoelastic properties by reinforcing the secondary wall with cellulose deposition and non-cellulosic components. Tightening which corresponds to a decrease in cell wall extensibility and/or turgor pressure is often accompanied by lignin biosynthesis in response to stress, except during flooding, when genes and/or enzymes involved in lignin biosynthesis are repressed/inhibited (Figure 10B). However, as the review reports, the stress effects on plant cell wall architecture depend on the plant species, plant genotype, and the age of the plant, as shown under ozone exposure or in response to cold. It also relies on the timing and intensity of the stress, as shown under flooding. Clearly, cell wall adjustment under abiotic stress is an important phenomenon in plant adaptation.

Although much progress has been made over the last decade in understanding the consequences of abiotic stress on cell wall metabolism, a multidisciplinary approach including physiology, biochemistry, proteomics and genetics based on one species for each given stress could provide more consistent results to elucidate the complex mechanism of the cell wall in its response to abiotic stress. Research needs to be pursued to engineer plants with cell walls adapted to global warming in the future.

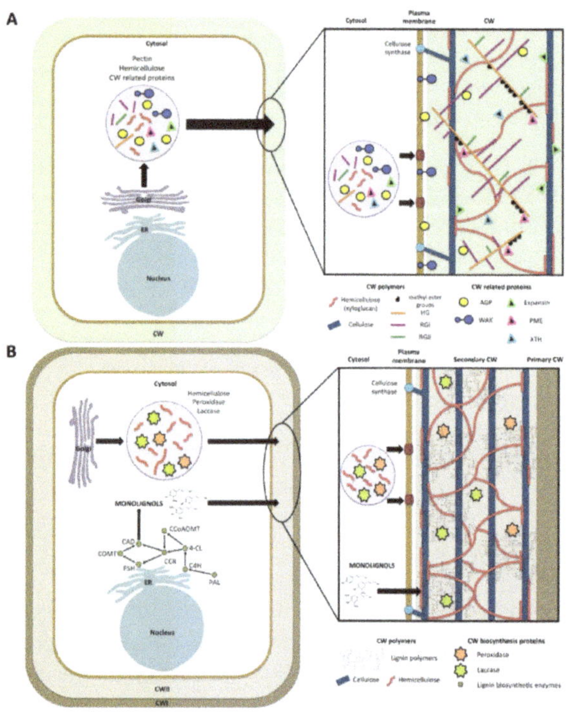

Figure 10. Model depicting changes in the plant cell wall mechanism in response to abiotic stress. (**A**) Key points of the dynamics of type I primary cell wall (CW) during abiotic stress exposure are shown. Cell wall polymer biosynthesis takes place at different locations within the cell. During cell wall formation, hemicelluloses and pectin are synthesized in the Golgi apparatus of the plant cell and then secreted to the apoplastic space while cellulose synthesis occurs at the plasma membrane. The organization and interactions of wall components is coordinated with dynamic assembly in the apoplast and rearrangement occurs during wall extension [328]. During growth under non-optimal environmental conditions, the primary cell wall composition is altered. In several abiotic stresses, cellulose, hemicellulose and pectin contents are changed. Different adaptive mechanisms can be observed, which depend on the species considered, the organ, the tissue, the age, and the exposure intensity of the stress. In most "tolerant" species, the biosynthesis of cellulose and xyloglucan, which is the most abundant non-cellulosic component of type I primary walls, is induced. This is associated with an up-regulation of the expression of genes encoding-*EXP* (expansin) and *XTH* (xyloglucan endo-β-transglucosylases/hydrolases). In parallel, levels of rhamnogalacturonan I (RGI) enriched in arabinan and/or galactan side chains are increased. The level of methylesterification of homogalacturonan (HG), regulated by PME (pectin methylesterase), is reduced. All these modifications of the cell wall architecture lead to a relative maintenance of cell wall extensibility, adapted to coping with a particular abiotic stress. For species that are "sensitive" to a given abiotic stress, cell wall degradation is observed, related to a decrease in cell wall polysaccharide content, an increase in cell wall hydrolases (EGase) and a decrease in cell wall biosynthesis and remodeling enzymes (PME, PG, SuSy). Abiotic stress can also alter wall-associated kinases (WAK), which are required for cell elongation and development. In plants that display tolerance to abiotic stress, the expression of genes encoding *WAK* are up-regulated, which suggests a perception of the stress at the cell wall/plasma membrane interface through the detection of released plant cell wall fragments [187,188,222]. Genes encoding other cell wall proteins, including arabinogalactan protein (AGP), proline-rich protein (PRP) and glycine-rich protein (GRP) are induced in response to abiotic stress, which could covalently link with pectin or hemicellulose thus contributing to the strengthening of the wall [329]. Another possibility is that HRGP is a signal molecule in plant defense [330]; (**B**) Key points of the secondary cell wall (CWII) during abiotic stress exposure are shown. The model is adapted from [78]. To cope with abiotic stress, some plants adopt a strategy that consists of increasing the secondary cell wall thickening by hemicellulose and cellulose synthesis [147,295]. This is often associated with a rigidification of the secondary wall by lignin deposition [23,182,282]. Monolignols, which are the building blocks of lignin, are synthesized from phenylalanine through the general phenylpropanoid and monolignol-specific pathways. This involves cytosolic enzymes including phenylalanine ammonia-lyase (PAL), 4 coumarate CoA-ligase (4CL), cinnamoyl-CoA reductase (CCR), caffeate *O*-methyltransferase (COMT), and caffeoyl-CoA 3-*O*-methyl-transferase (CCoAOMT) as well as ER membrane-anchored proteins including trans-cinnamate 4-hydroxylase (C4H), and ferulate 5-hydroxylase (F5H). The monolignols are then transported to the cell wall where they are polymerized by apoplastic peroxidase (PRX) and laccases into lignin [117,144]. This other adaptive mechanism, based on a decrease in cell wall expansion and cell extensibility, can limit water loss and prevent cell collapse due to dehydration.

Acknowledgments

This work was funded by grants from the Agence Nationale de la Recherche (GALAPAGOS, ANR-12-BSV5-0001 and NOSTRESSWALL, ANR-12-ADAP-0011) and the Conseil Regional de Picardie (PhD studentship awarded to F.P.). The funding of the Institut Universitaire de France (IUF) to J.P. is gratefully acknowledged.

Author Contributions

All authors contributed to this review.

Conflicts of Interest

The authors declare no conflict of interest.

REFERENCES

1. Shaik, R.; Ramakrishna, W. Machine learning approaches distinguish multiple stress conditions using stress-responsive genes and identify candidate genes for broad resistance in rice. *Plant Physiol.* **2014**, *164*, 481–495.
2. Wu, Q.; Shu, P.-C.; Cai, T.; Fan, X.-W.; Li, Y.-Z. Existence of abiotic stress-responsive genes within the regions of QTLs controlling maize grain yield: One of the root causes for unstability of the QTLs? *S. Afr. J. Bot.* **2014**, *93*, 231–241.
3. Wang, Y.; Frei, M. Stressed food—The impact of abiotic environmental stresses on crop quality. *Agric. Ecosyst. Environ.***2011**, *141*, 271–286.
4. Dolferus, R. To grow or not to grow: A stressful decision for plants. *Plant Sci.* **2014**, *229*, 247–261.

5. Potters, G.; Jansen, M.A.K.; Guisez, Y.; Pasternak, T. Stress drives plant cells to take the road towards embryogenesis. In *Floriculture, Ornamental and Plant Biotechnology, Advances and Topical Issues*; Teixeira da Silva, J.A., Ed.; Global Science Books Ltd.: London, UK, 2006; Volume 2, pp. 289–294.
6. Wrzaczek, M.; Brosché, M.; Kangasjärvi, J. ROS signaling loops—Production, perception, regulation. *Curr. Opin. Plant Biol.* **2013**, *16*, 575–582.
7. Gill, S.S.; Tuteja, N. Reactive oxygen species and antioxidant machinery in abiotic stress tolerance in crop plants. *Plant Physiol. Biochem.* **2010**, *48*, 909–930.
8. McCann, M.; Roberts, K. Architecture of the primary cell wall. In *The Cytoskeletal Basis of Plant Growth and Form*; Lloyd, C.W., Ed.; Academic Press: London, UK, 1991; pp. 109–129.
9. Carpita, N.C.; Gibeaut, D.M. Structural models of primary cell walls in flowering plants: Consistency of molecular structure with the physical properties of the walls during growth. *Plant J.* **1993**, *3*, 1–30.
10. Carpita, N.C.; McCann, M.C. Maize and sorghum: Genetic resources for bioenergy grasses. *Trends Plant Sci.* **2008**, *13*, 415–420.
11. Carpita, N.C. Structure and biogenesis of the cell walls of grasses. *Annu. Rev. Plant Physiol. Plant Mol. Biol.* **1996**, *47*, 445–476.
12. Grabber, J.H.; Ralph, J.; Lapierre, C.; Barrière, Y. Genetic and molecular basis of grass cell-wall degradability. I. Lignin–cell wall matrix interactions. *C. R. Biol.* **2004**, *327*, 455–465.
13. Sampedro, J.; Cosgrove, D.J. The expansin superfamily. *Genome Biol.* **2005**, *6*, 242.
14. Eklöf, J.M.; Brumer, H. The XTH gene family: An update on enzyme structure, function, and phylogeny in xyloglucan remodeling. *Plant Physiol.* **2010**, *153*, 456–466.
15. Sénéchal, F.; Wattier, C.; Rustérucci, C.; Pelloux, J. Homogalacturonan-modifying enzymes: Structure, expression, and roles in plants. *J. Exp. Bot.* **2014**, *65*, 5125–5160.
16. Liepman, A.H.; Wightman, R.; Geshi, N.; Turner, S.R.; Scheller, H.V. Arabidopsis a powerful model system for plant cell wall research. *Plant J.* **2010**, *61*, 1107–1121.
17. Vanholme, R.; Demedts, B.; Morreel, K.; Ralph, J.; Boerjan, W. Lignin Biosynthesis and Structure. *Plant Physiol.* **2010**,*153*, 895–905.
18. Tronchet, M.; Balagué, C.; Kroj, T.; Jouanin, L.; Roby, D. Cinnamyl alcohol dehydrogenases-C and D, key enzymes in lignin biosynthesis, play an essential role in disease resistance in Arabidopsis. *Mol. Plant Pathol.* **2010**, *11*, 83–92.
19. Hamann, T. Plant cell wall integrity maintenance as an essential component of biotic stress response mechanisms.*Front. Plant Sci.* **2012**, *3*, 77.
20. Seifert, G.J.; Blaukopf, C. Irritable walls: The plant extracellular matrix and signaling. *Plant Physiol.* **2010**, *153*, 467–478.
21. Sasidharan, R.; Voesenek, L.A.; Pierik, R. Cell wall modifying proteins mediate plant acclimatization to biotic and abiotic stresses. *Crit. Rev. Plant Sci.* **2011**, *30*, 548–562.
22. Moore, J.P.; Vicré-Gibouin, M.; Farrant, J.M.; Driouich, A. Adaptations of higher plant cell walls to water loss: Drought *vs.* desiccation. *Physiol. Plant* **2008**, *134*, 237–245.
23. Moura, J.C.M.S.; Bonine, C.A.V.; de Oliveira Fernandes Viana, J.; Dornelas, M.C.; Mazzafera, P. Abiotic and biotic stresses and changes in the lignin content and composition in plants. *J. Integr. Plant Biol.* **2010**, *52*, 360–376.
24. Cabane, M.; Afif, D.; Hawkins, S. Lignins and abiotic stresses. *Adv. Bot. Res.* **2013**, *61*, 220.
25. Zagorchev, L.; Kamenova, P.; Odjakova, M. The role of plant cell wall proteins in response to salt stress. *Sci. World J.***2014**. Article ID 764089.
26. Zhao, Q.; Zhang, H.; Wang, T.; Chen, S.; Dai, S. Proteomics-based investigation of salt-responsive mechanisms in plant roots. *J. Proteomics* **2013**, *82*, 230–253.
27. Krzeslowska, M. The wall cell in plant cell response to trace metals: Polysaccharide remodelling and its role in defense strategy. *Acta Physiol. Plant.* **2011**, *33*, 35–51.
28. Ovečka, M.; Takáč, T. Managing heavy metal toxicity stress in plants: Biological and biotechnological tools. *Biotechnol. Adv.* **2014**, *32*, 73–86.
29. Chen, X.Y.; Kim, J.Y. Callose synthesis in higher plants. *Plant Signal. Behav.* **2009**, *4*, 489–492.
30. Stass, A.; Horst, W.J. Callose in abiotic stress. In *Chemistry, Biochemistry and Biology of (1 3)- -Glucans and Related Polysaccharides*; Bacic, A., Fincher, G.B., Stone, B.A., Eds.; Academic: New York, NY, USA, 2009; pp. 499–524.
31. Guzmán, P.; Fernández, V.; Graça, J.; Cabral, V.; Kayali, N.; Khayet, M.; Gil, L. Chemical and structural analysis of*Eucalyptus globulus* and *E. camaldulensis* leaf cuticles: A lipidized cell wall region. *Front. Plant Sci.* **2014**, *5*.
32. Cosgrove, D.J.; Li, Z.C. Role of expansin in cell enlargement of oat coleoptiles (analysis of developmental gradients and photocontrol). *Plant Physiol.* **1993**, *103*, 1321–1328.
33. Tardieu, F.O.; Granier, C.; Muller, B. Water deficit and growth. Co-ordinating processes without an orchestrator? *Curr. Opin. Plant Biol.* **2011**, *14*, 283–289.
34. Zlatev, Z.; Cebola Lidon, F. An overview on drought induced changes in plant growth, water relations and photosynthesis. *Emir. J. Food Agric.* **2012**, *24*, 57–72.
35. Shao, H.-B.; Chu, L.-Y.; Jaleel, C.A.; Zhao, C.-X. Water-deficit stress-induced anatomical changes in higher plants. *C. R. Biol.* **2008**, *331*, 215–225.
36. Lisar, S.Y.S.; Motafakkerazad, R.; Hossain, M.M.; Rahman, I.M.M. Water stress in plants: Causes, effects and responses. In *Water Stress*; Rahman, M., Hasegawa, H., Eds.; InTech: Rijeka, Croatia, 2012; pp. 1–14.
37. Tardieu, F. Any trait or trait-related allele can confer drought tolerance: Just design the right drought scenario. *J. Exp. Bot.* **2011**, *63*, 25–31.

38. Wolf, S.; Greiner, S. Growth control by cell wall pectins. *Protoplasma* **2012**, *249*, 169–175.

39. Tardieu, F.; Parent, B.; Caldeira, C.F.; Welcker, C. Genetic and physiological controls of growth under water deficit.*Plant Physiol.* **2014**, *164*, 1628–1635.

40. Balducci, L.; Deslauriers, A.; Giovannelli, A.; Beaulieu, M.; Delzon, S.; Rossi, S.; Rathgeber, C.B.K. How do drought and warming influence survival and wood traits of *Picea mariana* saplings? *J. Exp. Bot.* **2014**, *4*, 1–13.

41. Hura, T.; Hura, K.; Dziurka, K.; Ostrowska, A.; Bączek-Kwinta, R.; Grzesiak, M. An increase in the content of cell wall-bound phenolics correlates with the productivity of triticale under soil drought. *J. Plant Physiol.* **2012**, *169*, 1728–1736.

42. Izanloo, A.; Condon, A.G.; Langridge, P.; Tester, M.; Schnurbusch, T. Different mechanisms of adaptation to cyclic water stress in two South Australian bread wheat cultivars. *J. Exp. Bot.* **2008**, *59*, 3327–3346.

43. Beikircher, B.; de Cesare, C.; Mayr1, S. Hydraulics of high-yield orchard trees: A case study of three *Malus domestica*cultivars. *Tree Physiol.* **2013**, *33*, 1296–1307.

44. Chartzoulakis, K.; Patakas, A.; Kofidis, G.; Bosabalidis, A.; Nastou, A. Water stress affects leaf anatomy, gas exchange, water relations and growth of two avocado cultivars. *Sci. Hortic.* **2002**, *95*, 39–50.

45. Saito, T.; Terashima, I. Reversible decreases in the bulk elastic modulus of mature leaves of deciduous *Quercus* species subjected to two drought treatments. *Plant Cell Environ.* **2004**, *27*, 863–875.

46. De Diego, N.; Sampedro, M.C.; Barrio, R.J.; Saiz-Fernandez, I.; Moncalean, P.; Lacuesta, M. Solute accumulation and elastic modulus changes in six radiata pine breeds exposed to drought. *Tree Physiol.* **2013**, *33*, 69–80.

47. Martínez, J.P.; Silva, H.; Ledent, J.F.; Pinto, M. Effect of drought stress on the osmotic adjustment, cell wall elasticity and cell volume of six cultivars of common beans (*Phaseolus vulgaris* L.). *Eur. J. Agron.* **2007**, *26*, 30–38.

48. Clifford, S.C.; Arndt, S.K.; Corlett, J.E.; Joshi, S.; Sankhla, N.; Popp, M.; Jones, H.G. The role of solute accumulation, osmotic adjustment and changes in cell wall elasticity in drought tolerance in *Ziziphus mauritiana* (Lamk.). *J. Exp. Bot.***1998**, *49*, 967–977.

49. Hessini, K.; Martínez, J.P.; Gandour, M.; Albouchi, A.; Soltani, A.; Abdelly, C. Effect of water stress on growth, osmotic adjustment, cell wall elasticity and water-use efficiency in *Spartina alterniflora*. *Environ. Exp. Bot.* **2009**, *67*, 312–319.

50. Iraki, N.M.; Bressan, R.A.; Hasegawa, P.M.; Carpita, N.C. Alteration of the physical and chemical structure of the primary cell wall of growth-limited plant cells adapted to osmotic stress. *Plant Physiol.* **1989**, *91*, 39–47.

51. Sweet, W.J.; Morrison, J.C.; Labavitch, J.M.; Matthews, M.A. Altered synthesis and composition of cell wall of grape (*Vitis vinifera* L.) leaves during expansion and growth inhibiting water deficit. *Plant Cell Physiol.* **1990**, *31*, 407–414.

52. Piro, G.; Leucci, M.R.; Waldron, K.; Dalessandro, G. Exposure to water stress causes changes in the biosynthesis of cell wall polysaccharides in roots of wheat cultivars varying in drought tolerance. *Plant Sci.* **2003**, *165*, 559–569.

53. Bray, E.A. Genes commonly regulated by water-deficit stress in *Arabidopsis thaliana*. *J. Exp. Bot.* **2004**, *55*, 2331–2341.

54. Zheng, M.; Meng, Y.; Yang, C.; Zhou, Z.; Wang, Y.; Chen, B. Protein expression changes during cotton fiber elongation in response to drought stress and recovery. *Proteomics* **2014**, *14*, 1776–1795.

55. Ricardi, M.M.; Gonzalez, R.M.; Zhong, S.; Dominguez, P.G.; Duffy, T.; Turjanski, P.G.; Salter, J.D.S.; Alleva, K.; Carrari, F.; Giovannoni, J.J. Genome-wide data (ChIP-seq) enabled identification of cell wall-related and aquaporin genes as targets of tomato ASR1, a drought stress-responsive transcription factor. *BMC Plant Biol.* **2014**, *14*, 29.

56. Yang, L.; Wang, C.C.; Guo, W.D.; Li, X.B.; Lu, M.; Yu, C.L. Differential expression of cell wall related genes in the elongation zone of rice roots under water deficit. *Russ. J. Plant Physiol.* **2006**, *53*, 390–395.

57. Cho, S.K.; Kim, J.E.; Park, J.A.; Eom, T.J.; Kim, W.T. Constitutive expression of abiotic stress-inducible hot pepper*CaXTH3*, which encodes a xyloglucan endotransglucosylase/hydrolase homolog, improves drought and salt tolerance in transgenic *Arabidopsis* plants. *FEBS Lett.* **2006**, *580*, 3136–3144.

58. Choi, J.Y.; Seo, Y.S.; Kim, S.J.; Kim, W.T.; Shin, J.S. Constitutive expression of *CaXTH3*, a hot pepper xyloglucan endotransglucosylase/hydrolase, enhanced tolerance to salt and drought stresses without phenotypic defects in tomato plants (*Solanum lycopersicum* cv. Dotaerang). *Plant Cell Rep.* **2011**, *30*, 867–877.

59. Zhu, J.; Alvarez, S.; Marsh, E.L.; LeNoble, M.E.; Cho, I.J.; Sivaguru, M.; Chen, S.; Nguyen, H.T.; Wu, Y.; Schachtman, D.P.; *et al.* Cell wall proteome in the maize primary root elongation zone. II. Region-specific changes in water soluble and lightly ionically bound proteins under water deficit. *Plant Physiol.* **2007**, *145*, 1533–1548.

60. Jones, L.; McQueen-Mason, S. A role for expansins in dehydration and rehydration of the resurrection plant*Craterostigma plantagineum*. *FEBS Lett.* **2004**, *559*, 61–65.

61. Wu, Y.; Sharp, R.E.; Durachko, D.M.; Cosgrove, D.J. Growth maintenance of the maize primary root at low water potentials involves increases in cell-wall extension properties, expansin activity, and wall susceptibility to expansins.*Plant Physiol.* **1996**, *111*, 765–772.

62. Wu, Y.; Thorne, E.T.; Sharp, R.E.; Cosgrove, D.J. Modification of expansin transcript levels in the maize primary root at low water potentials. *Plant Physiol.* **2001**, *126*, 1471–1479.

63. Wu, Y.; Cosgrove, D.J. Adaptation of roots to low water potentials by changes in cell wall extensibility and cell

wall proteins. *J. Exp. Bot.* **2000**, *51*, 1543–1553.

64. Guo, W.; Zhao, J.; Li, X.; Qin, L.; Yan, X.; Liao, H. A soybean β-expansin gene *GmEXPB2* intrinsically involved in root system architecture responses to abiotic stresses. *Plant J.* **2011**, *66*, 541–552.

65. Li, F.; Han, Y.; Feng, Y.; Xing, S.; Zhao, M.; Chen, Y.; Wang, W. Expression of wheat expansin driven by the *RD29*promoter in tobacco confers water-stress tolerance without impacting growth and development. *J. Biotechnol.* **2013**,*163*, 281–291.

66. Dai, F.; Zhang, C.; Jiang, X.; Kang, M.; Yin, X.; Lu, P.; Zhang, X.; Zheng, Y.; Gao, J. *RhNAC2* and *RhEXPA4* are involved in the regulation of dehydration tolerance during the expansion of rose petals. *Plant Physiol.* **2012**, *160*, 2064–2082.

67. Lü, P.; Kang, M.; Jiang, X.; Dai, F.; Gao, J.; Zhang, C. *RhEXPA4*, a rose expansin gene, modulates leaf growth and confers drought and salt tolerance to Arabidopsis. *Planta* **2013**, *237*, 1547–1559.

68. Wakabayashi, K.; Hoson, T.; Kamisaka, S. Changes in amounts and molecular mass distribution of cell-wall polysaccharides of wheat (*Triticum aestivum* L.) coleoptiles under water. *J. Plant Physiol.* **1997**, *151*, 33–40.

69. Peaucelle, A.; Braybrook, S.; Höfte, H. Cell wall mechanics and growth control in plants: The role of pectins revisited.*Front Plant Sci.* **2012**, *3*, 1–6.

70. Leucci, M.R.; Lenucci, M.S.; Piro, G.; Dalessandro, G. Water stress and cell wall polysaccharides in the apical root zone of wheat cultivars varying in drought tolerance. *J. Plant Physiol.* **2008**, *165*, 1168–1180.

71. Gribaa, A.; Dardelle, F.; Lehner, A.; Rihouey, C.; Burel, C.; Ferchichi, A.; Driouich, A.; Mollet, J.C. Effect of water deficit on the cell wall of the date palm (*Phoenix dactylifera* "Deglet nour", Arecales) fruit during development. *Plant Cell Environ.* **2013**, *36*, 1056–1070.

72. Moore, J.P.; Nguema-Ona, E.; Fagerström, A.; Fangel, J.U.; Willats, W.G.T.; Hugo, A.; Vivier, M.A. Profiling the main cell wall polysaccharides of grapevine leaves using high throughput methods. *Carbohydr. Polym.* **2014**, *99*, 190–198.

73. Pandey, A.; Rajamani, U.; Verma, J.; Subba, P.; Chakraborty, N.; Datta, A.; Chakraborty, S.; Chakraborty, N. Identification of extracellular matrix proteins of rice (*Oryza sativa* L.) involved in dehydration-responsive network: A proteomic approach. *J. Proteome Res.* **2010**, *9*, 3443–3464.

74. Atkinson, R.G.; Schröder, R.; Hallett, I.C.; Cohen, D.; MacRae, E. Overexpression of polygalacturonase in transgenic apple trees leads to a range of novel phenotypes involving changes in cell adhesion. *Plant Physiol.* **2002**, *129*, 122–133.

75. Tucker, G.A.; Robertson, N.G.; Grierson, D. The conversion of tomato fruit polygalacturonase isoenzyme 2 into isoenzyme 1 *in vitro*. *Eur. J. Biochem.* **1981**, *115*, 87–90.

76. Liu, H.; Ma, Y.; Chen, N.A.; Guo, S.; Liu, H.; Guo, X.; Chong, K.; Xu, Y. Overexpression of stress-inducible *OsBURP16*, the β subunit of polygalacturonase 1, decreases pectin content and cell adhesion and increases abiotic stress sensitivity in rice. *Plant Cell Environ.* **2013**, *37*, 1144–1158.

77. An, S.H.; Sohn, K.H.; Choi, H.W.; Hwang, I.S.; Lee, S.C.; Hwang, B.K. Pepper pectin methylesterase inhibitor protein CaPMEI1 is required for antifungal activity, basal disease resistance and abiotic stress tolerance. *Planta* **2008**, *228*, 61–78.

78. Wang, Y.; Chantreau, M.; Sibout, R.; Hawkins, S. Plant cell wall lignification and monolignol metabolism. *Front. Plant Sci.* **2013**, *4*, 220.

79. Terzi, R.; Güler, N.S.; Çaliskan, N.; Kadioglu, A. Lignification response for rolled leaves of *Ctenanthe setosa* under long-term drought stress. *Turk. J. Biol.* **2013**, *37*, 614–619.

80. Yoshimura, K.; Masuda, A.; Kuwano, M.; Yokota, A.; Akashi, K. Programmed proteome response for drought avoidance/tolerance in the root of a C3 xerophyte (wild watermelon) under water deficits. *Plant Cell Physiol.* **2007**, *49*, 226–241.

81. Fan, L. Progressive inhibition by water deficit of cell wall extensibility and growth along the elongation zone of maize roots is related to increased lignin metabolism and progressive stelar accumulation of wall phenolics. *Plant Physiol.***2006**, *140*, 603–612.

82. Moura-Sobczak, J.; Souza, U.; Mazzafera, P. Drought stress and changes in the lignin content and composition in Eucalyptus. *BMC Proc.* **2011**, *5*, 103.

83. Lin, C.C.; Kao, C.H. Osmotic stress-induced changes in cell wall peroxidase activity and hydrogen peroxide level in roots of rice seedlings. *Plant Growth Regul.* **2002**, *37*, 177–184.

84. Lee, B.R.; Kim, K.Y.; Jung, W.J.; Avice, J.C.; Ourry, A.; Kim, T.H. Peroxidases and lignification in relation to the intensity of water-deficit stress in white clover (*Trifolium repens* L.). *J. Exp. Bot.* **2007**, *58*, 1271–1279.

85. Bacon, M.A.; Thompson, D.S.; Davies, W.J. Can cell wall peroxidase activity explain the leaf growth response of*Lolium temulentum* L. during drought? *J. Exp. Bot.* **1997**, *48*, 2075–2085.

86. Hura, T.; Hura, K.; Ostrowska, A.; Grzesiak, M.; Dziurka, K. The cell wall-bound phenolics as a biochemical indicator of soil drought resistance in winter triticale. *Plant Soil Environ.* **2013**, *59*, 189–195.

87. Christianson, J.A.; Llewellyn, D.J.; Dennis, E.S.; Wilson, I.W. Globalgene expression responses to waterlogging in roots and leaves of cotton (*Gossypium hirsutum* L.). *Plant Cell Physiol.* **2010**, *51*, 21–37.

88. Kong, F.J.; Oyanagi, A.; Komatsu, S. Cell wall proteome of wheat roots under flooding stress using gel-based and LC MS/MS-based proteomics approaches. *Biochim. Biophys.* **2010**, *1804*, 124–136.

89. Salavati, A.; Khatoon, A.; Nanjo, Y.; Komatsu, S. Analysis of proteomic changes in roots of soybean seedlings during recovery after flooding. *J. Proteomics* **2012**, *75*, 878–893.

90. Nanjo, Y.; Nakamura, T.; Komatsu, S. Identification

of indicator proteins associated with flooding injury in soybean seedlings using label-free quantitative proteomics. *J. Proteome Res.* **2013**, *12*, 4758–4798.

91. Rauf, M.; Arif, M.; Fisahn, J.; Xue, G.P.; Balazadeh, S.; Mueller-Roeber, B. NAC transcription factor SPEEDY HYPONASTIC GROWTH regulates flooding-induced leaf movement in *Arabidopsis*. *Plant Cell.* **2013**, *25*, 4941–4955.

92. Vitorino, P.G.; Alves, J.D.; Magalhães, P.C.; Magalhães, M.M.; Lima, L.C.O.; Oliveira, L.E.M. Flooding tolerance and cell wall alterations in maize mesocotyl during hypoxia. *Pesq. Agropec. Bras.* **2001**, *36*, 1027–1035.

93. Ooume, K.; Inoue, Y.; Soga, K.; Wakabayashi, K.; Fujii, S.; Yamamoto, R.; Hoson, T. Cellular basis of growth suppression by submergence in azuki bean epicotyls. *Ann. Bot.* **2009**, *103*, 325–332.

94. Da Silva Ferreira, C.S.; Piedade, M.T.F.; Tine, M.A.S.; Rossatto, D.R.; Parolin, P.; Buckeridge, M.S. The role of carbohydrates in seed germination and seedling establishment of *Himatanthus sucuuba*, an Amazonian tree with populations adapted to flooded and non-flooded conditions. *Ann. Bot.* **2009**, *104*, 1111–1119.

95. Nanjo, Y.; Maruyama, K.; Yasue, H.; Yamaguchi-Shinozaki, K.; Shinozaki, K.; Komatsu, S. Transcriptional responses to flooding stress in roots including hypocotyl of soybean seedlings. *Plant Mol. Biol.* **2011**, *77*, 129–144.

96. Kreuzwieser, J.; Hauberg, J.; Howell, K.A.; Carroll, A.; Rennenberg, H.; Millar, A.H.; Whelan, J. Differential response of gray poplar leaves and roots underpins stress adaptation during hypoxia. *Plant Physiol.* **2009**, *149*, 461–473.

97. Komatsu, S.; Kobayashi, Y.; Nishizawa, K.; Nanjo, Y.; Furukawa, K. Comparative proteomics analysis of differentially expressed proteins in soybean cell wall during flooding stress. *Amino Acids* **2010**, *39*, 1435–1449.

98. Hall, A.E. Breeding for heat tolerance. *Plant Breed. Rev.* **1992**, *10*, 129–168.

99. Wahid, A.; Gelani, S.; Ashraf, M.; Foolad, M.R. Heat tolerance in plants: An overview. *Environ. Exp. Bot.* **2007**, *61*, 199–223.

100. Dolferus, R.; Ji, X.; Richards, R.A. Abiotic stress and control of grain number in cereals. *Plant Sci.* **2011**, *181*, 331–341.

101. Crafts-Brandner, S.J. Sensitivity of Photosynthesis in a C4 Plant, Maize, to Heat Stress. *Plant Physiol.* **2002**, *129*, 1773–1780.

102. Larkindale, J.; Knight, M.R. Protection against Heat Stress-Induced Oxidative Damage in Arabidopsis Involves Calcium, Abscisic Acid, Ethylene, and Salicylic Acid. *Plant Physiol.* **2002**, *128*, 682–695.

103. Kotak, S.; Larkindale, J.; Lee, U.; von Koskull-Döring, P.; Vierling, E.; Scharf, K.-D. Complexity of the heat stress response in plants. *Curr. Opin. Plant Biol.* **2007**, *10*, 310–316.

104. Hossain, A.; Sarker, M.; Saifuzzaman, M.; Teixeira da Silva, J.A.; Lozovskaya, M.V.; Akhter, M.M. Evaluation of growth, yield, relative performance and heat susceptibility of eight wheat (*Triticum aestivum* L.) genotypes grown under heat stress. *Int. J. Plant Prod.* **2013**, *7*, 615–636.

105. Liu, X.; Huang, B. Root physiological factors involved in creeping bentgrass response to high soil temperatures. *Environ. Exp. Bot.* **2005**, *53*, 233–245.

106. Wang, W.; Vinocur, B.; Altman, A. Plant responses to drought, salinity and extreme temperatures: Towards genetic engineering for stress tolerance. *Planta* **2003**, *218*, 1–14.

107. Suwa, R.; Hakata, H.; Hara, H.; El-Shemy, H.A.; Adu-Gyamfi, J.J.; Nguyen, N.T.; Kanai, S.; Lightfoot, D.A.; Mohapatra, P.K.; Fujita, K. High temperature effects on photosynthate partitioning and sugar metabolism during ear expansion in maize (*Zea mays* L.) genotypes. *Plant Physiol. Biochem.* **2010**, *48*, 124–130.

108. Lima, R.B.; Santos, dos, T.B.; Vieira, L.G.E.; de Lourdes Lúcio Ferrarese, M.; Ferrarese-Filho, O.; Donatti, L.; Boeger, M.R.T.; de Oliveira Petkowicz, C.L. Heat stress causes alterations in the cell-wall polymers and anatomy of coffee leaves (*Coffea arabica* L.). *Carbohydr. Polym.* **2013**, *93*, 135–143.

109. Guerra-Guimarães, L.; Vieira, A.; Chaves, I.; Pinheiro, C.; Queiroz, V.; Renaut, J.; Ricardo, C.P. Effect of greenhouse conditions on the leaf apoplastic proteome of *Coffea arabica* plants. *J. Proteomics* **2014**, *104*, 128–139.

110. Rakszegi, M.; Lovegrove, A.; Balla, K.; Láng, L.; Bedő, Z.; Veisz, O.; Shewry, P.R. Effect of heat and drought stress on the structure and composition of arabinoxylan and β-glucan in wheat grain. *Carbohydr. Polym.* **2014**, *102*, 557–565.

111. Yang, K.A.; Lim, C.J.; Hong, J.K.; Park, C.Y.; Cheong, Y.H.; Chung, W.S.; Lee, K.O.; Lee, S.Y.; Cho, M.J.; Lim, C.O. Identification of cell wall genes modified by a permissive high temperature in Chinese cabbage. *Plant Sci.* **2006**, *171*, 175–182.

112. Rienth, M.; Torregrosa, L.; Luchaire, N.; Chatbanyong, R.; Lecourieux, D.; Kelly, M.T.; Romieu, C. Day and night heat stress trigger different transcriptomic responses in green and ripening grapevine (*Vitis vinifera*) fruit. *BMC Plant Biol.***2013**, *14*, 108–108.

113. Xu, J.; Belanger, F.; Huang, B. Differential gene expression in shoots and roots under heat stress for a geothermal and non-thermal *Agrostis* grass species contrasting in heat tolerance. *Environ. Exp. Bot.* **2008**, *63*, 240–247.

114. Xu, Q.; Xu, X.; Shi, Y.; Xu, J.; Huang, B. Transgenic Tobacco Plants Overexpressing a Grass *PpEXP1* Gene Exhibit Enhanced Tolerance to Heat Stress. *PLoS One* **2014**, *9*, e100792.

115. Yu, E.; Fan, C.; Yang, Q.; Li, X.; Wan, B.; Dong, Y.; Wang, X.; Zhou, Y. Identification of Heat Responsive Genes in*Brassica napus* Siliques at the Seed-Filling Stage through Transcriptional Profiling. *PLoS One* **2014**, *9*, e101914.

116. Song, Y.; Ci, D.; Tian, M.; Zhang, D. Comparison of

the physiological effects and transcriptome responses of *Populus simonii* under different abiotic stresses. *Plant Mol. Biol.* **2014**, *86*, 139–156.

117. Han, F.; Chen, H.; Li, X.J.; Yang, M.F.; Liu, G.S.; Shen, S.H. A comparative proteomic analysis of rice seedlings under various high-temperature stresses. *Biochim. Biophys. Acta* **2009**, *1794*, 1625–1634.
118. Gulen, H.; Eris, A. Effect of heat stress on peroxidase activity and total protein content in strawberry plants. *Plant Sci.***2004**, *166*, 739–744.
119. Stitt, M.; Hurry, V. A plant for all seasons: Alterations in photosynthetic carbon metabolism during cold acclimation in *Arabidopsis*. *Curr. Opin. Plant Biol.* **2002**, *5*, 199–206.
120. Goulas, E.; Schubert, M.; Kieselbach, T.; Kleczkowski, L.A.; Gardeström, P.; Schröder, W.; Hurry, V. The chloroplast lumen and stromal proteomes of *Arabidopsis thaliana* show differential sensitivity to short- and long-term exposure to low temperature. *Plant J.* **2006**, *47*, 720–734.
121. Chinnusamy, V.; Zhu, J.; Zhu, J.K. Cold stress regulation of gene expression in plants. *Trends Plant Sci.* **2007**, *12*, 444–451.
122. Chang, M.-C.; Chien, W.-F.; Chao, C.-H.; Lu, M.-K. Effects of cold stress on alterations of physiochemical and chemical properties of rice polysaccharides. *Carbohydr. Polym.* **2010**, *80*, 373–376.
123. Mahajan, S.; Tuteja, N. Cold, salinity and drought stresses: An overview. *Arch. Biochem. Biophys.* **2005**, *444*, 139–158.
124. Sanghera, G.S.; Wani, S.H.; Hussain, W.; Singh, N.B. Engineering cold stress tolerance in crop plants. *Curr. Genomics***2011**, *12*, 30–43.
125. Noh, S.A.; Park, S.H.; Huh, G.H.; Paek, K.-H.; Shin, J.S.; Bae, J.M. Growth retardation and differential regulation of expansin genes in chilling-stressed sweet potato. *Plant Biotechnol. Rep.* **2009**, *3*, 75–85.
126. Stefanowska, M.; Kuras, M.; Kubacka-Zębalska, M.; Kacperska, A. Low Temperature Affects Pattern of Leaf Growth and Structure of Cell Walls in Winter Oilseed Rape (*Brassica napus* L., var *oleifera* L.). *Ann. Bot.* **1999**, *84*, 313–319.
127. Pearce, R.S. Molecular analysis of acclimation to cold. *Plant Growth Regul.* **1999**, *29*, 47–76.
128. Warren, G.J. Cold stress: Manipulating freezing tolerance in plants. *Curr. Biol.* **1998**, *8*, R514–R516.
129. Janská, A.; Maršík, P.; Zelenková, S.; Ovesná, J. Cold stress and acclimation - what is important for metabolic adjustment? *Plant Biol.* **2010**, *12*, 395–405.
130. Hajela, R.K.; Horvath, D.P.; Gilmour, S.J.; Thomashow, M.F. Molecular cloning and expression of cor (cold regulated) genes in *Arabidopsis thaliana*. *Plant Physiol.* **1990**, *93*, 1246–1252.
131. Kreps, J.A.; Wu, Y.; Chang, H.S.; Zhu, T.; Wang, X.; Harper, J.F. Transcriptome changes for Arabidopsis in response to salt, osmotic, and cold stress. *Plant Physiol.* **2002**, *130*, 2129–2141.
132. Zhu, J.; Dong, C.H.; Zhu, J.K. Interplay between cold-responsive gene regulation, metabolism and RNA processing during plant cold acclimation. *Curr. Opin. Plant Biol.* **2007**, *10*, 290–295.
133. Chen, L.; Zhong, H.; Ren, F.; Guo, Q.-Q.; Hu, X.-P.; Li, X.-B. A novel cold-regulated gene, *COR25*, of *Brassica napus* is involved in plant response and tolerance to cold stress. *Plant Cell. Rep.* **2011**, *30*, 463–471.
134. Janska, A.; Marsik, P.; Zelenkova, S.; Ovesna, J. Cold stress and acclimation-what is important for metabolic adjustment? *Plant Biol.* **2010**, *12*, 395–405.
135. Levitt, J. Chilling, freezing, and high temperature stresses. In *Responses of Plants to Environmental Stresses*, 2nd ed.; Academic Press: Orlando, FL, USA, 1980.
136. Kume, S.; Kobayashi, F.; Ishibashi, M.; Ohno, R.; Nakamura, C.; Takumi, S. Differential and coordinated expression of Cbf and Cor/Lea genes during long-term cold acclimation in two wheat cultivars showing distinct levels of freezing tolerance. *Genes Genet. Syst.* **2005**, *80*, 185–197.
137. Monroy, A.F.; Dryanova, A.; Malette, B.; Oren, D.H.; Farajalla, M.R.; Liu, W.; Danyluk, J.; Ubayasena, L.W.C.; Kane, K.; Scoles, G.J.; *et al.* Regulatory gene candidates and gene expression analysis of cold acclimation in winter and spring wheat. *Plant Mol. Biol.* **2007**, *64*, 409–423.
138. Winfield, M.O.; Lu, C.; Wilson, I.D.; Coghill, J.A.; Edwards, K.J. Plant responses to cold: Transcriptome analysis of wheat. *Plant Biotechnol. J.* **2010**, *8*, 749–771.
139. Fujikawa, S.; Kuroda, K. Cryo-scanning electron microscopic study on freezing behavior of xylem ray parenchyma cells in hardwood species. *Micron* **2000**, *31*, 669–686.
140. Pearce, R.S.; Fuller, M.P. Freezing of barley studied by infrared video thermography. *Plant Physiol.* **2001**, *125*, 227–240.
141. Yamada, T.; Kuroda, K.; Jitsuyama, Y.; Takezawa, D.; Arakawa, K.; Fujikawa, S. Roles of the plasma membrane and the cell wall in the responses of plant cells to freezing. *Planta* **2002**, *215*, 770–778.
142. Solecka, D.; Zebrowski, J.; Kacperska, A. Are involved in cold acclimation and de-acclimation of winter oil-seed rape plants? *Ann. Bot.* **2008**, *101*, 521–530.
143. Rajashekar, C.B.; Lafta, A. Cell-wall changes and cell tension in response to cold acclimation and exogenous abscisic acid in leaves and cell cultures. *Plant Physiol.* **1996**, *111*, 605–612.
144. Kubacka-Zebalska, M.; Kacperska, A. Low temperature-induced modifications of cell wall content and polysaccharide composition in leaves of winter oilseed rape (*Brassica napus* L. var *oleifera* L.). *Plant Sci.* **1999**, *148*, 59–67.
145. Weiser, R.L.; Wallner, S.J.; Waddell, J.W. Cell wall and extensin mRNA changes during cold acclimation of pea seedlings. *Plant Physiol.* **1990**, *93*, 1021–1026.
146. Carvalho, C.P.; Hayashi, A.H.; Braga, M.R.; Nievola, C.C. Biochemical and anatomical responses related to the *in vitro*survival of the tropical bromeliad *Nidularium*

minutum to low temperatures. *Plant Physiol. Biochem.* **2013**, *71*, 144–154.

147. Baldwin, L.; Domon, J.M.; Klimek, J.F.; Fournet, F.; Sellier, H.; Gillet, F.; Pelloux, J.; Lejeune-Hénaut, I.; Carpita, N.C.; Rayon, C. Structural alteration of cell wall pectins accompanies pea development in response to cold. *Phytochemistry***2014**, *104*, 37–47.
148. Thonar, C.; Liners, F.; van Cutsem, P. Polymorphism and modulation of cell wall esterase enzyme activities in the chicory root during the growing season. *J. Exp. Bot.* **2006**, *57*, 81–89.
149. Lee, J.Y. Use of serial analysis of gene expression technology to reveal changes in gene expression in arabidopsis pollen undergoing cold stress. *Plant Physiol.* **2003**, *132*, 517–529.
150. He, Z.H.; Fujiki, M.; Kohorn, B.D. A cell wall-associated, receptor-like protein kinase. *J. Biol. Chem.* **1996**, *271*, 19789–19793.
151. He, Z.H.; Cheeseman, I.; He, D.; Kohorn, B.D. A cluster of five cell wall-associated receptor kinase genes, Wak1–5, are expressed in specific organs of *Arabidopsis*. *Plant Mol. Biol.* **1999**, *39*, 1189–1196.
152. Meier, S.; Ruzvidzo, O.; Morse, M.; Donaldson, L.; Kwezi, L.; Gehring, C. The *Arabidopsis* wall associated kinase-like 10 gene encodes a functional guanylyl cyclase and is co-expressed with pathogen defense related genes. *PLoS One***2010**, *5*, e8904.
153. Kohorn, B.D.; Kohorn, S.L. The cell wall-associated kinases, WAKs, as pectin receptors. *Front. Plant Sci.* **2012**, *3*, 88–93.
154. Zhang, F.; Huang, L.; Wang, W.; Zhao, X.; Zhu, L.; Fu, B.; Li, Z. Genome-wide gene expression profiling of introgressed indica rice alleles associated with seedling cold tolerance improvement in a japonica rice background.*BMC Genomics* **2012**, *13*, 461.
155. Zabotin, A.I.; Barisheva, T.S.; Zabotina, O.A.; Larskaya, I.A.; Lozovaya, V.V.; Beldman, G.; Voragen, A.G.J. Alterations in cell walls of winter wheat roots during low temperature acclimation. *J. Plant Physiol.* **1998**, *152*, 473–479.
156. Ko, J.H.; Prassinos, C.; Keathley, D.; Han, K.H. Novel aspects of transcriptional regulation in the winter survival and maintenance mechanism of poplar. *Tree Physiol.* **2011**, *31*, 208–225.
157. Domon, J.-M.; Baldwin, L.; Acket, S.; Caudeville, E.; Arnoult, S.; Zub, H.; Gillet, F.; Lejeune-Hénaut, I.; Brancourt-Hulmel, M.; Pelloux, J.; *et al.* Cell wall compositional modifications of *Miscanthus* ecotypes in response to cold acclimation. *Phytochemistry* **2013**, *85*, 51–61.
158. Zabotin, A.I.; Barisheva, T.S.; Trofimova, O.I.; Toroschina, T.E.; Larskaya, I.A.; Zabotina, O.A. Oligosaccharin and ABA synergistically affect the acquisition of freezing tolerance in winter wheat. *Plant Physiol. Biochem.* **2009**, *47*, 854–858.
159. Imin, N.; Kerim, T.; Rolfe, B.G.; Weinman, J.J. Effect of early cold stress on the maturation of rice anthers. *Proteomics***2004**, *4*, 1873–1882.
160. Yamauchi, Y.; Ogawa, M.; Kuwahara, A.; Hanada, A.; Kamiya, Y.; Yamaguchi, S. Activation of gibberellin biosynthesis and response pathways by low temperature during imbibition of *Arabidopsis thaliana* seeds. *Plant Cell* **2004**, *16*, 367–378.
161. Seki, M.; Narusaka, M.; Ishida, J.; Nanjo, T.; Fujita, M.; Oono, Y.; Kamiya, A.; Nakajima, M.; Enju, A.; Sakurai, T. Monitoring the expression profiles of 7000 *Arabidopsis* genes under drought, cold and high-salinity stresses using a full-length cDNA microarray. *Plant J.* **2002**, *31*, 279–292.
162. Shi, H.; Ye, T.; Zhong, B.; Liu, X.; Jin, R.; Chan, Z. AtHAP5A modulates freezing stress resistance in Arabidopsis through binding to CCAAT motif of AtXTH21. *New Phytol.* **2014**, *203*, 554–567.
163. Dong, J.; J<em>iang, Y.; Chen, R.; Xu, Z.; Gao, X. Isolation of a novel xyloglucan endotransglucosylase (OsXET9) gene from rice and analysis of the response of this gene to abiotic stresses. *Afr. J. Biotechnol.* **2011**, *10*, 1–11.
164. Zhu, Y.-N.; Shi, D.-Q.; Ruan, M.-B.; Zhang, L.-L.; Meng, Z.-H.; Liu, J.; Yang, W.-C. Transcriptome analysis reveals crosstalk of responsive genes to multiple abiotic stresses in cotton (*Gossypium hirsutum* L.). *PLoS One* **2013**, *8*, e80218.
165. Evers, D.; Legay, S.; Lamoureux, D.; Hausman, J.F.; Hoffmann, L.; Renaut, J. Towards a synthetic view of potato cold and salt stress response by transcriptomic and proteomic analyses. *Plant Mol. Biol.* **2012**, *78*, 503–514.
166. Meng, M.; Geisler, M.; Johansson, H.; Mellerowicz, E.J.; Karpinski, S.; Kleczkowski, L.A. Differential tissue/organ-dependent expression of two sucrose- and cold-responsive genes for UDP-glucose pyrophosphorylase in *Populus.Gene* **2007**, *389*, 186–195.
167. Gong, S.Y.; Huang, G.Q.; Sun, X.; Li, P.; Zhao, L.L.; Zhang, D.J.; Li, X.B. GhAGP31, a cotton non-classical arabinogalactan protein, is involved in response to cold stress during early seedling development. *Plant Biol.* **2011**, *14*, 447–457.
168. Solecka, D.; Kacperska, A. Activity of L-phenylalanine ammonia-lyase in winter rape leaves treated with cold. *Acta Biochim. Pol.* **1993**, *40*, 113–118.
169. Solecka, D.; Boudet, A.M.; Kacperska, A. Phenylpropanoid and anthocyanin changes in low-temperature treated winter oilseed rape leaves. *Plant Physiol. Biochem.* **1999**, *37*, 491–496.
170. Solecka, D.; Kacperska, A. Phenylpropanoid deficiency affects the course of plant acclimation to cold. *Physiol. Plant***2003**, *119*, 253–262.
171. Stefanowska, M.; Kuras, M.; Kacperska, A. Low temperature-induced modifications in cell ultrastructure and localization of phenolics in winter oilseed rape (*Brassica napus* L. var *oleifera* L.) leaves. *Ann. Bot.* **2002**, *90*, 637–645.
172. Kim, Y.-H.; Jeong, J.C.; Lee, H.-S.; Kwak, S.-S. Comparative characterization of sweet potato antioxidant genes from expressed sequence tags of dehydration-treated fibrous roots under different abiotic stress conditions. *Mol. Biol. Rep.***2012**, *40*, 2887–2896.

173. Badowiec, A.; Swigonska, S.; Weidner, S. Changes in the protein patterns in pea (*Pisum sativum* L.) roots under the influence of long- and short-term chilling stress and post-stress recovery. *Plant Physiol. Biochem.* **2013**, *71*, 315–324.

174. Badowiec, A.; Weidner, S. Proteomic changes in the roots of germinating *Phaseolus vulgaris* seeds in response to chilling stress and post-stress recovery. *J. Plant Physiol.* **2014**, *171*, 389–398.

175. Munns, R.; Tester, M. Mechanisms of salinity tolerance. *Annu. Rev. Plant Biol.* **2008**, *59*, 651–681.

176. Szabolcs, I. *Salt-Affected Soils*; CRC Press: Boca Raton, FL, USA, 1998; p. 274.

177. Munns, R. Genes and salt tolerance: Bringing them together. *New Phytol.* **2005**, *167*, 645–663.

178. Verslues, P.E.; Agarwal, M.; Katiyar-Agarwal, S.; Zhu, J.; Zhu, J.-K. Methods and concepts in quantifying resistance to drought, salt and freezing, abiotic stresses that affect plant water status. *Plant J.* **2006**, *45*, 523–539.

179. Chinnusamy, V.; Zhu, J.; Zhu, J.-K. Salt stress signaling and mechanisms of plant salt tolerance. *Genet. Eng. (NY)* **2006**,*27*, 141–177.

180. Uddin, M.N.; Hanstein, S.; Faust, F.; Eitenmüller, P.T.; Pitann, B.; Schubert, S. Diferulic acids in the cell wall may contribute to the suppression of shoot growth in the first phase of salt stress in maize. *Phytochemistry* **2014**, *102*, 126–136.

181. De Lima, R.B.; dos Santos, T.B.; Vieira, L.G.E.; de Lourdes Lúcio Ferrarese, M.; Ferrarese-Filho, O.; Donatti, L.; Boeger, M.R.T.; de Oliveira Petkowicz, C.L. Salt stress alters the cell wall polysaccharides and anatomy of coffee (*Coffea arabica*L.) leaf cells. *Carbohydr. Polym.* **2014**, *112*, 686–694.

182. Neves, G.Y.S.; Marchiosi, R.; Ferrarese, M.L.L.; Siqueira-Soares, R.C.; Ferrarese-Filho, O. Root growth inhibition and lignification induced by salt stress in soybean. *J. Agron. Crop. Sci.* **2010**, *196*, 467–473.

183. Horie, T.; Karahara, I.; Katsuhara, M. Salinity tolerance mechanisms in glycophytes: An overview with the central focus on rice plants. *Rice* **2012**, *5*, 1–18.

184. Shabala, S.; Munns, R. Salinity Stress: Physiological constraints and adaptive mechanisms. In *Plant Stress Physiology*; Shabala, S., Ed.; CAB International: Oxford, UK; Oxford, MS, USA, 2012; pp. 59–93.

185. Deinlein, U.; Stephan, A.B.; Horie, T.; Luo, W.; Xu, G.; Schroeder, J.I. Plant salt-tolerance mechanisms. *Trends Plant Sci.***2014**, *19*, 371–379.

186. Golldack, D.; Li, C.; Mohan, H.; Probst, N. Tolerance to drought and salt stress in plants: Unraveling the signaling networks. *Front. Plant Sci.* **2014**, *5*, 151.

187. Kacperska, A. Sensor types in signal transduction pathways in plant cells responding to abiotic stressors: Do they depend on stress intensity? *Physiol. Plant.* **2004**, *122*, 159–168.

188. Hou, X.; Tong, H.; Selby, J.; Dewitt, J.; Peng, X.; He, Z.-H. Involvement of a cell wall-associated kinase, WAKL4, in*Arabidopsis* mineral responses. *Plant Physiol.* **2005**, *139*, 1704–1716.

189. Lippold, F.; Sanchez, D.H.; Musialak, M.; Schlereth, A.; Scheible, W.-R.; Hincha, D.K.; Udvardi, M.K. *AtMyb41*regulates transcriptional and metabolic responses to osmotic stress in *Arabidopsis*. *Plant Physiol.* **2009**, *149*, 1761–1772.

190. Schmidt, R.; Schippers, J.H.M.; Mieulet, D.; Obata, T.; Fernie, A.R.; Guiderdoni, E.; Mueller-Roeber, B. MULTIPASS, a rice R2R3-type MYB transcription factor, regulates adaptive growth by integrating multiple hormonal pathways.*Plant J.* **2013**, *76*, 258–273.

191. Cominelli, E.; Sala, T.; Calvi, D.; Gusmaroli, G.; Tonelli, C. Over-expression of the *Arabidopsis AtMYB41* gene alters cell expansion and leaf surface permeability. *Plant J.* **2008**, *53*, 53–64.

192. Han, Y.; Li, A.; Li, F.; Zhao, M.; Wang, W. Characterization of a wheat (*Triticum aestivum L.*) expansin gene, *TaEXPB23*, involved in the abiotic stress response and phytohormone regulation. *Plant Physiol. Biochem.* **2012**, *54*, 49–58.

193. Kwon, Y.R.; Lee, H.J.; Kim, K.H.; Hong, S.-W.; Lee, S.J.; Lee, H. Ectopic expression of Expansin3 or Expansinbeta1 causes enhanced hormone and salt stress sensitivity in Arabidopsis. *Biotechnol. Lett.* **2008**, *30*, 1281–1288.

194. Zhang, L.; Zhang, C.; Wu, P.; Chen, Y.; Li, M.; Jiang, H.; Wu, G. Global analysis of gene expression profiles in physic nut (*Jatropha curcas L.*) seedlings exposed to salt stress. *PLoS One* **2014**, *9*, e97878.

195. Kieliszewski, M.J.; Lamport, D.T.A.; Tan, L.; Cannon, M.C. Hydroxyproline-rich glycoproteins: Form and function. In*Annual Plant Reviews*; Wiley-Blackwell: Oxford, UK, 2010; pp. 321–342.

196. Lamport, D.T.A.; Kieliszewski, M.J.; Showalter, A.M. Salt stress upregulates periplasmic arabinogalactan proteins: Using salt stress to analyse AGP function. *New Phytol.* **2006**, *169*, 479–492.

197. Brinker, M.; Brosche, M.; Vinocur, B.; Abo-Ogiala, A.; Fayyaz, P.; Janz, D.; Ottow, E.A.; Cullmann, A.D.; Saborowski, J.; Kangasjarvi, J.; *et al.* Linking the salt transcriptome with physiological responses of a salt-resistant *Populus* species as a strategy to identify genes important for stress acclimation. *Plant Physiol.* **2010**, *154*, 1697–1709.

198. Ma, H.; Zhao, J. Genome-wide identification, classification, and expression analysis of the arabinogalactan protein gene family in rice (*Oryza sativa L.*). *J. Exp. Bot.* **2010**, *61*, 2647–2668.

199. Tseng, I.-C.; Hong, C.-Y.; Yu, S.-M.; Ho, T.-H.D. Abscisic acid- and stress-induced highly proline-rich glycoproteins regulate root growth in rice. *Plant Physiol.* **2013**, *163*, 118–134.

200. Long, R.; Yang, Q.; Kang, J.; Zhang, T.; Wang, H.; Li, M.; Zhang, Z. Overexpression of a novel salt stress-induced glycine-rich protein gene from alfalfa causes salt and ABA sensitivity in *Arabidopsis*. *Plant Cell Rep.* **2013**, *32*, 1289–1298.

201. Zhu, J.; Lee, B.-H.; Dellinger, M.; Cui, X.; Zhang, C.; Wu, S.; Nothnagel, E.A.; Zhu, J.-K. A cellulose synthase-like protein is required for osmotic stress tolerance in

Arabidopsis. *Plant J.* **2010**, *63*, 128–140.

202. Kav, N.N.V.; Srivastava, S.; Goonewardene, L.; Blade, S.F. Proteome-level changes in the roots of *Pisum sativum* in response to salinity. *Ann. Appl. Biol.* **2004**, *145*, 217–230.
203. Xu, C.; Sibicky, T.; Huang, B. Protein profile analysis of salt-responsive proteins in leaves and roots in two cultivars of creeping bentgrass differing in salinity tolerance. *Plant Cell Rep.* **2010**, *29*, 595–615.
204. Jiang, Y.; Yang, B.; Harris, N.S.; Deyholos, M.K. Comparative proteomic analysis of NaCl stress-responsive proteins in Arabidopsis roots. *J. Exp. Bot.* **2007**, *58*, 3591–3607.
205. Mustard, J.; Renault, S. Effects of NaCl on water relations and cell wall elasticity and composition of red-osier dogwood (*Cornus stolonifera*) seedlings. *Physiol. Plant.* **2004**, *121*, 265–271.
206. An, P.; Li, X.; Zheng, Y.; Matsuura, A.; Abe, J.; Eneji, A.E.; Tanimoto, E.; Inanaga, S. Effects of NaCl on root growth and cell wall composition of two soya bean cultivars with contrasting salt tolerance. *J. Agro. Crop Sci.* **2014**, *200*, 212–218.
207. Muszyńska, A.; Jarocka, K.; Kurczynska, E.U. Plasma membrane and cell wall properties of an aspen hybrid (*Populus tremula* × *tremuloides*) parenchyma cells under the influence of salt stress. *Acta Physiol. Plant* **2014**, *36*, 1155–1165.
208. Uddin, M.N.; Hanstein, S.; Leubner, R.; Schubert, S. Leaf cell-wall components as influenced in the first phase of salt stress in three maize (*Zea mays* L.) hybrids differing in salt resistance. *J. Agro Crop Sci.* **2013**, *199*, 405–415.
209. Sánchez-Aguayo, I.; Rodríguez-Galán, J.M.; García, R.; Torreblanca, J.; Pardo, J.M. Salt stress enhances xylem development and expression of S-adenosyl-L-methionine synthase in lignifying tissues of tomato plants. *Planta* **2004**,*220*, 278–285.
210. Guo, J.; Wang, M.-H. Characterization of the phenylalanine ammonia-lyase gene (*SlPAL5*) from tomato (*Solanum lycopersicum* L.). *Mol. Biol. Rep.* **2009**, *36*, 1579–1585.
211. Li, X.-J.; Yang, M.-F.; Chen, H.; Qu, L.-Q.; Chen, F.; Shen, S.-H. Abscisic acid pretreatment enhances salt tolerance of rice seedlings: Proteomic evidence. *Biochim. Biophys. Acta* **2010**, *1804*, 929–940.
212. Salekdeh, G.H.; Siopongco, J.; Wade, L.J.; Ghareyazie, B.; Bennett, J. Proteomic analysis of rice leaves during drought stress and recovery. *Proteomics* **2002**, *2*, 1131–1145.
213. Lee, S.; Lee, E.J.; Yang, E.J.; Lee, J.E.; Park, A.R.; Song, W.H.; Park, O.K. Proteomic identification of annexins, calcium-dependent membrane binding proteins that mediate osmotic stress and abscisic acid signal transduction in Arabidopsis. *Plant Cell* **2004**, *16*, 1378–1391.
214. Yeo, A.R.; Flowers, S.A.; Rao, G.; Welfare, K.; Senanayake, N.; Flowers, T.J. Silicon reduces sodium uptake in rice (*Oryza sativa L.*) in saline conditions and this is accounted for by a reduction in the transpirational bypass flow. *Plant Cell Environ.* **1999**, *22*, 559–565.
215. Yang, J.; Yen, H.E. Early salt stress effects on the changes in chemical composition in leaves of ice plant and Arabidopsis. A Fourier Transform Infrared Spectroscopy Study. *Plant Physiol.* **2002**, *130*, 1032–1042.
216. Hayatsu, M.; Suzuki, S.; Hasegawa, A.; Tsuchiya, S.; Sasamoto, H. Effect of NaCl on ionic content and distribution in suspension-cultured cells of the halophyte *Sonneratia alba vs.* the glycophyte *Oryza sativa*. *J. Plant Physiol.* **2014**, *171*, 1385–1391.
217. Chen, J.; Cheng, T.; Wang, P.; Liu, W.; Xiao, J.; Yang, Y.; Hu, X.; Jiang, Z.; Zhang, S.; Shi, J. Salinity-induced changes in protein expression in the halophytic plant *Nitraria sphaerocarpa*. *J. Proteomics* **2012**, *75*, 5226–5243.
218. Fan, P.; Nie, L.; Jiang, P.; Feng, J.; Lv, S.; Chen, X.; Bao, H.; Guo, J.; Tai, F.; Wang, J.; *et al.* Transcriptome Analysis of*Salicornia europaea* under saline conditions revealed the adaptive primary metabolic pathways as early events to facilitate salt adaptation. *PLoS One* **2013**, *8*, e80595.
219. D'Emilio, M.; Caggiano, R.; Macchiato, M.; Ragosta, M.; Sabia, S. Soil heavy metal contamination in an industrial area: Analysis of the data collected during a decade. *Environ. Monit. Asses.* **2012**, *185*, 5951–5964.
220. Kelepertzis, E. Accumulation of heavy metals in agricultural soils of Mediterranean: Insights from Argolida basin, Peloponnese, Greece. *Geoderma* **2014**, *221*, 82–90.
221. Doncheva, S.; Georgieva, K.; Vassileva, V.; Stoyanova, Z.; Popov, N.; Ignatov, G. Effects of succinate on manganese toxicity in pea plants. *J. Plant Nutr.* **2005**, *28*, 47–62.
222. Huang, T.L.; Nguyen, Q.T.T.; Fu, S.F.; Lin, C.Y.; Chen, Y.C.; Huang, H.J. Transcriptomic changes and signalling pathway induced by arsenic stress in rice roots. *Plant Mol. Biol.* **2012**, *80*, 587–608.
223. Feigl, G.; Kumar, D.; Lehotai, N.; Tugyi, N.; Molnor, A.; Ordog, A.; Szepesi, A.; Gemes, K.; Laskay, G.; Erdei, L.; *et al.* Physiological and morphological responses of the root system of Indian mustard (*Brassica jumcea* L. Czern.) and rapeseed (*Brassica napus* L.) to copper stress. *Ecotox. Environ. Safe* **2013**, *91*, 179–189.
224. Appenroth, K.J. What are "heavy metals" in plant sciences? *Acta Physiol. Plant* **2010**, *32*, 615–619.
225. Stohs, S.J.; Bagchi, D. Oxidative mechanisms in the toxicity of metal ions. *Free Radic. Biol. Med.* **1995**, *18*, 321–336.
226. Briat, J.F. Metal-ion-mediated oxidative stress and its control. In *Oxidative Stress in Plants*; Montagu, M., Inzé, D., Eds.; Taylor and Francis: London, UK, 2002; pp. 171–189.
227. Lander, H.M. An essential role for free radicals and derived species in signal transduction. *FASEB J.* **1997**, *11*, 118–124.
228. Opdenakker, K.; Remans, T.; Keunen, E.; Vangronsveld, J.; Cuypers, A. Exposure of *Arabidopsis thaliana* to Cd or Cu excess leads to oxidative stress mediated

alterations in MAPKinase transcript levels. *Environ. Exp. Bot.* **2012**, *83*, 53–61.

229. .Krzeslowska, M.; Lenartowska, M.; Mellerowicz, E.J.; Samardakiewicz, S.; Wosny, A. Pectinous cell wall thickenings formation-a response of moss protonema cells to Pb. *Environ. Exp. Bot.* **2009**, *65*, 119–131.
230. Krzeslowska, M.; Lenartowska, M.; Samardakiewicz, S.; Bilski, H.; Wosny, A. Lead deposited in the cell wall of*Funaria hygrometrica* protonema is not stable-a remobilization can occur. *Environ. Pollut.* **2010**, *158*, 325–338.
231. Vuletic, M.; Hadsi-Taskovic Sukalovic, V.; Markovic, K.; Kravic, N.; Vucinic, Z.; Maksimovic, V. Differential response of antioxidative systems of maize roots cell walls to osmotic and heavy metal stress. *Plant Biol.* **2014**, *16*, 88–96.
232. Jiang, X.; Wang, C. Zinc distribution and zinc-binding forms in *Phragmites australis* under zinc pollution. *J. Plant Physiol.* **2008**, *165*, 697–704.
233. Hossain, A.K.M.Z.; Koyama, H.; Hara, T. Growth and cell wall properties of two wheat cultivars differing in their sensitivity to aluminum stress. *J. Plant Physiol.* **2006**, *163*, 39–47.
234. Polec-Powlak, K.; Ruzik, R.; Lipiec, E.; Ciurzynska, M.; Gawronska, H. Investigation of Pb(II) binding to pectin in*Arabidopsis thaliana*. *J. Anal. Atom. Spectrom.* **2007**, *22*, 968–972.
235. Amenos, M.; Corrales, I.; Poschenrieder, C.; Illes, P.; Baluska, F.; Barcelo, J. Different effects of aluminum on the actin cytoskeleton and Brefeldin A-sensitive vesicle recycling in root apex cells of two maize varieties differing in root elongation rate and aluminum tolerance. *Plant Cell Physiol.* **2009**, *50*, 528–540.
236. Horst, W.J.; Wang, Y.; Eticha, D. The role of the root apoplast in Aluminum-induced inhibition of root elongation and in Aluminum resistance of plant. *Rev. Ann. Bot.* **2010**, *106*, 185–197.
237. Vaculík, M.; Lux, A.; Luxová, M.; Tanimoto, E.; Lichtscheidl, I. Silicon mitigates cadmium inhibitory effects in young maize plants. *Environ. Exp. Bot.* **2009**, *67*, 52–58.
238. Gu, J.; Yin, X.; Stomph, T.J.; Wang, H.; Struik, P.C. Physiological basis of genetic variation in leaf photosynthesis among rice (*Oryza sativa* L.) introgression lines under drought and well-watered conditions. *J. Exp. Bot.* **2012**, *63*, 5137–5153.
239. Eticha, D.; Stass, A.; Horst, W.J. Cell-wall pectin and its degree of methylation in the maize root apex: Significance for genotypic differences in aluminium resistance. *Plant Cell Env.* **2005**, *28*, 1410–1420.
240. Baluska, F.; Liners, F.; Hlavacka, A.; Schlicht, M.; van Cutsem, P.; McCurdy, D.W.; Menzel, D. Cell wall pectins and xyloglu-cans are internalised into dividing root cells and accumulate within cell plates during cytokinesis. *Protoplasma***2005**, *225*, 141–155.
241. Pelloux, J.; Rusterucci, C.; Mellerowicz, E.J. New insight into pectin methylesterase structure and function. *Trends Plant Sci.* **2007**, *12*, 1363–1372.
242. Willats, W.G.T.; Knox, J.P.; Mikkelsen, J.D. Pectin: New insights into an old polymer are starting to gel. *Trends Food Sci. Technol.* **2006**, *17*, 97–104.
243. Dronnet, V.M.; Renard, C.M.G.C.; Axelos, M.A.V.; Thibault, J.F. Heavy metal binding by pectins: Selectivity, quantification and characterization. *Carbohydr. Polym.* **1996**, *30*, 253–263.
244. Paynel, F.; Schaumann, A.; Arkoun, M.; Douchiche, O.; Morvan, C. Temporal regulation of cell wall pectin methylesterase and peroxidase isoforms in cadmium-treated flax hypocotyls. *Ann. Bot.* **2009**, *104*, 1363–1372.
245. Douchiche, O.; Rihouey, C.; Schaumann, A.; Driouich, A.; Morvan, C. Cadmium-induced alterations of the structural features of pectins in flax hypocotyls. *Planta* **2007**, *225*, 1301–1312.
246. Douchiche, O.; Driouich, A.; Morvan, C. Spatial regulation of cell wall structure in response to heavy metal stress: Cadmium-induced alteration of the methyl-esterification pattern of homogalacturonans. *Ann. Bot.* **2010**, *105*, 481–491.
247. Yang, Z-B.; Eticha, D.; Rao, I.M.; Horst, W.J. Alteration of Cell-wall Porosity is Involved in Osmotic Stress-induced Enhancement of Aluminium Resistance in Common Bean (*Phaseolus vulgaris* L.). *J. Exp. Bot.* **2010**, *61*, 3245–3258.
248. El-Moneim, D.A.; Contreras, R.; Silva-Navas, J.S.; Gallego, F.J.; Figueiras, A.M.; Benito, C. Pectin methylesterase gene and aluminum tolerance in *Secale cereale*. *Environ. Exp. Bot.* **2014**, *107*, 125–133.
249. Colzi, I.; Doumett, S.; Del Bubba, M.; Fornaini, J.; Arnetoli, A.; Gabbrielli, R.; Gonnelli, C. On the role of the cell wall in the phenomenon of copper tolerance in *Silene paradoxa* L. *Environ. Exp. Bot.* **2011**, *72*, 77–83.
250. Chandran, D.; Sharopova, N.; Ivashuta, S.; Gantt, J.S.; VandenBosch, K.A.; Samac, D.A. Transcriptome profiling identified novel genes associated with aluminium toxicity, resistance and tolerance in *Medicago truncatula*. *Planta* **2008**,*228*, 151–166.
251. Yang, Z.B.; Eticha, D.; Rotter, B.; Rao, I.M.; Horst, W.J. Physiological and molecular analysis of polyethylene glycol-induced reduction of aluminium accumulation in the root tips of common bean (*Phaseolus vulgaris*). *New Phytol.* **2011**,*192*, 99–113.
252. Han, Y.; Sa, G.; Sun, J.; Shen, Z.; Zhao, R.; Ding, M.; Deng, S.; Lu, Y.; Zhang, Y.; Shen, X.; Chen, S. Overexpression of*Populus euphratica* xyloglucan endotransglucosylase/hydrolase gene confers enhanced cadmium tolerance by the restriction of root cadmium uptake in transgenic tobacco. *Environ. Exp. Bot.* **2014**, *100*, 74–83.
253. Song, X.Q.; Liu, L.F.; Jiang, Y.J.; Zhang, B.C.; Gao, Y.P.; Liu, X.L.; Lin, Q.S.; Ling, H.Q.; Zhou, Y.H. Disruption of secondary wall cellulose biosynthesis alters cadmium translocation and tolerance in rice plants. *Mol. Plant.* **2013**, *6*, 768–780.
254. Liu, T.; Shen, C.; Wang, Y.; Huang, C.; Shi, J. New insights into regulation of proteome and polysaccharide in cell wall of *Elsholtzia splendens* in response to copper stress. *PLoS One* **2014**, *9*, e109573.

255. Gao, J.; Sun, L.; Yang, X.; Liu, J.X. Transcriptomic analysis of cadmium stress response in the heavy metal hyperaccumulator *Sedum alfredii* hance. *PLoS One* **2013**, *8*, e64643.

256. Tabbì, G.; Fry, S.C.; Bonomo, R.P. ESR study of the non-enzymic scission of xyloglucan by an ascorbate-H2O2-copper system: The involvement of the hydroxyl radical and the degradation of ascorbate. *J. Inorg. Biochem.* **2001**, *84*, 179–187.

257. Kärkönen, A.; Fry, S.C. Effect of ascorbate and its oxidation products on H_2O_2 production in cell-suspension cultures of *Picea abies* and in the absence of cells. *J. Exp. Bot.* **2006**, *57*, 1633–1644.

258. Claus, H. Laccases: Structure, reaction, distribution. *Micron* **2004**, *35*, 93–96.

259. Ali, M.B.; Singh, N.; Shohael, A.M.; Hahn, E.J.; Paek, K.Y. Phenolics metabolism and lignin synthesis in root suspension cultures of *Panax ginseng* in response to copper stress. *Plant Sci.* **2006**, *171*, 147–154.

260. Keuskamp, D.H.; Keller, M.M.; Ballaré, C.L.; Pierik, R. Blue light regulated shade avoidance. *Plant Signal. Behav.* **2012**,*7*, 514–517.

261. Franklin, K.A. Shade avoidance. *New Phytol.* **2008**, *179*, 930–944.

262. Sasidharan, R.; Keuskamp, D.H.; Kooke, R.; Voesenek, L.A.; Pierik, R. Interactions between auxin, microtubules and XTHs mediate green shade-induced petiole elongation in *Arabidopsis*. *PLoS One* **2014**, *9*, e90587.

263. Ma, L.; Li, J.; Qu, L.; Hager, J.; Chen, Z.; Zhao, H.; Deng, X.W. Light control of *Arabidopsis* development entails coordinated regulation of genome expression and cellular pathways. *Plant Cell.* **2001**, *13*, 2589–2607.

264. Sasidharan, R.; Chinnappa, C.C.; Voesenek, L.A.C.J.; Pierik, R. The regulation of cell wall extensibility during shade avoidance: A study using two contrasting ecotypes of *Stellaria longipes*. *Plant Physiol.* **2008**, *148*, 1557–1569.

265. Sasidharan, R.; Chinnappa, C.C.; Staal, M.; Elzenga, J.T.; Yokoyama, R.; Nishitani, K.; Voesenek, L.A.; Pierik, R. Light quality-mediated petiole elongation in Arabidopsis during shade avoidance involves cell wall modification by xyloglucan endotransglucosylase/hydrolases. *Plant Physiol.* **2010**, *154*, 978–990.

266. Masuda, Y.; Kamisaka, S.; Yanagisawa, H.; Suzuki, H. Effect of light on growth and metabolic activities in pea seedlings I. changes in cell wall polysaccharides during growth in the dark and in the light. *Biochem. Physiol. Pflanzen.***1981**, *176*, 23–34.

267. Wargent, J.J.; Gegas, V.C.; Jenkins, G.I.; Doonan, J.H.; Paul, N.D. UVR8 in *Arabidopsis thaliana* regulates multiple aspects of cellular differentiation during leaf development in response to ultraviolet B radiation. *New Phytol.* **2009**,*183*, 315–326.

268. Kigel, J.; Cosgrove, D.J. Photoinhibition of stem elongation by blue and red light. *Plant Physiol.* **1991**, *95*, 1049–1056.

269. Raffo, M.D.; Ponce, N.M.; Sozzi, G.O.; Vicente, A.R.; Stortz, C.A. Compositional changes in "Bartlett" pear (*Pyrus communis* L.) cell wall polysaccharides as affected by sunlight conditions. *J. Agric. Food Chem.* **2011**, *59*, 12155–12162.

270. Sharma, P.; Chatterjee, M.; Burman, N.; Khurana, J.P. Cryptochrome 1 regulates growth and development in *Brassica*through alteration in the expression of genes involved in light, phytohormone and stress signalling. *Plant Cell Environ.***2014**, *37*, 961–977.

271. Syros, T.D.; Yupsanis, T.A.; Economou, A.S. Expression of peroxidases during seedling growth in *Ebenus cretica* L. as affected by light and temperature treatments. *Plant Growth Regul.* **2005**, *46*, 143–151.

272. Ali, M.B.; Khatun, S.; Hahn, E.J.; Paek, K.Y. Enhancement of phenylpropanoid enzymes and lignin in *Phalaenopsis*orchid and their influence on plant acclimatisation at different levels of photosynthetic photon flux. *Plant Growth Regul.* **2005**, *49*, 137–146.

273. Kimura, M.; Yamamoto, Y.; Seki, M.; Sakurai, T.; Sato, M.; Abe, T.; Yoshida, S.; Manabe, K.; Shinozaki, K.; Matsui, M. Identification of *Arabidopsis* genes regulated by high light-stress using cDNA microarray. *Photochem. Photobiol.* **2003**,*77*, 226–233.

274. Ruhland, C.T.; Xiong, F.S.; Clark, W.D.; Day, T.A. The influence of ultraviolet-B radiation on growth, hydroxycinnamic acids and flavonoids of *Deschampsia antarctica* during Springtime ozone depletion in Antarctica.*Photochem. Photobiol.* **2005**, *5*, 1086–1093.

275. Lesniewska, E.; Adrian, M.; Klinguer, A.; Pugin, A. Cell wall modification in grapevine cells in response to UV stress investigated by atomic force microscopy. *Ultramicroscopy* **2004**, *100*, 171–178.

276. Pristov, J.B.; Spasić, M.; Spasojević, M. Converting low dose radiation to redox signaling. *Plant Signal Behav.* **2013**, *8*, e23151.

277. Messenger, D.J.; McLeod, A.R.; Fry, S.C. The role of ultraviolet radiation, photosensitizers, reactive oxygen species and ester groups in mechanisms of methane formation from pectin. *Plant Cell Environ.* **2009**, *32*, 1–9.

278. Pristov, J.B.; Jovanović, S.V.; Mitrović, A.; Spasojević, I. UV-irradiation provokes generation of superoxide on cell wall polygalacturonic acid. *Physiol. Plant.* **2013**, *148*, 574–581.

279. Barka, E.A.; Kalantari, S.; Makhlouf, J.; Arul, J. Impact of UV-C Irradiation on the cell wall-degrading enzymes during ripening of tomato (*Lycopersicon esculentum* L.) fruit. *J. Agric. Food Chem.* **2000**, *48*, 667–671.

280. Bu, J.; Yua, Y.; Aisikaera, G.; Yinga, T. Postharvest UV-C irradiation inhibits the production of ethylene andthe activity of cell wall-degrading enzymes during softening of tomato (*Lycopersicon esculentum* L.) fruit. *Postharvest Biol. Technol.***2013**, *86*, 337–345.

281. Yamasaki, S.; Noguchi, N.; Mimaki, K. Continuous UV-B irradiation induces morphological changes and the accumulation of polyphenolic compounds on the surface of cucumber cotyledons. *J. Radiat. Res.* **2007**, *48*, 443–454.

282. Zagoskina, N.V.; Dubravina, G.A.; Alyavina, A.K.;

Goncharuk, E.A. Effect of ultraviolet (UV-B) radiation on the formation and localization of phenolic compounds in tea plant callus cultures. *Russ. J. Plant Phys.* **2003**, *50*, 270–275.

283. Hilal, M.; Parrado, M.F.; Rosa, M.; Gallardo, M.; Orce, L.; Massa, E.D.; González, J.A.; Prado, F.E. Epidermal lignin deposition in quinoa cotyledons in response to UV-B radiation. *Photochem. Photobiol.* **2004**, *79*, 205–210.

284. Pontin, M.A.; Piccoli, P.N.; Francisco, R.; Bottini, R.; Martinez-Zapater, J.M.; Lijavetzky, D. Transcriptome changes in grapevine (*Vitis vinifera* L.) cv. Malbec leaves induced by ultraviolet-B radiation. *BMC Plant Biol.* **2010**, *10*, 224–237.

285. Yokawa, K.; Baluška, F. Pectins, ROS homeostasis and UV-B responses in plant roots. *Phytochemitry* **2014**. in press.

286. Lindroth, R.L. Impacts of elevated atmospheric CO_2 and O_3 on forests: Phytochemistry, trophic interactions, and ecosystem dynamics. *J. Chem. Ecol.* **2010**, *36*, 2–21.

287. Munné-Bosch, S.; Queval, G.; Foyer, C.H. The impact of global change factors on redox signaling underpinning stress tolerance. *Plant Physiol.* **2013**, *16*, 5–19.

288. Betzelberger, A.M.; Yendrek, C.R.; Sun, J.; Leisner, C.P.; Nelson, R.L.; Ort, D.R.; Ainsworth, E.A. Ozone exposure response for U.S. soybean cultivars: Linear reductions in photosynthetic potential, biomass and yield. *Plant Physiol.***2012**, *160*, 1827–1839.

289. Emberson, L.D.; Bueker, P.; Ashmore, M.R.; Mills, G.; Jackson, L.S.; Agrawal, M.; Atikuzzaman, M.D.; Cinderby, S.; Engardt, M.; Jamir, C.; *et al.* A comparison of North American and Asian exposure-response data for ozone effects on crop yields. *Atmos. Environ.* **2009**, *43*, 1945–1953.

290. Wilkinson, S.; Davies, W.J. Drought, ozone, ABA and ethylene: New insights from cell to plant to community. *Plant Cell Environ.* **2010**, *33*, 510–525.

291. Racherla, P.N.; Adams, P.J. The response of surface ozone to climate change over the eastern United States. *Atmos. Chem. Phys.* **2008**, *8*, 871–885.

292. Günthardt-Goerg, M.S.; Mc Quattie, C.J.; Scheidegger, C.; Rhiner, C.; Matyssek, R. Ozone-induced cytochemical and ultrastructural changes in leaf mesophyll cell walls. *Can. J. For. Res.* **1997**, *27*, 453–463.

293. Günthardt-Goerg, M.S. Different responses to ozone of tobacco, poplar, birch, and alder. *J. Plant Physiol.* **1996**, *148*, 207–214.

294. Paoletti, E.; de Marco, A.; Beddows, D.C.S.; Harrison, R.M.; Manning, W.J. Ozone levels in European and USA cities are increasing more than at rural sites, while peak values are decreasing. *Environ. Pollut.* **2014**, *192*, 295–299.

295. Paoletti, E.; Contran, N.; Bernasconi, P.; Günthardt-Goerg, M.S.; Wollenveider, P. Structural and physiological responses to ozone in Manna ash (*Fraxinus ornus* L.) leaves of seedlings and mature trees under controlled and ambient condition. *Sci. Total Environ.* **2010**, *408*, 2014–2024.

296. Richet, N.; Afif, D.; Huber, F.; Pollet, B.; Banvoy, J.; el Zein, R.; Lapierre, C.; Dizengremel, P.; Perré, P.; Cabané, M. Cellulose and lignin biosynthesis is altered by ozone in wood of hybrid poplar (*Populus tremula* × *alba*). *J. Exp. Bot.***2011**, *62*, 3575–3586.

297. Bussotti, F.; Agati, G.; Desotgiu, R.; Matteini, P.; Tani, C. Ozone foliar symptoms in woody plant species assessed with ultrastructural and fluorescence analysis. *New Phytol.* **2005**, *166*, 941–955.

298. Bussotti, F.; Pancrazi, M.; Matteucci, G.; Gerosa, G. Leaf morphology and chemistry in *Fagus sylvatica* L. (beech) trees as affected by site factors and ozone: Results from Level II permanent monitoring plots in Italy. *Tree Physiol.* **2005**, *25*, 211–219.

299. Keutgen, A.J.; Pawelzik, E. Contribution of amino acids to strawberry fruit quality and their relevance as stress indicators under NaCl salinity. *Food Chem.* **2008**, *111*, 642–647.

300. Kanter, U; Heller, W.; Durner, J.; Winkler, J.B.; Engel, M.; Behrendt, H.; Holzinger, A.; Braun, P.; Hauser, H.; Ferreira, F.; *et al.* Molecular and immunological characterization of ragweed (*Ambrosia artemisiifolia* L.) pollen after exposure of the plants to elevated ozone over a whole growing season. *PLoS One* **2013**, *8*, e61518.

301. Minas, I.S.; Vicente, O.R.; Dhanapal, O.P.; Manganaris, G.A.; Goulas, V.; Vasilakakis, M.; Crisosto, H.; Molassiotis, A. Ozone-induced kiwifruit ripening delay is mediated by ethylene biosynthesis inhibition and cell wall dismantling regulation. *Plant Sci.* **2014**, *29*, 76–85.

302. Oksanen, E.; Häikiö, E.; Sober, J.; Karnosky, D.F. Ozone-induced H_2O_2 accumulation in field-grown aspen and birch is linked to foliar ultrastructure and peroxisomal activity. *New Phytol.* **2003**, *161*, 791–799.

303. Cho, K.; Shibato, J.; Kubo, A.; Kohno, Y.; Satoh, K.; Kikuchi, S.; Sarkar, A.; Agrawal, G.K.; Rakwal, R. Comparative analysis of seed transcriptomes of ambient ozone-fumigated two different rice cultivars. *Plant Signal Behav.* **2013**, *8*, e26300.

304. Cabané, M.; Pireaux, J.C.; Leger, E.; Weber, E.; Dizengremel, P.; Pollet, B.; Lapierre, C. Condensed lignins are synthesized in poplar leaves exposed to ozone. *Plant Physiol.* **2004**, *134*, 586–594.

305. Sharma, Y.K.; Davis, K.R. Ozone-induced expression of stress-related genes in *Arabidopsis thaliana*. *Plant Physiol.* **1994**,*105*, 1089–1096.

306. Eckey-Kaltenbach, H.; Ernst, D.; Heller, W.; Sandermann, H. Biochemical plant responses to ozone: IV. Cross-induction of defensive pathways in parsley (*Petroselinum crispum* L.) plants. *Plant Physiol.* **1994**, *104*, 67–74.

307. Wiese, C.B.; Pell, E.J. Oxidative modification of the cell wall in tomato plants exposed to ozone. *Plant Physiol. Biochem.***2003**, *41*, 375–382.

308. Ranieri, A.; Castagna, A.; Pacini, J.; Baldan, B.; Mensuali Sodi, A.; Soldatini, G.F. Early production and scavenging of hydrogen peroxide in the apoplast of sunflowers plants exposed to ozone. *J. Exp. Bot.* **2003**, *54*, 2529–2540.

309. Ranieri, A.; Castagna, A.; Soldatini, G.F. Differential stimulation of ascorbate peroxidase isoforms by ozone

exposure in sunflower plants. *J. Plant Physiol.* **2000**, *156*, 266–271.

310. Keutgen, A.J.; Pawelzik, E. Apoplastic antioxidative system responses to ozone stress in strawberry leaves. *J. Plant Physiol.* **2008**, *165*, 868–875.

311. Meehl, G.A.; Stocker, W.D.; Collins, P.; Friedlingstein, A.T.; Gaye, J.M.; Gregory, A.; Kitoh, R.; Knutti, J.M.; Murphy, A.; Noda, S.C.B.; *et al.* Global Climate Projections. In *Climate Change 2007: The Physical Science Basis. Contribution of Working Group I to the Fourth Assessment Report of the Intergovernmental Panel on Climate Change*; Solomon, S., Qin, D., Manning, M., Chen, Z., Marquis, M., Averyt, K.B., Tignor, M., Miller, H.L., Eds.; Cambridge University Press: Cambridge, UK, New York, NY, USA, 2007.

312. Leakey, A.D.B.; Bernacchi, C.J.; Dohleman, F.G.; Ort, D.R.; Long, S.P. Will photosynthesis of maize (*Zea mays*) in the US Corn Belt increase in future CO_2 rich atmospheres? An analysis of diurnal courses of CO_2 uptake under free-air concentration enrichment (FACE). *Glob. Chang. Biol.* **2004**, *10*, 951–962.

313. De Souza, A.P.; Arundale, R.A.; Dohleman, F.G.; Long, S.P.; Buckeridge, M.S. Will the exceptional productivity of*Miscanthus x giganteus* increase further under rising atmospheric CO2? *Agricult. For. Meteorol.* **2013**, *171*, 82–92.

314. King, J.S.; Pregitzer, K.S.; Zak, D.R.; Karnosky, D.F.; Isebrands, J.G.; Dickson, R.E.; Hendrey, G.R.; Sober, J. Fine root biomass and fluxes of soil carbon in young stands of paper birch and trembling aspen as affected by elevated atmospheric CO_2 and tropospheric O_3. *Oecologia* **2001**, *128*, 237–250.

315. Oksanen, E.; Sôber, J.; Karnosky, D.F. Interactions of elevated CO_2 and ozone in leaf morphology of aspen (*Populus tremuloides*) and birch (*Betula papyrifera*) in aspen FACE experiment. *Environ. Pollut.* **2001**, *115*, 437–446.

316. Kubiske, M.E.; Quinn, V.S.; Marquart, P.E.; Karnosky, D.F. Effects of elevated atmospheric CO_2 and/or O_3 on intra- and inter-specific competitive ability of aspen. *Plant Biol.* **2007**, *9*, 342–355.

317. Cseke, L.J.; Tsai, C.J.; Rogers, A.; Nelsen, M.P.; White, H.L.; Karnosky, D.F.; P Odila, G.K. Transcriptomic comparison in the leaves of two aspen genotypes having similar carbon assimilation rates but different partitioning patterns under elevated (CO_2). *New Phytol.* **2009**, *182*, 891–911.

318. Wei, H.; Gou, J.; Yordanov, Y.; Zhang, H.; Thakur, R.; Jones, W.; Burton, A. Global transcriptome profiling of aspen trees under elevated (CO_2) for identifying potential molecular mechanisms responsible for enhanced growth. *J. Plant Res.* **2013**, *126*, 305–320.

319. Kostiainen, K.; Saranpää, P.; Lundqvist, S.O.; Kubiske, M.E.; Vapaavuori, E. Wood properties of *Populus* and *Betula* in long-term exposure to elevated CO_2 and O_3. *Plant Cell Environ.* **2014**, *37*, 1452–1463.

320. Kim, K.; Labbé, N.; Warren, J.M.; Elder, T.; Rials, T.G. Chemical and anatomical changes in *Liquidambar styraciflua* L. xylem after long term exposure to elevated CO_2. *Environ. Pollut.* **2015**, *17*, 179–185.

321. Gibeaut, D.M.; Cramer, G.R.; Seemann, J.R. Growth, cell walls, and UDP-Glc dehydrogenase activity of *Arabidopsis thaliana* grown in elevated carbon dioxide. *J. Plant Physiol.* **2001**, *158*, 569–576.

322. Duan, Z.; Homma, A.; Kobayashi, M.; Nagata, M.; Kaneko, Y.; Fujiki, Y.; Nishida, I. Photoassimilation assimilate translocation and plasmodesmal biogenesis in the source leaves of *Arabidopsis thaliana* grown under an increased atmospheric CO_2 concentration. *Plant Cell. Physiol.* **2014**, *55*, 358–369.

323. Perez-Lopez, U.; Robredo, A.; Lacuesta, M.; Munoz-Rueda, A.; Mena-Petite, A. Atmospheric CO_2 concentration influences the contributions of osmolyte accumulation and cell wall elasticity to salt tolerance in barley cultivars. *J. Plant Physiol.* **2010**, *167*, 15–22.

324. Huyskens-Keil, S.; Herppich, W. High CO_2 effects on postharvest biochemical and textural properties of white asparagus (*Asparagus officinalis* L.) spears. *Postharvest Biol. Technol.* **2013**, *75*, 45–53.

325. Chen, S.; Olbrich, A.; Langenfeld-Heyser, R.; Fritz, E.; Polle, A. Quantitative X-ray microanalysis of hydrogen peroxide within plant cells. *Microsc. Res. Techniq.* **2009**, *72*, 49–60.

326. Ookawara, R.; Satoh, S.; Yoshioka, T.; Ishizawa, K. Expression of α-expansin and xyloglucan endotransglucosylase/hydrolase genes associated with shoot elongation enhanced by anoxia, ethylene and carbon dioxide in arrowhead (*Sagittaria pygmaea* Miq.) tubers. *Ann. Bot.* **2005**, *96*, 693–702.

327. May, P.; Liao, W.; Wu, Y.; Shuai, B.; Mc Combie, W.R.; Zhang, M.Q.; Liu, Q.A. The effects of carbon dioxide and temperature on microRNA expression in Arabidopsis development. *Nat. Commun.* **2013**, *4*.

328. Cosgrove, D.J. Growth of the plant cell wall. *Nat. Rev. Mol. Cell Biol.* **2005**, *6*, 850–861.

329. Hijazi, M.; Velasquez, S.M.; Jamet, E.; Estevez, J.M.; Albenne, C. An update on post-translational modifications of hydroxyproline-rich glycoproteins: Toward a model highlighting their contribution to plant cell wall architecture.*Front. Plant Sci.* **2014**, *5*, 395.

330. Seifert, G.J.; Xue, H.; Acet, T. The Arabidopsis thaliana fasciclin like arabinogalactan protein 4 gene acts synergistically with abscisic acid signaling to control root growth. *Ann. Bot.* **2014**, *114*, 1125–1133.

CHAPTER 8

AIR-POLLUTANT-PHILIC PLANTS FOR AIR REMEDIATION

Misa Takahashi, Hiromichi Morikawa

Keywords: Air-Pollutant-Philic Plants; Air Remediation; Genetic Modification; Nitrogen Dioxide (NO_2); Plant Vitalization Effect of NO_2

INTRODUCTION

Over two decades ago, we started research investigating air-pollutant-philic plants that can grow with air pollutants as the sole nutrient source [1,2]. We thought (and still think) that such plants are critical for mitigation of air pollution. Our idea is a critical response to the need to cope with harsh air pollution worldwide and is a viable possibility due to plants' natural ability as a sink of various air pollutants. Although we still have much ground to cover and are not as far along as we had had hoped, we have made some progress in the research of air-pollutantphilic plants.

We have primarily focused on atmospheric nitrogen dioxide (NO_2) as a target pollutant. The weighted average of NO_2 concentration in cities from over 141 countries is reported to be 50.6 μg/m^3 (≈27 ppb) [3]. The current WHO air quality guideline value (annual mean) for NO_2 is 40 μg/m^3 (≈21 ppb) [4].

The air in polluted urban areas of Japan contains about 0.05 ppm of NO_2, a similar or even higher amount of nitrogen oxide (NO), and 0.1 ppm of non-methane hydrocarbons (NMHCs) [5]. Methane and other volatile organic hydrocarbons, including NMHCs, photochemically produce hydroxyl radicals (highly reactive molecules containing a single unpaired electron), which react with nitrogen oxides (NO_x) to produce photooxidative ozone. NO_2 and oxidants remain in the lower atmosphere (troposphere) and are very toxic gases for animals, including humans, as well as for plants [6]. Gaseous pollutants such as NO_x, once emitted, disperse into the atmosphere rapidly. Centralized treatment of these scattered pollutants is energy (equivalent to the entropy) intensive and costly. In contrast, plants are driven by solar energy, are self-reproducing, and can concentrate and detoxify pollutants. Therefore, plants should prove to be very useful in helping to clean up dispersed, ambient pollutants. In fact, we know that plants do uptake NO_2 [7-11] and incorporate its nitrogen into organic nitrogenous compounds [12-15].

Citation: M. Takahashi and H. Morikawa, " Air-Pollutant-Philic Plants for Air Remediation, Vol. 3 No. 10, 2012, pp. 1346-1352. doi: 10.4236/jep.2012.310153.

VARIATION IN PLANTS' ABILITIES TO MITIGATE AIR POLLUTANTS AMONG NATURALLY OCCURRING PLANTS

Learning to what extent naturally occurring plants are (or are not) air-pollutant-philic is important. We therefore investigated the uptake and assimilation of NO_2 in 217 taxa of naturally occurring plants that included 107 woody plants, 60 cultivated herbaceous plants, and 50 wild herbaceous plants collected from roadsides.

Plants were fumigated in a fumigation chamber at 22°C ± 0.3°C, 70% ± 4% relative humidity, and an atmospheric level of 0.03% to 0.04% CO_2 for 8 h during daytime (09:00 - 17:00) under fluorescent lights (70 μmol photons $m^{-2} \cdot s^{-1}$) with 4 ± 0.4 ppm $^{15}NO_2$ (51.6 atom% ^{15}N). Leaves were then cut from fumigated plants, washed with tap water and then with distilled water, and lyophilized. Lyophilized leaves were then analyzed for the ^{14}N plus ^{15}N content and the percentage of ^{15}N in the samples was determined using an EA-MS analyzer in which an elemental analyzer (EA/NA; Fisons Instruments, Milano, Italy) interfaced with a mass spectrometer (Delta C; Thermo Finnigan, Bremen, Germany), as described previously [16]. From these two values, total (^{14}N plus ^{15}N) nitrogen derived from NO_2 in the leaves of fumigated plants (mg N/g dry weight) was determined (designated $TNNO_2$). Parts of the lyophilized leaves were subjected to Kjeldahl digestion [17,18], and the reduced or Kjeldahl nitrogen fraction was obtained. The ^{14}N plus ^{15}N content and the percentage of ^{15}N in the Kjeldahl fraction were then analyzed. From these values, reduced nitrogen derived from NO_2 in the leaves of fumigated plants (mg N/g dry weight) was determined (designated $RNNO_2$). $TNNO_2$ (mg N/g dry wt), $RNNO_2$ (mg N/g dry weight), and the ratio of $RNNO_2$/RN (%) respectively give a measure of the ability of plants to act as a sink for NO_2, that of the ability of plants to metabolize (assimilate) NO_2, and of the contribution of NO_2 as a nitrogen source.

Details of the data on 217 taxa have been published [16,19], and here we briefly discuss some points in relation to the air-pollutant-philic plants. Among the 217 taxa, the $RNNO_2$ differed by a factor of 657 between the highest (Eucalyptus viminalis, 6.57) and lowest (Tillandsia ionantha and Tillandsia caput-medusae, 0.01). These two species belong to the Myrtaceae and Bromeliaceae family, respectively. The Myrtaceae seems to include plants with a high ability to assimilate NO_2(see below), and plants of the Bromeliaceae have less of an ability to assimilate NO_2. The other two species of the Bromeliaceae also showed low values for $RNNO_2$: 0.02 and 0.12 in Tillandsia usneoides and Tillandsia geminiflora, respectively.

The $RNNO_2$/RN differed about 85-fold between the highest (Magnolia kobus, 12.7%) and lowest (Codiaeum variegatum, 0.15%). These two species belong to the Magnoliaceae and Euphorbiaceae family, respectively. The 217 taxa examined included no other species of the Magnoliaceae. One species of the 217 taxa, belonging to the Euphorbiaceae, Sapium sebiferum, showed a particularly high value of $RNNO_2$/RN (10.1) [16].

Among the 217 taxa, six woody species, one cultivated herbaceous species, and two wild herbaceous species collected from roadsides were found to have $RNNO_2$ greater than 4 and $RNNO_2$/RN values greater than or close to 9 as shown in **Table 1**. These nine species may have more potential to be air-pollutant-philic specimens than those examined in the 217 taxa examined to date.

Since the fumigation in our study described above lasted for 8 h, $RNNO_2$/RN values reflect the turnover rate of reduced nitrogen in plants. Turnover of reduced nitrogen is considered to be first order as in the case of proteins [20]. A simple exponential curve was fitted to the $RNNO_2$/RN data to generate a first-order rated constant: k. The half-life of the reduced nitrogen was calculated as

Table 1. Nine species with high values of reduced nitrogen derived from NO_2 ($RNNO_2$) and the corresponding percentage of $RNNO_2$ in the total reduced nitrogen ($RNNO_2$/RN).

Species	$RNNO_2$ (mgN/g dry wt)	$RNNO_2$/RN (%)	Family
Woody species			
Eucalyptus viminalis	6.57	12.5	Myrtaceae
Populus nigra	5.14	10.7	Salicaceae
Magnolia kobus	4.92	12.7	Magnoliaceae
Robinia pseudoacacia	4.73	8.7	Leguminosae
Eucalyptus grandis	4.57	8.5	Myrtaceae
Eucalyptus globulus	4.08	9.4	Myrtaceae
Cultivated herbaceous species			
Nicotiana tabacum	5.72	11.4	Solanaceae
Wild herbaceous species			
Erechtites hieracifolia	5.72	10.1	Compositae
Crassocephalum crepidioides	5.07	9.2	Compositae

$t_{1/2} = \ln(2)/k$. In the aforementioned nine species, the half-life of reduced nitrogen for M. kobus with the highest $RNNO_2$/RN value (see **Table 1**) was calculated to be approximately 40 h. This value is comparable to the half-life of Cullin-associated and neddylation-dissociated 1 (CAND1; 30 h) and that of Transport inhibitor response 1 (TIR1; 10 h) proteins [20]. However, assuming a linear dependence of the turnover rate on the concentration of NO_2, the half-life of the reduced nitrogen at 0.04 ppm is estimated to be 4000 h. More recently, we found that 14% of $TNNO_2$ was derived from NO_2 in cucumber (Cucumis sativus) plants grown under natural lighting in the presence of 100 ± 50 ppb NO_2 for 5 weeks [21]. Using these values and the assumptions described above, the half-life of total nitrogen in cucumber was estimated to be 1000 h. Obviously, these calculations of half-lives of nitrogen-derived from NO_2 in the selected species indicates that no naturally occurring plants studied to date are air-pollutant-philic. However, the existence of great differences in both $RNNO_2$ and $RNNO_2/RN$ is encouraging for the fabrication of air-pollutant-philic plants. Investigations on the causes for the differences in these parameters between species or how the metabolism of NO_2 differs between species may provide an important clue to improve plants' abilities to assimilate NO_2 and hence to make plants air-pollutant-philic agents.

In the 217 taxa examined, the $RNNO_2$ value varied among the individuals of a species [16]. This variation was more pronounced in the wild herbaceous and woody plants than in the cultivated herbaceous plants. Among wild herbaceous plants (19 individuals), the variation was 26.2-fold in Solidago altissima, 21.4-fold in Rumex acetosella, and 17.6-fold in Crassocephalum crepidioides. In woody plants, the corresponding value was 11.9-fold in M. kobus, 10.6-fold in Euonymus japonicus, and 6.7-fold in Prunus cerasoides [16]. Understanding differences among individuals of the same species in their abilities to assimilate NO_2 is also important in terms of genetically constructing air-pollutant-philic plants.

We attempted selection of "mutant" lines of Rhododendron mucronatum, which is widely used as roadside trees and can be propagated by cuttings [22]. R. mucronatum was ranked 105th among the 217 taxa in $RNNO_2$. In 1996, 1452 different cutting lines of R. mucronatum (referred to as the "96-plants") were fumigated with 4 ppm $^{15}NO_2$ for 8 h, and $TNNO_2$ in the leaves was determined. A 56-fold difference was observed between the individuals with the highest and lowest capacities. We propagated approximately 100 lines (designated the "97- plants") each from the high-, middle-, and low-capacity groups, and assayed them for $TNNO_2$ in the following year. The maximum variation in $TNNO_2$ was 5.6 times. In 1998, 227 lines propagated from the 96-plants and 95 lines from the 97-plants (the "98-plants") were assayed for $TNNO_2$. The variations in $TNNO_2$ in the 97- and 98- plants were maximally 4.9 and 5.2 times, respectively. The reason(s) for the disappearance of the initial large variation in $TNNO_2$ remains unclear.

GENETIC IMPROVEMENT OF PLANTS' ABILITIES TO MITIGATE AIR POLLUTANTS

Foliar uptake of NO_2 is believed to occur via stomatal and nonstomatal pathways as a gas or as nitrite or nitrate [11]. However, the precise mechanisms remain unclear. Most NO_2 molecules taken up into plants are converted, either chemically or enzymatically, to nitrate and nitrite, assimilated to ammonia, and incorporated into amino acids and other organic nitrogenous compounds through the primary nitrate assimilation pathway [12-15]. In the primary nitrate pathway, nitrate is converted to nitrite by a two-electron reduction catalyzed by the nitrate reducetase (NR), the first enzyme. The second enzyme, nitrite reductase (NiR), catalyzes a six-electron reduction to convert nitrite to ammonia. Ammonia thus formed is incorporated into glutamic acid to form glutamine by the glutamine synthetase (GS), the third enzyme. Therefore, engineering of these genes should be instrumental to improve metabolism of NO_2 and help make plants air-pollutant-philic.

To date we have produced transgenic plants of Arabidopsis thaliana (Cruciferae), a model species that overexpresses any one of the three genes of the primary nitrate reduction pathway [17]. We have also produced transgenic plants of Rhaphiolepis umbellata (Rosaceae) that overexpress the NiR gene [23]. R. umbellata is a very popular roadside shrub in Japan. Finally, we produced mutants of Ficus pumila (Moraceae) using an ion-beam-irradiation technique [24], an emerging method to produce various mutants [25]. F. pumila is an evergreen climber that is sometimes used to cover the walls of houses and buildings. Data on the uptake and assimilation of NO_2 by these three genetically modified plants are shown in **Table 2**.

The uptake of NO_2 in transgenic R. umbellata and ion-beam-irradiated F. pumila was 1.8 to 1.9 times greater compared with their respective control plants (**Table 2**). Although these increases in the uptake of NO_2 may not be enough to make these plants air-pollutantphilic, note that engineering of the NiR gene and ionbeam irradiation in F. pumila did improve the plants' abilities to uptake NO_2. Future studies investigating what causes the increase in uptake of NO_2 in these modified plants may provide some information that is helpful in making plants air-pollutant-philic.

The assimilation of NO_2 in transgenic A. thaliana and R. umbellata, and ion-beam-irradiated F. pumila, was 1.6 – 1.8 times greater compared with their respective control plants (**Table 2**). Thus, gene engineering of the NiR gene and ion-beam irradiation improved plants' abilities to assimilate NO_2. This finding was in contrast to the effects of overexpressing NR and GS (see below). In transgenic A. thaliana, a 1.5-time increase in NiR enzyme activity was observed as compared to the control [17]. The increase in the ability of assimilating NO_2 in transgenic A. thaliana can therefore be attributed to the increase in NiR activity. However, no such increase in NiR activity was observed in ion-beam-irradiated F. pumila [24].

Table 2. Total nitrogen derived from NO_2 ($TNNO_2$) and reduced nitrogen derived from NO_2 ($RNNO_2$) in transgenic Arabidopsis thaliana and Rhaphiolepis umbellata plants, and plants regenerated from ion-beam-irradiated explants of Ficus pumila.

	NO_2 uptake ($TNNO_2$ μgN/gDW)					NO_2 assimilation ($RNNO_2$ μgN/gDW)				
	Transgenic or irradiated	*n*	*Wild type*	*n*	*Fold*	*Transgenic or irradiated*	*n*	*Wild type*	*n*	*Fold*
Arabidopsis thaliana	-	-	-	-	-	1650 ± 140**	3	1018 ± 80	3	1.6
Rhaphiolepis umbellate	333	1	174.2 ± 4.0	3	1.9	282	1	170.1	1	1.7
Ficus pumila	635 ± 160*	9	457 ± 108	10	1.8	519 ± 83**	6	293 ± 72	3	1.8

A. thaliana, *R. umbellate*, and *F. pumila* were fumigated with 4 ppm NO_2 for 8 h under artificial light [17] with 0.2 ppm NO_2 for 1 week under natural light [23], or with 1 ppm ^{15}N-labeled NO_2 for 8 h under artificial light [24], respectively. Data represent the mean ± SD. *$P < 0.05$, **$P < 0.01$.

This implies that genetic change(s) other than manipulation of the NiR gene may be capable of improving the assimilation of NO_2. Future studies on the molecular causes of improved NO_2 assimilation in ion-beam-irradiated F. pumila may provide a clue to help create airpollutant-philic plants.

In the case of the overexpression of NR and GS in A. thaliana, no significant increases in the uptake or assimilation of NO_2 was observed despite respective enzyme activities having been increased by 1.8- and 1.5-fold compared to the controls [17]. Taken together with the above results on NiR overexpresssion, NiR, but not NR or GS, is a controlling enzyme of NO_2 assimilation in plants. This conclusion was supported by the determination of the flux control coefficient, a measure of the effect of change in a single enzyme activity on flux [26,27] as described previously [17].

PLANT VITALIZATION EFFECT OF NO_2

During our study on air-pollutant-philic plants, we unexpectedly discovered that prolonged exposure of plants to a sufficient level of NO_2 activates the uptake and metabolism of nutrients that fuel plant growth and development [28]. We named this phenomenon the plant vitalization effect of NO_2 (PVEON).

In the literature, atmospheric NO_2 has been regarded as being either detrimental [15,29] or beneficial [30,31] to plants depending on the plant's nitrogen nutrient conditions. Recent reports that NO_2 stimulates or fails to stimulate the growth of a particular plant [32,33] suggest that this debate is still active. We speculate that a cause for this discrepancy may be that the response of plant growth to NO_2 is highly dependent on the target species, and thus both the concentration of NO_2 and duration of exposure to NO_2 need to be optimized for each plant species to observe the PVEON.

The PVEON was first found in Nicotiana plumbaginifolia [28], and then in other species such as cucumber, kenaf, lettuce, pumpkin, and tomato (**Table 3**) [21,28,34, 35]. Shoot biomass in NO_2-exposed plants (+NO_2) was 1.4- to 2.4-fold greater than that in non-exposed (–NO_2) plants. A. thaliana showed PVEON and its increase in shoot biomass in response to this effect differed between the accessions Columbia and C24 (1.6- and 2.3-fold, respectively) compared to no-exposure plants when plants were exposed to 50 ppb NO_2 for 3 weeks [35]. The optimal NO_2 concentration for the PVEON varied from 50 to 200 ppb depending on the species. The duration of exposure to NO_2 was changed from 3 to 12 weeks as a function of target species (**Table 3**). The increase in shoot biomass in response to PVEON was always accompanied by an increase in total leaf area in all plant species studied to date. In Nicotiana plumbaginifolia, the contents per shoot of C, N, S, P, K, Ca, Mg, free amino acids and crude proteins were almost doubled in +NO_2 plants compared with –NO_2plants [28].

In tomato (cv. Micro-Tom), the PVEON significantly increases the yield of fruits (**Table 3**). This increase in fruit yield was accompanied by acceleration of flowering time by 3.2 days and an increase in flower number per plant of up to 60%. Not surprisingly, the average weight of a tomato fruit was not significantly different between +NO_2 plants and –NO_2 plants, but the number of fruits on a plant was distinctly increased in the former [35]. Thus, the PVEON increases fruit yield via stimulation of flowering in tomato. This is in sharp contrast to Spierings' [36] report that exposure of Moneymaker tomato plants to NO_2 at 250 ppb during the entire growth period of about 4 months caused a 22% decrease in fruit yield. In addition, Spierings [36] found that NO_2 was detrimental to the vegetative growth of Moneymaker; fumigation with NO_2 increased the height of very young plants, but resulted in smaller leaves and shorter petioles, and

Table 3. Effects of exposure to NO_2 on the yield of shoot biomass or fruit.

Species	NO_2 (ppb)	Exposure duration (weeks)	Shoot biomass (mg/plant)	Fruit yield (g/plant)	*n*	Fold
Cucumis sativus (cucumber)[1]						
–NO_2	<5	5	370 ± 90	-	3	
+NO_2	100 ± 50	5	620 ± 110*	-	3	1.7 ± 0.3
Cucurbita moschata (pumpkin)[2]						
–NO_2	<5	5	5300 ± 100	-	3	
+NO_2	200 ± 50	5	8300 ± 1000*	-	3	1.6 ± 0.2
Helianthus annuus (sunflower)[3]						
–NO_2	<5	6	1760 ± 810	-	3	
+NO_2	200 ± 50	6	3550 ± 30*	-	3	2.0 ± 0.02
Hibiscus cannabinus (kenaf)[4]						
–NO_2	<5	3	586 ± 284	-	5	
+NO_2	100 ± 50	3	793 ± 119	-	5	1.4 ± 0.2
Lactuca sativa (lettuce)[5]						
–NO_2	<5	6	140 ± 70	-	3	
+NO_2	50 ± 10	6	340 ± 90*	-	3	2.4 ± 0.7
Nicotiana plumbaginifolia[6]						
–NO_2	<5	10	1044 ± 87	-	6	
+NO_2	150 ± 50	10	1767 ± 119***	-	6	1.7 ± 0.1
Solanum lycopersicum (tomato)[7]						
–NO_2	<5	12	-	50.9 ± 8.7	10	
+NO_2	50 ± 10	12	-	70.0 ± 14.6**	10	1.4 ± 0.2

[1-3,5][21]; [4][35]; [6][28]; [7][36]. ***$P < 0.001$, **$P < 0.01$, *$P < 0.05$.

caused a loss of leaves. These findings are also contrary to our results for Micro-Tom, where no detrimental effect of NO_2 on vegetative growth was observed, and the biomass yield of Micro-Tom plants grown with or without 50 ppb NO_2 was not significantly different. The reason for these discrepancies remains unclear.

Based on the mass spectrometric analysis on the $^{15}N/^{14}N$ ratio, it was found that NO_2-derived N (NO_2-N) comprised less than 3% - 14% of the total plant nitrogen [21,28,34], indicating that PVEON is due to the effect of NO_2 as a signal rather than as a nitrogen source. This speculation is in line with the common understanding that NO_2 is a reactive nitrogen species acting as a cellular signal in both plants and animals.

The molecular mechanisms of PVEON and the genes involved in PVEON are currently being investigated using Arabidopsis plants. Histological analyses examining the causes of an approximately two times increase in leaf size in Arabidopsis plants indicates that an increase in cell size rather than cell number is the most probable cause. Based on transcriptome analysis, we found that destruction of a gene (Vita 1) causes loss of PVEON. Thus, this gene is a PVEON-responsive gene in Arabidopsis plants. These genes will prove useful in helping to make plants air-pollutant-philic in future studies.

Acknowledgements

This work was supported in part by a Grant-in-Aid for Scientific Research (nos.18770037 and 21580403) from the Japan Society for the Promotion of Science and by the Research for Promoting Technological Seeds Program of the Japan Science and Technology Agency.

REFERENCES

1. H. Morikawa, A. Higaki, M. Nohno, M. Kamada, M. Nakata, G. Toyohara, K. Fujita and K. Irifune, "Air-Pollutant-Philic Plants from Nature," In: N. Murata, Ed., Research in Photosynthesis, Vol. IV, Kluwer Academic Publishers, Dordrecht, 1992, pp. 79-82.
2. M. Kamada, A. Higaki, Y. Jin, M. Ayabe, M. Seki, T. Sawasaki, S. Ida, G. Toyohara, K. Irifune and H. Morikawa, "Transgenic 'Air-Pollutant-Philic Plants' Produced by Particle Bombardment," In: N. Murata, Ed., Research in Photosynthesis, Vol. IV, Kluwer Academic Publishers, Dordrecht, 1992, pp. 83-86.
3. NationMaster, "Urban NO_2 Concentration (Most Recent) by Country," http://www.nationmaster.com/graph/env_urb_no2_con-environment-urban-no2- WHO, "Air Quality and Health," 2011. http://www.who.int/mediacentre/factsheets/fs313/en/index.html
4. Ministry of the Environment of Government of Japan, "White Paper," http://www.env.go.jp/
5. A. R. Wellburn, "Air Pollution and Climate Change: The Biological Impact," Longman Scientific & Technical, Harlow, 1994.
6. A. J. Zeevaart, "Some Effects of Fumigating Plants for Short Periods with NO_2," Environmental Pollution, Vol. 11, No. 2, 1976, pp. 97-108. doi:10.1016/0013-9327(76)90022-7
7. Y.-N. Lee and S. E. Schwartz, "Reaction Kinetics of Nitrogen Dioxide with Liquid Water at Low Partial Pressure," Journal of Physical Chemistry, Vol. 85, No. 7, 1981, pp. 840-848.doi:10.1021/j150607a022
8. A. Rowland, A. J. S. Murray and A. R. Wellburn, "Oxides of Nitrogen and Their Impact upon Vegetation," Reviews on Environmental Health, Vol. 5, No. 4, 1985, pp. 295-342.
9. P. Ramge, F.-W. Badeck, M. Plöchl and G. H. Kohlmaier, "Apoplastic Antioxidants as Decisive Elimination Factors within the Uptake Process of Nitrogen Dioxide into Leaf Tissues," New Phytologist, Vol. 125, No. 4, 1993, pp. 771- 785. doi:10.1111/j.1469-8137.1993.tb03927.x
10. W. Larcher, "Physiological Plant Ecology," Springer-Verlag, Berlin, 1995. doi:10.1007/978-3-642-87851-0
11. S. V. Durmishidze and N. N. Nutsubidze, "Absorption and Conversion of Nitrogen Dioxide by Higher Plants," Doklady Biochemistry and Biophysics, Vol. 227, 1976, pp. 104- 107.
12. H. H. Rogers, J. C. Campbell and R. J. Volk, "Nitrogen-15 Dioxide Uptake and Incorporation by Phaseolus vulgaris (L.)," Science, Vol. 206, No. 4416, 1979, pp. 333- 335. doi:10.1126/science.206.4416.333
13. T. Yoneyama and H. Sasakawa, "Transformation of Atmospheric Nitrogen Dioxide Absorbed in Spinach Leaves," Plant and Cell Physiology, Vol. 20, 1979, pp. 263-266.
14. A. R. Wellburn, "Why Are Atmospheric Oxides of Nitrogen Usually Phytotoxic and Not Alternative Fertilizers?" New Phytologist, Vol. 115, No. 3, 1990, pp. 395- 429.doi:10.1111/j.1469-8137.1990.tb00467.x
15. H. Morikawa, A. Higaki, M. Nohno, M. Takahashi, M. Kamada, M. Nakata, G. Toyohara, Y. Okamura, K. Matsui, S. Kitani, K. Fujita, K. Irifune and N. Goshima, "More than a 600-Fold Variation in Nitrogen Dioxide Assimilation among 217 Plant Taxa," Plant, Cell and Environment, Vol. 21, 1998, pp. 180-190. doi:10.1046/j.1365-3040.1998.00255.x Paper reference:4
16. M. Takahashi, Y. Sasaki, S. Ida and H. Morikawa, "Enrichment of Nitrite Reductase Gene Improves the Ability of Arabidopsis thaliana Plants to Assimilate Nitrogen Dioxide," Plant Physiology, Vol. 126, No. 2, 2001, pp. 731- 741. doi:10.1104/pp.126.2.731 Paper reference:4
17. H. Morikawa, M. Takahashi, A. Sakamoto, T. Matsubara, G. Arimura, Y. Kawamura, K. Fukunaga, K. Fujita, N. Sakurai, T. Hirata, H. Ide, N. Nonoyama and H. Suzuki, "Formation of Unidentified Nitrogen in Plants: An Implication for a Novel Nitrogen Metabolism," Planta, Vol. 219, No. 1, 2004, pp. 14-22. doi:10.1007/s00425-003-1200-7
18. M. Takahashi, A. Higaki, M. Nohno, M. Kamada, Y. Okamura, K. Matsui, S. Kitani and H. Morikawa, "Differential Assimilation of Nitrogen Dioxide by 70 Taxa of Roadside Trees at an Urban Pollution Level," Chemosphere, Vol. 61, No. 5, 2005, pp. 633-639. doi:10.1016/j.chemosphere.2005.03.033
19. X.-Y. Yang, W.-P. Chen, A. K. Rendah, A. D. Hegeman,

W. M. Gray and J. D. Cohen, "Measuring the Turnover Rates of Arabidopsis Proteins Using Deuterium Oxide: An Auxin Signaling Case Study," The Plant Journal, Vol. 63, No. 4, 2010, pp. 680-695.doi:10.1111/j.1365-313X.2010.04266.x Paper reference:2

20. S. E. H. Adam, J. Shigeto, A. Sakamoto, M. Takahashi and H. Morikawa, "Atmospheric Nitrogen Dioxide at Ambient Levels Stimulates Growth and Development of Horticultural Plants," Botany, Vol. 86, No. 2, 2008, pp. 213-217. doi:10.1139/B07-129

21. Y. Kawamura, K. Fukunaga, A. Umehara, M. Takahashi and H. Morikawa, "Selection of Rhododendron mucronatum Plants that Have a High Capacity for Nitrogen Dioxide Uptake," Acta Biotechnologica, Vol. 22, No. 1-2, 2002, pp. 113-120. doi:10.1002/1521-3846(200205)22:1/2<113::AID-ABIO113>3.0.CO;2-B

22. J. Shigeto, S. Yoshihara, S. E. H. Adam, K. Sueyoshi, A. Sakamoto, H. Morikawa and M. Takahashi, "Genetic Engineering of Nitrite Reductase Gene Improves Uptake and Assimilation of Nitrogen Dioxide by Rhaphiolepis umbellata (Thunb.) Makino," Plant Biotechnology, Vol. 23, 2006, pp. 111-116. doi:10.5511/plantbiotechnology.23.111

23. M. Takahashi, S. Kohama, J. Shigeto, Y. Hase, A. Tanaka and H. Morikawa, "Mutants of Ficus pumila Produced by Ion Beam Irradiation with an Improved Ability to Uptake and Assimilate Atmospheric Nitrogen Dioxide," International Journal of Phytoremediation, Vol. 14, No. 3, 2012, pp. 275-281. doi:10.1080/15226514.2011.604694 Paper reference:2

24. A. Tanaka, N. Shikazono, Y. Yokota, H. Watanabe and S. Tana, "Effects of Heavy Ions on the Generation and Survival of Arabidopsis thaliana," International Journal of Radiation Biology, Vol. 72, No. 1, 1997, pp. 121-127. doi:10.1080/095530097143608

25. H. Kacser and J. W. Porteous, "Control of Metabolism: What Do We Have to Measure?" Trends in Biochemical Sciences, Vol. 12, No. 1, 1987, pp. 5-14. doi:10.1016/0968-0004(87)90005-3

26. M. Stitt and U. Sonnewald, "Regulation of Metabolism in Transgenic Plants," Annual Review of Plant Physiology and Plant Molecular Biology, Vol. 46, 1995, pp. 341-368.doi:10.1146/annurev.pp.46.060195.002013

27. M. Takahashi, M. Nakagawa, A. Sakamoto, C. Ohsumi, T. Matsubara and H. Morikawa, "Atmospheric Nitrogen Dioxide Gas Is a Plant Vitalization Signal to Increase Plant Size and the Contents of Cell Constituents," New Phytologist, Vol. 168, No. 1, 2005, pp. 149-154.doi:10.1111/j.1469-8137.2005.01493.x Paper reference:3

28. T. M. Capron and T. M. Mansfield, "Inhibition of Growth in Tomato by Air Polluted with Nitrogen Oxides," Journal of Experimental Botany, Vol. 28, No. 1, 1977, pp. 112- 116.doi:10.1093/jxb/28.1.112

29. H. Saxe, "Relative Sensitivity of Greenhouse Pot Plants to Long-Term Exposures of NOand NO_2-Containing Air," Environmental Pollution, Vol. 85, No. 3, 1994, pp. 283-290.doi:10.1016/0269-7491(94)90049-3

30. R. Sandhu and G. Gupta, "Effects of Nitrogen Dioxide on Growth and Yield of Black Turtle Bean (Phaseolus vulgaris L.) cv. 'Domino,'" Environmental Pollution, Vol. 59, No. 4, 1989, pp. 337-344. doi:10.1016/0269-7491(89)90160-7

31. Q. Xu, B. Zhou, C. Ma, X. Xu, J. Xu, Y. Jiang, C. Liu, G. Li, S. J. Herbert and L. Hao, "Salicylic Acid-Altering Arabidopsis Mutants Response to NO_2 Exposure," Bulletin of Environmental Contamination and Toxicology, Vol. 84, No. 1, 2010, pp. 106-111. doi:10.1007/s00128-009-9913-3

32. D. M. Vallano and J. P. Sparks, "Quantifying Foliar UpTake of Gaseous Nitrogen Dioxide Using Enriched Foliar ∂^{15}N Values," New Phytologist, Vol. 177, No. 4, 2008, pp. 946-955.doi:10.1111/j.1469-8137.2007.02311.x

33. M. Takahashi, S. E. Adam, D. Konaka and H. Morikawa, "Nitrogen Dioxide at an Ambient Level Improves the Capability of Kenaf (Hibiscus cannabinus) to Decontaminate Cadmium," International Journal of Phytoremediation, Vol. 10, No. 1, 2008, pp. 73-76. doi:10.1080/15226510701827085

34. M. Takahashi, A. Sakamoto, H. Ezura and H. Morikawa, "Prolonged Exposure to Atmospheric Nitrogen Dioxide Increases Fruit Yield of Tomato Plants," Plant Biotechnology, Vol. 28, No. 5, 2011, pp. 485-487. doi:10.5511/plantbiotechnology.11.0819a Paper reference:2

35. F. H. F. G. Spierings, "Influence of Fumigations with NO_2 on Growth and Yield of Tomato Plants," Netherlands Journal of Plant Pathology, Vol. 77, No. 6, 1971, pp. 194-200.doi:10.1007/BF01977278

CHAPTER 9

EXPOSURE TO ELEVATED OZONE LEVELS DIFFERENTIALLY AFFECTS THE ANTIOXIDANT

Fernanda Freitas Caregnato, Rafael Calixto Bortolin, Armando Molina Divan Junior, José Cláudio Fonseca Moreira

Keywords: Antioxidant capacity; Ozone; *Phaseolus vulgaris*; Reactive oxygen species; Redox balance

INTRODUCTION

Tropospheric ozone (O_3) is a widespread secondary air pollutant which is produced through the reactions between primary pollutants (nitric oxides, sulfur oxides, carbon oxides and hydrocarbons), and sunlight (Finlayson et al., 1997). Although O_3 is produced on areas with intense primary pollution, meteorological and topographic conditions can move its precursors from these areas towards less polluted ones, such as rural zones, with detrimental effects on natural and cultivated plant species (Crutzen and Lelieveld, 2001). Over the past few years, the ground-level O_3 concentrations has become a major threat to vegetation at a global scale, and experimental studies have confirmed that exposure to elevated O_3 levels results in plant damage specially because it can suppress photosynthesis, accelerate senescence, decrease growth and reduce crop yields (Agrawal et al., 2003 and Booker et al., 2009). The threat imposed by O_3 has brought important implications for agriculture across many parts of the world and a large amount of data indicate that substantial yield losses of important crops, such as wheat, rice and soybean, arise under current O_3 concentrations (Feng et al., 2003, Fishman et al., 2010 and Sarkar and Agrawal, 2010). Predictive models indicate that global surface O_3 levels are expected to rise significantly throughout the 21st century. According to Vingarzan (2004), modeling studies using the Intergovernmental Panel on Climate Change (IPCC, 2001) emission scenarios predict that the average global surface O_3 concentration is expect to be in the range of 35–48 ppb by 2040, 38–71 ppb by 2060, 41–87 by 2080 and 42–84 ppb by 2100. Such increases exceed the internationally accepted environmental criteria and can have severe implications on human health, crops and vegetation.

Numerous studies report that elevated O_3 concentrations often results in reduced photosynthesis and whole-plant growth, decreased stomatal conductance, altered antioxidant system, and accelerated senescence (Kobayashi, 1995, Calatayud and Barreno, 2004, Ashmore, 2005, Black et al., 2007, Calatayud et al., 2007 and Feng et al., 2008). The effects of O_3 depend on a number of events, starting with the gas uptake through leaf stomata. In the sub-stomatal chamber, O_3 can directly

Citation: Fernanda Freitas Caregnato, Rafael Calixto Bortolin, Armando Molina Divan Junior, José Cláudio Fonseca Moreira, Exposure to elevated ozone levels differentially affects the antioxidant, Volume 93, Issue 2, September 2013, Pages 320-330, ISSN 0045-6535, http://dx.doi.org/10.1016/j.chemosphere.2013.04.084.

react with plasma membrane through "ozonolysis" stimulating lipid peroxidation and impairing membrane fluidity, or it can be spontaneously converted into reactive oxygen species (ROS) such as superoxide radical (O_2^-), hydroxyl radical (OH^-) and hydrogen peroxide (H_2O_2) (Kanofsky and Sima, 1995 and Halliwell and Gutteridge, 2007). Increased ROS levels may create a pro-oxidant environment inside the cell, resulting in proteins modification, both in their structure and activity (Iriti and Faoro, 2008). In particular, O_3 and the derived ROS can react with the exposed proteins sulphydryl groups presented on the amino acids resides, mainly cysteine, tryptophan, tyrosine, methionine and histidine, which might contribute directly to lower the overall plant vigor (Mudd et al., 1969 and Foyer and Noctor, 2008). Furthermore, photosynthetic capacity and photosynthetic efficiency are decreased under high O_3 levels, which has greatly contribute to yield loss in U.S. soybean (*Glycine max*) cultivars that are particularly sensitive to O_3 (Betzelberger et al., 2012). Some studies report that O_3-mediated decline in photosynthetic pigments, chlorophylls *a* and *b*, arises as a consequence of the oxidative destruction of these molecules (Leitao et al., 2008). Changes on the levels of photosynthetic pigments modify the chlorophyll fluorescence parameters, specifically affecting the photosystem II (PSII) activity (Flowers et al., 2007).

Plant cells possess an array of antioxidants (both enzymatic and non-enzymatic) to reduce the stress elicited by the ROS. Enzymatic scavenging mechanisms in plants include important antioxidant enzymes such as superoxide dismutase (SOD), ascorbate peroxidase (APX), glutathione reductase (GR), and catalase (CAT). Among these enzymes, catalase plays an important role in keeping the levels of H_2O_2 under control, and some authors suggest that catalase functions as a cellular sink for H_2O_2 (Willekens et al., 1997). In contrast to superoxide and hydroxyl radical, H_2O_2 is relatively stable, and thus is capable to function as a molecular signal for cell-to-cell communication (Desikan, 2001 and Vandenabeele et al., 2004). Thus, sustained increases in H_2O_2 may provoke transcriptional responses that include expression of hypersensitive response and programmed cell death genes (Vandenabeele et al., 2003). In guard cells, for example, H_2O_2 participates in the activation of membrane-localized anion channels that lead to stomatal closure (Schroeder et al., 2001 and Zhang and Klessig, 2001). The importance of catalase has long been discussed and experiments performed with catalase-deficient mutant of barley (*Hordeum vulgare* L.) showed that transgenic plants present severe leaf bleaching, specifically under conditions that H_2O_2 production is favored through photorespiration (Kendall et al., 1983).

Non-enzymatic ROS scavengers include a number of compounds with high reducing power, such as the ascorbic acid and glutathione. According to several authors, the apoplastic AsA is the first line of defense against O_3-derived ROS, and O_3 sensitivity can be generally correlated to the AsA redox status of the leaf tissue (Kanofsky and Sima, 1995, Ranieri et al., 1999 and Castagna and Ranieri, 2009). It has been reported that AsA levels are increased in plants challenged with O_3, suggesting that AsA plays an important role in providing resistance to oxidative stress imposed by O_3 (Menser, 1964 and Sharma et al., 1997). However, some authors have recently reported that elevated apoplastic AsA levels are apparently not sufficient to explain O_3 tolerance (D'Haese et al., 2005 and Dizengremel et al., 2008). Although the resulting increment on AsA levels may be a predictor of resistance to O_3 exposure (Eltayeb et al., 2006), other antioxidant molecules, and an improved cellular ability to regenerate these antioxidants could be more relevant to protect the cells against O_3 toxic effects. In plants, the physiological significance of AsA is strictly dependent on total thiol groups redox state, especially the glutathione (GSH) molecule which is the predominant non-protein thiol. Thus, the AsA–GSH system (also known as Halliwell–Asada cycle) is responsible for the maintenance of the cellular the redox homeostasis (Smirnoff, 1996 and Horemans et al., 2000). In addition to GSH, thiol-redox proteins also play a important role as antioxidants and, together with GSH, are responsible for keeping the cellular redox balance (Foyer and Noctor, 2008). Analysis of total thiol content may thus be important for understanding how O_3-derived ROS changes the thiol-redox state of cellular components, such as enzymes activity modulation trough redox modifications of active site residues required for catalysis and/or ligand binding, oligomerization (Paget and Buttner, 2003).

Most studies and evidences that assess the harmful effects of O_3 on crop plants health and productivity are currently performed in countries which are experiencing a rapid increase in population and industrialization levels, such as China and India (Sarkar et al., 2010 and Wang et al., 2012). Besides that, extensive studies have been conducted on the United Sates that helped to improve the scientists' knowledge about the biochemical, genetics and molecular aspects ruling O_3 toxicity in native and agronomic plants (Krupa et al., 2000, Booker et al., 2009 and Betzelberger et al., 2010).

However, O_3 trend data for the Southern Hemisphere is sparse, and regional characterization is still necessary to determine the levels of O_3 observed on the industrialized southern countries (Vingarzan, 2004). Furthermore, investigations and evidences with respect to vegetation damage and human health effects of O_3 in South America countries are scarce. In Brazil, relative little information concerning the levels and the impacts of O_3 pollution on plants health and crop productivity are available and, therefore, it should be some of the urgent research priorities for the future of the country. Over the past few decades, Brazilian commodities production has expanded and the country has observed a rapid rise in exports of crops, like soybean and wheat. Crop area has expanded rapidly and is expected to continue to do so, shifting the crop cultivation to new regions of the country. This scenario has brought an increment on the production of the major cultivated crops.

According to the US DA (2010), Brazil is currently the largest world bean producer and consumer. The National Food Supply Agency (CONAB) had forecasted Brazilian dry bean production for 2010/11 at 3.45 million metric tons, up 5% when compared to 2009, with a cultivated area estimated at 3.6 million hectares, up 1% percent from 2009. Yield was forecasted at 0.965 tons/hectare, 6% greater than 2009 and well above the 5-year average of 0.857 tons/hectare. Dry beans are cultivated in nearly every Brazilian state, but most of the production is concentrated in 8 states, which are responsible for approximately 80% of Brazil's production. Family farmers produce about 70% of total bean production. Feng and Kobayashi (2009) conducted a quantitative meta-analysis study to assess the effects of rising O_3 concentrations on yield and yield components of major food crops (potato, barley, wheat, rice, bean and soybean) and observed that bean was the crop with the largest yield loss (19%). The authors indicate that bean is very sensitive to O_3, which is in agreement with the study of Mills et al. (2007), who ranked bean as the most sensitive among agricultural crops according to O_3 critical levels (AOT40 for 5% yield reduction).

Keeping in view that O_3 levels are increasing worldwide and Brazil is a developing country that still suffer from a general lack of information concerning the negative impacts of O_3 over the major agricultural crops, the aim of the present work was to analyze the effects of O_3 exposure on the redox homeostasis balance on leaf tissue of two Brazilian bean varieties (Fepagro 26 and Irai). These varieties were developed by the Founding State of Agricultural Research of Rio Grande do Sul (FEPAGRO-RS) and are exclusively cultivated in the southern states of Brazil. In spite of that, no information describing the physiology or the chemical profile of these varieties was done before. Thus, our laboratory decided to start the investigation on how O_3 affects the physiological and the biochemical aspects of these varieties (information on physiological aspects seeClebsch et al., 2009). In a previous analysis we observed that Fepagro 26 and Irai have very different oxidative responses to O_3, suggesting that these varieties might present biochemical constitutive differences regarding their redox status and these differences could be associated with a higher tolerance or susceptibility to O_3 stress (see Caregnato et al., 2010). To further test whether the constitutive antioxidant biochemical differences between the varieties could account for their distinct responses to O_3 stress we decided to assess the following biochemical traits on the foliar tissue: (1) changes in the levels of ROS, (2) modifications on the immunocontent of the antioxidant enzyme catalase, (3) concentrations of non-enzymatic antioxidant (ascorbate and total thiol groups), and (4) and changes on the levels of the photosynthetic pigments chlorophylls *a* and *b*. In addition, *in situ* histochemical techniques were used to determine the type of ROS that was prevalently being produced during O_3 exposure.

MATERIALS AND METHODS

Plant Material and Growing Conditions

Seeds from two different varieties of *Phaseolus vulgaris* L. (Fepagro 26 and Irai) developed by FEPAGRO-RS were germinated in 2 L pots containing 2:1:1 of coarse vermiculite, washed coarse sand and peat. The plants were grown under controlled conditions in a glasshouse, at an average temperature of 17.7 °C, and relative humidity of 88.6%. The plants were irrigated once a day with 100 mL of water. One day before O_3treatment each plant was irrigated with 100 mL of Vitaplan Nutriverde 13-13-15 fertilizer (Nutriplan Products Company, Parana, Brazil). Eight days after the seeds were planted, 12 *P. vulgaris* seedlings from each variety were selected for the experiment. Six seedlings from each variety were subjected to O_3 exposure (n = 6), and six seedlings were used as control (n = 6).

Ozone Exposure

Open top chambers were used for O_3 exposure experiments as previously reported byCaregnato et al. (2010) and Clebsch et al. (2009). Ozone fumigation was done using an ozonizer (ozone generator GHR150B, OZ Engenharia, Porto Alegre, Brazil) equipped with three O_3 production cells (corona effect) with a nominal production of 85 mg h^{-1}each. Fumigation was performed daily from 10:00 to 16:00 h during one week. Control plants were maintained in identical chambers in which normal air was introduced. Ozone monitoring was done on a 6 h d^{-1} basis (10:00–16:00 h) because this period is considered to be the peak of O_3 production found under ambient conditions. Ozone concentration inside the chambers was monitored by the iodometric method (APHA, 1992). Gases were sampled in a fresh washer of the impinger type, containing 75 mL of absorbent solution of KI 2% with the help of a gas sampler (LaMotte, model BD, Chestertown, USA) at a flow of 1.5 L min^{-1}. The sampling point was located in the central position of the chambers in relation to their diameters, and at the height of the unifoliolated primary leaves. The O_3 doses fumigated in each chamber (control and O_3) were calculated as the accumulated hourly O_3 exposure over a threshold concentration of 40 ppb (AOT_{40}) during daylight hours (10:00–16:00 h). Exposure index for O_3, i.e. AOT_{40}, was calculated and the mean value of O_3 concentration obtained on our experiments was 122.6 ppb h. During the experiment period each seedling was irrigated with 200 mL of tap water every day. At the end of the fumigation period, control and treated seedlings were harvested and the second leaf pair from each plant was sampled. Leaf samples were weighed and immediately stored at – 80 °C for biochemical analysis. The O_3 exposure experiments were performed on three independent sets, and the results presented are representative from one of the three biological replications. Results are presented as the average values from 6 seedlings for each treatment (control and O_3).

Determination of Reactive Oxygen Species Levels

About 50 mg of frozen tissue was used for the ROS level measurement assay. The assay was performed as described by Myhre et al. (2003). Leaf tissue was homogenized in 1 mL of Tris HCl 10 mM buffer (pH 7.2), and centrifuged at 14 000*g* at 4 °C for 5 min. About 100 μL of this homogenate was mixed with 100 μL of 20 μM 2′,7′-dichlorodihydrofluorescein diacetate (H_2DCFDA) (Sigma Chemicals, St. Louis, MO, USA) and incubated in the dark for 10 min at room temperature. An aliquot from these homogenate was used to measure the sample fluorescence in fluorometer equipment (SpectraMax M5 model; Molecular Device, CA, USA). The results are expressed as relative fluorescence units (RFUs) per milligram of protein.

Histochemical Staining

In situ histochemical assays were conducted for determining the nature of ROS as described by Thordal-Christensen et al. (1997) and Fryer et al. (2002) followed by minor modifications.

Superoxide Radical Detection

For detection of superoxide radical, 5 discs (diameter of 6.7 mm each) from all seedlings (O_3 and control) and from different leaves were incubated with 10 mM phosphate-buffered saline (pH 6.4) containing 0.01% nitroblue tetrazolium (NBT) during

60 min in the dark. Disc samples were illuminated for 2 h, when the reaction of NBT with superoxide anion stabilizes and the result appears as dark blue spots, which is a characteristic of blue formazan precipitates generation.

Hydrogen Peroxide Detection

To detect the H_2O_2 molecule, five leaf discs (diameter of 6.7 mm each) from all seedlings (O_3 and control) were immersed in diaminobenzidine (DAB) solution of 1 mg mL^{-1} (pH 3.8) during 8 h. As negative control, DAB solution was supplemented with 10 mM ascorbic acid, an antioxidant that prevents H_2O_2 generation.

Before visualization, disc samples from both NBT and DAB staining assays were bleached by immersing the samples in boiling solution of ethanol (96%) – glycerin (9:1 v/v) for 5 min to remove the chlorophyll. The destained discs were placed on a microscope slide and observed under a microscope (Nikon Eclipse TE300, Nikon, Melville, NU, USA), with a digital camera (Nikon Digital Camera DMX1200, Nikon, Melville, NU, USA) attached to it that was used to capture the images.

Determination of Ascorbate Concentration

Ascorbate concentration was assessed according to Keller and Schwager (1977). About 50 mg of frozen leaf tissue was homogenized in 1 mL of aqueous solution of EDTA-Na_2(0.07%) and oxalic acid (0.5%). Samples were centrifuged at 14 000 *g* during 30 min at 4 °C. An aliquot of supernatant was add 0.02% of 2,6-*Dichloroindophenol* Sodium Salt Dihydrate (DCPiP) (1:1 v/v). Samples were divided in two groups; to one of the groups were added 20 μL of ascorbic acid (1%). This group was considered the positive control. The absorbance of each sample was measured at $\lambda = 520$ nm in a micro-plate reader and the difference in the absorbance between the two groups (group without ascorbic acid – group with ascorbic acid) were used to quantify the total amount of ascorbate. Values are expressed as mg of ascorbate per milligram of protein.

Total reduced Thiol Content

Leaf samples were analyzed for their content of total reduced thiol, which is considered an estimation of intracellular redox balance alterations. As previously described byEllman (1959), an aliquot of the sample (30–80 μg protein) was diluted in 10 mM phosphate-buffered saline (pH 7.4) Then, 0.01 M 5,5-dithiobis-2-nitrobenzoic acid (DTNB) dissolved in ethanol, was added. After 60 minutes an intense yellow color was developed and samples were read in a spectrophotometer at 412 nm. Values are expressed as nmol of thiol groups (SH) per milligram of protein.

Catalase Immunoblotting

To perform immunoblot experiments leaf samples were homogenized in Laemmli-sample buffer (62.5 mM Tris–HCl, pH 6.8, 1% (w/v) SDS, 10% (v/v) glycerol) and equal amounts of protein (60 μg/lane) were fractionated by SDS–polyacrylamide gel electrophoresis (PAGE) and electroblotted onto polyvinyledilene difluorife (PVDF) membranes. Protein loading and electroblotting efficiency were verified through Ponceau S staining, and the membrane was blocked in Tween–Tris buffered saline (TTBS:100 mM Tris–HCl, pH 7.5, containing 0.9% NaCl and 0.1% Tween-20) containing 0.5% albumin and incubated overnight with the primary antibody to be tested. The membrane was washed and incubated with Horseradish peroxidasecoupled anti-rabbit IgG antibodies, and the immunoreactivity was detected by enhanced chemiluminescence using the West Pico detection Kit (Pierce, Rockford, IL, USA). Densitometric analysis of the X-ray films exposed to the chemiluminescence was performed with the IMAGE J® software. Blots were developed to be linear in the range used for densitometry. All results are expressed as relative CAT units, and Comassie-blue staining was used as the gel internal loading control.

Chlorophyll Content

Chlorophylls *a* and *b* content was measured by UV–VIS Spectroscopy. Ten foliar discs of 6.7 mm in diameter were incubated in 10 mL of ethanol 96% during 1 week in darkness. The absorbance of leaf pigment extracted in the ethanol was measured at 649 and 665 nm. Levels of chlorophylls *a* and *b*, total chlorophylls ($a + b$) and chlorophyll ratio (a/b) were calculated according to extinction coefficients and equations reported byLichtenthaler (1987). Data from the absorbance were averaged and chlorophyll content was expressed on basis of final volume of ethanol and the dry weight of leaf discs after the period of incubation. Results are expressed as mg of chlorophyll per gram of dry weight.

Protein Content

Protein content was measured spectrophotometrically at 595 nm by the protein–dye binding method of Bradford (1976), using bovine serum albumin as standard.

Statistical Analysis

Data shown in figures and tables are means ± SEM for each variety. Differences between the varieties were analyzed employing two-way analysis of variance (ANOVA) and the student *t*-test. Statistically significant differences found between controls and ozone exposed on the ANOVA test were assessed via Bonferroni post-hoc test, with a significant level set at $p < 0.05$. The relationship between chlorophylls a/b ratio and protein concentration in leaf tissue were assessed using linear bivariate regression analysis within GraphPad software version 5.01 (Graphpad Software Inc., 2007).

RESULTS

Reactive Oxygen Species Generation and Histochemical Staining

The two *P. vulgaris* varieties showed different patterns of ROS production in leaf tissue when growing under the presence of O_3. Fepagro 26 exposed seedlings presented lower constitutive levels of ROS content, but significantly increased the foliar ROS concentration in response to O_3 treatment. After the exposure period of 1 week, ROS levels on Fepagro 26 leaves were 1.8-fold higher than control plants. Conversely, no significant changes on Irai leaf ROS levels could be observed after the same exposure time period (Fig. 1). However, the endogenous levels of ROS on Irai leaves were 1.2-times higher than Fepagro 26 ($p < 0.0001$).

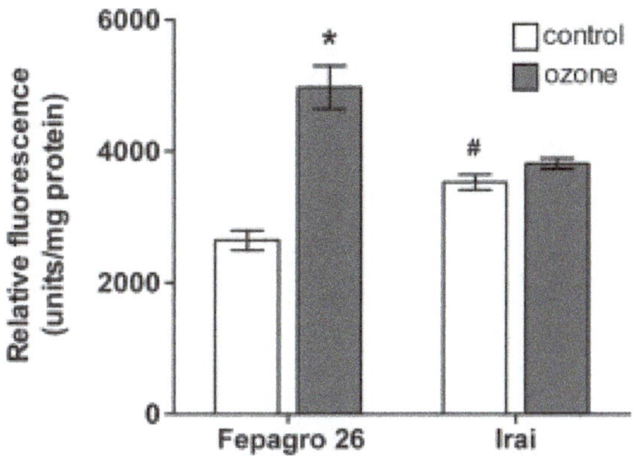

Figure 1. Reactive oxygen species levels on control and O_3-exposed leaves from Fepagro 26 and Irai seedlings. Values represent mean ± SEM ($n = 6$) of three independent experiments. * Indicate difference from control ($p < 0.05$); # indicate difference between varieties ($p < 0.05$) as analyzed by ANOVA.

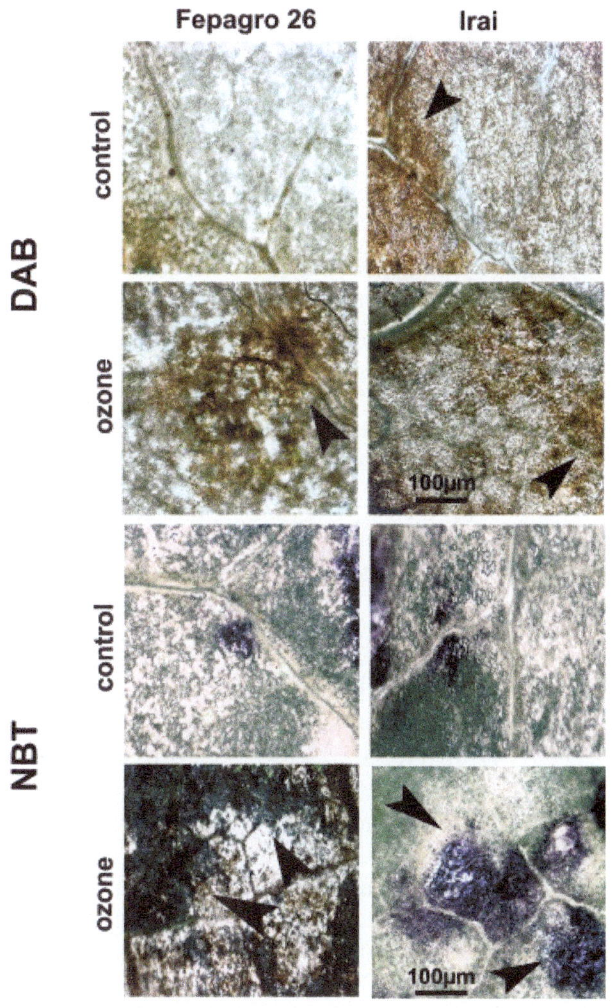

Figure 2. *In situ* localization of O_2^- and H_2O_2 respectively by NBT and DAB staining in control and O_3-exposed leaves from Fepagro 26 and Irai seedlings. Arrows are indicating the H_2O_2 accumulation areas (brown-stained spots), and O_2^- accumulation areas (purple spots) (magnification 10×). (For interpretation of the references to color in this figure legend, the reader is referred to the web version of this article.)

To assess the type of ROS that was prevalently being produced during the O_3 exposition we used two *in situ* histochemical techniques. Diaminobenzidine (DAB) reacts with $H_2O_2$to form reddish-brown deposits that can be examined with a light microscope. Image analyses showed that Fepagro 26 fumigated leaves present more intense spots for DAB staining than control leaves (Fig. 2, arrows). On the other hand, when analyzing Irai leaves no substantial changes in the amount of DAB precipitates formed during the same time period could be noticed. When comparing Irai control and exposed leaves the same pattern of brow spots intensity could be observed. To ensure that the results we were seeing were really from the reaction between H_2O_2 and DAB, we tested the samples with a positive control. Ascorbate (10 mM) was added to the DAB staining solution and the incubation of leaf discs was performed. Almost all DAB deposits were completely reduced on these samples, indicating that DAB staining was actually due to local differences in H_2O_2 generation (data not shown).

The production of superoxide radical was imaged in leaves infiltrated with NBT, which specifically reacts with superoxide anions resulting in purple formazan deposits, which can easily be visualized in the microscopy. Within 1 week of treatment visible heavier staining spots were observed on Fepagro 26 exposed leaves, indicating that superoxide radical production was higher on these leaves than control (Fig. 2, arrows). When comparing exposed and non-exposed Irai leaves, we observed that generation of purple formazam precipitates occurred on both groups, and when comparing exposed Irai and Fepagro 26 leaves we could notice that a lighter staining was found on Irai leaves.

Non-enzymatic Antioxidant Defense System

Ascorbate Levels

Ozone sensitivity is generally correlated with the oxidative status of AsA on leaf tissue, and the antioxidant role played by AsA molecule is considered to be strictly dependent on the cell ability to maintain it in a reduced state (Castagna and Ranieri, 2009; Conklin and Barth, 2004). Comparing the AsA levels of Irai and Fepagro 26 leaves it was evident that the endogenous levels of AsA on Fepagro 26 are significantly lower than Irai ($p < 0.0001$), in fact AsA concentration of Fepagro 26 are 1.26 times lower than Irai (Fig. 3a). After 1 week of O_3 treatment the AsA pool in Fepagro 26 leaves was changed, and a significant increase of nearly 50% on the AsA concentration was recorded on exposed Fepagro 26 plants. On the opposite, AsA levels on Irai leaves were not changed in response to O_3 exposure (Fig. 3a).

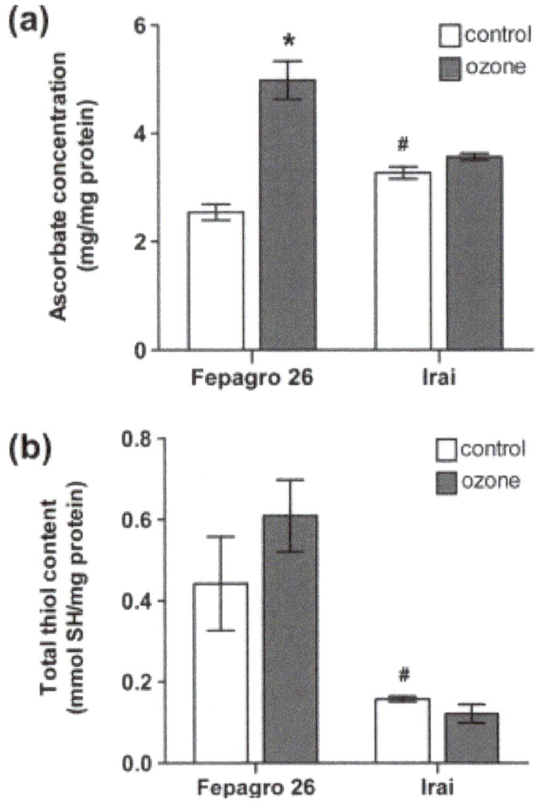

Figure 3. (a) Ascorbate concentration and (b) total thiol content on control and O_3-exposed leaves from Fepagro 26 and Irai seedlings. Values represent mean ± SEM (n = 6) of three independent experiments. * Indicate difference from control ($p < 0.05$); # indicate difference between varieties ($p < 0.05$) as analyzed by ANOVA.

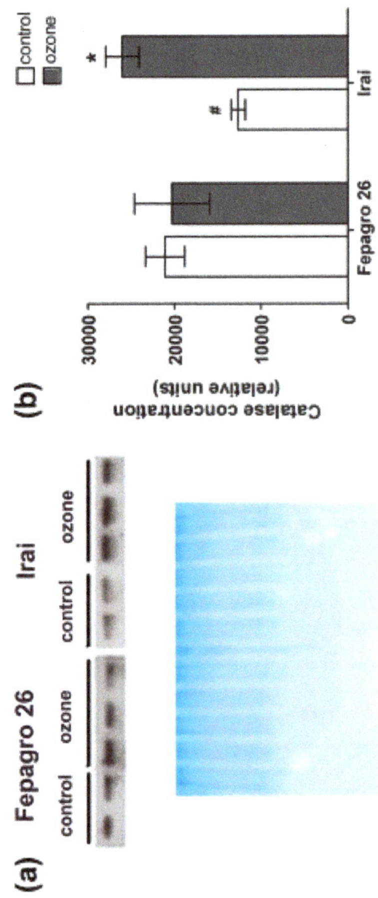

Figure 4. (a) Catalase immunocontent on control and O_3-exposed leaves from Fepagro 26 and Irai seedlings and loading control with Comassie Blue staining. (b) Catalase immunocontent expressed as relative units. Values represent mean ± SEM (n = 3) of three independent experiments. * Indicate difference from control ($p < 0.05$); # indicate difference between varieties ($p < 0.05$) as analyzed by student t-test.

Total Reduced Thiol Content

The physiological significance of thiol metabolism in plants is specially related to sulfur assimilation and antioxidant defense. Here we could notice that significant differences on the physiological levels of total reduced thiol content when comparing the two *P. vulgaris*varieties. Leaves from Fepagro 26 had a baseline level of total reduced thiol content 4.2-fold higher than Irai (p = 0.0007). After 1 week of O_3 fumigation no significant changes on the levels of total thiol concentration were observed neither for Fepagro 26 nor Irai leaves (Fig. 3b).

Catalase Immunoblotting Analysis

To better understand how O_3 exposure was affecting the leaves enzymatic antioxidant apparatus at the level of enzyme amount, we decided to perform a quantitative analysis of the catalase enzyme. As previously reported, the two varieties have a very different response of CAT activation when exposed to O_3 (Caregnato et al., 2010). Here we demonstrated that the differences observed on CAT activity are based on the amount of CAT protein. Irai leaves showed a significant lower trend on the baseline CAT immunocontent than Fepagro 26 (p = 0.02). Catalase immunocontent on Fepagro 26 leaf tissue was 1.66 times higher than Irai leaves. But differently from Fepagro 26, Irai seedlings challenged with O_3 were capable of increasing the amount of CAT protein. Catalase immunocontent on treated Irai plants were found to be 2 times higher when compared to control Irai plants (p = 0.01) (Fig. 4).

Photosynthetic Pigments Content

The levels of photosynthetic pigments were differently affected by O_3 exposure on each*P. vulgaris* variety (Table 1). As suggested by the loss of leaves green color, a significant declining trend under O_3 exposure was observed in both chlorophylls *a* and *b* (53% and 40%, respectively) on O_3 fumigated Fepagro 26 leaves. Total chlorophyll (Chl *a* + Chl *b*) content was also affected by O_3, and a significant reduction of approximately 50% of total Chl was recorded in the O_3-exposed Fepagro 26. Furthermore, a significant decrease was also observed on the Chl *a*/*b* ratio, which indicates that Chl *a* levels was probably more affected by the O_3 treatment than Chl *b*. On the opposite, no modifications in the levels of the foliar pigmentation of O_3-treated Irai leaves could be observed, indicating that chlorophyll pigments levels were not affected when this cultivar is growing under the presence of elevated O_3 concentration. In this variety neither Chl *a* nor Chl *b* amounts were found to be modified by O_3 treatment, and thus no differences could be assessed either on total Chl content or on the Chl *a*/*b* ratio.

Table 1. Chlorophyll a and b content, and protein concentration on control and O_3-exposed *Phaseolus vulgaris*seedlings. Values represent mean ± SEM ($n = 6$) of three independent experiments.

	Fepagro 26			Irai		
	Control	O_3 treated	*p* Value	Control	O_3 treated	*p*Value
Chlorophyll a (mg g^{-1})	7.7 ± 0.41	3.6 ± 0.31*	<0.0001	7.8 ± 0.71	6.5 ± 0.44	ns
Chlorophyll b (mg g^{-1})	3.5 ± 0.20	2.1 ± 0.17*	<0.001	3.7 ± 0.34	3.2 ± 0.14	ns
Total chlorophyll (mg g^{-1})	11.3 ± 0.59	5.7 ± 0.49*	<0.001	11.62 ± 1.06	9.8 ± 0.60	ns
Chlorophyll *a*/*b* ratio	2.1 ± 0.040	1.7 ± 0.062*	<0.0005	2.1 ± 0.030	2.0 ± 0.04	ns
Protein (μg mL^{-1})	3227.9 ± 183	1722.1 ± 130*	<0.001	2353.6 ± 77	2219.5 ± 48*	<0.05

ns – Indicates no significant differences between control and exposed leaves as analyzed by ANOVA.

*Indicate difference between control and exposed leaves ($p < 0.05$).

Relationship between Chlorophyll and Protein Concentration

Levels of total protein content were significantly decreased in response to O_3 treatment in both varieties. However, the reduction on protein content was significantly higher in Fepagro 26 (approximately 50%), when compared to total protein content loss found on Irai plants (5.6%) ($p < 0.05$) (Table 1). In order to assess if there was a significant relationship between the chlorophyll content and total protein concentration we performed a linear regression analysis. To the test this relationship we choose to use the ratio of photosynthetic pigments, chlorophylls *a* and *b*, once their concentration are tightly regulated in the plant photosynthesis apparatus and their proportional changes may affect the entire photosynthesis process.

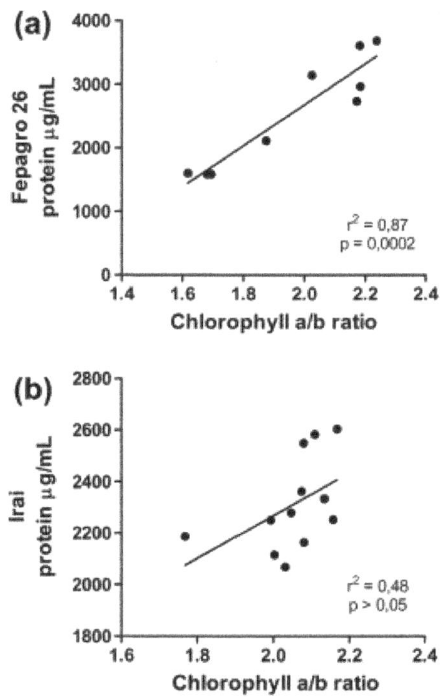

Figure 5. Relationship between chlorophyll content and total protein levels on control and O_3-exposed leaves from (a) Fepagro 26 and (b) Irai. Values represent mean ± SEM ($n = 6$) of three independent experiments. Data were analyzed using linear bivariate regression analysis. The significance of the *p*-values are shown, and a significant relationship was considered for a $p < 0.05$.

We found a significant positive linear relationship between the protein content and Chl *a*/*b* ratio only for Fepagro 26 seedlings (Fig. 5a). Regression analysis results indicate thus that a decrease in the Chl *a*/*b* ratio is significantly associated with a decrease in total protein concentration when Fepagro 26 plants grown under the presence of toxic O_3 concentrations ($r^2 = 0.87$; $p < 0.0005$;$F = 49.63$). Since no differences on chlorophylls *a* and *b* content could be found on fumigated Irai leaves, no association between this parameter and protein content were observed ($r^2 = 0.48$; $p = 0.11$; $F = 3.07$) (Fig. 5b).

DISCUSSION

The present study reveals that the antioxidant capacity on leaf tissue of two Brazilian *P. vulgaris* varieties exposed to high O_3 concentration during the period of 1 week was significantly affected. Our results showed that integrated biochemical mechanisms related to a better capacity to counterbalance the O_3-mediated oxidative process are behind the differences on the O_3-susceptibility observed between the two studied bean varieties. Not differently from the most up-to-date studies we demonstrated here that the detrimental effects of O_3 on plant metabolism occur due to its high oxidative power and ability to be converted into different ROS in the leaf internal space. However, as plant species possess a wide range of responses to cope with O_3-induced oxidative stress and, as suggested by other experiments with different crops worldwide, variation on these responses can be attributed to innate and inducible antioxidant biochemical features that are distinct among plant species and varieties (Burkey et al., 2005, Scebba et al., 2006, Tosti et al., 2006 and Matyssek et al., 2008). Differential sensitivity to O_3exhibited by different varieties within a plant species has been proven to be especially dependent on the efficiency of ROS scavenging system to maintain the cellular redox steady-state of leaf tissue (Puckette et al., 2007, Castagna and Ranieri, 2009 and Giacomo et al., 2010).

The evaluation of ROS levels through DCF analysis revealed that Irai variety present higher constitutive levels of ROS than Fepagro 26 (Fig. 1). Ozone exposure did not alter the ROS generation on Irai leaves, but significantly increased the ROS production in Fepagro 26 plants. Using histochemical *in situ* localization techniques we could detect and visualize the two major ROS, superoxide radical and H_2O_2, on leaf tissue from both varieties (Fig. 2). On Fepagro 26 leaf samples we observed that both superoxide radical and

H_2O_2 accounted for the significant enhancement on ROS levels after O_3 treatment. The strong staining pattern observed after the reaction of superoxide radical with NBT on O_3-exposed Fepagro 26 leaves suggest that the levels of this ROS was pronouncedly increased on this bean variety. With the *in situ* localization techniques we were also able to notice that Irai control leaves showed stronger stained spots for DAB reaction suggesting that endogenous levels of H_2O_2 on this variety are constitutively higher than those levels found on Fepagro 26 control leaves (Fig. 2). Previous studies have demonstrated that distinct patterns of ROS accumulation occurs in differentially O_3-sensitive plants as shown for tobacco (*Nicotiana tabacum* L. cv Bel W3) and clover plants (*Trifolium repens* L.) (Pasqualini et al., 2003 and Scebba et al., 2006). In sensitive varieties both superoxide radical and H_2O_2 accumulate in leaf margins following O_3exposure, and the sites of ROS generation (particularly H_2O_2) has been correlated with the appearance of leaf injury signals, such as necrotic lesions (Di Baccio et al., 2008; for review see Castagna and Ranieri, 2009). Furthermore, Guidi et al. (2010) when comparing the responses of two *P. vulgaris* cultivars to acute O_3 treatment (165 nL L^{-1}) showed that after 3 h of exposure the levels of H_2O_2 accumulated to a markedly greater extent in the leaves of the sensitive cultivar Cannellino than in tolerant Top Crop, whereas the H_2O_2 deposits were only become detectable for Top Crop after 24 h from the beginning of the fumigation treatment. In the O_3-sensitive cultivar Cannellino this early accumulation of H_2O_2 was found to be correlated to the almost complete disruption of cell structure, and irreversible damages to the photosynthetic apparatus. In addition, interesting inverse correlation between the levels of ROS in control and O_3 sensitivity plants were reported for 38 different accessions of *Medicago truncatula* fumigated with 300 nmol mol^{-1} of O_3 for 6 h. Accessions with higher levels of endogenous ROS were more tolerant compared to accessions with lower ROS levels, suggesting that plants with elevated ROS levels may be 'intrinsically primed' to tolerate extreme oxidative stress, a fact that strengths the role of ROS as important signaling molecules during normal growth and stress conditions (Puckette et al., 2007).

Nevertheless, the degree of sensitivity of a plant to O_3 not only depends on the quantity of gaseous molecules entering the leaf and that are subsequently converted to ROS. Over the past years, several studies have pointed out that the main level of O_3 defense relies both on the existing content of cellular antioxidants (e.g. ascorbate and glutathione) and the intensity of the detoxifying pathways that are responsible for regenerating these metabolites (Calatayud et al., 2001, Dizengremel et al., 2008, Dizengremel et al., 2009 and Luwe et al., 1993). The protective role of AsA as ROS-scavenger was first supported by the enhanced O_3-sensitivity shown by *Arabidopsis thaliana* mutants deficient in AsA synthesis (Conklin et al., 1996). Even so, the relationship between O_3sensitivity and apoplastic AsA concentration remains controversial and some studies have postulated that elevated apoplastic AsA levels cannot always be sufficient to render a plant tolerant to O_3 (Ranieri et al., 1999, D'Haese et al., 2005 and Di Baccio et al., 2008). Here we found that Irai plants have higher constitutive AsA levels than Fepagro 26, and O_3 exposure does not modify the levels of this antioxidant molecule on Irai leaf tissue (Fig. 3a). Although it is well-known that apoplastic AsA is quickly oxidized when in presence of O_3 and ROS, our results may indicate that Irai plants are capable to maintain a higher AsA-mediated antioxidant capacity, a feature that is probably related to the ability of the foliar cells to keep both extra and intracellular supply of reduced AsA at levels that are effective enough to detoxify ROS generated under the presence of O_3. Conversely, Fepagro 26 baseline levels of AsA were significantly lower, and when this variety was fumigated with O_3 the levels of foliar AsA were significantly enhanced. These findings suggest that O_3 exposure of Fepagro 26 plants might stimulate AsA production and/or its movement from the intracellular to apoplastic space in attempt to increase the amount of ROS that can be scavenged by the cells. The rate of AsA regeneration controls the O_3 influx to leaf interior, which means that non-detoxified O_3 molecules are subsequently free to promote its oxidative action, especially trough its conversion to ROS in the leaf apoplast (Smirnoff, 1996, Van Hove et al., 2001 and Di Baccio et al., 2008). In spite of the fact that O_3-treated Fepagro 26 increases the concentration of AsA, its antioxidant capacity to respond O_3 influx is probably less effective than that of Irai. Together with the results from foliar ROS accumulation we can infer that the AsA antioxidant capacity per unit of O_3 flux is decreased in Fepagro 26, and thus more O_3molecules are free to being decomposed into ROS (H_2O_2 and superoxide radical) inside the foliar tissue.

The antioxidant role played by AsA is, however, strictly dependent on the cell ability to maintain it in a reduced state which occurs trough the action of AsA–GSH cycle (Smirnoff, 1996, Noctor and Foyer, 1998 and Di Baccio et al., 2008). Using high O_3concentrations (300 ppb) Luwe et al. (1993) observed a time-dependent relationship between oxidation of both extracellular AsA and intracellular GSH pool, while the cellular AsA redox state was unaltered during fumigation. As reported by numerous other studies, AsA regeneration is tightly coupled to GSH within the cell and transport activity was responsible for replenish the reduced apoplastic AsA pool (for review see Horemans et al., 2000b and Smirnoff, 1996). Thus, trough a GSH-dependent regeneration mechanism, the extracellular AsA oxidation is supposed to affect the intracellular thiol signature. Analysis on the levels of total reduced thiol indicated that in both varieties the antioxidant thiol pool was unaffected by O_3 exposure (Fig. 3b); nonetheless, there was a significant difference in the endogenous pool size between Irai and Fepagro 26, being the values for total reduced thiol content on Fepagro 26 leaves 4.2 times higher than Irai. Unlike the apoplasm, the cytoplasm is a cellular compartment where innumerous molecules involved in different metabolic pathways are localized. Modifications on the intracellular redox balance can directly modify proteins structures, and changes in thiol-redox state can modulate enzymes activity by altering active site residues required for catalysis and/or ligand binding, oligomerization, and cellular localization (Paget and Buttner, 2003 and Yi et al., 2010). Thus, besides GSH other reducing co-factors for several enzymes involved in ROS detoxification are also conjugate to proteins to avoid their oxidation, and this process is catalyzed by enzymes such as glutaredoxins, thioredoxins and peroxiredoxins (Foyer et al., 1997 and Halliwell, 2006) . Here we analyzed leaf total reduced thiol content, not only the GSH levels, and maybe this why we have not observed a positive relationship between the levels of total reduced thiol and AsA during O_3 detoxification process on Fepagro 26 plants. Based on recently proteomic studies that support that thiol-based regulation play a critical role in plant responses to O_3exposure (Gillespie et al., 2011 and Galant et al., 2012), we can suggest that in an effort to avoid deleterious oxidative changes on proteins, Fepagro 26 leaf cells try to maintain the levels of total reduced thiol as a mechanism of redox-protection of enzymes involved in major metabolic processes.

In addition to the non-enzimatic antioxidant components, the balance of antioxidant enzymes activities is crucial for suppressing the toxic effects of elevated ROS levels (Apel and Hirt, 2004). Modulation of the major antioxidant enzymes activities (SOD, CAT, APX, GR) have been observed in different O_3 exposure experiments (Pell et al., 1999,Calatayud et al., 2003 and Heath, 2007). The changes on enzymes activities are often linked to parallel changes in gene expression in the intracellular compartment (Sharma and Davis, 1994, Willekens et al., 1994, Pellinen et al., 2002, Pasqualini et al., 2007 and Torres et al., 2007). On a previous work, when comparing the results for CAT activity, we observed that the two varieties had a very different pattern of CAT stimulation in response to O_3 fumigation. Irai seedlings significantly increased CAT activity after O_3exposure, whereas Fepagro 26 did not change the enzyme activity when challenged with the same O_3 concentrations ($AOT_{40=}$212 ppb h, during one week) (Caregnato et al., 2010). Trough western blot analysis we demonstrated that, even though Irai leaves have lower endogenous levels of CAT (Fig. 4), when this variety is subjected to O_3 treatment the concentration of the antioxidant enzyme is significantly increased. Altogether, our results indicate that in Irai leaves the increment on CAT activity elicited by the presence of O_3 is probably a consequence of the increased concentration of the enzyme found on the intracellular space, a fact that might improve the antioxidant capacity of Irai foliar cells. Catalase in the one of the most important enzymes needed for H_2O_2 detoxification (Willekens et al., 1995). Plants contain several types of H_2O_2-degrading enzymes (peroxidases), however CAT are unique as it do not require cellular reducing equivalents to convert H_2O_2 to water (Sharma et al., 2012). The most important reducing substrate used for peroxidases is the AsA (Mehlhorn et al., 1996), an antioxidant molecule that is easily oxidized under stressful conditions such as O_3 exposure. Stress analysis on transgenic CAT-deficient *Arabidopsis thaliana* revealed that reduced peroxissomal CAT activity increased sensitivity towards O_3 and triggers cell death by photorespiratory H_2O_2(Vandenabeele et al., 2004). Transgenic tobacco plants with approximately 10% of wild-type CAT activity showed accumulation of oxidized GSH (GSSG) and a 4-fold decrease in AsA; moreover, CAT deficiency renders tobacco transgenic plants markedly more vulnerable to O_3 stress (Willekens et al., 1997). In sum, in a pro-oxidant environment an effective CAT action is critical for maintaining the redox balance to avoid further oxidative stress.

It is known that one of the main detrimental effects of O_3 that occur at leaf subcellular level is the impairment of the photosynthesis process (Pell et al., 1994). Ozone reduction in net photosynthesis is attributed to be a consequence of degradation of both structural and functional components of the chloroplasts (Pääkkönen et al., 1996). Trough the oxidation of membrane lipids, O_3 and its derived ROS can cause the disruption of biological membranes, leading to irreversible modifications of chloroplast components, such as the tylakoids (Calatayud et al., 2001, Calatayud et al., 2003 and Anderson et al., 2003). Tylakoids disintegration results in the loss of membrane attached chlorophyll molecules, which in free solution are prone to be oxidatively damaged by ROS present in the intracellular environment (Sakaki et al., 1983 and Ranieri et al., 2001). Moreover, recent analyses of transcriptome changes in O_3-sensitive and resistant *Medicago truncatula* accessions identified that genes involved in chlorophyll biosynthesis were down regulated after 6 h post-fumigation on sensitive accession Jemalong (Puckette et al., 2008). When analyzing leaf chlorophyll content we found that chlorophylls *a* and *b*, total chlorophyll and the ratio of Chl *a*/*b* levels on O_3-exposed Fepagro 26 plants were significantly reduced, a change that was not observed on Irai leaves (Table 1). Considering that Fepagro 26 leaves are unable to properly detoxify the ROS, we can suggest that chlorophyll loss can be either associated to increased chlorophyll oxidation or to an altered regulation of pigments biosynthesis. A comparative study with 12 bean varieties demonstrated that, after 75–135 min of O_3 exposure, a high stomatal conductance was accompanied by chlorophyll loss and/or inhibition of photosynthesis was a common feature of sensitive varieties (Guzy and Heath, 1993). Furthermore,Leitao et al. (2008) observed that *P. vulgaris* plants exposed to non-filtered ambient air supplied with 80 nL L^{-1} of O_3 had a loss of 43% on total chlorophyll content. Similarly to our results, reductions in both Chl *a* and *b* were evident, and as underlined by the authors, a similar trend of Chl *a*/*b* ratio reduction indicate that Chl *a* tended to be decreased a little more by O_3 than Chl *b*.

Using correlation analysis we could show that when chlorophylls *a*/*b* ratio is decreased on Fepagro 26 exposed plants, the amount of total protein is also reduced. Such significant relationship was not observed for Irai. This result may indicate that ROS accumulation is leading to chlorophyll loss, which affects the photosynthetic process and the biomass acquisition of Fepagro 26 exposed plants. Clebsch et al. (2009) studying different subtropical bean varieties, including Fepagro 26 and Irai, observed that O_3treatment decrease the carbon net assimilation on all studied varieties, but Irai. Carboxylation efficiency, which is considered to be the main factor involved in photosynthesis performance (Pell et al., 1994), was found to be significantly reduced in Fepagro 26 seedlings. Disruption of photosynthetic apparatus might be responsible for reduced carboxylation efficiency by the Calvin cycle enzymes, which finally result in growth impairment on this variety. Experimental studies have recently described that high O_3-mediated impairment on growth and yield of two important Indian rice cultivars (*Oryza sativa* L. cvs Malviya dhan 36 and Shivani) are associated to reductions of plant height, leaf area, total chlorophyll content, photosynthetic rate, stomatal conductance, and chlorophyll fluorescence (Sarkar and Agrawal, 2012). In addition, studies done with a particular O_3-sensitive crop, soybean (*Glycine max*), demonstrated that O_3 exposure negatively impacts the plants efficiency to convert solar energy into biomass energy by reducing photosynthetic efficiency and by increasing respiratory costs, which would contribute to yield loss (Betzelberger et al., 2012).

In fact, numerous studies throughout the world that performed similar experiments with a wide range of crops exposed to O_3 have showed that, on sensitive plants, the changes on redox-related biochemical parameters are related with yield losses and, recently, with altered grain quality and nutritional traits (Flowers et al., 2007, Feng and Kobayashi, 2009, Iriti et al., 2009, Sarkar and Agrawal, 2010 and Wang et al., 2012). Considering that the risk of the negative effects of O_3 on crop productivity has been creating the need for new crop varieties that are better adapted to O_3 stress, it is of particular importance to further study how the O_3-induced changes on the redox status on the sensitive Fepagro 26 can be associated to modifications on grain quality and yield on this variety. In addition, *P. vulgaris* L. is considered by some authors to be one of the most sensitive agricultural crops to O_3 toxic effects, and experiments to better predict the impacts of increasing O_3 levels are urgently needed. Thus, it must be noted that improved understanding of the exposure–response mechanisms underlying O_3sensitivity/tolerance are still necessary for major Brazilian bean varieties.

CONCLUSIONS

Based on our results we can suggest that Fepagro 26 is more likely to suffer the negative impacts of high O_3 levels than Irai, which seems to be a more robust variety. The greater sensitivity of Fepagro 26 variety to O_3 stress seems to be associated with an increased ROS production (superoxide radical and H_2O_2) which creates a pro-oxidative environment inside the cells. Apparently, both enzymatic (CAT) and non-enzymatic (AsA) leaf antioxidants are unable to maintain the intracellular redox homeostasis, which leads to oxidative damages to structural molecules such as the membrane lipids (seeCaregnato et al., 2010). Disruption of membrane integrity can cause photosynthetic pigments loss (chlorophylls *a* and *b*), which reduces the photosynthesis rate and affects the biomass acquisition of Fepagro 26 plants (protein levels reduction). On the opposite, Irai presents higher endogenous levels of ROS, and under the presence of O_3 its antioxidant apparatus is stimulated to avoid oxidative stress. Similarly to the conclusions achieved by Guidi et al. (2010), the O_3-tolerance in bean depends more on the specific potential ability of each cultivar than the stress-induced physiological and biochemical adjustments to avoid and counter stress-induced oxidative damage. Although the parameters assessed in this study could arise as a consequence of other environmental stresses, our data provide reliable results concerning some of the major biochemical differences that are behind the responses of the two subtropical *P. vulgaris* varieties to the O_3 exposure. To better characterize these bean varieties according to their O_3susceptibility an assessment of dose dependent sensitivity with respect to growth, plant yield parameters and grain quality should be further conducted.

Acknowledgments

Funding support for this project was generously provided by the Brazilian agencies CNPq, CAPES, FAPERGS and PROPESQ-UFRGS.

REFERENCES

1. Agrawal, M., Singh, B., Rajput, M., Marshall, F., Bell, J.N., 2003. Effect of air pollution on periurban agriculture: a case study. Environ. Pollut. 126, 323–329.
2. Anderson, P.D., Palmer, B., Houpis, J.L.J., Smith, M.K., Pushnik, J.C., 2003. Chloroplastic responses of ponderosa pine (Pinus ponderosa) seedlings to ozone exposure. Environ. Int. 29, 407–413.
3. Apel, K., Hirt, H., 2004. Reactive oxygen species: metabolism, oxidative stress, and signal transduction. Annu. Rev. Plant Biol. 55, 373–399.
4. Ashmore, M.R., 2005. Assessing the future global impacts of ozone on vegetation. Plant Cell Environ. 28, 949–964.
5. Betzelberger, A.M., Gillespie, K.M., McGrath, J.M., Koester, R.P., Nelson, R.L., Ainsworth, E.A., 2010. Effects of chronic elevated ozone concentration on antioxidant capacity, photosynthesis and seed yield of 10 soybean cultivars. Plant Cell Environ. 33, 1569–1581.
6. Betzelberger, A.M., Yendrek, C.R., Sun, J., Leisner, C.P., Nelson, R.L., Ort, D.R., Ainsworth, E.A., 2012. Ozone exposure response for US soybean cultivars: linear reductions in photosynthetic potential, biomass, and yield. Plant Physiol. 160, 1827–1839.
7. Booker, F., Muntifering, R., McGrath, M., Burkey, K., Decoteau, D., Fiscus, E., Manning, W., Krupa, S., Chappelka, A., Grantz, D., 2009. The ozone component of global change: potential effects on agricultural and horticultural plant yield, product quality and interactions with invasive species. J. Integr. Plant Biol. 51, 337–351.
8. Bradford, M.M., 1976. Rapid and sensitive method for quantitation of microgram quantities of protein utilizing principle of protein-dye binding. Anal. Bioch. 72, 248–254.
9. Calatayud, A., Barreno, E., 2004. Response to ozone in two lettuce varieties on chlorophyll a fluorescence, photosynthetic pigments and lipid peroxidation. Plant Physiol. Biochem. 42, 549–555.
10. Calatayud, A., Alvarado, J.W., Barreno, E., 2001. Changes in chlorophyll a fluorescence, lipid peroxidation, and detoxificant system in potato plants grown under filtered and non-filtered air in open-top chambers. Photosynthetica. 39, 507–513.
11. Calatayud, A., Iglesias, D.J., Talon, M., Barreno, E., 2003. Effects of 2-month ozone exposure in spinach leaves on photosynthesis, antioxidant systems and lipid peroxidation. Plant Physiol. Biochem. 41, 839–845.
12. Desikan, R., 2001. Regulation of the Arabidopsis transcriptome by oxidative stress. Plant Physiol. 127, 159–172.
13. Di Baccio, D., Castagna, A., Paoletti, E., Sebastiani, L., Ranier, A., Ranieri, A., 2008. Could the differences in O(3) sensitivity between two poplar clones be related to a difference in antioxidant defense and secondary metabolic response to O(3) influx? Tree Physiol. 28, 1761–1772.
14. Dizengremel, P., Le Thiec, D., Bagard, M., Jolivet, Y., 2008. Ozone risk assessment for plants: central role of metabolism-dependent changes in reducing power. Environ. Pollut. 156, 11–15.
15. Dizengremel, P., Le Thiec, D., Hasenfratz-Sauder, M.-P., Vaultier, M.-N., Bagard, M., Jolivet, Y., 2009. Metabolic-

dependent changes in plant cell redox power after ozone exposure. Plant Biol. (Stuttgart) 1, 35–42.

16. Eltayeb, A.E., Kawano, N., Badawi, G.H., Kaminaka, H., Sanekata, T., Morishima, I., Shibahara, T., Inanaga, S., Tanaka, K., 2006. Enhanced tolerance to ozone and drought stresses in transgenic tobacco overexpressing dehydroascorbate reductase in cytosol. Physiol. Plant. 127, 57–65. Feng, Z., Kobayashi, K., 2009. Assessing the im

17. Feng, Z.-W., Jin, M.-H., Zhang, F.-Z., Huang, Y.-Z., 2003. Effects of ground-level ozone (O3) pollution on the yields of rice and winter wheat in the Yangtze River Delta. J. Environ. Sci. (China) 15, 360–362.

18. Fishman, J., Creilson, J.K., Parker, P.A., Ainsworth, E.A., Vining, G.G., Szarka, J., Booker, F.L., Xu, X., 2010. An investigation of widespread ozone damage to the soybean crop in the upper Midwest determined from ground-based and satellite measurements. Atmos. Environ. 44, 2248–2256

19. Galant, A., Koester, R.P., Ainsworth, E.A., Hicks, L.M., Jez, J.M., 2012. From climate change to molecular response: redox proteomics of ozone-induced responses in soybean. New Phytol. 194, 220–229.

20. Gillespie, K.M., Rogers, A., Ainsworth, E.A., 2011. Growth at elevated ozone or elevated carbon dioxide concentration alters antioxidant capacity and response to acute oxidative stress in soybean (Glycine max). J. Exp. Bot. 62, 2667–2678.

21. Halliwell, B., 2006. Reactive species and antioxidants. redox biology is a fundamental theme of aerobic life. Plant Physiol. 141, 312–322.

22. IPCC (Intergovernmental Panel on Climate Change), 2001. Climate change 2001: the scientific basis. In: Contribution of Working Group I to the Third Assessment Report of the Intergovernmental Panel on Climate Change. Cambridge University Press, Cambridge, UK, 881pp (ISBN 0521 807670).

23. Iriti, M., Di Maro, A., Bernasconi, S., Burlini, N., Simonetti, P., Picchi, V., Panigada, C., Gerosa, G., Parente, A., Faoro, F., 2009. Nutritional traits of bean (Phaseolus vulgaris) seeds from plants chronically exposed to ozone pollution. J. Agri. Food Chem. 57, 201–208.

24. Kanofsky, J.R., Sima, P.D., 1995. Singlet oxygen generation from the reaction of ozone with plant leaves. J. Biol. Chem. 270, 7850–7852.

25. Kobayashi, K., 1995. Effects of ozone on dry matter partitioning and yield of Japanese cultivars of rice (Oryza sativa L.). Agric. Ecosyst. Environ. 53, 109–122.

26. Leitao, L., Dizengremel, P., Biolley, J.-P.P., 2008. Foliar CO2 fixation in bean (Phaseolus vulgaris L.) submitted to elevated ozone: distinct changes in Rubisco and PEPc activities in relation to pigment content. Ecotox. Environ. Saf. 69, 531–540.

27. Matyssek, R., Sandermann, H., Wieser, G., Booker, F., Cieslik, S., Musselman, R., Ernst, D., 2008. The challenge of making ozone risk assessment for forest trees more mechanistic. Environ. Pollut. 156, 567–582.

28. Noctor, G., Foyer, C.H., 1998. Ascorbate and glutathione: keeping active oxygen under control. Annu. Rev. Plant Physiol. Plant Mol. Biol. 49, 249–279.

29. Pääkkönen, E., Vahala, J., Holopainen, T., Karjalainen, R., Kärenlampi, L., 1996. Growth responses and related biochemical and ultrastructural changes of the photosynthetic apparatus in birch (Betula pendula) saplings exposed to low concentrations of ozone. Tree Physiol. 16, 597–605.

30. Pasqualini, S., Paolocci, F., Borgogni, A., Morettini, R., Ederli, L., 2007. The overexpression of an alternative oxidase gene triggers ozone sensitivity in tobacco plants. Plant Cell Environ. 30, 1545–1556.

31. Ranieri, A., Castagna, A., Padu, E., Moldau, H., Rahi, M., Soldatini, G.F., 1999. The decay of O-3 through direct reaction with cell wall ascorbate is not sufficient to explain the different degrees of O-3-sensitivity in two poplar clones. J. Plant Physiol. 154, 250–255.

32. Ranieri, A., Giuntini, D., Ferraro, F., Nali, C., Baldan, B., Lorenzini, G., Soldatini, G.F., Soldatini, F., 2001. Chronic ozone fumigation induces alterations in thylakoid functionality and composition in two poplar clones. Plant Physiol. Biochem. 39, 999–1008.

33. Sarkar, A., Rakwal, R., Agrawal, S.B., Shibato, J., Ogawa, Y., Yoshida, Y., Agrawal, G.K., Agrawal, M., 2010. Investigating the impact of elevated levels of ozone on tropical wheat using integrated phenotypical, physiological, biochemical, and proteomics approaches. J. Proteome Res., 4565–4584.

34. Thordal-Christensen, H., Zhang, Z., Wei, Y., Collinge, D.B., 1997. Subcellular localization of H2O2 in plants. H2O2 accumulation in papillae and hypersensitive response during the barley—powdery mildew interaction. Plant J. 11, 1187–1194.

35. Torres, N.L., Cho, K., Shibato, J., Hirano, M., Kubo, A., Masuo, Y., Iwahashi, H., Jwa, N.S., Agrawal, G.K., Rakwal, R., 2007. Gel-based proteomics reveals potential novel protein markers of ozone stress in leaves of cultivated bean and maize species of Panama. Electrophoresis 28, 4369–4381.

36. Van Hove, L.W., Bossen, M.E., San Gabino, B.G., Sgreva, C., 2001. The ability of apoplastic ascorbate to protect poplar leaves against ambient ozone concentrations: a quantitative approach. Environ. Pollut. 114, 371–382.

37. Wang, Y., Yang, L., Han, Y., Zhu, J., Kobayashi, K., Tang, H., Wang, Y., 2012. The impact of elevated tropospheric ozone on grain quality of hybrid rice: a freeair gas concentration enrichment (FACE) experiment. Field Crops Res. 129, 81– 89.

38. Willekens, H., Van Camp, W., Van Montagu, M., Inze, D., Langebartels, C., Sandermann, H., 1994. Ozone, sulfur dioxide, and ultraviolet b have similar effects on mRNA accumulation of antioxidant genes in Nicotiana plumbaginifolia L. Plant Physiol. 106, 1007–1014.

39. Yi, H., Galant, A., Ravilious, G.E., Preuss, M.L., Jez, J.M., 2010. Sensing sulfur conditions: simple to complex protein regulatory mechanisms in plant thiol metabolism. Mol. Plant. 3, 269–279.

40. Zhang, S., Klessig, D.F., 2001. MAPK cascades in plant defense signaling. Trends Plant Sci. 6, 520–527.

CHAPTER 10

EFFECT OF CO, NOX AND SO2 ON ROS PRODUCTION, PHOTOSYNTHESIS AND ASCORBATE–GLUTATHIONE PATHWAY TO INDUCE FRAGARIA×ANNASA AS A HYPERACCUMULATOR

Sowbiya Muneer, Tae Hwan Kim, Byung Chul Choi, Beom Seon Lee, Jeong Hyun Lee

Keywords: Ascorbate–glutathione pathways, *Fragaria annasa*, Hyperaccumulator, Photosynthesis, ROS production

INTRODUCTION

Environmental pollution from industrial effluents and other organic pollutants has become a major environmental and human concern worldwide [1] and [2]. The major hazardous gases released from industries are abundant in number, among which carbon monoxide (CO) is a piousness gas to mammals, and has many similar qualities and effects to those of carbon dioxide (CO_2), except for its flammability at high concentration[3]. As an important signaling molecule in physiological activity, CO has a similar toxic effect to that of cyanide, the inhibitory effect of a higher ratio of CO to O_2 (above 4:1) on cytochrome *c* oxidase in the mitochondrial electron transport chain at the cellular level or the effect on the detoxification enzyme P-450 in the monoxygenase system and multiprotein complex proteins in chloroplasts. CO can have important signaling roles as well as toxic effects, but this depends upon the amount produced and context, as plants also contain a variety of heme moiety-containing proteins (catalase, peroxidase, cytochromes, etc.), and produce a number of important, biliverdin-related tetrapyrrole pigments, such as phycocyanobilin [4] and phytochromes [5].

On the other hand nitrogen oxides (NO*x*) are produced by combustion processes and NO*x* is usually >90% nitric oxide (NO), with the balance being composed of nitrogen dioxide (NO_2). In the atmosphere, chemical reactions convert NO to NO_2, which can further react to produce O_3. While NO*x* emission rates vary from plant to plant, largely according to the design of the plant and the characteristics of the fuel, an uncontrolled emission rate of 1.0 b mmBtu^{-1} is typical for coal-fired generators. NO*x* is a free radical gas which transfers electrons across biological membranes [6]. In general the electron acceptor is oxygen and the product of the electron transfer reaction is superoxide. The biological function of NO*x* is therefore the generation of reactive oxygen species (ROS)[6]. The physiological generation of ROS can occur as a byproduct of other biological reactions. ROS generated as a byproduct occurs within mitochondria, peroxisomes and other cellular elements [7].

Sulfur dioxide (SO_2) is one of the most common harmful air pollutants. The phytotoxicity of SO_2 depends on

Citation: Sowbiya Muneer, Tae Hwan Kim, Byung Chul Choi, Beom Seon Lee, Jeong Hyun Lee, Effect of CO, NOx and SO2 on ROS production, photosynthesis and ascorbate–glutathione pathway to induce Fragaria×annasa as a hyperaccumulator, Volume 2, 2014, Pages 91-98, ISSN 2213-2317, http://dx.doi.org/10.1016/j.redox.2013.12.006.

its concentration and on the duration of exposure [1] and is influenced by sulfur status in plants [8]. The low dose of sulfur even can be useful to plants since sulfur is important for plants, and can help to alleviate other nutrient deficiencies in plants, such as that of Fe [9]. Sulfur is also a structural component of amino acids, proteins, vitamins and chlorophyll [8], [10] and [11]. Sulfur enhances the development of nodules and nitrogen fixation, and also affects carbohydrate metabolism. However, exposure to high doses can lead to leaf chlorosis and necrosis, growth inhibition and plant death [12]. The concentration of SO_2 in the atmosphere has increased in many countries during the past few decades. It is estimated that parts of global plants exposed to SO_2 have prompted widespread attempts to evaluate defense responses.

The overall toxicity of these hazardous gases CO, NO*x* and SO_2 is derived from toxic molecular species such as those from the combustion of CO_2 to CO, NO*x* to NO_2, as well as O_3, sulphite (SO_3^{2-}) and bisulphite (HSO^{3-}). The detoxification reaction of these compounds further leads to the production of reactive oxygen species, such as hydroxyl radical (OH·) and hydrogen peroxide. Increased production of ROS can attack bio macromolecules and results in oxidative damage to nucleic acids, proteins and lipids[13] and [14]. However, plants can scavenge excess ROS by invoking the antioxidant defense system to avoid oxidative damage. Many studies have shown that some antioxidant enzymes are involved in plant response under various abiotic stresses. The induction of antioxidant enzymes is thought to be a protective reaction of plants against abiotic stress, but the exact defense mechanism is not clear.

Strawberry plants are the richest grown plants in greenhouses which are often fertilized by direct CO_2 enrichments fertilizers for heating. The CO_2 boilers/direct heating systems used in greenhouse however, often lead to incomplete combustion which results in formation of hazardous gases such as carbon monoxide (CO), nitroxide (NO*x*) and sulfur dioxide (SO_2) and other hydrocarbons. In the present study, we selected hazardous gases to conduct an investigation of alterations of ROS production, photosynthetic measurements and the ascorbate–glutathione pathway in strawberry plants exposed to different concentrations of CO, NO*x* and SO_2, assessing strawberry plants as a hyper-accumulator of these gasses.

METHODS

Plant Material and Treatments

Young strawberry (*Fragaria×anansa*) plants were propagated on rock–wool cubes (10×10×6.5 cm) after germination at 25 °C under fluorescent light (300 μmol m^{-2} s^{-1} of photosynthetic photon flux density) for three weeks. The rooted strawberry plants were grown on the cubes with a hydroponic nutrient solution containing (m mole per litter for the macro elements): N 9.5, P 1.5, K 5.5, Ca 5.0, S 1.5, P 1.5 Mg 1.5 and (μmole per litter for the micro elements): Fe 33, Mn 22, B 10, Zn 5, Cu 1.0, Mo 0.5. The nutrient solution supplied was regularly determined for its electrical conductivity (EC) (1.0 dS m−1) and pH was maintained at 6.5–7.0. Fluorescent light was supplied at 150 μmol m^{-2} s^{-1} at the canopy height for 16 h day^{-1}. Six-week-old plants were transferred to air tight glass bottles (3 l) and were supplied with 133 (low dose), 267 (medium dose) and 533 (high dose) ppm of carbon monoxide (CO), 25 (low dose), 50 (medium dose) and 199 (high dose) ppm of nitroxide (NO*x*), and sulfur dioxide (SO_2). Upon 24 h of treatment, the leaves were separated from the order of antigenic appearance and young leaves were used for the study. Plant samples were immediately frozen in liquid N_2 and were stored in a deep—freezer (−80 °C) prior to further analysis.

Determination of H_2O_2 and O_2^{-1}

H_2O_2 concentration was measured calorimetrically as described by Lin and Kao [15]using titanium sulfate. H_2O_2 concentration was calculated using the extinction coefficient 0.28 mM^{-1} cm^{-1} and was expressed as nmol g^{-1} tissue fresh weight. For O_2^{-1}determination fresh tissue was extracted in phosphate buffer (pH 7.5) and centrifuged at 10,000*g* for 5 min. The resultant extract was added to 1 ml of 10 mM NH_2OH–HCl and phosphate buffer (pH 7.8) and incubated at 25° for 1 h. After incubation the mixture was added with 0.5 ml of 17 mM alpha aminobenzene sulphonic acid and 0.5 ml of 7 mM alpha naphthylamine solution and was incubated again for 20 min at 25° and absorbance was recorded at 530 nm.

Measurement of net Photosynthetic Activity and F_V/Fm ratio

Photosynthetic rate was measured using a portable photosynthesis measurement system (LI-6400XT. LI-COR, Inc., Lincoln, NE). Chlorophyll fluorescence (*FV*/*Fm*) was measured by using a chlorophyll fluorescence meter (PAM 2000; Heinz Walz GmbH, Effeltrich, Germany). The leaves were adapted to dark conditions for 30 min before the measurement. The maximum fluorescence (*Fm*) and minimum fluorescence (F_0) were determined by applying a saturating light pulse (20 kHz) of 1100 μmol m^{-2} s^{-1} PPF for 3 μs. The maximum PS II quantum yield (*FV*/*Fm*) was calculated as $FV/Fm=(Fm-F_0)/Fm$ for 24 h at different concentrations of CO, NO*x* and SO_2.

Photosynthetic Pigments and Stomatal Observation

The content of chlorophyll and carotenoid was estimated by the method of Hiscox and Israclstam [16]. The fresh leaves were collected in glass vials, to which 10 ml dimethyl sulfoxide (DMSO) was added, after which samples were kept in an oven at 65 °C for complete leeching of pigments for 1 h. Optical density was recorded at 480, 645, 520 and 663 nm. The content of total chlorophyll and carotenoid was calculated using the formulae given by Arnon [17].

For stomatal observation, thin layers of leaf tissues were carefully cut and were laid on glass slides, covered with cover slips by adding few drops of water, and were observed under a light microscope (Leica DM4000 M) at 40× magnification.

RuBisCO Determination by SDS-PAGE

Leaf tissues were homogenized at 4°C in 100 mM Tris buffer (pH 7.5) containing 5 mM of DTT, 2 mM iodoacetate and 5% (v/v) glycerol at a leaf; buffer ratio of 1:5–10 (g:ml). For this extraction, a buffer without sodium or potassium ion was recommended for SDS-PAGE analysis because those cations reduce the solubility of DS (dodecyl sulfate). Before centrifugation, a TritonX100 (25%, v/v) was added to a portion of leaf homogenate to make a final concentration of 0.1% (v/v). An addition of TritonX100 was effective for the extraction of RuBisCO bound to the membrane fraction [18]. The homogenates were centrifuged at 5000*g* for 3 min at 4 °C. A lithium DS solution (25% w/v) and 2-mercaptoethanol were added to the supernatant fluid to a final concentration of 1.0% (w/v) and 1% (v/v), respectively. This preparation was immediately treated at 100 °C for 1 min, and was then stored at −30 °C until the analysis of SDS-PAGE. The samples containing 2–10 µg RuBisCO were loaded onto 12% polyacrylamide gel. After electrophoresis, the gels were stained in 0.25% (w/v) CBB-R. The stained bands corresponding to larger and smaller subunits of RuBisCO were cut out of the gels with a razor blade and were eluted in 1–2.5 ml of formamide in a stoppered amber test tube at 50 °C for 5 h with shaking. The absorbance of the resultant solution was read at 595 nm with a spectrophotometer. RuBisCO content was determined by using the standard curve calculated from the absorbance of a known amount of purified RuBisCO.

Total Carbohydrate and Sucrose Content

The carbohydrate content was determined by anthrone method [19]. One hundred milligrams of dried leaf tissue were added to a boiling tube. The tissues were hydrolyzed in a boiling tube by keeping them in a water bath for 3 h with 5 ml of 2.5 N-HCl, and were then cooled at room temperature. The solution was then neutralized with solid sodium carbonate until effervescence ceased, and samples were then centrifuged. After centrifugation 1 ml of supernatant had added to it 4 ml of anthrone reagent. The solution was then heated for 8 min in a boiling water bath and absorbance at 630 nm were recorded after cooling down at room temperature. The corresponding concentration of carbohydrate was determined against a standard curve prepared by using glucose solution.

For sucrose content 0.1 g of dried leaf was taken in 10 ml of ethanol (90%). The vials were incubated at 60 °C for 1 h and volume was made up to 25 ml by adding 15 ml of 95% ethanol. A 1 ml aliquot had added to it 1 ml of 5% phenol, and the solution was thoroughly mixed. Five ml of concentrated H_2SO_4 was added and the solution was allowed to cool at room temperature, after which the absorbance was recorded at 485 nm using a Beckman DU 640 spectrophotometer. The corresponding concentration of sucrose was determined against a standard curve which was prepared using glucose solution.

Antioxidant Enzyme Assays

Superoxide dismutase (SOD) activity was determined by the method of Dhindsa et al.[20] with minor modifications. Fresh tissues (200 mg) were extracted in phosphate buffer (pH 7.3) and were then centrifuged at 15,000*g*. SOD activity in the supernatant was assayed for its ability to inhibit photochemical reduction of nitro blue tetrazolium (NBT) by reading the absorbance at 560 nm. One unit of enzyme activity was defined as the amount of enzyme required to inhibit 50% of the NBT photo reduction in comparison with tubes lacking the plant extract.

Catalase (CAT) activity was determined by the method of Aebi [21]. About 500 mg of grounded tissues was extracted in phosphate buffer (pH 7.3), and were then centrifuged at 15,000*g* for 20 min. The decrease in absorbance at 240 nm was recorded as a result of H_2O_2 degradation (extinction coefficient of 36 mM^{-1} cm^{-1}). One unit of enzyme is defined as the amount necessary to decompose 1 µm of H_2O_2 per min.

Glutathione reductase (GR) activity was determined by the method of Rao [22]. About 500 mg of fresh tissues was extracted in phosphate buffer (pH 7.0), and were centrifuged at 15,000*g* for 10 min. The supernatant was immediately assayed for GR activity through the glutathione-dependent oxidation of NADPH at 340 nm. One unit of enzyme is defined as the amount necessary to decompose 1 µmol of NADPH per min.

Ascorbate peroxidise (APX) activity was estimated by measuring the decrease in absorbance at 290 nm (extinction coefficient of absorbance 2.8 mM^{-1} cm^{-1} for ascorbate) according to Nokano et al. [23]. One hundred milligrams of fresh tissue were extracted in 100 mM K-phosphate (pH 7.0), and were then centrifuged at 15,000*g* for 10 min. One unit of enzyme was expressed as the amount necessary to decompose 1 µmol of ascorbate in 1 min.

Statistical Analysis

A completely randomized design was used with three biological replicates for five treatments. An individual Student's *t* test was employed to compare the means of separate replicates by using software SAS (version 9.1, USA).

RESULTS AND DISCUSSION

CO, NO*x* and SO_2 induced ROS generation

Reactive oxygen species generation was observed by determination of hydrogen peroxide (H_2O_2) and singlet oxygen (O_2^{-1}). Strawberry leaves which were exposed to CO, NO*x*, and SO_2 gases exhibited high concentration of H_2O_2 and O_2^{-1} (Fig. 1A and B). The concentration of H_2O_2 and O_2^{-1} was observed highest by more than 90 and 80% respectively under high dose of CO, NO*x* and SO_2 compared to control.

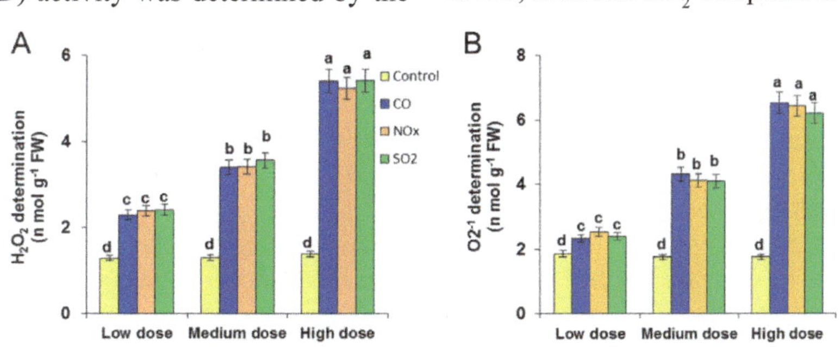

Figure 1. Production of reactive oxygen species (ROS) (A) H_2O_2 and (B) O2^{-1} determination, as affected by low, medium and high dose of

CO, NO*x*, and SO_2. Vertical bars indicate ±SE by means with $n=3$. Means denoted by different letters are significantly different at $p<0.05$ according to the Tukey's studentized range test.

In plant cells, ROS are unavoidable by-products of aerobic metabolism. Under normal growth conditions, the amount of ROS is modest, and cells experience only mild oxidative stress, whereas, many stresses enhance ROS production [13], [24] and [25]. The results of our study clearly show that CO, NO*x* and SO_2 triggered the rapid induction of O_2^{-1} and H_2O_2 generation in strawberry leaves. This enhanced production of ROS under stress can pose a threat to cells and can also act as a signal to activate stress response pathways such as the ascorbate–glutathione cycle [13] and [26].

Photosynthetic Parameters

The net photosynthetic rate in plants exposed under low doses of hazardous gases (CO, NO*x* and SO_2) showed a slight reduction of net photosynthesis, however, at the medium dose of gas stress the net photosynthetic rate decreased (Fig. 2A) and the reduction of photosynthesis was observed to be highest in plants under high doses of CO, NO*x* and SO_2.

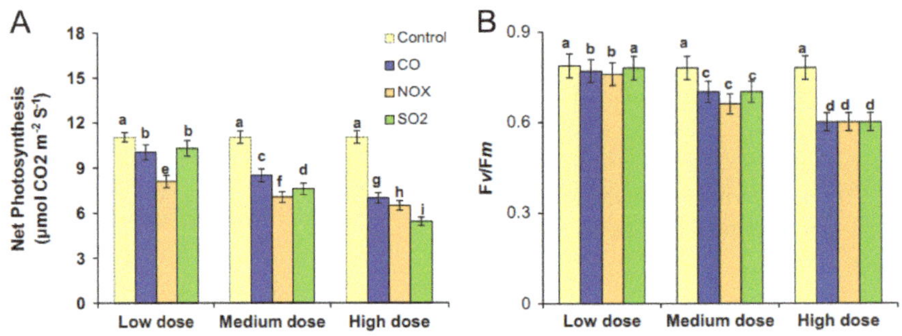

Figure 2. Changes in photosynthetic parameters (A) photosynthesis rate, (B) *FV/Fm* ratio, as affected by low, medium and high dose of CO, NO*x*, and SO_2. Vertical bars indicate ±SE by means with $n=3$. Means denoted by different letters are significantly different at $p<0.05$ according to the Tukey's studentized range test.

Similarly, the ratio between *FV/Fm* under CO, NO*x* and SO_2 treatment was not different between control plants, and those exposed to a low dose of gas stress (Fig. 2B). After a medium dose of gas stress, the ratio between *FV/Fm* decreased, and the lowest ratio was observed under a high dose of gas stress by 13%, 24% and 11%, respectively, when compared to the control.

We also observed that photosynthetic pigments (total chlorophyll and carotenoid) under CO, NO*x* and SO_2 were reduced slightly under a low dose of gas stress, whereas under a medium dose of hazardous gas stress the photosynthetic pigments reduced significantly and the highest reduction was observed in plants under a high dose of gas stress, reducing total chlorophyll by 54%, 60% and 59% (Fig. 3A) and carotenoid content by 66%, 80% and 70% (Fig. 3B) compared to control.

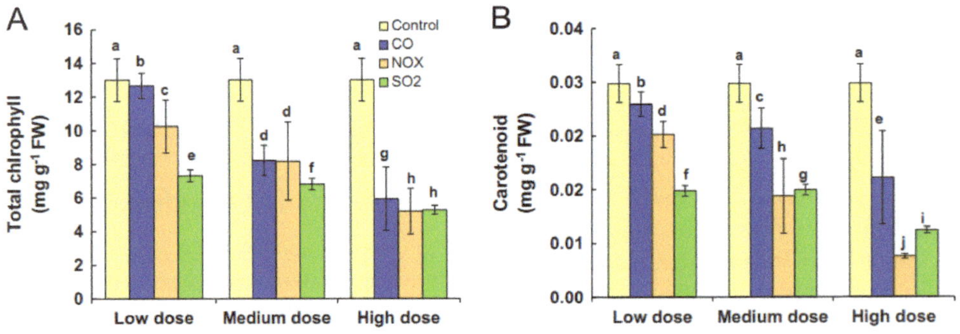

Figure 3. Changes in the content of photosynthetic pigments (A) total chlorophyll (B) carotenoid, as affected by low, medium and high dose of CO, NO*x*, and SO_2. Vertical bars indicate ±SE by means with $n=3$. Means denoted by different letters are significantly different at $p<0.05$, according to Tukey's studentized range test.

The present results show that photosynthetic parameters such as net-photosynthesis,*FV/Fm* ratio and photosynthetic pigments (chlorophyll and carotenoid) were reduced under CO, NO*x* and SO_2, which represents that there might be a transfer of lesser electrons and lesser energy investment in the biochemical machinery of strawberry. The reduction of net photosynthesis and ratio between *FV/Fm* has been also observed under salt stress in green algae [27] and in *Vigna radiata* under mineral deficiency [24] and [25]. The decline in net photosynthesis and *FV/Fm* under hazardous gas stress might be due to the inactivation of reaction centers of photosystems which receive an initial amount of light energy and cannot be exploited efficiently due to the presence of oxidative stress. The ineffective energy exploitation [28] and [29] leads to an extreme increase of dissipated energy, and therefore, decreases photosynthesis. In addition to this the loss of carotenoids might be due to the loss of chlorophyll, because carotenoids absorb light energy for use in photosynthesis, and they protect chlorophyll from photo-damage. Such a decrease in chlorophyll content may also be attributed to a reduction of hazardous gases for the formation of precursor molecules, δ-aminolevulinic acid and protochlorophyllide [30].

Observation of Stomata under Microscope

Microscopic observations (40×) have shown that stomata did not show any change in behavior in plants under low dose of CO, NO*x*, or SO_2 treatment (Fig. 4 and Fig. 5). With increased doses of gas exposure the stomata gradually closed and guard/subsidiary cells were damaged compared to those of the control, which was found to be exaggerated at high dose of gas exposed after 24 h.

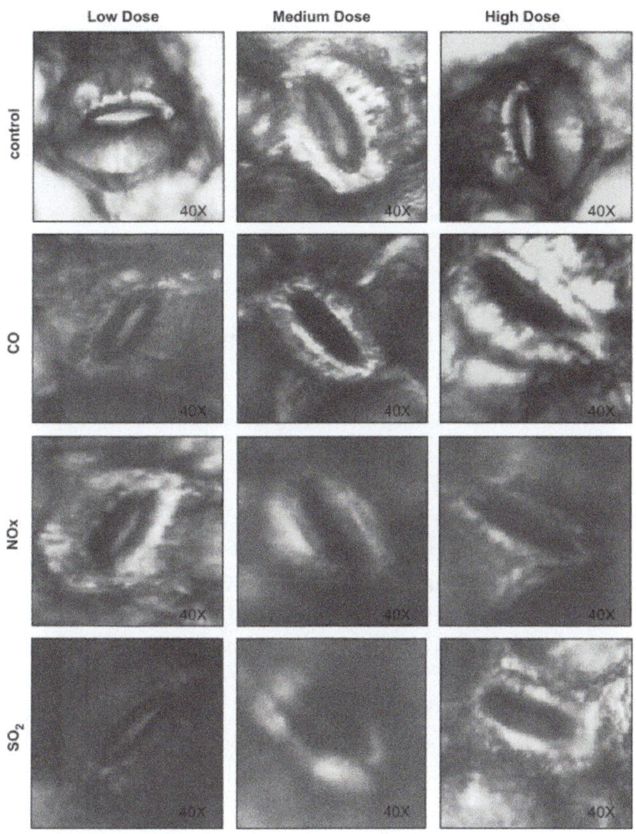

Figure 4. Representative images of stomata, as affected by low, medium and high doses of CO, NO*x*, and SO_2. Thin layer of leaf outer covering were peeled off carefully and laid on a glass slide, covered with a cover slip, and were observed under a light microscope (Leica CME) at 40× magnification.

As we know, stomata play an important regulatory role in leaf physiological processes, as the primary pathway for the exchange of gases between internal leaf surfaces and the atmosphere [31]. The closing of stomata as a result of exposure under hazardous gas has been reported in several species such as *Vicia faba* [32] and *Pisum sativum* [31]under SO_2 stress. The closure of stomata under high concentrations of CO, NO*x* and SO_2 demonstrated that certain key enzymes involved in stomatal metabolism might have been affected, which are also involved in several other metabolic activities in plants such as malate formation and metabolism in epidermis, and turgor pressure of guard cells. The destructive damage of guard cells and density might also occur due to defoliation of leaves during growth conditions of strawberry, and indeed, due to high temperatures under hazardous gas conditions [33], which might have resulted in the induction of palisade and sponge parenchyma cell length and thickness.

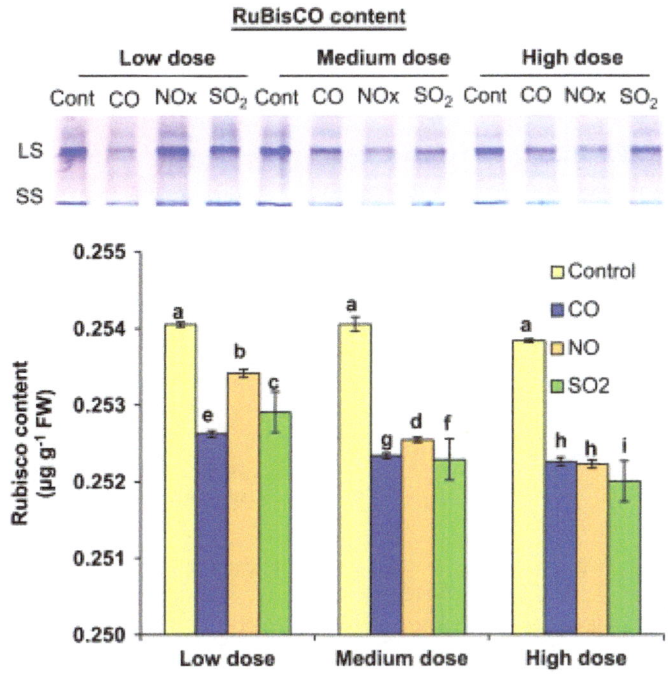

Figure 5. Regulation of RuBisCO larger and smaller subunits to low, medium and high doses of CO, NO*x* and SO_2(A) SDS-PAGE of leaf proteins from control and gas stressed plants, (B) quantification of the large subunit protein band. Vertical bars indicate ±SE by means with n=3. Means denoted by different letters are significantly different at p<0.05 according to Tukey's studentized range test.

RuBisCO contents

Fig. 4 shows that the RuBisCO content determined by SDS-PAGE, the RuBisCO content in plants under low dose CO, NO*x* or SO_2 showed great change in expression, and statistically significant results, as generated by a spectrophotometer, have shown a huge change even early stages of hazardous gas treatment. With increasing hazardous gas doses the reduction of RuBisCO content was observed to be highest in plants exposed to low doses of gas stress when compared to that of the control.

The changes in RuBisCO content in response to various abiotic stresses has been reported in many species, such as in *Trifolium alexandrinum* [34] under salt stress conditions. The maintenance of a higher RuBisCO content in plants has been correlated with a higher rate of photosynthesis in certain genotypes, and a lower content of RuBisCO is correlated with lower rates of photosynthesis under abiotic stresses [35]. In the present study, the reduction of RuBisCO content under hazardous gas stress (CO, NO*x* and SO_2) coincides with a reduction of photosynthesis (Fig. 2). In recent studies we observed the proteomic data of *Brassica napus* under Fe-deficiency to reduce most proteins classified as photosynthetic precursors, including RuBisCO content [10]. The loss of RuBisCO protein might be due to the progressive depletion of biochemical pathways associated with signal transduction and gene regulation, and in particular, might be involved in protein synthesis [36] and [37], and also might be associated with an excessive production of ROS which leads to incorrect folding or assembly of proteins, and consequent protein degradation [38].

Carbohydrate and Sugar Content

The total carbohydrate in plants under low dose of CO, NO*x* or SO_2 was reduced compared to control (Fig. 6A) and more reduction was observed in plants under medium dose of hazardous gas stress by 54%, 81% and 72% respectively. The reduction was observed to be highest under high dose of CO, NO*x* and SO_2, by 65%, 87% and 70% respectively.

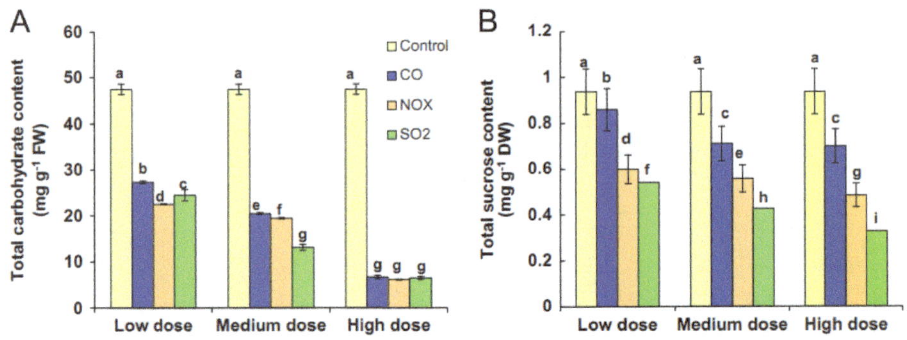

Figure 6. Changes in content of (A) total carbohydrate and (B) sucrose content, as affected by low, medium and high dose of CO, NO*x*, and SO_2. Vertical bars indicate ±SE by means with n=3. Means denoted by a different letter are significantly different at p<0.05 according to Tukey's studentized range test.

Sucrose content in plants under low dose of CO, NO*x* and SO_2 showed a slight reduction, which was greater (and significant) in medium dose plants (Fig. 6B), however, a high dose of CO, NO*x* and SO_2 reduced the sucrose content by 23%, 48% and 65%, respectively, when compared to control.

The decrease in total carbohydrates and sucrose content of damaged leaves probably corresponded with photosynthetic inhibition, or with the stimulation of a lower respiratory rate. Higher starch accumulation in damaged leaves may have resulted both in the higher resistance of their photosynthetic apparatus [39] and low starch export from the mesophyll. The negative effect of hazardous gases on carbon metabolism is a result of their possible interaction with the reactive center of ribulose bisphosphate carboxylase[40].

CO, NOx and SO_2-induced Antioxidative defense Response

We analyze the evolution of stress responses in strawberry plants exposed to low, medium and high doses of CO, NO*x* or SO_2 for 24 h. The treatment evokes physiological changes in the form of ascorbate glutathione cycle. There was a stark contrast between the responses of all antioxidative enzymes examined in this study. CO, NO*x* and SO_2increased SOD, CAT, GR and APX, activity at low and medium doses of gas stress conditions (Fig. 7A–D), while at higher doses, SOD activity was decreased by 55%, 55.1% and 78%, catalase activity was decreased by 81%, 89% and 86%, GR activity was decreased by 69%, 30% and 47% and APX activity was decreased by 32%, 62% and 25% compared with control.

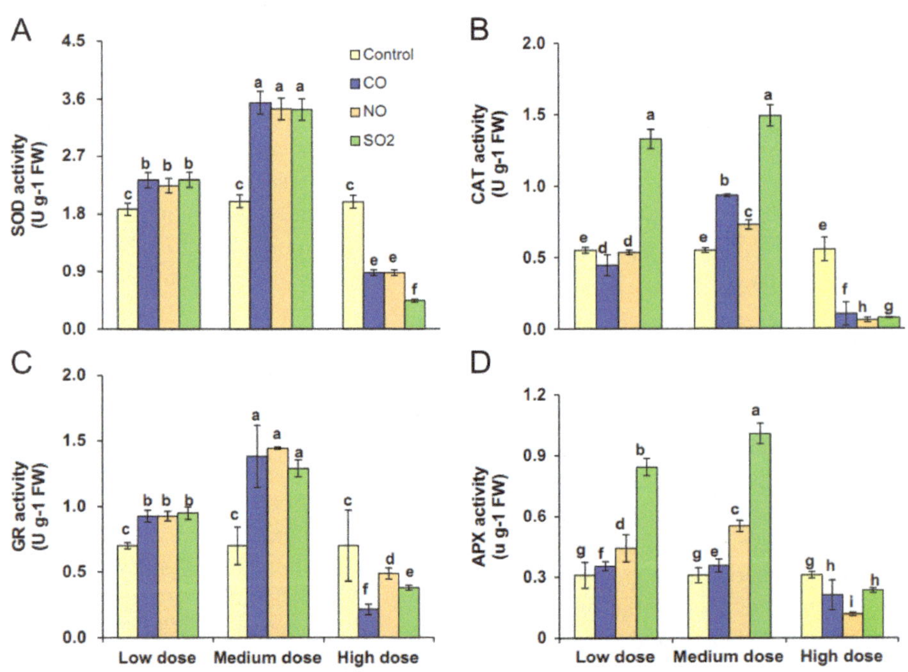

Figure 7. Changes in antioxidant enzyme activities (A) SOD (B) CAT (C) GR and (D) APX as affected by low, medium and high dose of CO, NO*x*, and SO_2. Vertical bars indicate ±SE by means with n=3. Means are denoted by different letters are significantly different at p<0.05 according to Tukey's studentized range test.

In this study, the activities of antioxidative enzymes (APX, SOD, CAT and GR) were found to be increased at low and medium doses of CO, NO*x* and SO_2 stress, whereas they were significantly decreased at high doses of gas stress. Several reviews have been published on different aspects of ascorbate and glutathione, ranging from their biosynthesis [41] and [42], to their roles in the transport system [43] and stress defense[44] in plants. It has been generally established that glutathione and ascorbate play a prominent role in non-enzymatic mechanisms to prevent the oxidation of cellular compounds [44], and are involved in the induction of enzymes and gene expression. SOD catalyzes the dismutation reaction of O_2^{-1} and H_2O_2 for the initial period of stress conditions, as induction of these enzymes was observed in our study, whereas, after prolonged exposure the SOD could not dismutase these ROS into H_2O_2 and H_2O, which resulted in the reduction of enzyme activities of ascorbate glutathione cycle.

Glutathione is involved as a substrate for glutathione peroxidase and is therefore necessary for the removal of lipid peroxidases, because GR reduces the oxidized glutathione, which is generated by glutathione peroxidase [42] and [45]. The observed increase in GR at initial period of stress resulted in the detoxification of oxidative damage, while a high dose of CO, NO*x* and SO_2 resulted in a disturbance in ascorbate glutathione pathways. CAT and APX are the primary H_2O_2 scavenging enzymes in plants. In this study, CO, NO*x* and SO_2 pronouncedly an increased APX activity and CAT activity in the initial period of stress, which indicates that APX and CAT are efficient enzymes for detoxifying ROS arising from hazardous gas treatments. Moreover the increased activity of these enzymes can contribute to resistance and enhancement since they participate in many other physiological processes in plant defense reactions.

CONCLUSIONS

In conclusion, this is a systematic study which provided a specific insight into the changes in ROS production, photosynthesis physiology and antioxidant status in strawberry plants under CO, NO*x* and SO_2 stresses, which contribute the mechanism of plant adaptation and promote that strawberry plant, can act as hyper-accumulator to mitigate oxidative damage. Our results suggest that CO, NO*x* and SO_2 caused a high level of ROS production in the form of O_2^{-1} and H_2O_2, as well as a reduction of photosynthetic physiology.

Conversely, antioxidant enzymes of ascorbate glutathione pathways were induced to scavenge the negative effects of gas stresses. However, increased levels of lipid peroxidation products in strawberry plants exposed to CO, NO*x*and SO_2 for high dose demonstrated that the antioxidant defense system was no longer used to mitigate free radicals. Ascorbate glutathione cycle status may play a critical role in defense against CO, NO*x*, and SO_2, as well as the maintenance of photosynthesis metabolism in plants, and will help to mitigate oxidative damage under abiotic stress. These results suggest that strawberry plants can act as a hyperaccumulator for detoxifying these gases.

Acknowledgements

This study was supported by iPET (Korea Institute of Planning and Evaluation for Technology in Food, Agriculture, Forestry and Fisheries), Ministry for Food, Agriculture, Forestryand Fisheries, Republic of Korea.

REFERENCES

1. Long Li., Huilan Yi., Effect of sulfur dioxide on ROS production, gene expression and antioxidant enzyme activity in Arabidopsis plants, J. Plant Physiol 58 (2012) 46–53.
2. Z.Y. Dong, W.H. Huang, D.F. Xing, H.F. Zhang, Remediation of soil co-contaminated with petroleum and heavy metals by the integration of electrokinetics and biostimulation, J. Hazard. Mater. 260 (2013) 399–408.
3. J.Z. Wang, L.J. Cui, Y. Wang, J.L. Li, Growth, lipid peroxidation and photosynthesis in two tall fescue cultivars differing in heat tolerance, Biol. Plant 53 (2009) 237–242.
4. R.F. Troxler, Synthesis of bile pigments in plants. Formation of carbon monoxide and phycocyanobilin in wild-type and mutant strains of the alga, Cyanidium caldarium, Biochemistry 11 (1972) 4235–4242.
5. T.D. Elich, A.F. McDonagh, L.A. Palma, J.C. Lagarias, Phytochrome chromophore biosynthesis. treatment of tetrapyrrole-deficient Avena explants with natural and non-natural bilatrienes leads to formation of spectrally active holoproteins, J. Biol. Chem 264 (1989) 183–189.
6. J. Xu, H. Yin, X. Liu, Nitric oxide is associated with long term zinc tolerance in Solnum nigrum, Plant Physiol 154 (3) (2010) 1319–1334.
7. K. Bedard, K.H. Krause, The NOx family of ROS generating NADPH oxidase: physiology and pathophysiology, Phys. Rev. 87 (2010) 245–313.
8. L.J. De-Kok, Sulfur metabolism in plants exposed to atmospheric sulfur, in: M. Rennenberg, C. Brunold, L.J. De-Kok, I. Stulen (Eds.), Sulfur Nutrition and Sulfur Assimilation in Higher Plants, SPB Academic Publishing, The Hague, 1990, p. 111e130. (1990).
9. S. Muneer, B.R. Lee, D.W. Won, T.H. Kim, Changes in expression of proteins involved in alleviation of Fe deficiency by sulfur nutrition in Brassica napus L, Acta Physiol. Plant (2013), http://dx.doi.org/10.1007/s11738-013-1336-4.
10. T.H. Maugh II, SO2 pollution may be good for plants, Science 205 (1979) 383.
11. L.P. Yang, I. Stulen, L.J. De-Kok, Sulfur dioxide: relevance of toxic and nutritional effects for Chinese cabbage, Environ. Exp. Bot 57 (2006) 236–245.
12. M. Agrawal, B. Singha, M. Rajputa, F. Marshallb, J.N.B. Bellb, Effect of air pollution on peri-urban agriculture: a case study, Environ. Pollut. 126 (2003) 323–329.
13. R. Mittler, S. Vanderauwera, M. Gollery, F. Van-Breusegem, Reactive oxygen gene network of plants, Trends Plant Sci. 9 (2004) 490–498.
14. C.H. Foyer, G. Noctor, Oxidant and antioxidant signaling in plants: a reevaluation of the concept of oxidative stress in a physiological context, Plant Cell Environ 28 (2005) 1056–1071.
15. C.C. Lin, C.H. Kao, Abscisic acid induced changes in cell wall peroxidase activity and hydrogen peroxide level

in roots of rice seedlings, Plant Sci 160 (2001) 323–329.

16. J.D. Hiscox, G.F. Israclstam, A method for interaction of chlorophyll from leaf tissue without maceration, Can. J. Bot 57 (1979) 1332–1334.
17. J.D. Arnon, G.F. Israelstam, Copper enzymes in isolated chloroplast oxidase in Beta vulgaris, Plant Physiol. 42 (1979) 287–292.
18. A. Makino, B. Osmond, Effects of nitrogen nutrition on nitrogen partitioning between chloroplasts and mitochondria in pea and wheat, Plant Physiol 96 (1991) 355–362.
19. J.E. Hedge, B.T. Hofreiter, Carbohydrate Chemistry, in: R.L. Whistler, J.N. Be Miller (Eds.), Academic Press, New York, 1962.
20. R.H. Dhindsa, P. Plumb-Dhindsa, T.A. Thorpe, Leaf senescence correlated with increased level of membrane permeability, lipid peroxidation and decreased level of SOD and CAT, J. Exp. Bot 32 (1981) 93–101.
21. H. Aebi, Catalase in vitro, Methods Enzymol. 105 (1984) 121–126.
22. M.V. Rao, Cellular detoxification mechanisms to determine age dependent injury in tropical plant exposed to SO2, J. Plant Physiol 140 (1992) 733–740.
23. M. Nakano, K. Nobuta, K. Vemaraju, S.S. Tej, J.W. Skogen, B.C. Meyers, Plant MPSS databases: signature-based transcriptional resources for analyses of mRNA and small RNA, Nucleic Acids Res. 34 (2006) 731–735.
24. S. Muneer, T.H. Kim, M.I. Qureshi, Fe modulates Cd-induced oxidative stress and the expression of stress responsive proteins in the nodules of Vigna radiata, Plant Growth Regul. 68 (2012) 421–443.
25. S. Muneer, J. Ahmad, M.I. Qureshi, Involvement of Fe nutrition in modulating oxidative stress and the expression of stress responsive proteins in leaves of Vigna radiata L, Aust. J. Crop Sci 7 (9) (2013) 1333–1342.
26. K. Apel, H. Hirt, Reactive oxygen species: metabolism, oxidative stress, and signal transduction, Ann. Rev. Plant Biol 55 (2004) 373–399.
27. G. Demetriou, N. Christina, N. Eleni, K. Kiriakos, Salt stress impact on the molecular structure and function of the photosynthetic apparatus—the protective role of polyamines, Biochim. Biophys. Acta 1767 (2007) 272–280.
28. K. Satoh, C.M. Smith, D.C. Fork, Effects of salinity on primary processes of photosynthesis in the red alga (Porphyra perforate), Plant Physiol. 73 (1983) 643–647.
29. C. Lu, A. Vonshak, Effects of salinity stress on photosystem II function in cyanobacterial Spirulina platensis cells, Physiol. Plant 114 (2002) 405–413.
30. H. Marschner, Mineral Nutrition in Higher Plants, Academic Press, London, 1986.
31. M.I. Rao, R.B. Amundson, R. Herman-Alscher, L.E. Anderson, Effect of SO2 on stomatal metabolism in Pisum sativum L, Plant Physiol. 72 (1983) 573–577.
32. V.J. Black, M.H. Unsworth, Effects of low concentration of sulfur dioxide on net photosynthesis and dark respiration of Vicia faba, J. Exp. Biol 30 (1979) 473–483.
33. V. Fernandez, T. Eichert, V. Del Rio, G. Lopez-Casado, J.A. Heredia-Guerrero, A. Anunciacion, A. Heredia, A. Abadia, Leaf structural changes associated with iron deficiency chlorosis in field-grown pear and peach: physiological implications, Plant Soil 311 (2008) 161–172.
34. G.M. Abogadallah, Sensitivity of Trifolium alexandrinum L. to salt stress is related to the lack of long-term stress-induced gene expression, Plant Sci 178 (2010) 491–500.
35. C. Lu, A. Vonshak, Effects of salinity stress on photosystem II function in cyanobacterial Spirulina platensis cells, Physiol. Plant 114 (2002) 405–413.
36. A. Pandey, M.K. Choudhary, D. Bhushan, A. Chattopadhyay, S. Chakraborty, A. Datta, N. Chakroborty, The nuclear proteome of chick pea (Cicer arietinum L.) reveals predicted and un-expected proteins, J. Proteome Res 5 (12) (2006) 3301–3311.
37. M.K. Choudhary, D. Basu, A. Datta, N. Chakraborty, S. Chakroborty, Dehydration-responsive nuclear proteome of rice (Oryza sativa L.) illustrates protein network, novel regulators of cellular adaptation, and evolutionary, Mol. Cell. Proteomics 8 (7) (2009) 1579–1598.
38. S. Luo, H. Ishida, A. Makino, T. Mae, Fe2þ-catalyzed site specific cleavage of the large subunit of ribulose 1,5-bisphosphatecarboxylase close to the active site, J. Biol. Chem 277 (2002) 12382–12387.
39. E., Prokopiev Afforestation of Industrial Areas. Sofia, Zemizdat, 208 p. (In Bulg.). 1978.
40. M. Stiborova, M. Ditrichova, A. Brezinova, Effect of heavy metal ions on growth and biochemical characteristics of photosynthesis of barley and maize seedlings, Biol. Plant 29 (1987) 453–467.
41. N. Smirnoff, P.L. Conklin, F.A. Loewus, Biosynthesis of ascorbic acid in plants: a renaissance, Annu. Rev. Plant Physiol. Plant Mol. Biol. 52 (2001) 437.
42. G. Potters, L. De Gara, H. Assad, N. Horemans, Ascorbate and glutathione: guardians of the cell cycle, partners in crime? Plant Physiol. Biochem 40 (2002) 537–548.
43. N. Horemans, C.H. Foyer, G. Potters, H. Asard, Ascorbate functions and associated transport systems in plants, Plant Physiol. Biochem 38 (2000) 531–540.
44. G. Noctor, C.H. Foyer, Ascorbate and glutathione: keeping active oxygen under control, Annu. Rev. Plant Physiol. Plant Mol. Biol 49 (1998) 249–279.
45. M.B. Ali, E.J. Hahn, K.Y. Paek, Effects of temperature on oxidative stress defense systems, lipid peroxidation and lipoxygenase activity in Phalaenopsis, Plant Physiol. Biochem 43 (2005) 213–223.

CHAPTER 11

CHANGES IN ANTI-OXIDATIVE ENZYMES ACTIVITY, PROTEIN CONTENT AND ASCORBIC ACID LEVEL IN PROSOPIS JULIFLORA EXPOSED TO INDUSTRIAL AIR POLLUTION

Seyyednejad, S.M., Koochak, H., Vaezi, J

Keywords: Air pollution, Prosopis juliflora, anti-oxidative enzymes activity, protein, ascorbic acid.

INTRODUCTION

Urban air pollution is a serious problem in both developing and developed countries [1]. The increasing number of industries and automobile vehicles are continuously adding toxic gases and other substances to the environment [2]. Vegetation is an effective indicator of the overall impact of air pollution, and the effect observed is a time-averaged result that is more reliable than the one obtained from direct determination of the pollutant in air over a short period. Although a large number of trees and shrubs have been identified and used as dust filters to check the rising urban dust pollution level [3]. All combustion release gases and particles into the air. These can include sulphur and nitrogen oxides, carbon monoxide and soot particles, as well as smaller quantities or toxic metals, organic molecules and radioactive isotope [4]. The atmospheric SO_2 adversely affects various morphological and physiological characteristics of plants. High soil moisture and high relative humidity aggravated SO_2 injury in plants [5]. One of the biochemical changes which occur in trees subjected to environmental stress is the increased production of reactive oxygen species (ROS) such as superoxide, hydrogen peroxide and hydroxyl radicals [6]. ROS causes damage to the biomolecules such as membrane lipids, proteins, chloroplast pigments, enzymes, nucleic acid, and etc. lipid peroxidation is a biochemical marker for the free radical mediated injuries [7]. In general, plants produce O_2^- and H_2O_2 in chloroplasts and mitochondria during their normal aerobic metabolism, as well as under environmentally induced oxidative stress [1]. High level exposure to pollutants causes chloroplasts to function at a very high level of excitation energy, which increases the generation of ROS and induces oxidative stress. Such elevated ROS production is correlated with a rise in the activities of anti-oxidative enzymes, which then enables the trees to reduce the oxidative damage triggered by those ROS [8]. In fact, high ROS level initiate signaling responses that include enzyme activation, gene expression, programmed cell death and cellular damage [9]. Anthropogenic pollution causes oxidative stress in plants and reactive oxygen species (ROS) are diminished by anti-oxidative enzymes and small molecular antioxidants [10]. An efficient anti-oxidative defense system is present in plants to counteract oxidative stress. It is composed by enzymatic and non-enzymatic

Citation: Seyyednejad, S.M., Koochak, H., Vaezi, J, "Changes in anti-oxidative enzymes activity, protein content and ascorbic acid level in Prosopis juliflora exposed to industrial air pollution," ISSN 2322-3308.

mechanisms. The former includes superoxide dismutase, catalase, peroxidase, ascorbate peroxidase, dehydroascorbate reductase, monodehydroascorbate reductase and glutathione reductase. The non-enzymatic mechanisms consist of antioxidants such as ascorbate, glutathione, α-tocopherol and carotenoids [9]. In fact, all plants have evolved anti-oxidative defense systems consisting of low molecular weight antioxidants in aqueous and lipid phases, enzymes, and photoprotective pigments – to detoxify ROS or repair their adverse effects. If the capacity of these systems is too low to keep up with the generation rates of ROS, cell structures are destroyed and tissue damages emerge [11]. In plants, CAT is one of the main H_2O_2-scavenging enzymes that dismutates H_2O_2 into water and O_2, and it is a constitutive component of peroxisomes. APX is a specific peroxidase that catalyzes the elimination of the toxic product H_2O_2 at the expense of oxidizing ascorbate to monodehydroascorbate. Peroxidases are enzymes that catalyze the H_2O_2-dependent oxidation of a wide variety of substrates, mainly phenolics and they are often found in multiple molecular forms [9]. Ascorbic acid, a natural antioxidant in plants has been shown to play an important role in pollution tolerance [12]. Additional environmental stresses as air and soil pollution, UV-B radiation, can modify antioxidant level in plant cells [13]. The genus *Prosopis* is in the family Leguminosae (Fabaceae), sub-family Mimosoideae. The native range of *Prosopis* species can be approximately divided into five regions, simply defined as Asia, Africa, North America, Central America and South America. The genera *Prosopis* contain some of the most widespread and important tree species in the arid and semiarid zones of the tropical and sub-tropical world. The genus *Prosopis* contains 44 species and a large number of varieties. The genetic variation of *P. juliflora* is high for all characters tested in the species. Prosopis is able to improve the soil in which it is growing by means of biological nitrogen fixation, leaf litter accumulation, nutrient pumping from deeper soil layers, loosening of a hard soil structure, stabilizing of loose sands, and an increase of the fauna above and below the ground [14]. In order to understand the effects due to air pollutants on plants and the adaptation mechanisms of plant under stress conditions, present study examines the impact of air pollution on *P. juliflora* around one of the oil factory in Iran.

MATERIALS AND METHODS

Area of Study and Sample Collection

The area of study is situated in industrial region (around one of the oil fields in south west of Iran). This is designated as polluted site. A site with similar ecological conditions was selected as the control (unpolluted) site. Three replicates of the plant sample leaves in two regions were collected. After collection, leaf samples immediately were taken to the laboratory in 4°C and stored in -70°C until analyzed. The plants studied in two regions were 10 to 15 years old. The climate in this land is warm and humid.

Measurement of Protein and Enzyme Assay

Preparation of Extract

Frozen leaves (0.5 g) were homogenized in 10 ml of 0.1 M potassium phosphate buffer (pH 7.4) at 4°C and the homogenates were centrifuged at 15000 rpm for 20 min at 4°C. The supernatants were then stored at -70°C. This obtained extracts were used for measurement of protein and enzyme assay.

Measurement of Protein

Protein concentration of the samples was evaluated spectrophotometrically at 660 nm according to the method of Lowry et al. [15] with bovine serum albumin as standard.

Enzyme Assay

The activity of peroxidase was determined spectrophotometrically using the method of Hoyle [16]. The reaction mixture contained 0.2 M acetate buffer (pH 5), 0.3% H_2O_2, 0.02 M benzidine soluble in 50% methanol and 0.1 ml enzyme extract. Absorbance was recorded at 530 nm in 3-5 min. Ascorbate peroxidase activity was analyzed spectrophotometrically by measuring decrease of ascorbate at 240 nm for 1 min according to Chen and Asada [17] method. The reaction mixture contained 0.1 ml enzyme extract, 0.5 mM ascorbate, 0.1 M potassium phosphate buffer (pH 7.0), 0.1 mM EDTA and 1% H_2O_2. Catalase activity was determined spectrophotometrically by measuring decomposition of hydrogen peroxide (H_2O_2) at 240 nm for 2-3 min using Aebi [18] method. The reaction mixture contained 0.05 M potassium phosphate buffer (pH 7.0), 10 mM H_2O_2 and 0.2 ml enzyme extract. The activity of all tested enzymes was expressed as $\Delta OD.min^{-1}$ per mg protein.

Measurement of Ascorbic Acid Content

0.5 g fresh leaves were homogenized in 10 mm metaphosphoric acid (5%), then were centrifuged at 15000 rpm for 20 min at 4°C. The supernatants were used for measurement of ascorbic acid content according to Smirnoff [19] method. The absorbance was determined spectrophotometrically at 520 nm and ascorbic acid concentration was expressed as mg/g fresh weight. The results were analyzed statistically by using the statistical software MSTATC in order to determine the significant differences between polluted and control stands. Analysis of variance was done by Completely Randomized Design.

RESULTS

The results of experiment exhibited that the plants exposed to air pollution showed changes in peroxidase, Ascorbate peroxidase and Catalase activity, as well as protein and ascorbic acid content, in comparison to control. According to the results, protein content in *P. juliflora* leaves, at polluted site (17.574) was lower as compared to control samples (37.771). A reduction of 53.47% in protein content was observed in leaves collected from polluted site (Fig. 1). This change was significant at $p < 0.01$. Ascorbic acid content in leaves increased from 0.100 in control to 0.111 in polluted site (Fig. 2).

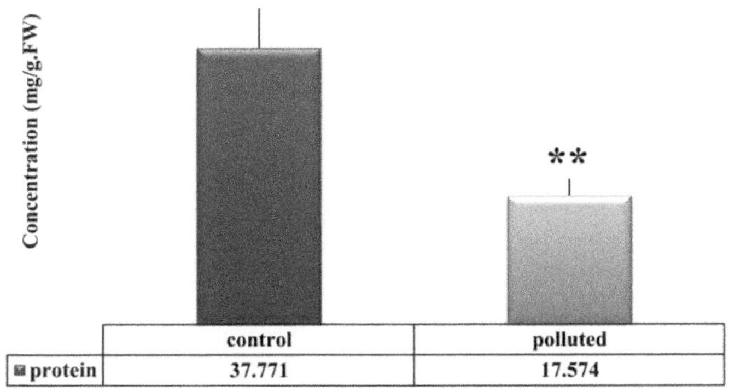

Figure 1. Protein content of *P. juliflora* grown in control and polluted sites. Bars represent standard deviation. * = $p < 0.05$; ** = $p < 0.01$.

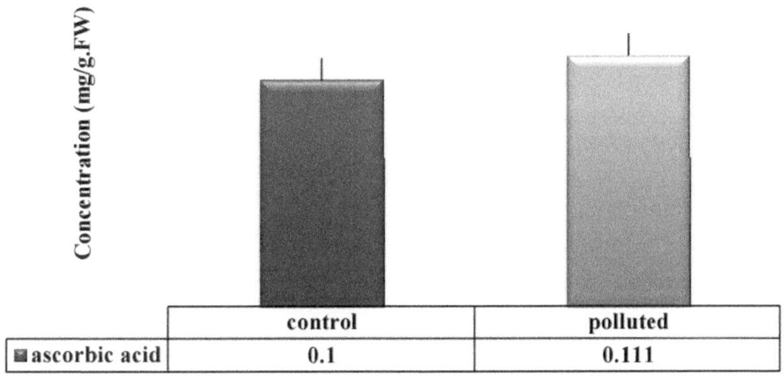

Figure 2. Ascorbic acid content of *P. juliflora* grown in control and polluted sites. Bars represent standard deviation. * = $p < 0.05$; ** = $p < 0.01$.

In fact, under air pollution, an increase of 11% in ascorbic acid concentration was exhibited at polluted leaves. The activity of peroxidase, Ascorbate peroxidase and Catalase from *P. juliflora* leaves was higher in polluted site (169.63%, 153.28% and 270.47%, respectively) as compared to control site (Figs. 3, 4, 5). These increases were significant ($p < 0.01$) about peroxidase and Catalase.

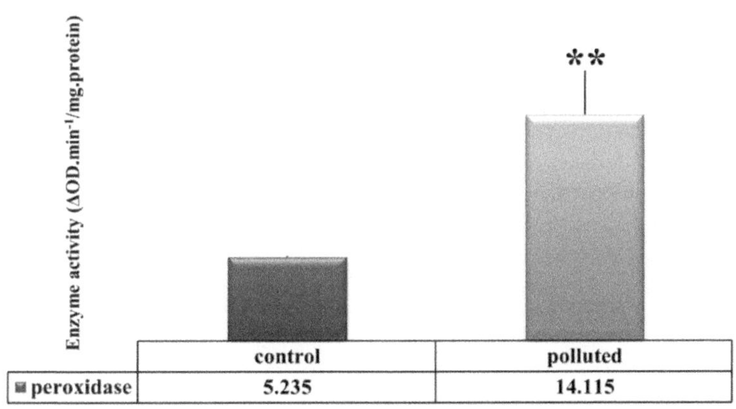

Figure 3. Peroxidase activity of *P. juliflora* grown in control and polluted sites. Bars represent standard deviation. * = $p < 0.05$; ** = $p < 0.01$.

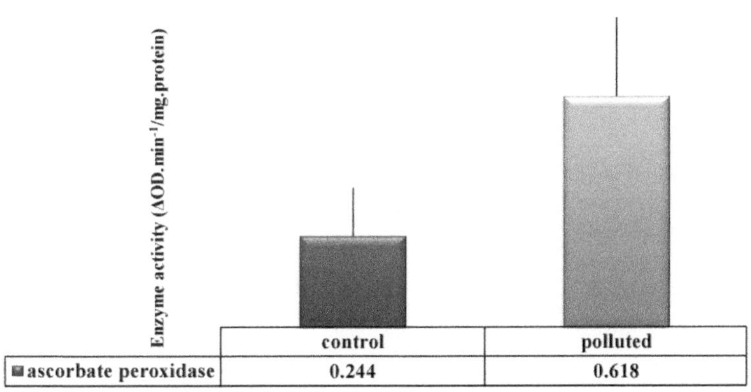

Figure 4. ascorbate peroxidase activity of *P. juliflora* grown in control and polluted sites. Bars represent standard deviation. * = $p < 0.05$; ** = $p < 0.01$.

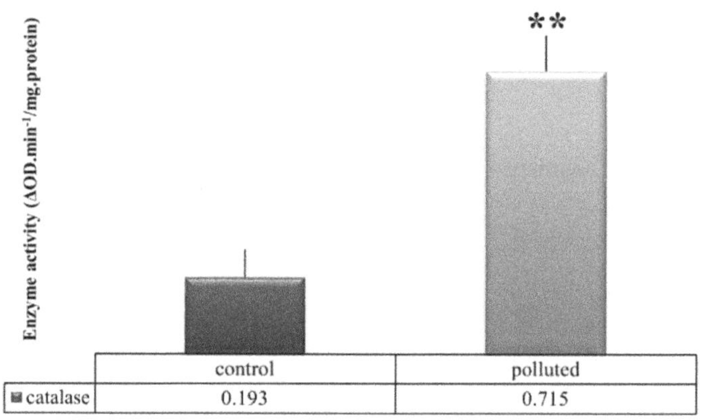

Figure 5. Catalase activity of *P. juliflora* grown in control and polluted sites. Bars represent standard deviation. * = $p < 0.05$; ** = $p < 0.01$.

DISCUSSION

Air pollution can directly affect plants via leaves or indirectly via soil acidification. When exposed to airborne pollutants, most plants experienced physiological changes before exhibiting visible damage to leaves [20]. Pollution may induce defense cascades such as the production of jasmonates, which may further lead to the activation of defense-related genes such as genes encoding peroxidases. ROS oxidize membrane lipids, proteins and DNA. Plants protect themselves against oxidative stress with antioxidant enzymes such as superoxide dismutase (SOD) that converts O_2^- to H_2O_2 and O_2, catalase (CAT) and peroxidase (POD) that both detoxify H_2O_2. Increased activities of these enzymes and levels of non-enzymatic antioxidants are considered as indicators of oxidative stress in plants [10]. Among antioxidant enzymes, peroxidase activity in plants was found to be a sensitive indicator of pollutant exposure and has been suggested as a marker to evaluate urban air pollution [1]. Ascorbate peroxidase, considered the most important peroxidase for H_2O_2 detoxification, is distributed throughout the cell. It catalyzes the reduction of H_2O_2 to H_2O, using ascorbate as an electron donor in the first step of the ascorbate-glutathione cycle [8]. Catalase, located in peroxisomes, mitochondrial and cytosol, can scavenge H_2O_2 with co-substrates. It is another important enzyme against oxidative stress, being able to scavenging H_2O_2, which is the major product produced by superoxide dismutase [21]. The reports have showed that the increase of SOD, POD, CAT and APX was induced by the accumulation of O_2^- and H_2O_2 in plants exposed to ozone [22]. Similarly, in this study, it has been shown that in response to air pollution, the activity of CAT, POD, and APX from *P. juliflora* leaves increased. The changes in POD and CAT activities were significant ($p < 0.01$). Lignin in cell walls provides both a mechanical support as well as a plant production from the chemical and biological stress. Lignin production is considered to be catalyzed by peroxidase. Lignin and POD are known to have a considerable part in the mechanism of resistance to the external stress in plants [23]. Phenolic compounds also have antioxidative properties and they are substrates of plant oxidases, POD. POD-catalyzed oxidation of phenolics to quinines consumes H_2O_2, thereby suppressing oxidative stress, and the oxidation of phenolics is suggested to act synergistically with antioxidant, ascorbate, in the detoxification of H_2O_2 [10]. The increase in peroxidase activity in plant cells under a variety of stresses, such as mechanical injury and attack by pathogen or an influence of environmental pollution has been reported. The increase in peroxidase activity varies with the plant species and the concentration of pollutants. It has been reported that leaves of the resistant plants might have high peroxidase activity [24]. The increase in total POD activity has been used as an indicator of pollution and other stresses. POD_s are induced by various abiotic stresses such as acid rain and heavy metal pollution and by biotic stresses such as pathogen infection, wounding and herbivory [10]. Several researches have suggested that peroxidase activity may not be used as a specific indicator of single air pollutant and can only be considered as a general indicator of oxidative stress [1]. The increase in anti-oxidative enzymes activity under stress conditions has been reported in several studies [6, 8, 9, 10, 21, 22, 25]. The mechanism of the influence of different pollutants on plant organism is associated with oxidative damage at cellular level and the plants response to this oxidative stress by changes in the activities of ROS scavenging enzymes [13]. Thus, increased anti-oxidative enzymes activity in the leaves is strongly associated with their resistance to air pollution [8]. But, the sensitivity to different pollutants differs between plants and even between clones of the same species [13]. It has been also reported that ageing leaves contain lower antioxidant levels than younger leaves [22]. In the present study, the protein content significantly ($p < 0.01$) decreased in polluted leaves that is in agreement with those of Tripathi and Gautam [24] and Tiwari et al. [25] that shown decreased protein level in leaves subjected to ambient air pollution. Reduction in protein content might be due to the enhanced rate of protein denaturation. It has been shown that enhanced protein denaturation and breakdown of existing protein to amino acid is as the main causes of reduction in protein content [24]. Ascorbic acid is known for its function as a powerful antioxidant and for its role in photosynthesis and photoprotection against oxidative stress, including SO_2 [26]. Numerous studies indicate that oxidative stress evoked by atmospheric pollutants enhances the quantity of low molecular antioxidants such as ascorbic acid in cells. Ascorbic acid is a very important primary antioxidant which reacts with hydroxyl radical, superoxide and singlet oxygen as well as secondary antioxidant, reducing the oxidized form of α-tocopherol [13]. Ascorbic acid is known to provide stability to the plant cell membranes during pollution stress and scavenges cytotoxic free radicals, which can otherwise cause lipid peroxidation and destruction of membranes [25]. In this study, ascorbic acid level in polluted leaves of *P. juliflora* increased in response to air pollution stress. It has been reported that this increase might be a consequence of substantial oxidative stress. Many studies exhibited accumulation of ascorbic acid content in plant cells after pollutant exposure. Similarly, it has been reported that resistant species showed increase in ascorbic acid content, whereas sensitive species showed decline [25]. The increase in ascorbic acid content may be due to the more rate of production of reactive oxygen species (ROS) such as SO_3^-, HSO_3^-, OH^-, O_2^- etc. during photooxidation of SO_3^- to SO_4^- where sulfites are generated SO_2 absorbed. The free radical production under SO_2 exposure would increase the free radical scavengers, such as ascorbic acid based on dosage and physiological status of plant.

Increased level of ascorbic acid may be due to the defense mechanism of the plant [24]. Tripathi and Gautam [24] and Pukacka and Pukacki [13] have also reported increased ascorbic acid content in plants exposed to air pollution. Species differed in the magnitude of response to pollutant exposure, although differences were not consistent within taxonomic or functional groups [27]. The present knowledge of effects of air pollution on plants is mostly based on experiments where plants have been exposed to high concentrations of air pollutants for short periods under experimental conditions. However, less is known about responses of plants to air pollutants at environmentally relevant concentrations and for long durations in field conditions [1]. The present study was intended to increase the understanding of the adaptation and survival mechanisms of plants in response to air pollution. Many important and widespread air pollutants, such as SO_2, NO_X and O_3, can produce an oxidative stress in plants and adversely affect physiological processes of different plants. In addition to morphological damage, responses of plants to oxidative stresses include a number of physiological and biochemical changes which are involved in both enzymatic and non-enzymatic antioxidant defense systems. Today, it is well known that oxidative stresses increase activities of antioxidant enzymes [1]. In summary, many changes in plant physiology and growth, such as those caused by air pollution, are biological compensatory responses to environmental stress. The main stress compensatory strategy in plants is to minimize damage from stress [6]. In fact, these changes help plants minimize stress and maximize use of internal and external resources [28].

ACKNOWLEDGEMENT

The authors wish to thank vice president for Research and Technology of Shahid Chamran University of Ahvaz for financial support.

REFERENCES

1. Li, M.H, 2003. Peroxidase and superoxide dismutase activities in fig leaves in response to ambient air pollution in a subtropical city. Arch Environ Contam Toxicol 45: 168-176.
2. Jahan, S. and Iqbal, M.Z, 1992. Morphological and anatomical studies of leaves of different plants affected by motor vehicles exhaust. Journal of Islamic Academy of Sciences 5(1): 21-23.
3. Rai, A., Kulshreshtha, K., Srivastava, P.K. and Mohanty, C.S, 2009. Leaf surface structure alterations due to particulate pollution in some common plants. Environmentalist 30: 18-23.
4. Agbaire, P.O. and Esiefarienrhe, E, 2009. Air pollution tolerance indices (apti) of some plants around Otorogun gas plant in Delta State, Nigeria. J Appl Sci Environ Manage 13(1): 11-14.
5. Tankha, K. and Gupta, R.K, 1992. Effect of water deficit and sulphur dioxide on total soluble proteins, nitrate reductase activity and free proline content in sunflower leaves. Biologia Plantarum 34(3-4): 305-310.
6. Woo, S.Y., Lee, D.K. and Lee, Y.K, 2007. Net photosynthetic rate, ascorbate peroxidase and glutathione reductase activities of Erythrina orientalis in polluted and non-polluted areas. Photosynthetica 45(2): 293-295.
7. Wang, F., Zeng, B., Sun, Z. and Zhu, C, 2009. relationship between proline and Hg2+ - induced oxidative stress in a tolerant rice mutant. Arch Environ Contam Toxicol 56: 723-731.
8. Woo, S.Y. and Je, S.M, 2006. photosynthetic rates and antioxidant enzyme activity of Platanus occidentalis growing under two levels of air pollution along the streets of Seoul. Journal of Plant Biology 49(4): 315-319.
9. Yannarelli, G.G., Gallego, S.M. and Tomaro, M.L, 2006. Effect of UV-B radiation on the activity and isoforms of enzymes with peroxidase activity in sunflower cotyledons. Environmental and Experimental Botany 56: 174-181.
10. Ruuhola, T., Rantala, L.M., Neuvonen, S., Yang, S. and Rantala, M.J, 2009. Effects of long-term simulated acid rain on a plant-herbivore interaction. Basic and Applied Ecology 10: 589-596.
11. Grill, D., Pfanz, H., Lomsky, B., Bytnerowicz, A., Grulke, N.E. and Tausz, M, 2005. Plant responses to air pollution and global change. In: Omasa K, Nouchi I, Kok DLJ (ed.) Physiological responses of trees to air pollutants at high elevation sites, Springer Japan, pp 37-44.
12. Joshi, P.C. and Swami, A, 2007. Physiologigal responses of some tree species under roadside automobile pollution stress around city of Haridwar, India. Environmentalist 27: 365-374.
13. Pukacka, S. and Pukacki, P.M, 2000. Seasonal changes in antioxidant level of Scots pine (Pinus sylvestris L.) needles exposed to industrial pollution. I. Ascorbate and thiol content. Acta Physiologiae Plantarum 22: 451-456.
14. Kaushik, N. and Kumar, V, 2003. Khejri (Prosopis cineraria)-based agroforestry system for arid Haryana, India. Journal of Arid Environments 55: 433-440.
15. Lowry, O.H., Rosebroiugh, N.J., Farr, A.L. and Randall, R.J, 1951. Protein measurement with the Folin phenol reagent. J Biol Chem 193: 265-275.
16. Hoyle, M.C, 1972. Indoleacetic acid oxidase: A dual catalytic enzyme? Plant Physiol 50: 15-18.
17. Chen, G.N. and Asada, K, 1989. Ascorbate peroxidase in tea leaves: occurrence of two isozymes and differences in their enzymatic and molecular properties. Plant and Cell Physiology 30: 987-998.
18. Aebi, H, 1984. Catalase in vitro. Methods Enzymol 105:

121–126.

19. Smirnoff, N, 2000. Ascorbic acid: metabolism and functions of a multifaceted molecule. Curr Opin Plant Biol 3: 229-235.

20. Liu, Y.J. and Ding, H, 2008. Variation in air pollution tolerance index of plants near a steel factory: implication for landscape-plant species selection for industrial areas. Wseas Transactions on Environment and Development 4: 24-32.

21. Zhang, L.Z., Wei, N., Wu, Q.X. and Ping, M.L, 2007. Anti-oxidant response of Cucumis sativus L. to fungicide carbendazim. Pesticide Biochemistry and Physiology 89: 54-59.

22. Wu, Y.X. and Tiedemann, A.V, 2002. Impact of fungicides on active oxygen species and antioxidant enzymes in spring barley (Hordeum vulgare L.) exposed to ozone. Environmental pollution 116: 37-47.

23. Bogdanovic, J., Dikanovic, D., Maksimovic, V., Tufegdzic, S., Dokovic, D., Isajev, V. and Radotic, K, 2006. Phenolics, lignin content and peroxidase activity in Picea omorika lines. Biologia Plantarum 50(3): 461-464.

24. Tripathi, A.K. and Gautam, M, 2007. Biochemical parameters of plants as indicators of air pollution. Journal of Environmental Biology 28(1): 127-132.

25. Tiwari, S., Agrawal, M. and Marshall, F.M, 2006. Evaluation of ambient air pollution impact on carrot plants at a sub urban site using open top chambers. Environmental Monitoring and Assessment 119: 15-30.

26. Caviglia, A.M. and Modenesi, P, 1999. Oxidative stress and ascorbic acid contents in Parmotrema reticulatum and Parmelia sulcata thalli. Lichenologist 31(1): 105-110.

27. Honour, S.L., Bell, J.N.B., Ashenden, T.W., Cape, J.N. and Power, S.A, 2009. Responses of herbaceous plants to urban air pollution: effects on growth, phenology and leaf surface characteristics. Environmental pollution 157: 1279-1286.

28. Dineva, S.B, 2004. Comparative studies of the leaf morphology and structure of white ash Fraxinus americana L. and London plane tree Platanus acerifolia Willd growing in polluted area. Dendrobiology 52: 3-8.

CHAPTER 12

POTENTIAL OF AQUEOUS OZONE TO CONTROL AFLATOXIGENIC FUNGI IN NUTS

Otniel Freitas-Silva, Héctor Morales-Valle, and Armando Venâncio

INTRODUCTION

The Brazil nut (Bertholietia excelsa) is an important nontimber forest product (NTFP) from the Amazon forest. Brazil produces approximately 24% of the total world supply of this nut. Brazil nuts exploitation is an important social and economic activity for people living in forest areas. Besides this, it is an important factor in forestry conservation and environmental sustainability [1].

Despite the positive nutrition and health-related aspects of Brazil nuts, they are susceptible to colonization by mycotoxin-producing fungi and consequently to contamination with mycotoxins. There are reports of the presence of Aspergillus flavus, and A. nomius, which are the main producers of aflatoxins (AF) [2], and more recently two other aflatoxigenic species A. bombycis and A. arachidicola [3] and a new nonaflatoxigenic speciesA. bertholletius [4], all of them from this same section, were isolated and identified in Brazil nut samples.

Citation: Otniel Freitas-Silva, Héctor Morales-Valle, and Armando Venâncio, "Potential of Aqueous Ozone to Control Aflatoxigenic Fungi in Nuts, vol. 2013, Article ID 859830, 6 pages, 2013. doi:10.5402/2013/859830.

EU countries and the United States have been the major importers of in-shell and shelled Brazil nuts, respectively. AF contamination constitutes not only an economic problem for Brazil nut producing countries but also a serious health risk for consumers all over the world [5].

Both industries and producers have been making considerable efforts in the past 15 years to minimize fungal growth and AF contamination of tree nuts. Particularly in the case of Brazil nuts, the climate conditions in the Amazon environment and the characteristics of the exploitation activity (collecting and primary handling) cannot be controlled, exerting direct or indirect effects on the toxigenic fungi and on AF or multitoxin production [6].

Processes and/or treatments that have been proven to reduce AF levels in Brazil nuts include shelling and sorting by size, specific gravity, color, or damage [1]. Furthermore, several strategies have been reported for the management of fungi and mycotoxins already present in raw material and food [7]. Of the available methods to ensure food safety, ozone application is one of the most promising ones. Ozonation is a chemical means of food processing that involves exposing food or an intermediate product to O_3 [7, 8]. Ozone can be applied as a gas or dissolved in an aqueous solution. The first method is more useful, but both have been used successfully on many food products to reduce (a) postharvest diseases [9],

(b) viability of potential mycotoxin-producing fungi, and (c) mycotoxin accumulation [7].

The fact that ozonation is a nonresidual treatment, an advantage in light of the increasing concern related to the use of chemicals on food chains, makes it a promising method to treat Brazil nuts. The assays performed in the present work are focused on the use of aqueous ozone since it is known to be a strong antimicrobial agent [7,10] and its use for mycotoxin destruction has been studied earlier [11]. Also, ozone is generally recognized as a safe (GRAS) product, and it has been already used in many agricultural products, including organically labeled ones [12], like Brazil nuts. In view of this, the present work aimed to investigate the potential of aqueous ozone to control the fungi associated with Brazil nuts, focusing on Aspergillus flavus.

MATERIALS AND METHODS

The samples used were collected from the 2009 Brazil nuts harvest, obtained from an agroforestry production area in Pará state, Brazil. The initial sample weighed about 40 kg and was homogenized to obtain a working sample.

The working samples of Brazil nuts were divided into two groups:

- naturally contaminated nuts (without sterilization) and
- sterilized nuts intended for studies with artificially contaminated nuts.

For the latter group, the sterilization was carried out by autoclaving at 121°C during 20 min at 1 atm, and the contamination was with the MUM 9201 strain of A. flavus. This strain is a well-known producer of both AF and cyclopiazonic acid (CPA) [13, 14].

To obtain the stock solution, water saturated with ozone was prepared by bubbling gas, generated by the passage of extra dry oxygen through a reactor (Anseros, Model CDCOM-HF-4) for 15 minutes (power generator at 100%; gas flow of 25 Lh^{-1}) in a bottle with 1000 mL of water, at 3°C. The final aqueous ozone concentration was determined by the colorimetric method with a JASCO V560 spectrophotometer (λmax = 258 nm and ε = 2900 M^{-1} cm^{-1}) [15]. The ozone concentration in the stock solution ranged from 25 to 40 mg/L.

The different ozone concentrations required for the assays were obtained from the stock solution diluted with ozone demand-free water as necessary, until obtaining the ozone working dilution. Nonozonated water was used for control treatments. Only purified water obtained by successive reverse osmosis by Milli-Q plus system (Millipore, Molsheim, France) was used in the tests.

A MUM 9201 suspension of 1 × 10^7 conidia × mL^{-1} was prepared in peptone water (0.1%). From that, a 10x diluted conidial suspension was prepared and the number of viable conidia of both suspensions was assessed by inoculation on malt extract agar (MEA: malt extract, 20 g/L; glucose, 20 g/L; peptone, 1 g/L, agar, 20 g/L).

The high-concentrate conidial solution (HC) yielded about 400 viable conidia, whereas the low-concentrate suspension (LC) yielded about 40. From now on all the assays are based on the number of viable conidia.

Effect of Ozonation Time on Viable Conidia

Suspensions of A. flavus (MUM 9201) with 0.1 mL of either HC or LC conidia were washed in 20 mL of ozone solution for 2.5 min, 5 min, 10 min, 20 min, 30 min, or 60 min. Subsequently, the ozone solution was filtered in a Millipore filtration system. Filters (0.2 µm) with washed conidia were plated directly on MEA and incubated at 25°C for 3 days for assessment of colony-forming units (cfu). Control tests were carried out by filtration of each conidial suspension in a solution of Tween 80 in Milli-Q water (0.005% v/v). The duration of the ozone treatment was controlled by the addition of 1 mL of 1 M reagent-grade sodium formate solution to neutralize the remaining dissolved ozone after the desired contact time [16]. Three replications of each treatment were carried out, and the results underwent full factorial testing at $\alpha < 0.05$. In order to assess whether the factor time was affected by the ozone concentration (interaction between effects), the whole assay was carried out at four different ozone concentrations (0, 1, 10, and 20 mg L^{-1}). In summary, the assay consisted of two different conidial concentrations (HC or LC), six submersion times (2.5, 5, 10, 20, 30, and 60 min), four ozone concentrations (0, 1, 10, and 20 mg L^{-1}), and three replications (2 × 6 × 4 × 3 = 144 experiments).

Effect of Ozone Concentration on Recovery of Viable Conidia from Brazil Nuts

Six sterilized nuts were inoculated with 0.1 mL of either the LC or HC conidial solution of A. flavus MUM 9201. The nuts were stored overnight at room temperature. Then the nuts were submerged in 20 mL of ozone solution at different concentrations (1, 5, 10, 15, 20, 25, and 30 mg L^{-1}). Control tests were carried out by submerging nuts in a solution of Tween 80 in Milli-Q water (0.005% v/v). After 15 min, 20 mL of the ozone or Tween 80 solution was filtered (0.2 µm filter) in a Millipore filtration system, and the filtrate was plated directly on MEA and incubated at 25°C for 3 days for determination of cfu. The whole experiment was carried out twice, with three replications per assay. The effect of both the ozone concentration and initial conidial concentration on the viable spores after ozonation was studied by full factorial tests at .

Effect of Ozonation on Shell Color

The effect of the ozonation on the color of the nutshell was tested. For that purpose, 20 nuts were submerged in aqueous ozone during 30 min. Five different ozone concentrations were tested (0, 0.1, 1.0, 10, and 20 mg L^{-1}). The control sample consisted on 20 untreated nuts chosen randomly. Chromaticity values of the shells in the $L^*a^*b^*$ space coordinates were recorded [17]. The values of $L*$ (lightness or brightness) varied from black (0) to white (100), and the chroma values of $a*$ ranged from green (−60) to red (+60) and b^* ranged from blue (−60) to yellow (+60). The hue [defined as arctan(b^*/a^*)] was calculated. An Avantes spectrometer (AvaMouse 2.0, Eerbeek, Netherland) was used. The change in shell color was evaluated by comparing the parameters hue and L of the control samples with those of the treated ones. The significance of the results was assessed by t-tests for independent samples at $\alpha < 0.05$.

Assays with Inoculated and Natural Contaminated Nuts

In order to apply the correct ozone concentration during the treatment, it was necessary to know the degree of contamination of the nutshell. For that purpose, a previous assay with artificially contaminated nuts was carried out to learn the percentage of viable conidia that migrate from the nutshell to an aqueous solution. For this purpose, 100 µL of either the LC or HC conidial suspension was inoculated onto the sterilized nut surface. After one hour at room temperature, the nut was submerged in 20 mL of a solution of Tween 80 (0.005% v/v) in water. After 1 hour of shaking, this solution was filtered (0.2 µm filter) in a Millipore filtration system, and the filtrate was plated directly on MEA and incubated at 25°C for 3 days for cfu determination. The assay was carried out three times with three replications (three nuts) each time.

In parallel, to know the degree of contamination of the naturally contaminated nuts, they were submerged in 20 mL of a solution of Tween 80 (0.005% v/v) in water. Just as in the previous assay, after 1 hour shaking, 20 mL of the aqueous solution was filtered (0.2 µm filter), and the filtrate was plated directly on MEA and incubated at 25°C for 3 days for cfu determination. Three tests were performed with three nuts each. The percentage of migration obtained in the previous assay is used to assess the total initial viable conidia on the nutshell.

Next, three more sterilized nuts were inoculated with 100 µL of either the LC or HC conidial solution and left 1 hour at room temperature. In parallel, three naturally contaminated nuts were randomly picked. Each sample was submerged in 20 mL of ozone solution (20 mg L^{-1}). Each ozone solution was filtered (0.2 µm filter) in a Millipore filtration system and the filtrate was plated directly on MEA. The determination of cfu was carried out after 3 days at 25°C. The whole assay was carried out three times. The fungi isolated from the naturally contaminated samples were identified by genus according to the identification key of Pitt and Hocking [18].

The significance of the efficiency of ozonation was assessed by comparison of viable conidia in aqueous solution to viable conidia in ozone solution by means of t-test for separate samples at $\alpha < 0.05$.

RESULTS

Neither the duration of aqueous ozone treatment nor the interaction with the ozone concentration affected the efficiency of the treatment. Table 1 shows the percentage of viable spores after different treatment durations. However, the ozone concentration factor as a single effect was highly significant ($P < 0.001$) and so was further tested (see below). Therefore, the effect of the ozone against the spores was instantaneous at all ozone concentrations. This was corroborated by comparing the results of an assay in which sodium formate solution was used with the results of another assay in which the treatment duration was determined only by the time of submersion of the nuts in the ozone solution (data not shown). Thereafter, the sodium formate solution was no longer used, and the duration of the treatment (which influences the shell color) was studied

Table 1: Viable conidia of A. flavus (MUM 9201) after different treatment duration and aqueous ozone concentration

	Time (min)	Viable conidia (%) after different time exposure (min) to ozone					
		2.5 (±sd)	5 (±sd)	10 (±sd)	20 (±sd)	30 (±sd)	60 (±sd)
Ozone (mg L^{-1})	0	96 ± 7	94 ± 2	92 ± 0	93 ± 5	97 ± 5	94 ± 3
	1	51 ± 1	69 ± 2	64 ± 20	59 ± 4	58 ± 7	61 ± 3
	10	7 ± 4	5 ± 2	6 ± 6	5 ± 3	3 ± 2	2 ± 2
	20	2 ± 3	1 ± 0	2 ± 1	2 ± 1	1 ± 1	3 ± 1

The effect of the ozone concentration on the treatment efficiency was highly significant ($P < 0.001$). In general, whatever the initial charge of spores, the higher the ozone concentration, the lower the final concentration of viable conidia (Figure 1).

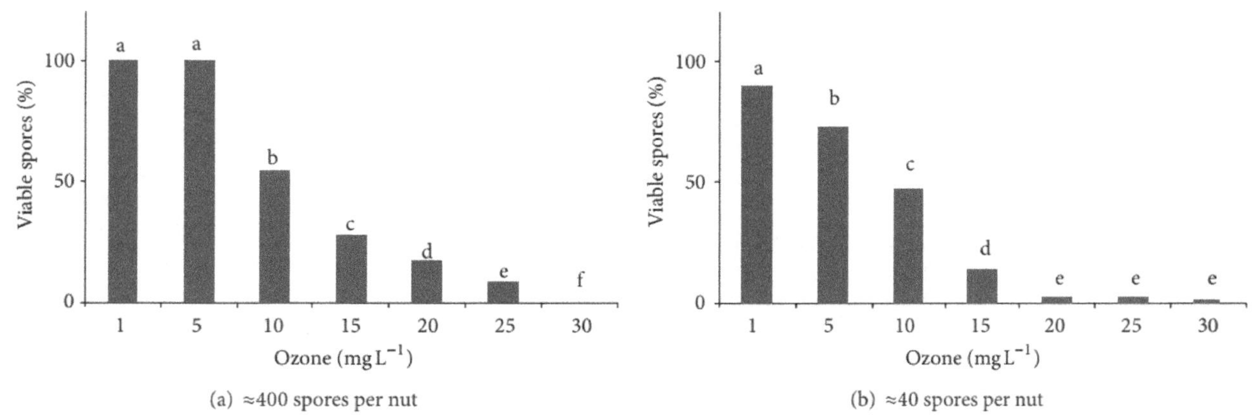

(a) ≈400 spores per nut (b) ≈40 spores per nut

Figure 1: Percentiles of A. flavus viable conidia recovered after aqueous ozone exposure up to 30 mg L^{-1} in high (a) and low (b) concentration of conidia per nut.

However, the interaction of ozone concentration with initial viable spores was also highly significant ($P < 0.001$). This interaction showed that for lower initial viable spore charge it was enough to use an ozone concentration of 20 mg L^{-1}, as no significantly better results were observed with higher concentrations.

The submersion of LC or HC inoculated nuts in a solution of Tween 80 showed that, whatever the initial number of viable spores (HC or LC), about 60% of them migrated to the water solution. The submersion of naturally contaminated nuts in the Tween 80 solution yielded 32 ± 5 conidia per nut. This means the initial contamination was 53 ± 9 conidia per nut. According to the results obtained with different ozone concentrations (Figure 1), a concentration of 20 mg L−1 of ozone was chosen to assay with HC and LC inoculated nuts and naturally contaminated nuts. The most frequently isolated filamentous fungi from the naturally contaminated nuts were, in descending order, Aspergillus section Nigri, Cunninghamella sp, Penicillium spp., and Aspergillus section Flavi, with incidences of 26%, 21%, 16%, and 5%, respectively. The rest of the fungal isolates (33%) were yeasts.

The results of submersion of (a) nuts artificially inoculated with LC solution of A. flavus MUM 9201, (b) naturally contaminated nuts, and (c) nuts artificially contaminated with HC suspension of A. flavus MUM 9201 in a 20 mg L^{-1} ozone solution showed almost 0 viable spores after treatment in cases (a) and (b) and about 3 viable spores per nut in case (c) (Figure 2). In all cases, the reduction of viable spores was highly significant ($P < 0.001$).

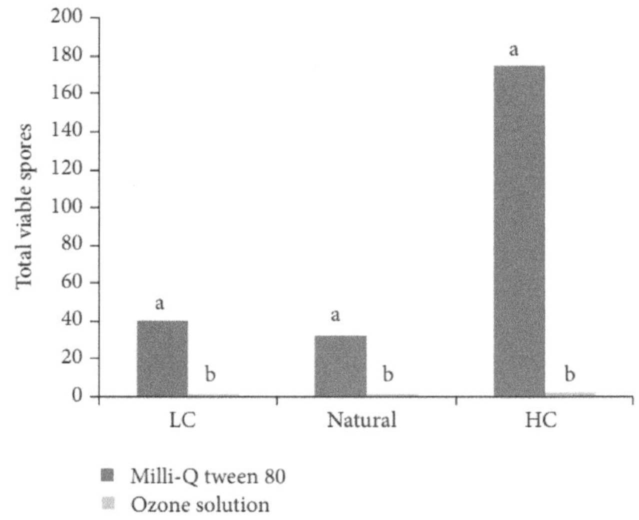

Figure 2: Reduction of viable spores on Brazil nuts from natural population and inoculated A. flavus (MUM 9201) after aqueous ozone 20 mg mL^{-1}. The low-concentrate suspension (LC) and the high-concentrate conidial solution (HC).

The influence of the aqueous ozone on the shell color was evident at higher ozone concentrations. The values of the parameter L increased with rising ozone concentration and significantly differed from the control ones at concentrations higher than 10 mg L^{-1}. On the other hand, the hue values decreased with increasing ozone concentration and significantly differed from control ones at concentrations of 10 mg L^{-1} and higher (Figure 3).

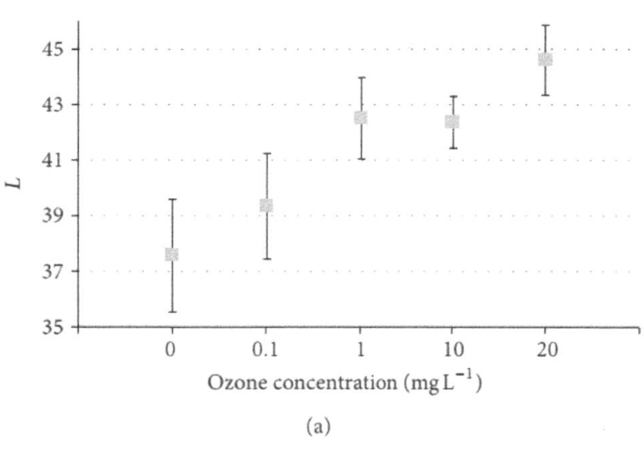

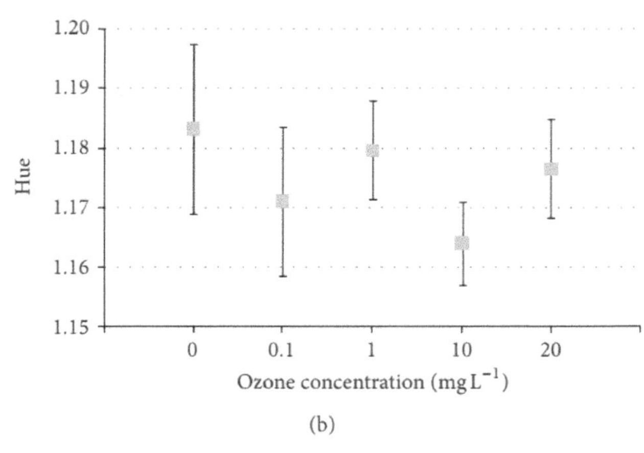

Figure 3: Effect of aqueous ozone concentration on color parameters (a) L and (b) hue.

DISCUSSION

The genus Aspergillus is mainly responsible for fungal contamination and subsequent production of AF in the field, at harvest, and during postharvest operations and storage of nuts [13]. Due to high potential risk of nut contamination, decontamination methods are of great interest from economic, public health, and environmental aspects. Improving postharvest processing followed by further prevention of fungal growth with ozone treatment is one of the most effective ways to restrict AF contamination and would go a long way to reduce health-related risks from eating contaminated nuts and economic losses during production [19].

This work focused on fungal decontamination of Brazil nuts using different ozone concentrations. The use of ozone in dried fruits and nuts has been studied before. The contamination of dried figs using 0.01 to 0.02 g $m^3$of gaseous O_3 for 3 and 4 h resulted in the reduction of total mycobiota counts [20]. When the efficiency of gaseous ozone was evaluated for the degradation of AF on pistachio kernels and ground pistachios [21], pistachios were contaminated with known concentrations of AFB_1, AFB_2, AFG_1, and AFG_2, and then samples were exposed to gaseous ozone at 5.0, 7.0, and 9.0 g m^{-3} concentrations for 140 and 420 min at 20°C and 70% RH. This works although presented a significant reduction in total AF and in AFB_1 for pistachio kernels, no significant change in the fatty acid composition was observed. Likewise, no significant changes were found in sweetness, rancidity, flavor, appearance and overall palatability between ozonated and nonozonated pistachio kernels.

In Brazil nuts, other authors have studied the potential of gaseous ozone. In research conducted by Giordano and coauthors [22] they had shown that O_3 treatment for 5 h at 31 mg L^{-1} inhibited the viability of fungi of the nut-contaminating mycobiota, including the aflatoxigenic Aspergillus. In another work [23] also was reported the potential of gaseous ozone to treat Brazil nuts and considered the best method, especially when associated with vacuum. According to these authors, ozone treatment was able to (a) degrade AF (15 μg kg^{-1}), (b) reduce fungi and yeast spores from 1.8×10^4 cfu g^{-1} to no growth, (c) keep the fatty acid oxidation indicator (i.e., malondialdehyde) stable, and (d) even improve the sensory attributes for consumer acceptance. Such features are of great importance for the gaseous ozone treatment to be considered a good sanitizer. Unfortunately, the use of gaseous ozone in some cases could cause toxicity since the concentrations of gaseous O_3 necessary to inactivate the conidia of Penicillium digitatum, P. italicum, and Botrytis cinerea from table grapes were relatively high. The need for such high O_3 concentration makes it hard to achieve complete containment of the gas, and thus the worker safety may be jeopardized [24]. Gaseous ozone treatment needs some adjustments in its application to comply with regulations since O_3 concentrations cannot exceed 0.075 μLL^{-1} during an 8 h workday [25].

According to our results, the reaction to the ozone is almost instantaneous. The application of aqueous ozone treatments in industries does not seem to be very complicated. It is necessary to learn the degree of contamination of the nuts in advance, because if it is not very high, lower ozone concentrations can be used. For the same reason it is necessary to learn which degree of contamination is acceptable after treatment, because higher ozone concentrations will be required if lower levels of spores per nut are to be achieved.

Although the effect of gaseous ozone on red pepper treated with concentrations up to 66 g·m^3 for up to 1 hour did not significantly change the product's color [26], it seems that some changes could happen in the case of Brazil nuts. Such effect could depend on the ozone concentration as well as the submersion time. Taking into account that the effect of ozone against conidia seems to be immediate, it seems feasible to reduce color changes by minimizing as much as possible the submersion time and adjusting the ozone concentration to the minimum needed for the initial degree of contamination.

When the effectiveness of gaseous and aqueous ozone on mycobiota and AFB_1 content of dried figs [27] was studied, in both treatments, degradation of AFB_1 increased with increasing ozonation time. Long gaseous ozone treatment, as mentioned previously, is hard to control and can jeopardize worker safety. These authors [27] stated that gaseous ozone was more effective than ozonated water for reduction of AFB_1, whereas ozonated water was more effective for decreasing microbial counts.

Our results show that the use of aqueous ozone is a viable alternative technique to control fungi on Brazil nuts, showing optimum control of a large spectrum of fungi and good applicability to complex matrices such as nuts. The aqueous O_3 solution can also be recommended to control other mycotoxin-producing Aspergilli. However, additional experiments are needed to adjust the method for use in Brazil nut packing houses.

Food-related application of ozone is currently restricted in packing houses, especially for leaf and fruit sterilization [7]. The food processing industry constantly searches for new technologies to improve commercial sterilization of food commodities, especially those associated with mycotoxins and other hazards.

As shown in this paper, aqueous ozone application with accurate O_3 dosification can be an efficient method for disinfection of dry food surfaces. O_3 disinfection of food surfaces opens up an alternative for the elimination of fungal contamination since ozone application on food is easy and, according to the results shown here, effective at low doses and short durations, yielding no residues on the product or in the environment, for which reason it has GRAS certification [12].

More research must be undertaken to evaluate this technique to become common in the food industry, to determine the ideal concentration and projected cost of large-scale treatment of Brazil nuts.

ACKNOWLEDGMENTS

H. Morales-Valle is grateful to the Portuguese Government (Ministério da Ciência, Tecnologia e Ensino Superior; Fundação para a Ciência e a Tecnologia FCT) Grant ref. SFRH/BPD/38011/2007).

REFERENCES

1. O. Freitas-Silva and A. Venâncio, "Brazil nuts: benefits and risks associated with contamination by fungi and mycotoxins," Food Research International, vol. 44, no. 5, pp. 1434–1440, 2011. ·
2. M. Olsen, P. Johnsson, T. Möller, R. Paladino, and M. Lindblad, "Aspergillus nomius, an important aflatoxin producer in Brazil nuts?" World Mycotoxin Journal, vol. 1, pp. 123–126, 2008.
3. J. S. Gonçalves, L. M. Ferracin, M. L. C. Vieira, B. T. Iamanaka, M. H. Taniwaki, and M. H. P. Fungaro, "Molecular analysis of Aspergillus section Flavi isolated from Brazil nuts," World Journal of Microbiology and Biotechnology, vol. 28, no. 4, pp. 1817–1825, 2012.
4. M. H. Taniwaki, J. I. Pitt, B. T. Iamanaka et al., "Aspergillus bertholletius sp. nov. from Brazil Nuts," PLoS ONE, vol. 7, no. 8, Article ID e42480, 2012.
5. O. Freitas-Silva, A. Teixeira, F. Q. da Cunha, R. L. O. de Godoy, and A. Venâncio, "Predominant mycobiota and aflatoxin content in Brazil nuts," Journal for Consumer Protection and Food Safety, vol. 6, no. 4, pp. 465–472, 2011.
6. O. Freitas-Silva, M. D. L. M. de Souza, and A. Venâncio, "Tracing fungi secondary metabolites in Brazil nuts using LC-MS/MS," Drug Metabolism Letters, vol. 5, no. 3, pp. 150–155, 2011. ·
7. O. Freitas-Silva and A. Venâncio, "Ozone applications to prevent and degrade mycotoxins: a review,"Drug Metabolism Reviews, vol. 42, no. 4, pp. 612–620, 2010.

8. Z. B. Guzel-Seydim, A. K. Greene, and A. C. Seydim, "Use of ozone in the food industry," Lebensmittel-Wissenschaft und-Technologi, vol. 37, no. 4, pp. 453–460, 2004.

9. J.-G. Kim, A. E. Yousef, and S. Dave, "Application of ozone for enhancing the microbiological safety and quality of foods: a review," Journal of Food Protection, vol. 62, no. 9, pp. 1071–1087, 1999.

10. M. A. Khadre, A. E. Yousef, and J.-G. Kim, "Microbiological aspects of ozone applications in food: a review," Journal of Food Science, vol. 66, no. 9, pp. 1242–1252, 2001.

11. K. S. McKenzie, A. B. Sarr, K. Mayura et al., "Oxidative degradation and detoxification of mycotoxins using a novel source of ozone," Food and Chemical Toxicology, vol. 35, no. 8, pp. 807–820, 1997.

12. F. M. Gabler, J. L. Smilanick, M. F. Mansour, and H. Karaca, "Influence of fumigation with high concentrations of ozone gas on postharvest gray mold and fungicide residues on table grapes,"Postharvest Biology and Technology, vol. 55, no. 2, pp. 85–90, 2010.

13. P. Rodrigues, A. Venâncio, Z. Kozakiewicz, and N. Lima, "A polyphasic approach to the identification of aflatoxigenic and non-aflatoxigenic strains of Aspergillus section Flavi isolated from Portuguese almonds," International Journal of Food Microbiology, vol. 129, no. 2, pp. 187–193, 2009.

14. C. Soares, P. Rodrigues, O. Freitas-Silva, L. Abrunhosa, and A. Venâncio, "HPLC method for simultaneous detection of aflatoxins and cyclopiazonic acid from Aspergillus section Flavi," World Mycotoxin Journal, vol. 3, no. 3, pp. 225–231, 2010.

15. H. Bader and J. Hoigne, "Determination of ozone in water by the indigo method," Water Research, vol. 15, no. 4, pp. 449–456, 1981.

16. K. Biswas, S. Craik, D. W. Smith, and M. Belosevic, "Synergistic inactivation of Cryptosporidium parvumusing ozone followed by free chlorine in natural water," Water Research, vol. 37, no. 19, pp. 4737–4747, 2003.

17. R. G. McGuire, "Reporting of objective colour measurements," HortScience, vol. 27, pp. 1254–1255, 1992.

18. J. I. Pitt and A. D. Hocking, Fungi and Food Spoilage, Blackie Academic and Professional, London, UK, 3rd edition, 2007.

19. P. Basaran, N. Basaran-Akgul, and L. Oksuz, "Elimination of Aspergillus parasiticus from nut surface with low pressure cold plasma (LPCP) treatment," Food Microbiology, vol. 25, no. 4, pp. 626–632, 2008.

20. S. Öztekin, B. Zorlugenç, and F. K. Zorlugenç, "Effects of ozone treatment on microflora of dried figs,"Journal of Food Engineering, vol. 75, pp. 396–399, 2006.

21. M. Y. Akbas and M. Ozdemir, "Effect of different ozone treatments on aflatoxin degradation and physicochemical properties of pistachios," Journal of the Science of Food and Agriculture, vol. 86, no. 13, pp. 2099–2104, 2006.

22. B. N. E. Giordano, V. Simao, D. Manfio, et al., "Reduction of in-shell Brazil nut (Bertholletia excelsaH.B.K.) aflatoxin contamination by ozone gas application during storage," Julius-Kühn-Archiv, vol. 425, pp. 566–572, 2010.

23. V. M. Scussel, B. N. Giordano, V. Simao, et al., "Effect of oxygen-reducing atmospheres on the safety of packaged shelled Brazil nuts during storage," International Journal of Analytical Chemistry, vol. 2011, Article ID 813591, 9 pages, 2011.

24. R. Ozkan, J. L. Smilanick, and O. A. Karabulut, "Toxicity of ozone gas to conidia of Penicillium digitatum,Penicillium italicum, and Botrytis cinerea and control of gray mold on table grapes," Postharvest Biology and Technology, vol. 60, no. 1, pp. 47–51, 2011.

25. USEPA, "National ambient air quality standards for ozone, final rule," Federal Register, vol. 73, pp. 16436–16514, 2008.

26. F. Inan, M. Pala, and I. Doymaz, "Use of ozone in detoxification of aflatoxin B_1 in red pepper," Journal of Stored Products Research, vol. 43, no. 4, pp. 425–429, 2007. ·

27. B. Zorlugenç, F. Kirolu Zorlugenç, S. Öztekin, and I. B. Evliya, "The influence of gaseous ozone and ozonated water on microbial flora and degradation of aflatoxin B_1 in dried figs," Food and Chemical Toxicology, vol. 46, no. 12, pp. 3593–3597, 2008.

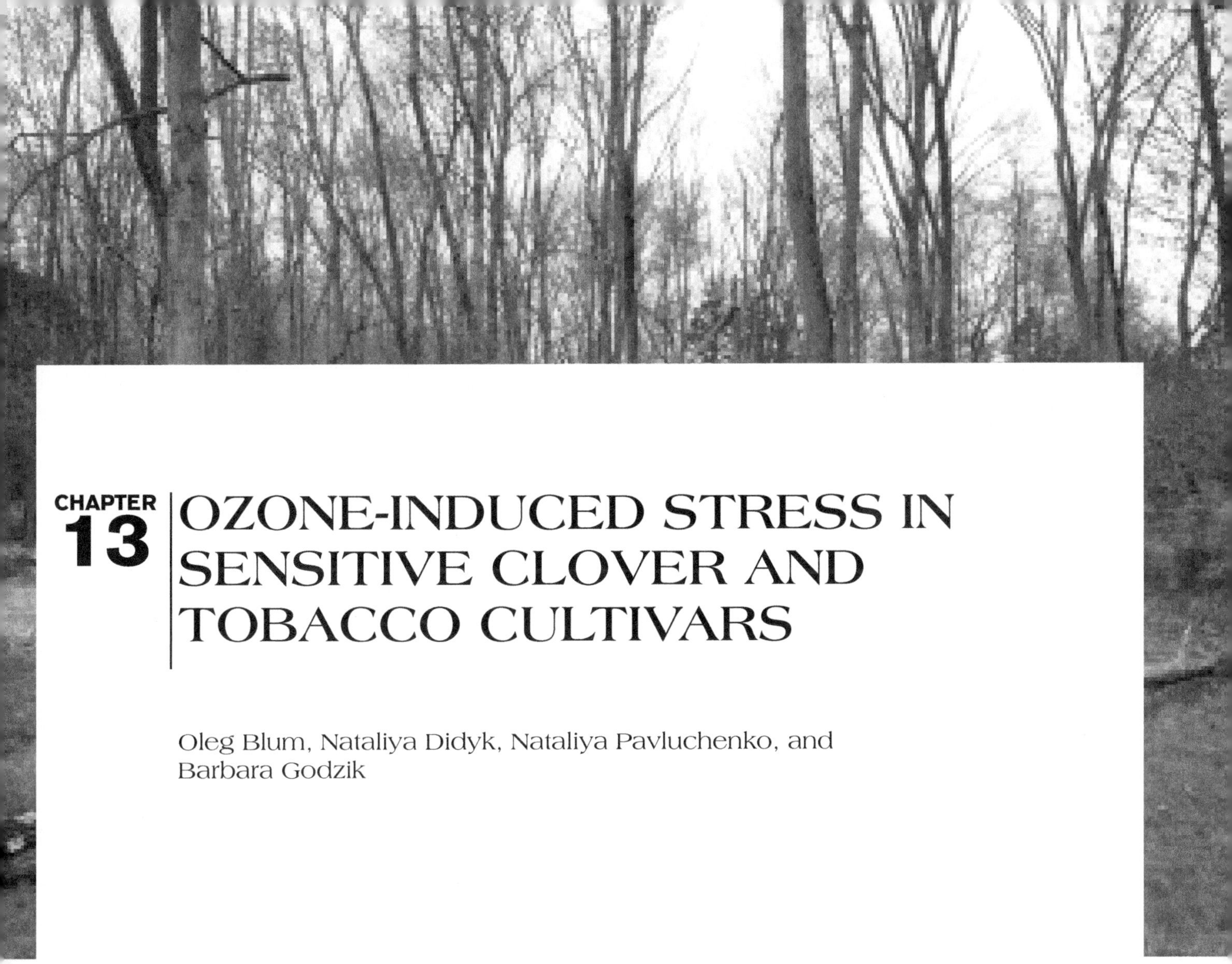

CHAPTER 13 OZONE-INDUCED STRESS IN SENSITIVE CLOVER AND TOBACCO CULTIVARS

Oleg Blum, Nataliya Didyk, Nataliya Pavluchenko, and Barbara Godzik

INTRODUCTION

Ozone is one of the major toxic to vegetation and human health gaseous air pollutants. It contributes to crop losses and forest decline. It was estimated that the present day global relative yield losses range between 7% and 12% for wheat, between 6% and 16% for soybean, between 3% and 4% for rice, and between 3% and 5% for maize (range resulting from different metrics used) [1]. For the year 2000 global crop yield loss due to ambient ozone was estimated to be worth \$14–26 billion. About 40% of this damage is occurring in China and India [1]. Crop yield loss due to ambient ozone in the U.S.A. was estimated to be worth \$3–5 billion annually [2].

The ozone-induced crop losses could be mitigated by two ways:

- selection of O_3-tolerant cultivars;
- application of chemical protectants.

So far there has been only limited success in producing genetically transformed plants with increased tolerance to elevated ozone concentrations because plant resistance to ozone is a very complicated phenomenon involving multiple signaling pathways and defense responses [3, 4].

The phytotoxicity of ozone results primarily from the oxidative stress imposed by the pollutant on sensitive components of the plasmalemma [3, 5]. In connection to this, application of antioxidants for protection of crops from ozone injury has been extensively studied over the last four decades [6, 7]. A large number of antioxidants (fungicides, insecticides, herbicides, growth regulators, etc.) have been evaluated [6]. Among them the systemic antioxidant ethylene diurea—N-[2-(2-oxo-1-imidazolidinyl) ethyl]-N′ phenylurea (EDU) was found to be the most efficient [6]. EDU, applied as a foliar spray, soil drench, or stem infusion agent, was shown to prevent acute ozone injury and inhibit plant senescence [6, 8]. Physiological effects of EDU associated with its protective properties are still unclear. Yet, there is some evidence that EDU may confer tolerance to ozone through the induction of enzyme systems involved in the elimination of activated oxygen species and free radicals [9].

Presently, modern agrochemicals containing antioxidant compounds, such as fungicides belonging to strobilurins and triazoles (azoxystrobin, epoxiconazole, penconazole, etc.), are gaining increasing attention as possible ozone protectants [10]. The above-mentioned fungicides, especially strobilurins, which

Citation: Oleg Blum, Nataliya Didyk, Nataliya Pavluchenko, and Barbara Godzik, "Ozone-Induced Stress in Sensitive Clover and Tobacco Cultivars, vol. 2011, Article ID 308598, 4 pages, 2011. doi:10.1155/2011/308598.

were developed on the basis of natural substances extracted from the fungus Strobilurus tenacellus (Pers.) Singer, have low toxicity for human health and environment. Therefore they are more preferable for application in agriculture as protectants, than other synthetic agrochemicals such as EDU. In spite of the above-said a rather limited number of studies have been devoted to the protective effects of strobilurins against oxidative stress from the environment. Yet it was shown that application of strobilurins to plants caused marked enhancement of antioxidative enzymes and enhanced scavenging of potentially harmful reactive oxygen species.

Another promising group of agrochemicals is growth regulators containing phytohormones or/and others antioxidants. Apart from their effects on plant development, some phytohormones, that is, cytokinins, gibberellins, salicylic acid and their synthetic analogs were shown to enhance tolerance of crops to abiotic stresses including ambient ozone [6, 7]. The aim of this study was to assess protective properties of some modern agrochemicals with antioxidant potential including three fungicides belonging to strobilurins and some new growth regulators in comparison with EDU.

MATERIALS AND METHODS

The experiments were conducted simultaneously on two monitoring stations located in semiurban (National Botanic Garden in Kyiv city, Ukraine) and rural (Field Station of Institute of Botany Polish Academy of Sciences in Szarów, 30 km from Krakow, Poland) sites. In Kyiv plants of subterranean clover (Trifolium subterraneum L., O_3-sensitive cv. Geraldton) and tobacco (Nicotiana tabacum L.) cvs. Bel-W3 (O_3-sensitive) and Bel-B (O_3-resistant) were used. In Szarów only subterranean clover was used. Prior to exposure to ozone, tobacco and clover were cultivated in ozone free air until they reach 4 leaves. In Kyiv plants were cultivated in the special chamber, made from organic glass, with the volume of 1.1 m^3, at 25–28°C temperature, 65–70% RH, light illumination of 66 $\mu mol\,m^{-2}{\cdot}s^{-1}$ PAR and 14:10 (L:D) h photoperiod. The air was pumped up through charcoal filter with the rate of 0.3 m^3 per min. In Szarów plants were cultivated in the special ozone free greenhouse at temperature of 25 ± 2°C with a relative humidity of 75% under natural photoperiod. Afterwards, the plants were sprayed with distilled water (control) or solutions of one of the following agrochemicals: (1) EDU (ethylenediurea), in concentrations of 150 $mg\,L^{-1}$; (2) "Emistym C", a natural growth regulator containing exometabolites of micorhizal fungi (flavonoids, phytohormones and organic acids), in concentration of 30 and 150 $mg\,L^{-1}$; (3) "Agrostimulin," containing synthetic growth regulator—ivin, in concentration of 30 and 150 $mg\,L^{-1}$; modern fungicides belonging to the group of strobilurins such as (4) "Quadris" (containing 25% of azoxystrobin) in concentration of 150 and 300 $mg\,L^{-1}$; (5) "Flint" (containing 50% of trifloxystrobin) in concentration of 150 and 300 $mg\,L^{-1}$; (6) "Strobi" (containing 50% of kresoxim-methyl) in concentration of 150 and 300 $mg\,L^{-1}$. Ozone protective effect of "Emistym C" and "Agrostimulin" has been studied for the first time.

Two modes of agrochemical application (except for EDU) were studied. In Kyiv higher concentrations of the chemicals were applied only once (one day before exposure). In Sharów lower concentrations of agrochemicals were applied twice (one day before exposure and on 16th day of exposure). We used EDU as a standard ozone protectant, in concentration, whose effectiveness is well established on many plant species [6, 8].

The test plants were exposed into the field on the following day after application of agrochemicals. The duration of exposure of the test plants to ambient ozone was 2 and 4 weeks for tobacco and clover, respectively. On the last day of exposure ozone-induced foliar injury (%) and content of photosynthetic pigments (chlorophylls a and b, carotenoids) in leaves of test plants were evaluated. Photosynthetic pigments were extracted by 100% acetone and determined spectrophotometrically using "Spekol 11" (Carl zeiss, Jena, Germany) [11]. Dry biomass of test plants was also determined. For clover eight replications of ten plants each per treatment were used in both locations. Thus, values reported are the means of 80 plants. For tobacco ten replications of one plant each per treatment were used. Thus, values reported are the means of 10 plants.

During the period of the experiments ambient ozone concentrations were continuously monitored at the both locations. In Kyiv and Szarów ozone monitoring was conducted using the Thermo Environmental UV Photometric Ozone Analyzer Model 49-003 and Model 49C, correspondingly. Average daily ozone and temperature concentrations as well as the doses of ozone, which test plants received during the period of experimentation were calculated using index of AOT 40 (accumulated exposure over a threshold of 40 ppb).

Statistical analysis of the results obtained was conducted with the usage of descriptive statistics and analysis of variance. Treatment means were compared using ANOVA and LSD-test (Statistica 6.0 software) [12].

RESULTS AND DISCUSSION

For the most part of the periods of observations the average daily ambient ozone concentrations exceeded the threshold of subterranean clover and tobacco Bel-W3 cv. sensitivity to ozone of 25 ppb [13, 14] in both experimental sites. The dose of ozone (calculated using index of AOT 40), which test plants received during the exposition periods, was 828 ppb·h for clover and 700 ppb·h for tobacco in Kyiv, and 1662 ppb·h for clover in Sharów. Despite the fact that in Szarów ambient ozone concentrations during the period of experimentation were higher than in Kyiv, clover plants in control (without application of agrochemicals) displayed similar degree of visible foliar injury (about 28%) in both sites (Figure 1). However, dry phytomass of clover plants in Sharów was almost two times lower than that in Kyiv. O_3-sensitive tobacco plants demonstrated higher sensitivity to ozone than clover in terms of visible foliar injuries. All the applied agrochemicals (with the exception of "Quadris", which have no significant effect on foliar injuries and biomass, accumulation in tobacco) showed partial protective effect to O_3-sensitive plants. The plants treated with the solutions of the mentioned agrochemicals had less foliar injuries, higher biomass and higher content of photosynthetic pigments (chlorophylls a and b) in leaves as compared to control (sprayed with distil water) (Figures 1, 2, and3). Tobacco plants treated with solutions of "Emistym C" had even higher biomass than plants of O_3-resistant Bel-B cv. treated with distill water, which was evidently due to the growth promoting effect of this agrochemical rather than its ozone protective effect. Among the studied agrochemicals the most effective as protectant against ozone-induced injuries was fungicide "Strobi". Natural growth regulator "Emistym C" was slightly less effective. Fungicide "Quadris" demonstrated the lowest effectiveness.

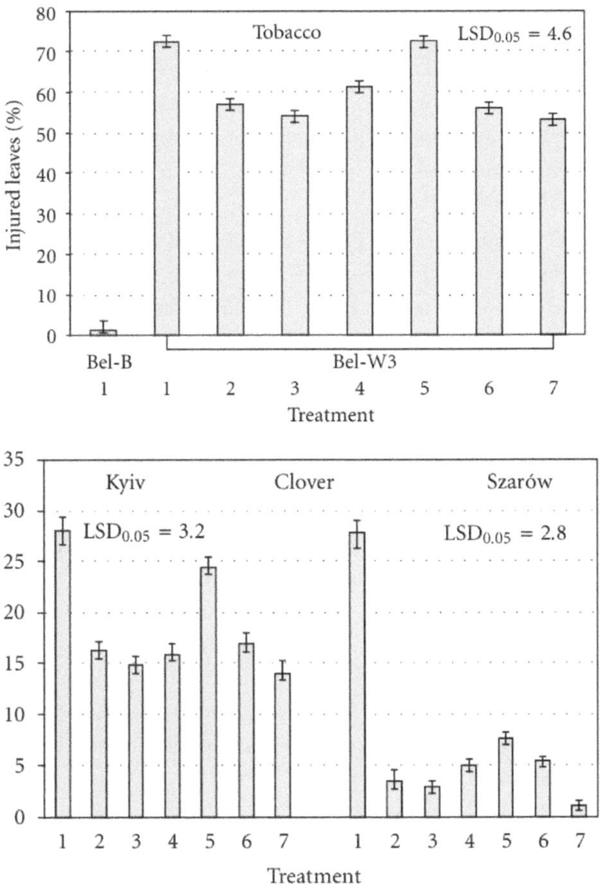

Figure 1: Ozone-induced visible injuries on leaves of clover and tobacco sprayed with distilled water (1), solutions of EDU, 150 mg L^{-1} (2), "Emistym C", 150 mg L^{-1} in Kyiv and 30 mg L^{-1} in Szarów (3), "Agrostymulin", 150 mg L^{-1} in Kyiv and 30 mg L^{-1} in Szarów (4), "Quadris", 300 mg L^{-1} in Kyiv and 150 mg L^{-1} in Szarów (5), "Flint", 300 mg L^{-1} in Kyiv and 150 mg L^{-1} in Szarów (6), "Strobi", 300 mg L^{-1} in Kyiv and 150 mg L^{-1} in Szarów (7) on the last day of exposure to ambient ozone. Vertical bars represent standard error. $LSD_{0,05}$: least significant difference at $P < 0.05$.

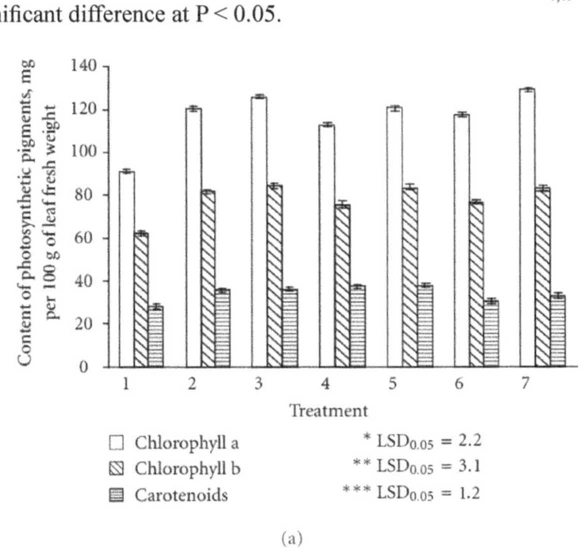

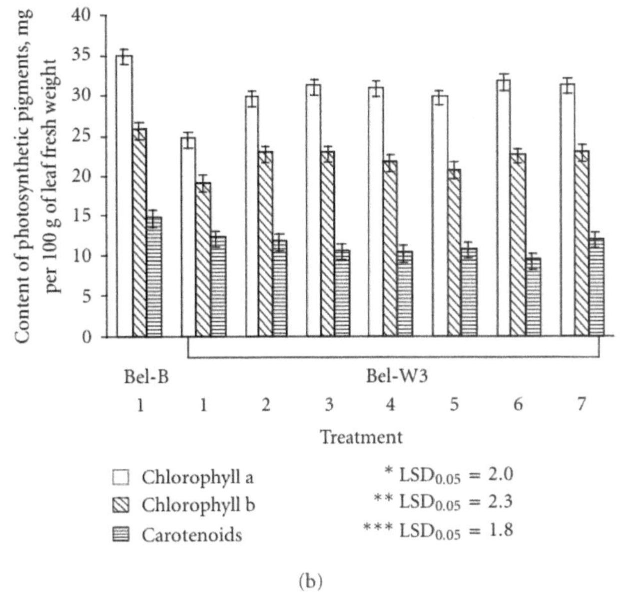

Figure 2: Content of photosynthetic pigments in leaves of clover (a) and tobacco (b) sprayed with distilled water (1), solutions of EDU, 150 mg L^{-1} (2), "Emistym C", 150 mg L^{-1} (3), "Agrostymulin" 150 mg L^{-1} (4), "Quadris", 300 mg L^{-1} (5), "Flint", 300 mg L^{-1} (6), "Strobi", 300 mg L^{-1} (7) on the last day of exposure to ambient ozone in Kyiv. Vertical bars represent standard error. $LSD_{0.05}$: least significant difference at $P < 0.05$ for *chlorophyll a; **chlorophyll b; ***carotenoids.

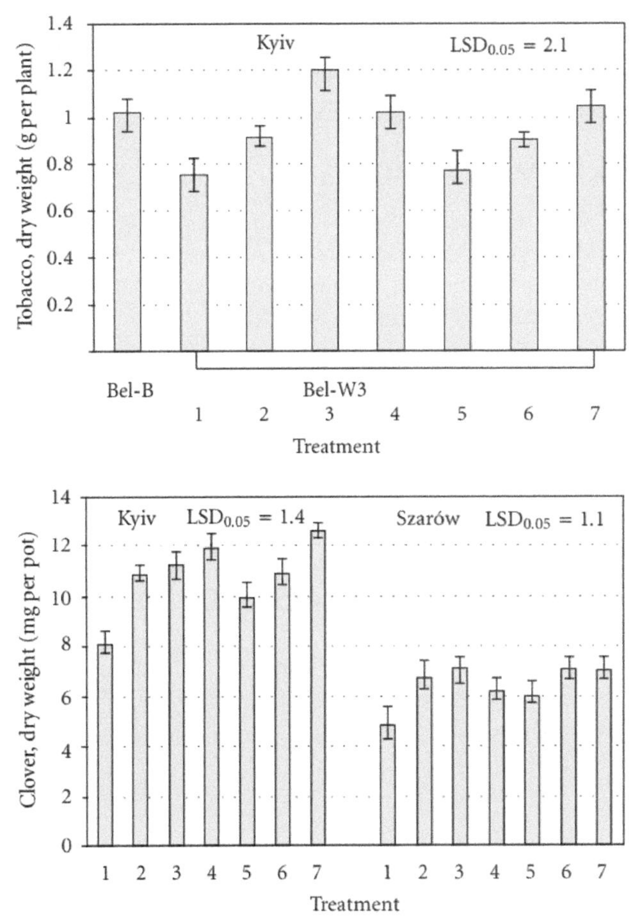

Figure 3: Mean dry weight of clover and tobacco sprayed with distilled water (control) (1), solutions of EDU, 150 mg L^{-1} (2), "Emistym C", 150 mg L^{-1} in Kyiv and 30 mg L^{-1} in Szarów (3), "Agrostymulin",

150 mg L^{-1} in Kyiv and 30 mg L^{-1} in Szarów (4), "Quadris", 300 mg L^{-1} in Kyiv and 150 mg L^{-1} in Szarów (5), "Flint", 300 mg L^{-1} in Kyiv and 150 mg L^{-1} in Szarów (6), "Strobi", 300 mg L^{-1} in Kyiv and 150 mg L^{-1} in Szarów (7) on the last day of exposure to ambient ozone. Vertical bars represent standard error. $LSD_{0.05}$— least significant difference at $P < 0.05$.

In terms of visible foliar injuries and biomass accumulation, all the applied agrochemicals more effectively protected clover plants when they were applied two times (one day before exposure and on 16th day of exposure) in lower concentration (in Szarów) than when they were applied once (one day before exposure) in higher concentration (in Kyiv). For biomass accumulation, this tendency was less defined. It could be supposed that repeated application of the tested agrochemicals even in lower concentrations is more effective as it allows protection of newly emerged leaves. Though, the differences in ozone doses, which plants received in both locations, could also contribute to the observed effects. The degree of protective effect for all tested agrochemicals was higher in clover plants as compared to tobacco. Evidently it was a consequence of higher degree of ozone injuries in tobacco as compared to clover.

Thus, conducted studies confirmed the effectiveness of ozone protective effect of EDU and fungicides belonging to the group of strobilurins [6, 8, 10]. The protective effect of growth regulators "Emistym C" and "Agrostimulin" against ozone damage of plants has been shown for the first time. Out of the studied agrochemicals fungicide "Strobi" and natural growth regulator "Emistym C" demonstrated the best protective effects. These agrochemicals present promise for further studies of their possible utilization for enhancement of ozone tolerance of sensitive crops.

ACKNOWLEDGMENTS

This study was funded by STCU (Project N3894) and by the "M. M. Gryshko" National Botanical Garden of the National Academy of Sciences of Ukraine and by the W. Szafer Institute of Botany of Polish Academy of Sciences.

REFERENCES

1. R. Van Dingenen, F. J. Dentener, F. Raes, M. C. Krol, L. Emberson, and J. Cofala, "The global impact of ozone on agricultural crop yields under current and future air quality legislation," Atmospheric Environment, vol. 43, no. 3, pp. 604–618, 2009. · ·
2. E. L. Fiscus, F. L. Booker, and K. O. Burkey, "Crop responses to ozone: uptake, modes of action, carbon assimilation and partitioning," Plant, Cell and Environment, vol. 28, no. 8, pp. 997–1011, 2005. · ·
3. Z. Chen and D. R. Gallie, "Increasing tolerance to ozone by elevating foliar ascorbic acid confers greater protection against ozone than increasing avoidance," Plant Physiology, vol. 138, no. 3, pp. 1673–1689, 2005. · ·
4. N. Tosti, S. Pasqualini, A. Borgogni et al., "Gene expression profiles of O_3-treated Arabidopsis plants,"Plant, Cell and Environment, vol. 29, no. 9, pp. 1686–1702, 2006. · ·
5. M. C. Puckette, Y. Tang, and R. Mahalingam, "Transcriptomic changes induced by acute ozone in resistant and sensitive Medicago truncatula accessions," BMC Plant Biology, vol. 8, article 46, 2008. · ·
6. D. Archambault, D. J. Slaski, and J. J. Li, "Ozone protection in plants. The potential use of chemical protectants to measure oxidant damage in Alberta crops," Report prepared for the Air Research Users Group, Alberta Environment, Edmonton, Canada, 2000.
7. N. P. Didyk and O. B. Blum, "Natural antioxidants of plant origin against ozone damage of sensitive crops," Acta Physiologiae Plantarum, vol. 33, no. 1, pp. 25–34, 2011. · ·
8. B. Godzik and W. J. Manning, "Relative effectiveness of ethylenediurea, and constituent amounts of urea and phenylurea in ethylenediurea, in prevention of ozone injury to tobacco," Environmental Pollution, vol. 103, no. 1, pp. 1–6, 1998. · ·
9. B. D. Whitaker, E. H. Lee, and R. A. Rowland, "EDU and ozone protection: foliar glycerolipids and steryl lipids in snapbean exposed to O_3," Physiologia Plantarum, vol. 80, no. 2, pp. 286–293, 1990.
10. Y. X. Wu and A. Von Tiedemann, "Impact of fungicides on active oxygen species and antioxidant enzymes in spring barley (Hordeum vulgare L.) exposed to ozone," Environmental Pollution, vol. 116, no. 1, pp. 37–47, 2002. · ·
11. M. M. Musiyenko, T. B. Parshikova, and P. C. Slavnyi, Spectrophotometric Methods in the Practice of Physiology, Biochemistry and Ecology of Plants, Phytosociotsentr, Kyiv, Ukraine, 2001.
12. G. N. Zaytsev, Mathematical Statistics in Experimental Botany, Nauka, Moscow, Russia, 1984.
13. G. P. Karlsson, H. Pleijel, E. Sild et al., "Clover Sweden—a national three-year study of the effects of tropospheric ozone on Trifolium subterraneum, L.," Water, Air, and Soil Pollution, vol. 85, no. 3, pp. 1503–1508, 1995. · ·
14. H. E. Heggestad, "Origin of Bel-W3, Bel-C and Bel-B tobacco varieties and their use as indicators of ozone," Environmental Pollution, vol. 74, no. 4, pp. 264–291, 1991.

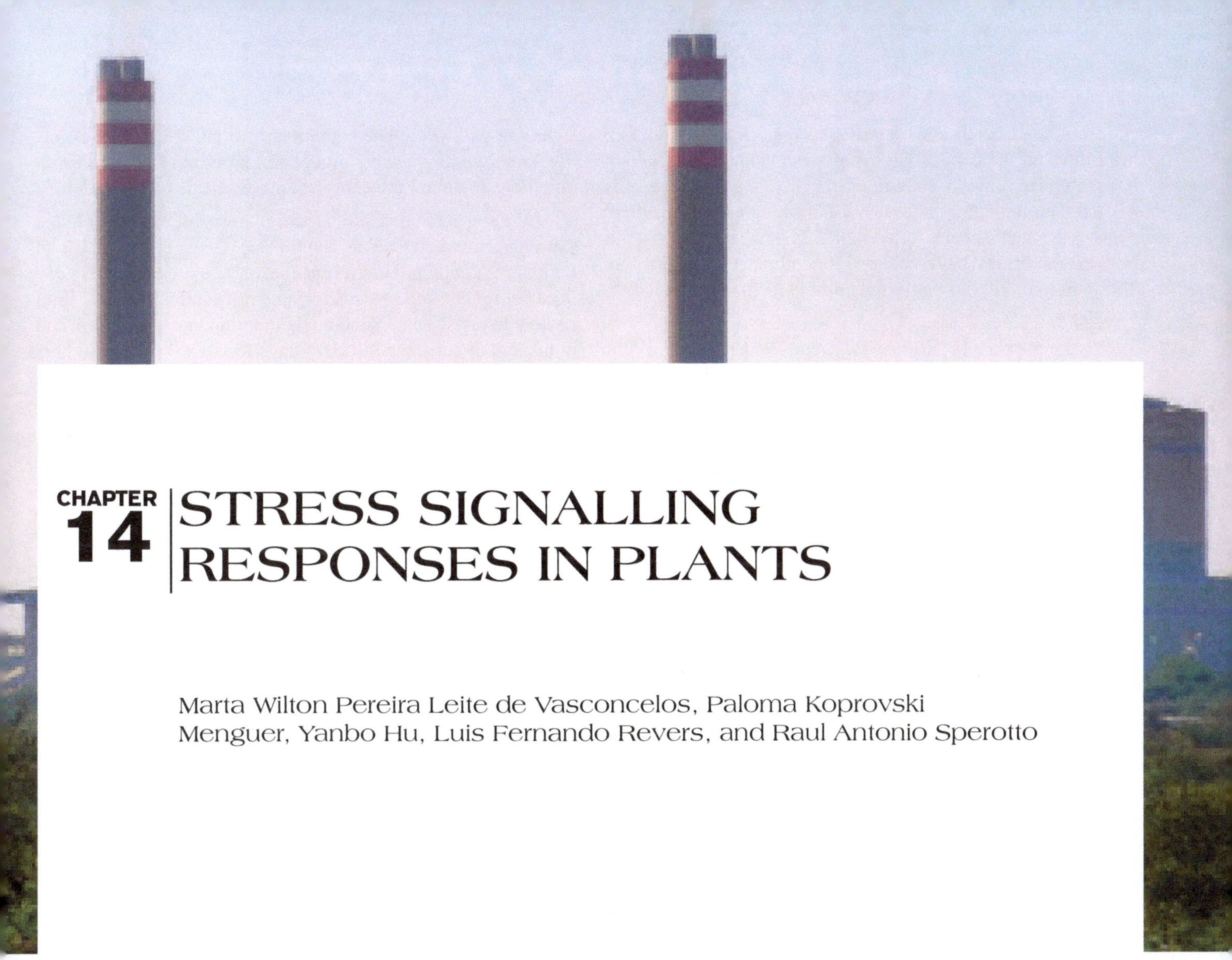

CHAPTER 14 STRESS SIGNALLING RESPONSES IN PLANTS

Marta Wilton Pereira Leite de Vasconcelos, Paloma Koprovski Menguer, Yanbo Hu, Luis Fernando Revers, and Raul Antonio Sperotto

Plants undergo continuous exposure to various biotic and abiotic stresses in their natural environment. To survive under such conditions, plants exhibit stress tolerance or stress avoidance through acclimation and adaptation mechanisms that ultimately reestablish cellular and organismal homeostasis or reduce episodic shock effects. These abilities involve intricate and complex mechanisms of perception, transduction, and transmission of stress stimuli, allowing optimal response to environmental conditions. The perception of stimuli and their expansion in cells involves signaling molecules such as intracellular calcium and reactive oxygen species, which intensify the action of particular signaling pathways. To date, the molecular mechanisms that are involved in each stress have been revealed comparatively independently, and so our understanding of convergence points between biotic and abiotic stress signaling pathways remains rudimentary. However, recent studies have revealed several molecules as promising candidates for common players that are involved in crosstalk between stress signaling pathways.

This special issue aimed to join original research and review articles related to understanding of plant responses to abiotic and biotic stress conditions, identifying novel players involved in plant responses to stress conditions, biotechnological strategies to increase plant tolerance to abiotic and biotic stresses, and understanding molecular interactions and crosstalks among different stress conditions.

M. Nourbakhsh-Rey and M. Libault in "Decipher the Molecular Response of Plant Single Cell Types to Environmental Stresses" explore omic studies to understand the response of single cell types to environmental stresses in order to clearly depict the contribution of each cell type composing the sample in response to stress. Cellular complexity of entire organs masks cell-specific responses to environmental stresses and logically leads to the dilution of the molecular changes occurring in each cell type composing the tissue/organ/plant in response to the stress. Specifically, the authors highlight that combining one or two omic analyses to look at single cell system biology can provide more precise molecular characterization and more dynamic models of the interactions between the plant and its environment.

J. Ren et al. in "Drought Tolerance Is Correlated with the Activity of Antioxidant Enzymes in Cerasus humilisSeedlings" describe the correlation between the activities of antioxidant enzymes and drought tolerance. By exploring drought-resistant

Citation: Marta Wilton Pereira Leite de Vasconcelos, Paloma Koprovski Menguer, Yanbo Hu, Luis Fernando Revers, and Raul Antonio Sperotto, "Stress Signaling Responses in Plants," BioMed Research International, vol. 2016, Article ID 4720280, 2 pages, 2016. doi:10.1155/2016/4720280.

and drought-susceptible C. humilis accessions, they compare the abilities of the contrasting genotypes to induce antioxidant defense under drought conditions. Their manuscript presents original data indicating that plants exhibiting a more efficient reactive oxygen species scavenging system in response to drought conditions have enhanced membrane protection, an ability that could be directly linked to their higher adaptation to drought.

X. Tang et al. in "Reference Gene Selection for qPCR Normalization of Kosteletzkya virginica under Salt Stress" employed RT-qPCR to select the most stable reference gene of the perennial halophytic plant K. virginica under salinity stresses. The stable reference gene selected in this study will be very helpful for revealing the gene expression profiles of K. virginica under salt stress, allowing a better understanding of the salt-resistant mechanisms in halophyte plants.

Y. Oono et al. in "Genome-Wide Transcriptome Analysis of Cadmium Stress in Rice" used RNAseq strategy to elucidate the molecular basis of the rice response to cadmium (Cd) stress, a widespread heavy metal pollutant that is highly toxic to living cells, negatively affecting nutrient uptake and homeostasis in plants. In this work, rice plants were hydroponically treated with low Cd concentrations, revealing novel Cd-responsive transporters by analyzing gene expression under different Cd concentrations. This study could help to develop novel strategies for improving tolerance to Cd exposure in rice and other cereal crops.

J. S. Rohila et al. in "Leaf Proteome Analysis Reveals Prospective Drought and Heat Stress Response Mechanisms in Soybean" investigate the effect of drought, heat, and co-occurring drought and heat stresses in the leaf proteome of two contrasting soybean genotypes. The authors identified genes involved in photosynthesis that were differentially expressed during drought and heat stress conditions, suggesting that photosynthesis-related proteins could be affecting RuBisCO regulation, electron transport, and Calvin cycle during abiotic stress, which ultimately alter the carbon fixation in leaves. The authors discuss the role of heat shock-related proteins and ROS detoxification capacity via carbonic anhydrase as heat and drought tolerance mechanisms, respectively.

J. M. Garcia-Mina et al. in "Involvement of Hormone- and ROS-Signaling Pathways in the Beneficial Action of Humic Substances on Plants Growing under Normal and Stressing Conditions" review our current knowledge about the mechanisms by which soil humus affects soil fertility. In particular, they discuss the relationships between two main signaling pathway families that are affected by humic substances within the plant: hormone- and ROS-mediated signaling pathways. The authors aim to integrate these events in a more comprehensive model and point out future research directions to unveil the complete mechanisms of regulation.

W. Fang et al. in "Cloning and Expression Analysis of One Gamma-Glutamylcysteine Synthetase Gene (Hbγ-ECS1) in Latex Production in Hevea brasiliensis" isolated a γ-ECS gene from the rubber tree and investigated its function linked to thiol content in latex. To understand the relation between γ-ECS and thiols and to correlate these findings to latex flow, the authors conducted RT-qPCR analysis and found that the expression levels of Hbγ-ECS1 were induced by tapping and ethrel stimulation, and the expression was related to thiols content in the latex. When looking at long-term flowing latex, the gene was related to the duration of latex flow. This work may have important biotechnological applications, since rubber tree is a major commercial source of latex, which is used for rubber production.

A. Vian et al. in "Plant Responses to High Frequency Electromagnetic Fields" conducted an original review, looking at the possible effects of HF-EMFs, which are increasingly present in the environment (due to, e.g., cell phones, Wi-Fi, and different types of connected devices) on different essential plant metabolic activities, such as reactive oxygen species metabolism, Krebs cycle, pentose phosphate pathway, chlorophyll content, and terpene emission. They found that not only are most of these pathways indeed modified by HF-EMF exposure, but also radiation brings about alterations in gene expression and plant growth. The authors propose to consider nonionizing HF-EMF radiation as a noninjurious, albeit influential environmental factor that induces significant changes in plant metabolism.

ACKNOWLEDGMENTS

We sincerely thank the editorial board for their approval of this concept and continuous help in the successful publication of this special issue. We would also like to thank contributors to this special issue for their scientifically sound papers. We extend our thanks to the reviewers for critical assessment of each paper, their constructive criticisms, and timely responses that made this special issue possible. We hope that this special issue will contribute to a more thorough understanding of the diverse aspects of plant stress signaling and response.

Marta Wilton Pereira Leite de Vasconcelos

Paloma Koprovski Menguer

Yanbo Hu

Luis Fernando Revers

Raul Antonio Sperotto

CHAPTER 15

PLANT RESPONSES TO HIGH FREQUENCY ELECTROMAGNETIC FIELDS

Alain Vian, Eric Davies, Michel Gendraud and Pierre Bonnet

INTRODUCTION

High frequency electromagnetic fields (HF-EMF, i.e., frequencies from 300 MHz to 3 GHz, wavelengths from 1 m to 10 cm) are mainly human-produced, nonionizing electromagnetic radiations that do not naturally occur in the environment, excluding the low amplitude VHF (very high frequency) cosmic radiation. HF-EMF are increasingly present in the environment [1] because of the active development of wireless technology, including cell phones, Wi-Fi, and various kinds of connected devices. Since living material is not a perfect dielectric, it readily interferes with HF-EMF in a way that depends upon several parameters, including (but not restricted to) its general shape, the conductivity and density of the tissue, and the frequency and amplitude of the EMF. The interaction between the living material and the electromagnetic radiation may (or not) induce an elevation of the tissue temperature, thus defining the thermal (versus nonthermal) associated metabolic responses. In the case of a thermal response, the resulting heat dissipation is normalized with the specific absorption rate (SAR) index. This has led to considerable research efforts to study the possible biological effects due to exposure to HF-EMF. While the vast majority of these studies have focused on animals and humans because of health concerns, with contradictory or nonconclusive results [2], numerous experiments have also been performed on plants. Plants are outstanding models compared to animals to conduct such investigations: they are immobile and therefore keep a constant orientation in the EMF and their specific scheme of development (high surface area to volume ratio) makes them ideally suited to efficiently intercept EMF [3]. It is also quite easy in plants to achieve genetically stable plant lines through the selection of species that favor asexual reproduction [4] or self-pollination [5]. Furthermore, metabolic mutants are easily available for several species and constitute invaluable tools to understand the way the EMF signal is transduced [6]. Indeed, several reports have pointed out that plants actually perceive HF-EMF of even small amplitudes and transduce them into molecular responses and/or alterations of their developmental scheme [3–9]. The way that HF-EMF interact with plants remains essentially unanswered. However, since EMF evoke a multitude of responses in plants, they might be considered as a genuine environmental stimulus. Indeed, EMF exposure alters the activity of several enzymes, including those of reactive oxygen species (ROS) metabolism [7], a well-known

Citation: Alain Vian, Eric Davies, Michel Gendraud, and Pierre Bonnet, "Plant Responses to High Frequency Electromagnetic Fields vol. 2016, Article ID 1830262, 13 pages, 2016. doi:10.1155/2016/1830262.

marker of plant responses to various kinds of environmental factors. EMF exposure also evokes the expression of specific genes previously implicated in plant responses to wounding [5, 8] and modifies the development of plants [9]. Furthermore, these responses are systemic insofar as exposure of only a small region of a plant results in almost immediate molecular responses throughout the plant [6]. These responses were abolished in the presence of calcium chelators [6] or inhibitors of oxidative phosphorylation [10] which implies the involvement of ATP pools. In the present review, we describe exposure devices, SAR determination methods, and biological responses (at both the cellular/molecular and whole plant levels) observed after plant exposure to EMF. We focused this review on radiated (i.e., EMF that are emitted through an antenna) HF-EMF (mainly within the range of 300 MHz–3 GHz) and consequently will not address the biological effects of static magnetic fields (SMF), extremely low frequency electromagnetic fields (ELF), or HF current injection, since their inherent physical properties are dramatically different from those of high frequencies. Therefore, the HF-EMF we consider should be viewed through the prism of classical electromagnetism: macroscopic electrodynamics phenomena described in terms of vector and scalar fields.

EXPOSURE SYSTEMS AND DOSIMETRY

HF-EMF are a combination of an electric field and a magnetic field governed by Maxwell's equations. At high frequency, these vector quantities are coupled and obey wave equations whether for propagating waves or for standing waves. In vacuum, the former travel at the speed of light ($\approx 3\ 10^8\,\mathrm{m\,s^{-1}}$) and have the structure of a plane wave (Figure 1(a)). In other media, the speed decreases and the spatial distribution for the electric and the magnetic fields are generally arbitrary (thus not being a plane wave). The latter, which do not propagate but vibrate up and down in place, appear in some particular conditions (e.g., bounded medium like metallic cavity) and play important roles in many physical applications (resonator, waveguide, etc.).

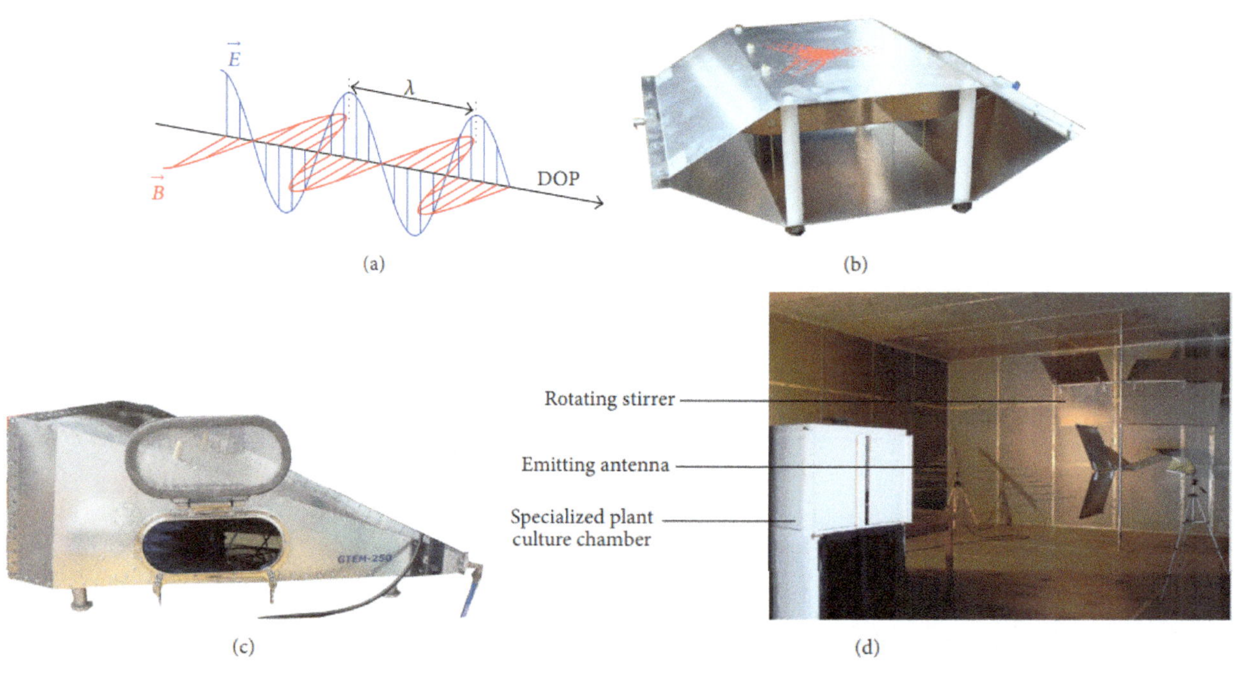

Figure 1: Electromagnetic wave and experimental set-up. (a) Schematic representation of an electromagnetic plane wave showing the transverse and space varying electric (E) and magnetic field (B). The wavelength (λ) is the distance between two crests. DOP: direction of propagation. (b) A TEM cell (transverse electromagnetic cell). (c) A GTEM cell (gigahertz transverse electromagnetic cell). (d) MSRC (mode stirred reverberation chamber). Note the double-sided metallic walls, the emitting antenna, the rotating stirrer, and the specialized culture chamber that stands in the "working volume" where the electromagnetic field characteristics have been extensively characterized.

In both cases, HF-EMF are characterized by an amplitude of the electric (E) or magnetic (H) components (measured in volts or amperes per meter), a frequency f (number of cycles per second of the wave quantity, measured in hertz), and a wavelength λ (distance between wave crests, measured in meters). These properties are related through the following equation:

$$\lambda = \frac{c}{f} = c \times T, \qquad (1)$$

where c is the speed of the wave in the considered medium and T is the period of the wave (time between successive wave crests, measured in seconds). The wavelength λ is then the distance traveled by the wave during a period T. The electromagnetic power density associated with an electromagnetic wave (measured in watts per square meter) is obtained by a vector product between the electric and magnetic field vectors (namely, the Poynting vector) for every point in space. The total HF-EMF power crossing any given surface is derived from Poynting's theorem [11]. For an incident plane wave in vacuum, the time-averaged electromagnetic power P_i (measured in watts) illuminating a surface of 1 m^2 orthogonal to the direction of propagation is given by the following equation:

$$P_i = \frac{E^2}{2 \times Z_0}, \qquad (2)$$

where Z_0 is the characteristic impedance of free vacuum space (377 Ω).

The absorbed electromagnetic power (*Pd*), converted to heat by Joule effect in a volume (*V*) and averaged over a time period, is given by (3) for an electrically and magnetically linear material that obeys Ohm's law (conductivity σ in siemens per meter):

$$Pd = \iiint_v \frac{\sigma \times E^2}{2} dV. \quad (3)$$

Diversity of Exposure Devices

Due to the wide variety of electromagnetic waves, physicians developed a lot of electromagnetic exposure facilities, mainly for electromagnetic compatibility (EMC) test purposes. Some of these devices are used for plant exposure to HF-EMF.

HF-EMF exposure set-up is usually made up with the following two basic elements:

- HF source (radio frequency generator, Gunn oscillator) associated with a radiating element (antenna, strip-line) and
- a structure that allows the propagation of EM waves and the exposure of the sample.

The simplest exposure set-up relies on the use of standard cell phones as a source of HF-EMF [12, 13] radiating in an open-area test site. While this apparatus has the advantage of being simple and economical, it poses many limitations that may compromise the quality of the exposure. Indeed, these communication devices are operated with different protocols that may modify or even interrupt the emitted power. Also, the biological samples are placed in the immediate neighborhood of the antenna, which is a region where the electromagnetic field is not completely established (near-field conditions) and therefore is difficult to measure; this situation may constitute an issue for bioelectromagnetics studies. These apparatuses are nowadays used only in a small proportion of studies. Moreover, the use of open-area test sites exposes the biological samples to the uncontrolled electromagnetic ambient environment. The use of shielded rooms is a good solution to overcome this issue. Indeed, anechoic chambers provide shielded enclosures, which are designed to completely absorb reflected electromagnetic waves. However, these facilities are often large structures requiring specific equipment and costly absorbers to generate an incident plane wave (far-field illumination) and are consequently seldom used for plant exposure [14, 15].

In contrast, numerous studies are based upon dedicated apparatus of relatively small volume (Figure 1(b)), namely, the transverse electromagnetic (TEM) cell [16]. TEM cells are usually quite small (about 50 cm long × 20 cm wide) and therefore only allow the use of seeds or seedlings as plant models. Many TEM cells are based upon the classic "Crawford cell" [17]. They consist of a section of rectangular coaxial transmission line tapered at each end to adapt to standard coaxial connectors. A uniform plane wave of fixed polarization and direction is generated in the sample space for experiments between the inner conductor (septum) and the upper metallic wall. Because this cost-efficient device is enclosed, high amplitude EMF can be developed with relatively little injected power. Under some conditions, two parallel walls of the TEM cell can be removed (therefore constituting the so-called open TEM cell) without dramatically compromising the performances. This configuration is adequate to allow plant lighting. Special attention must still be paid to the relative position of the samples in the system since the disposition of the different organs within the EMF could severely affect the efficiency of the plant samples' coupling with the electromagnetic field. The main TEM cell limitation is that the upper useful frequency is bound by its physical dimensions limiting the practical size of samples at high frequency.

The gigahertz transverse electromagnetic (GTEM) cell has emerged as a more recent EMF emission test facility (Figure 1(c)) [18]. It is a hybrid between an anechoic chamber and a TEM cell and could therefore be considered as a high frequency version of the TEM cell. The GTEM cell comprises only a tapered section, with one port and a broadband termination. This termination consists of a 50 Ω resistor board for low frequencies and pyramidal absorbers for high frequencies. This exposure device removes the inherent upper frequency limit of TEM cell while retaining some of its advantages (mainly the fact that no antenna set-up is required and the fact that high field strength could be achieved with low injected power).

Waveguides are another kind of screened enclosures that are seldom used in plant exposure [19, 20]. These classical and easy to use exposure devices generate traveling waves along the transmission coordinate and standing waves along the transverse coordinates. In contrast to the TEM cell, waveguides do not generate uniform plane waves but rather allow the propagation of more complex EMF, namely, propagation modes. Each mode is characterized by a cutoff frequency below which the mode cannot propagate. When the ends of the waveguide are short-circuited, a so-called resonant cavity is constituted, from which a recent large facility, originally designed for EMC studies, namely, the mode stirred reverberation chamber (MSRC, Figure 1(d)), is based. While this equipment is expensive and technically difficult to set up, it is the state of the art in terms of electromagnetic field characteristics, allowing the establishment of an isotropic and homogeneous field in a volume large enough to hold a dedicated plant culture chamber (either transparent or shielded toward EMF [6]). This latter characteristic permits experiments on large plants that are kept in an adequate controlled environment [6]. Our group pioneered the use of this facility, based on judicious combinations of standing waves patterns in a complex screened enclosure, in plant bioelectromagnetics studies [8] and extensively described the MSRC functionality [21]. Finally, each exposure set-up may differ in concept, polarization, frequency, or incident power but these setups always need to be optimally designed and based on well-understood physical concepts in order to assess well-controlled HF-EMF exposure conditions (homogeneity, repeatability, reproducibility, etc.).

Different Types of Exposure Signals

From each of the previous exposure devices, two very different types of EMF can be used to expose plants. The most commonly encountered mode is the continuous wave (CW) mode, in which the biological samples are continuously exposed for a specific duration to an EMF of given frequency and amplitude (rarely more than a few dozen $V\,m^{-1}$). The second mode is the pulsed electromagnetic field (PEMF) mode, in which the biological samples are subjected to several series of discontinuous pulses of ultrashort duration EMF (within the range of μs to ns) and usually of very high amplitude (up to several hundred $kV\,m^{-1}$). This last kind of exposure [22, 23] is seldom used because of the scarcity and great complexity of the equipment needed to generate the EMF and the difficulty to design the dedicated antennae able to deliver such ultrashort power surges [24].

The HF-EMF could also be modulated (i.e., varied in time at a given, usually much lower frequency). Only a few studies explicitly addressed modulation effect on biological responses. Răcuciu et al. [25] exposed maize caryopses to low levels (7 dBm), 900 MHz RF field, for 24 h in either continuous wave (CW), amplitude modulated (AM), or frequency modulated (FM) modes. They found that 12-day-old plant lengths were reduced by about 25% in modulated EMF (AM or FM type) compared to control (unexposed samples), while CW exposure had an opposite (growth stimulation) effect, suggesting that EMF modulation actually modifies biological responses.

Dosimetry

In order to compare the biological effects observed in different exposure conditions, the National Council on Radiation Protection and Measurements officially introduced in 1981 an EMF exposure metric, the specific absorption rate (SAR). The formal definition of this basic dosimetry (the amount of dose absorbed) is "the time derivative of the incremental energy absorbed (dW) by (dissipated in) an incremental mass contained in a volume (dV) of a given density ρ." From this definition and (3), the SAR (measured in $W\,kg^{-1}$) is given by the following equation:

$$\mathrm{SAR} = \frac{d}{dt}\left(\frac{dW}{\rho \times dV}\right) = \frac{\sigma \times E^2}{2 \times \rho}. \tag{5}$$

SAR is the power absorbed by living tissue during exposure to CW-EMF (this quantity does not apply to PEMF mode because of the very short duration of the pulses that do not cause temperature increase in the samples). SAR can be calculated from the dielectric characteristics of plant tissues at the working frequencies, using (4). While ρ could be easily determined, the value of σ is dependent upon the frequency and is difficult to assess in the range of GHz. It is usually evaluated from the literature [40], since the experimental set-up to measure this parameter at a given frequency (waveguide, open waveguide, and coaxial line technique, e.g., D-Line) is rarely used because of its complex set-up. From the biological heat-transfer equation, the SAR can also be determined using the temperature increase evoked in plant tissue after exposure to EMF, using the following equation:

$$\mathrm{SAR} = C \times \left(\frac{dT}{dt}\right)_{t \to 0}, \tag{6}$$

where C is the heat capacity ($J\,K^{-1}\,kg^{-1}$, which is available for some tissues in the literature) and dT (measured in Kelvin) is the sample temperature increase corresponding to the elapsed time dt (measured in second) since the beginning of HF-EMF exposure. Either for animals or plants, the SAR measurement is subject to uncertainty [46]. Since the specific heat is frequency independent and the temperature distribution is usually more uniform than the internal electric field, (5) provides, for detectable temperature increases, a better way for SAR estimation.

In animal and human tissue, SAR is determined using dedicated phantoms [47] filled with a special liquid that mimics the dielectric properties of biological fluids. While this approach is adequate in animals, in which the developmental scheme produced volumes, it could not be adapted to most plant organs (e.g., leaves) that have a high surface area to volume ratio [3] but could be used in fruits and tuberous structures. In contrast, surface temperature can be easily assessed with dedicated instruments (e.g., Luxtron® fiber optic temperature probe) and used to feed (5) [45]. The SAR can also be determined using the differential power method based on the measurement of power absorption (reviewed in [48]) that takes place in the absence or presence of biological samples [39]. The SAR is then calculated by dividing the absorbed power by the mass of the living material.

BIOLOGICAL RESPONSES

Biological responses should be considered as reporters of, and evidence for, the plant's ability to perceive and interact with EMF. These responses can take place at the subcellular level, implying molecular events or modification of enzymatic activities, or at the level of the whole plant, taking the form of growth modification. Tables 1–3 summarize some work reporting HF-EMF effects observed at the scale of the whole plant, biochemical processes, or gene regulation, respectively.

Table 1: Metabolic pathways affected after plant exposure to HF-EMF radiations

Enzymes or metabolites	Metabolic pathways	Organisms	Exposure conditions	Response to EMF
Phenylalanine ammonia-lyase	Phenylpropanoids	Phaseolus vulgaris	N/A (PEMF)	Synergistic action with growth regulators in cultured cells [26]
Polyphenol oxidase	Polyphenols	Vigna radiata	900 MHz, up to 4 h, 8.55 $\mu W\,cm^{-2}$	8.5-fold increase [27]
α- and β-amylases	Starch metabolism	Vigna radiata	900 MHz, up to 4 h, 8.55 $\mu W\,cm^{-2}$	2.5- and 15-fold increase for α- and β-amylases, respectively [27]

α- and β-amylases, starch phosphorylases	Starch metabolism	Zea mays	1800 MHz, up to 4 h, 332 mW m^{-2}	2-fold increase for amylases. −73% for starch phosphorylases [28]
Water soluble sugars	Sugar metabolism	Phaseolus vulgaris	900 MHz, 4 h	2-fold reduction in soluble sugars [12]
Acid and alkaline invertases	Sucrose metabolism	Zea mays	1800 MHz, up to 4 h, 332 mW m^{-2}	1.8- and 2.6-fold increase for acid and alkaline forms, respectively [28]
Malate and NADP isocitrate dehydrogenases, glucose-6P dehydrogenase	Krebs cycle, pentose phosphate pathway	Plectranthus	900 MHz, 1 h	Lower activity (−10 to −30%) at the end of the stimulus and then a 2-fold increase 24 h later [29]
ATP content and adenylate energy charge (AEC)	Energetic metabolism	Solanum lycopersicon	900 MHz, 10 min, 5 V m^{-1}	Drop of ATP content (30%) and AEC (0.8 to 0.6) 30 min after the stimulus [10]
MDA content, H_2O_2, superoxide dismutase, catalase, guaiacol peroxidase, glutathione reductase, ascorbate peroxidase	Lipid peroxidation-oxidative metabolism	Vigna radiata	900 MHz, 8.55 μW cm^{-2}	All oxidative metabolism markers increased (2-fold to 5-fold) [7]
MDA and H_2O_2 content, catalase, ascorbate peroxidase	Lipid peroxidation	Lemna minor	400 and 900 MHz, 2 to 4 h, 10 to 120 V m^{-1}	MDA and H_2O_2 content, catalase and ascorbate peroxidase activities increased (10–30%) [30]
Peroxidases	Oxidative metabolism	Vigna radiata, Lemna minor	900 MHz, 1 to 4 h, 8.55 μW cm^{-2} or 41 V m^{-1}	Peroxidase activities increased [18, 27]
MDA, oxidized and reduced glutathione, NO synthase	Oxidative metabolism-NO metabolism	Triticum aestivum	2.45 GHz, 5 to 25 s, 126 mW mm^{-2}concomitantly with NaCl treatment	Exposure to EMF reduced the oxidative response of plants to high salt treatment [31]
Protein metabolism-DNA damage	Oxidative protein and DNA damage (comet assay)	Nicotiana tabacum	900 MHz, 23 V m^{-1}	Carbonyl content and tail DNA value increased (1.8-fold and 30%, resp.) [30]
Protein metabolism	Protein content	Phaseolus vulgaris, Vigna radiata, Triticum aestivum	Cell phone, 4 h	Drop in protein content inPhaseolus (71%) and Vigna(57%) [27, 32] and Triticum [13]
Amino acid metabolism	Proline accumulation	Zea mays, Vigna radiata	940 MHz, 2 days Cell phone, 2 h, 8.55 μW cm^{-2}	1.8- and 5-fold increase in Z. mays [33] and V. radiata [7], respectively
Global terpene emission	Monoterpene metabolism	Petroselinum crispum, Apium graveolens, Anethum graveolens	900–2400 MHz, 70–100 mW m^{-2}	Enhanced emission of terpene compounds [34]

Table 2: Genes whose expression is altered after plant exposure to HF-EMF

Gene	Organism	Function	Exposure conditions	Response to EMF exposure
lebZIP1	Solanum lycopersicon, whole plant	Transcription factor	900 MHz, 5 V m^{-1}, CW in a MSRC	Increase (3-fold to 4-fold) [6,8]

lebZIP1	Solanum lycopersicon, whole plant	Transcription factor	Cell phone	Increase (3-4-fold) [35]
cam	Solanum lycopersicon, whole plant	Ca^{2+} signal transduction	900 MHz, 5 V m^{-1}, CW in a MSRC	Increase (5-fold) [5, 10]
cdpk	Solanum lycopersicon, whole plant	Ca^{2+} signal transduction	900 MHz, 5 V m^{-1}, CW in a MSRC	Increase (5-fold) [10]
cmbp	Solanum lycopersicon, whole plant	mRNA metabolism	900 MHz, 5 V m^{-1}, CW in a MSRC	Increase (6-fold) [5]
pin2	Solanum lycopersicon, whole plant	Proteinase inhibitor	900 MHz, 5 V m^{-1}, CW in a MSRC	Increase (4.5-fold [5] and 2.5-fold) [6]
pin2	Solanum lycopersicon, whole plant	Proteinase inhibitor	900 MHz, cell phone	Increase (2-fold) [35]
At4g26260	Arabidopsis thaliana, cell suspension culture	Similar to myo-inositol oxygenase	1.9 GHz, 8 mW cm^{-2}	Decrease (0.3-fold) [36]
At3g47340	Arabidopsis thaliana, cell suspension culture	Glutamine-dependent asparagine synthetase	1.9 GHz, 8 mW cm^{-2}	Decrease (0.4-fold) [36]
At3g15460	Arabidopsis thaliana, cell suspension culture	Brix domain protein	1.9 GHz, 8 mW cm^{-2}	Decrease (0.5-fold) [36]
At4g39675	Arabidopsis thaliana, cell suspension culture	Expressed protein	1.9 GHz, 8 mW cm^{-2}	Increase (1.5-fold) [36]
At5g10040	Arabidopsis thaliana, cell suspension culture	Expressed protein	1.9 GHz, 8 mW cm^{-2}	Increase (1.4-fold) [36]
AtCg00120	Arabidopsis thaliana, cell suspension culture	ATPase alpha subunit (chloroplast)	1.9 GHz, 8 mW cm^{-2}	Increase (1.4-fold) [36]

Table 3: Morphogenetic responses observed after plant exposure to HF-EMF

Plant species	Exposure conditions	Responses to HF-EMF exposure and references
Raphanus sativus	Gunn generator 10.5 GHz, 14 mW, exposure of seeds and hypocotyls	Germination inhibition (45%), reduction of hypocotyl elongation (40%) [37]
Lens culinaris	Cell phone, 1800 MHz (1 mW), exposure of dormant seeds	Reduction of seedlings' root growth (60%) and mitotic index (12%). Abnormal mitosis increased (52%) [38]
Vigna radiata	Cell phone, 900 MHz, 8.55 μW cm^{-2}	Rhizogenesis (root number and length) severely affected [7]
Vigna radiata	Cell phone, 900 MHz, 8.55 μW cm^{-2}	Inhibition of germination (50%), hypocotyl (46%), and root growth (59%). Dry weight reduced by 43% [27]
Phaseolus aureus, Vigna radiata	Cell phone, 4 h exposure	Root and stem elongations severely affected (−44 and −39%, resp.) [12, 32]
Vigna radiata, Lablab purpureus	1.8 GHz, 0.48–1.45 mW cm^{-2}	Reduction of height and fresh weight [39]
Zea mays	1 GHz, 1 to 8 h, 0.47 W cm^{-2}	Reduced growth of 12-day-old plants (about 50% after 8 h of exposure) [40]
Zea mays	1800 MHz, 4 h, 332 mW m^{-2}	Reduced growth of roots and coleoptiles (16 and 22%, resp.) [28]
Vigna radiata, Triticum aestivum	Cell phone, 900 MHz, 4 h exposure	Growth reduction (21 and 50%) in Vigna and Triticum, respectively [13]
Triticum aestivum, Cicer arietinum, Vigna radiata, Vigna aconitifolia	Klystron-based EMF generator, 9.6 GHz, 1 dBm to 3.5 dBm	Growth and biomass reduction [41]
Vigna radiata, Ipomoea aquatica	425 MHz, 2 h, 1 mW	Growth stimulation of primary root [16]
Glycine max	900 MHz, 5.7 to 41 V m^{-1}	Inhibition of epicotyl and/or root growth, depending on exposure set-up [9]
Lemna minor	400–1900 MHz, 23 to 390 V m^{-1}, whole plant exposure	Growth slowed down, at least in the first days following exposure [18]
Trigonella foenum-graecum, Pisum sativum	900 MHz, 0.5–8 h	Increased root size, nodule number, and size [42]
Hibiscus sabdariffa	Resulting field from a GSM base antenna (not measured)	Reduction of flower bud abscission with increasing distances from the antenna [43]

Linum usitatissimum	Cell phone or Gunn generator (105 GHz), 2 h	Production of epidermic meristems under calcium deprivation condition [44]
Rosa hybrida	900 MHz, 5–200 V m^{-1}, whole plant exposure in MSRC	Delayed and reduced (45%) growth of secondary axes [45]

Cellular and Molecular Level

Numerous reports [4, 7, 33] indicate an increase in the production of malondialdehyde (MDA, a well-known marker of membrane alteration) along with ROS metabolism activation after exposing plants to HF-EMF (Table 1). Membrane alteration and ROS metabolism activation are likely to establish transduction cascades that enable specific responses. Indeed, the critical role of calcium, a crucial second messenger in plants, has long been pointed out [6, 10]: the responses (e.g., changes in calm-n6, lecdpk-1, and pin2 gene expression) to EMF exposure are severely reduced when plants are cultivated with excess of calcium or in the presence of calcium counteracting agents (Figure 2) such as chelators (EGTA and BAPTA) or a channel blocker ($LaCl_3$). The importance of calcium in the establishment of the plant response is also highlighted by the fact that early gene expression associated with EMF exposure involves at least 2 calcium-related products (calmodulin and calcium-dependent protein kinase) [5, 10]. This response is also energy-dependent: an important drop (30%, Figure 3) in ATP content and adenylate energy charge (AEC) occurs after HF-EMF exposure [10]. It is not clear for now if the AEC drop is the consequence of altered membranes allowing passive ATP exit or if higher consumption of ATP occurred because of increased metabolic activity. Indeed, it is well known that a drop in AEC stimulates the catabolic enzymatic pathways through allosteric modulations. Nevertheless, inhibiting ATP biosynthesis with the decoupling agent carbonyl cyanide m-chlorophenyl hydrazone (CCCP) abolished plant responses to EMF exposure [10]. Nitric oxide (NO) is another signaling molecule that is tightly related to environmental factors' impact on plants [49]. NO rapidly increases after various kinds of stimuli including drought stress or wounding. Chen et al. [50] recently demonstrated the increased activity of nitric oxide synthase and accumulation of NO after exposing caryopses of wheat for 10 s to high power 2.45 GHz EMF. Similarly, Qiu et al. [51] showed in wheat that the tolerance to cadmium evoked by microwave pretreatment was abolished by the addition of 2-(4-carboxyphenyl)-4,4,5,5-tetramethylimidazoline-1-oxyl-3-oxide (carboxy-PTIO), an NO scavenger, suggesting that microwave-induced NO production was involved in this mechanism. Taken together, these results advocate for the EMF induction of NO synthase. However, these studies used high power EMF (modified microwave oven) as stimulating tool and the fact that a temperature increase of the sample was the cause of NO increase is not excluded. To our knowledge, the involvement of NO has not yet been demonstrated after low power (i.e., nonthermal) EMF exposure. Furthermore, well-known actors of plant responses to environmental stimuli are also involved: the tomato mutants sitiens and JL-5 for abscisic (ABA) or jasmonic (JA) acids biosynthesis, respectively, display normal responses (accumulation of stress-related transcripts) when whole plants are exposed to EMF [6]. In contrast, very rapid distant responses to local exposure that occur in the wild plants (Figure 4(a)) are impaired in sitiens ABA mutant (Figure 4(b)) and JL-5mutants, highlighting the existence of a transmitted signal (whose genesis and/or transmission is dependent on ABA and JA) in the whole plant after local exposure [6]. The nature of this signal is still unknown, but very recent work has demonstrated that membrane potential is affected after exposure to EMF [14]. It could therefore be hypothesized that electrical signals (action potential and/or variation potential) could be the transmitted signal, strongly implying that HF-EMF is a genuine environmental factor.

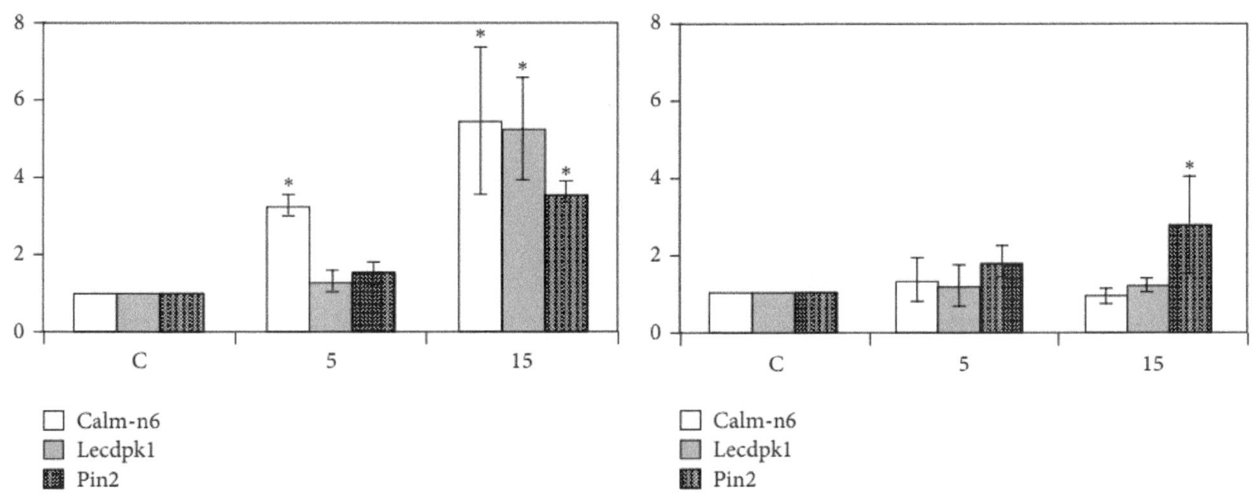

(a) Standard (0.73 mM calcium)

(b) Standard + calcium (7.3 mM)

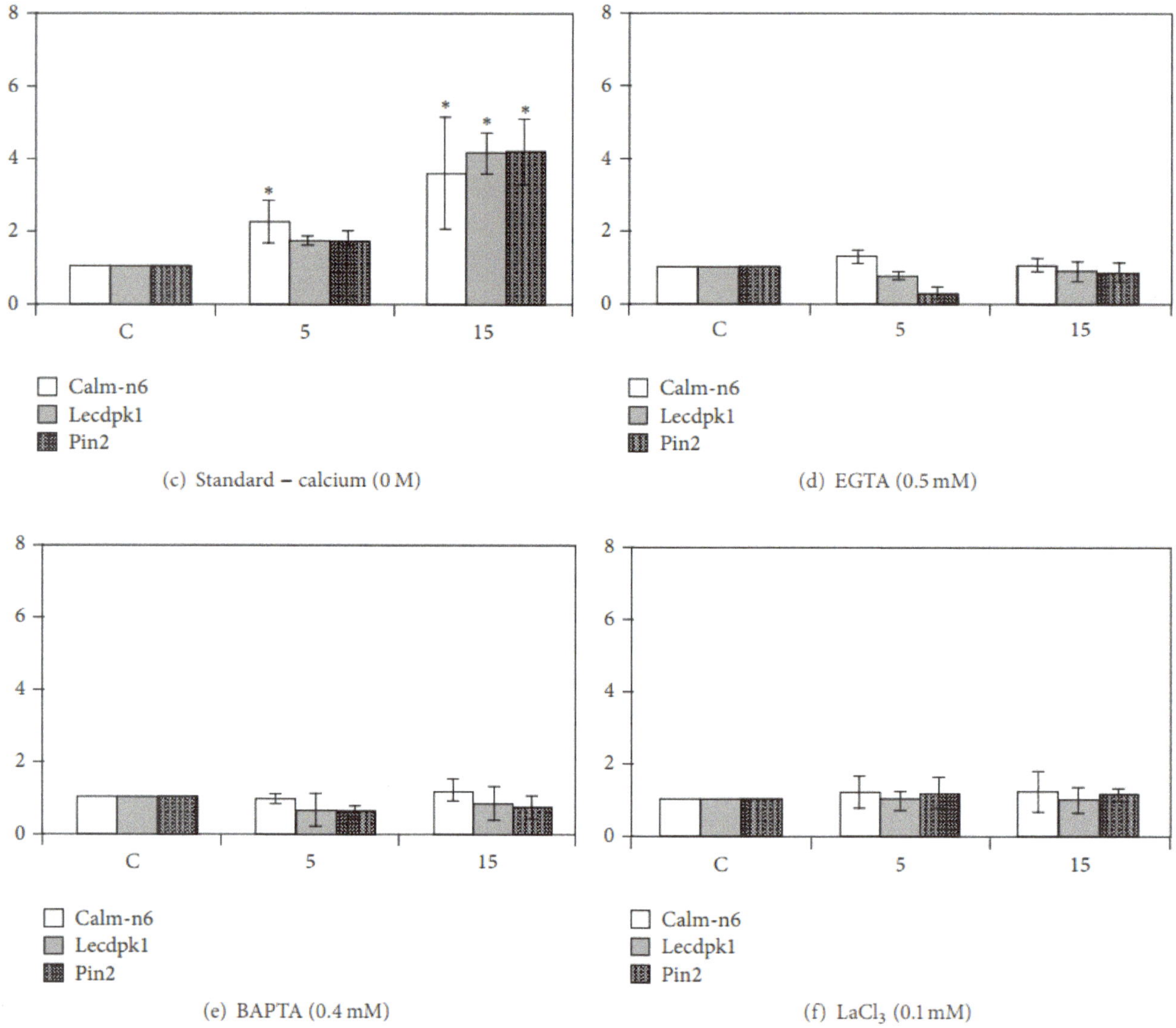

Figure 2: Effect of calcium concentration in culture medium on calm-n6 (calmodulin), pin2 (proteinase inhibitor), and lecdpk1 (calcium-dependent protein kinase) transcript accumulation in response to the HF-EMF exposure. (a) Standard medium (0.73 mM of calcium). (b) Ten-fold extra calcium (7.3 mM). (c) No calcium (0 mM). (d) No calcium (0 mM) with 0.5 mM of EGTA. (e) No calcium (0 mM) with 0.4 mM of BAPTA (1,2-bis(o-aminophenoxy)ethane-N,N,N′,N′-tetraacetic acid), a specific Ca^{2+} chelator. (f) No calcium (0 mM) with 0.1 mM $LaCl_3$. Bars represent mean values ± SE from at least three independent experiments. An asterisk over the bars states the significant differences according to the one-sided Mann-Whitney U test. Reproduced from [10], with permission.

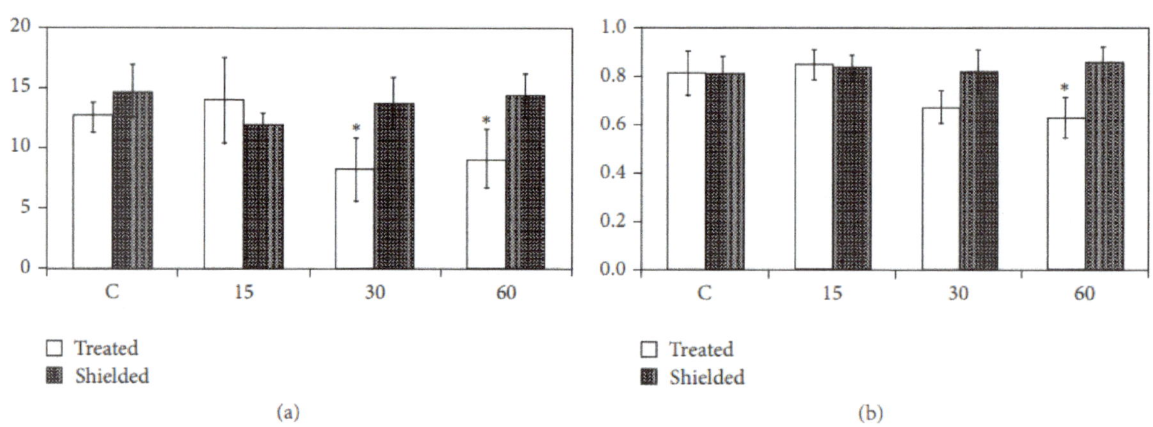

Figure 3: ATP concentration and adenylate energy charge (AEC) changes after HF-EMF exposure ($5\,V\,m^{-1}$, 10 min) in a mode stirred reverberation chamber. C: control, unexposed plants. 15, 30, and 60: time (min) after the end of HF-EMF exposure. (a) ATP concentration ($pmol\,mg^{-1}$ Prot.). (b) Adenylate energy charge (ratio). Bars represent mean values ± SE from at least three independent experiments. An asterisk over the bars states the significant differences according to the one-sided Mann-Whitney U test. Reproduced from [10], with permission.

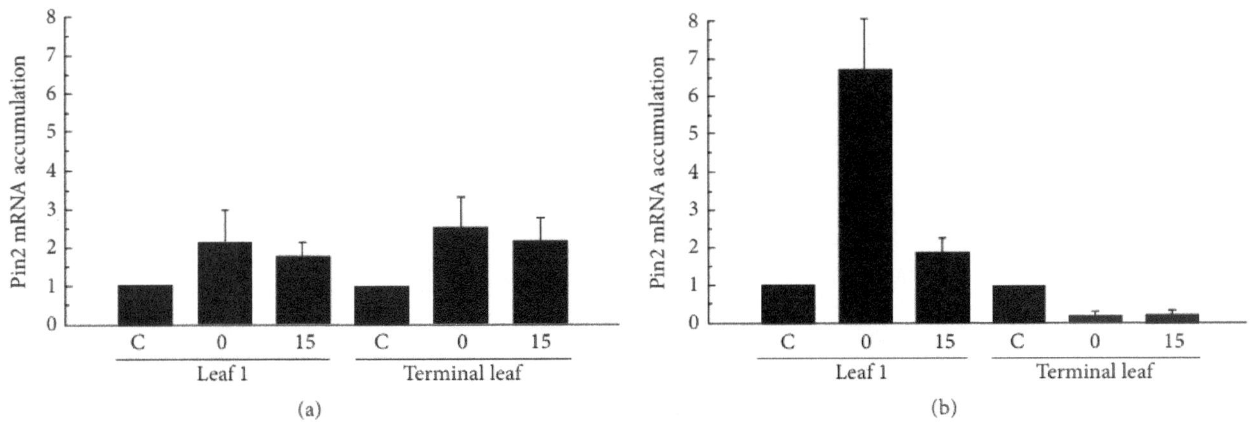

Figure 4: Local and systemic responses after HF-EMF exposure. (a) Systemic response after local exposure to HF-EMF in wild type. Local and distant responses after stimulation of leaf 1 (with the rest of the plant being protected from the EMF). The stimulated tissue (leaf 1) and distant one (terminal leaf) both displayed responses (accumulation of pin2 transcript). HF-EMF exposure: $5\,V\,m^{-1}$, 10 min. (b) Impairment of distant response after exposure to HF-EMF in sitiens (ABA deficient) mutant. The stimulated tissue (leaf 1) displays the response to EMF exposure (accumulation of pin2 transcript), while the response in the distant tissue (terminal leaf) is impaired. HF-EMF exposure: $5\,V\,m^{-1}$, 10 min. Reproduced from [6], with permission.

Alterations of Enzymatic Activities

Table 1 summarizes some of the enzymatic activities that are modified after exposing plants to HF-EMF. As previously noted, ROS metabolism is very often activated after plant exposure to EMF. Enzymatic activities such as peroxidase, catalase, superoxide dismutase, and ascorbate peroxidase have twofold to fourfold increase [4, 7, 18, 27, 33]. The question remains open to determine if this could be the consequence of a direct action of EMF on living tissue. Indeed, the very low energy that is associated with the EMF at these frequencies makes them nonionizing radiations. Side effects of elevated ROS metabolism are also noted: H_2O_2 production [4, 7], MDA increases [4, 7, 33], and protein damage [30]. An increase in polyphenol oxidase [27] and phenylalanine ammonia-lyase [26] may indicate stress responses linked to an increased lignification, a common response of plants to environmental stress.

Protein content is reduced in Vigna and Phaseolus [27, 32] as well as in Triticum [13]. It is not yet known if the decrease in protein content results from an increase in protein degradation and/or a decrease in protein synthesis, but this may constitute a stimulating field of investigation, since evidence shows that mRNA selection from translation occurs after plant exposure to HF-EMF [10]. Hydrolytic enzymatic activities (α- and β-amylases and invertases) responsible for the production of soluble sugar increase in germinating seeds after exposure to HF-EMF [12, 28, 32], while the starch phosphorylase activity, phosphorolytic and potentially reversible, is diminished. In contrast, HF-EMF exposure causes a drop of soluble sugar that may be related to the inhibition of Krebs cycle and pentose phosphate pathway in Plectranthus (Lamiaceae) leaves after exposure to 900 MHz EMF [29], suggesting that seeds and adult leaves respond in a different way to HF-EMF exposure. The accumulation of proline, reported by several authors [7, 33], and an increase in terpenoid emission and content in aromatic plants [34] are also classical responses of plants to environmental stresses.

Modification of Gene Expression

While numerous reports focused on enzymatic activities alterations after exposure to EMF, only a few studies concentrate on gene expression modifications (Table 2). Tafforeau et al. [44] demonstrated using Gunn generator (105 GHz) several reproducible variations in 2D gel electrophoresis profiles, showing that gene expression is likely to be altered by the exposure treatment. Jangid et al. [52] provided indirect proof (RAPD profiles) suggesting that high power microwave irradiation (2450 MHz, $800\,W\,cm^{-2}$) modifies gene expression in Vigna aconitifolia, while these results do not exclude a possible thermal effect of microwave treatment. Arabidopsis thaliana suspension-cultured cells exposed to HF-EMF (1.9 GHz, $8\,mW\,cm^{-2}$) showed differential expression of several genes (p values < 0.05) compared to the control (unexposed) condition in microarray analysis [36]. Most of them are downregulated (while At4g39675, At5g10040, and AtCg00120 displayed a slight increase; see Table 2). However, the RT-PCR p value lowers the significance of these variations and these authors consequently concluded the absence of HF-EMF effect on plant gene expression. In contrast, short duration, high frequency, low amplitude EMF exposure (10 min, 900 MHz, $5\,V\,m^{-1}$) performed on whole 3-week-old tomatoes in MSRC [5, 6, 8, 10] demonstrated altered expressions of at least 5 stress-related genes (Table 2), suggesting that whole plants are more sensitive to HF-EMF than cultured cells. These experiments have been independently replicated by Rammal et al. [35], using a longer exposure period and a far less sophisticated exposure set-up (cell phone). Stress responses of plants quite often display a biphasic pattern [53]: a very rapid increase in transcript accumulation that lasts 15–30 min, followed by a brief return to basal level, and then a second increase (after 60 min). This pattern was observed after tomato exposure to EMF so we questioned the meaning of the early and late population of transcripts in terms of physiological significance by measuring their association to polysomes (which reflects their putative translation to proteins). We found that the early (0–15 min) mRNA population was only faintly associated with polysomes, yet being poorly translated, while the late mRNA population (60 min) is highly associated with polysomes [10]. This result strongly suggests that only the late mRNA population may have a physiological importance since it is the only one to be efficiently translated into proteins.

Whole Plant Level

The biochemical and molecular modifications observed after plant exposure to EMF and described in the previous paragraphs might induce morphogenetic alterations of plant development. Indeed, an increasing number of studies report modifications of plant growth after exposure to HF-EMF (Table 3). These treatments are effective at different stages of plant development (seeds, seedlings, or whole plants) and may affect different organs or developmental processes including seeds germination and stem and root growth, indicating that biological samples of even small sizes (a few mm) are able to perceive HF-EMF. Seed exposure to EMF generally results in a reduced germination rate [27, 37, 39], while in other cases germination is unaffected [42] or even stimulated [16]. The seedlings issued from EMF-exposed seeds displayed reduced growth of roots and/or stem [13, 28, 32, 37–39, 41] but rarely a stimulatory effect [16]. This point strongly differs from exposure to static magnetic fields or extremely low frequency EMF, in which the stimulatory effects on growth are largely predominant [54]. Ultrashort pulsed high power EMF (PEMF, 4 µs, 9.3 GHz, 320 kV m^{-1}) also tends to stimulate germination of seeds of radish, carrot, and tomato and increase plant height and photosynthetic surface area in radish and tomato [20] and roots of tobacco seedlings [22]. These different effects of PEMF compared to HF-EMF on plants may be related to their fundamental difference in terms of physical properties. Exposure to HF-EMF of seedlings or plants (rather than seeds) also generally resulted in growth inhibition [9,18, 27, 28, 39]. Singh et al. [7] showed that rhizogenesis (root number and length) is severely affected in mung bean after exposure to cell phone radiation, possibly through the activation of several stress-related enzymes (peroxidases and polyphenol oxidases). Akbal et al. [38] showed that root growth was reduced by almost 60% in Lens culinaris seeds exposed in the dormant state to 1800 MHz EMF radiation. Concomitantly, these authors reported an increase in ROS-related enzymes, lipid peroxidation, and proline accumulation, with all of these responses being characteristic of plant responses to stressful conditions. Afzal and Mansoor [13] investigated the effect of a 72 h cell phone exposure (900 MHz) on both monocotyledonous (wheat) and dicotyledonous (mung bean) plants seeds: germination was not affected, while the seedlings of both species displayed growth inhibition, protein content reduction, and strong increase in the enzymatic activities of ROS metabolism. It is however worth noting that growth of mung bean and water convolvulus seedlings exposed at a lower frequency (425 MHz, 2 h, 1 mW) is stimulated because of higher elongation of primary root [11], while duckweed (Lemna minor, Araceae) growth was significantly slowed down not only by exposure at a similar frequency (400 MHz, 4 h, 23 V m^{-1}) but also after exposure at 900 and 1900 MHz for different field amplitudes (23, 41, and 390 V m^{-1}) at least in the first days following the exposure [18]. Surducan et al. [15] also found stimulation of seedling growth in bean and maize after exposure to EMF (2.452 GHz, 0.005 mW cm^{-2}). Senavirathna et al. [55] studied real-time impact of EMF radiation (2 GHz, 1.42 W m^{-2}) on instantaneous growth in the aquatic plant, parrot's feather (Myriophyllum aquaticum, Haloragaceae), using nanometer scale elongation rate fluctuations. These authors demonstrated that EMF-exposed plants displayed reduced fluctuation rates that lasted for several hours after the exposure, strongly suggesting that plants' metabolism experienced a stressful situation. It is worth noting that the exposure did not cause any plant heating (as measured using sensitive thermal imaging). Some other kind of morphological changes also occurred after plant exposure to HF-EMF: induction of epidermal meristems in flax [44], flower bud abscission [43], nitrogen-fixation nodule number increase in leguminous [42], or delayed reduced growth of secondary axis in Rosa [45].

These growth reductions may be related to a lower photosynthetic potential since Răcuciu et al. [40] showed that exposing 12-day-old maize seedlings to 0.47 W kg^{-1} 1 GHz EMF induces a drop in photosynthetic pigment content: the diminution was especially important in chlorophyll a, which was reduced by 80% after 7 h of exposure. Ursache et al. [56] showed that exposure of maize seedlings to microwave (1 mW cm^{-2}, 10.75 GHz) also caused a drop in chlorophyll a and b content. Similarly, Hamada [57] found a decrease in chlorophyll content in 14-day-old seedlings after exposing the caryopses for 75 min at 10.5 GHz. Kumar et al. showed a 13% decrease in total chlorophyll after 4 h exposure of maize seedlings to 1800 MHz (332 mW m^{-2}). These modifications may be related to abnormal photosynthetic activity, which relies on many parameters, including chlorophyll and carotenoid content. Senavirathna et al. [58] showed that exposing duckweeds to 2–8 GHz, 45–50 V m^{-1} EMF induced changes in the nonphotosynthetic quenching, indicating a potential stressful condition. Three aromatic species belonging to Apiaceae family (Petroselinum crispum, Apium graveolens, and Anethum graveolens) strongly respond to global system for mobile communications radiation (GSM, 0.9 GHz, 100 mW cm^{-2}) or wireless local area network (WLAN, 2.45 GHz, 70 mW cm^{-2}) exposure by decreasing the net assimilation rate (over 50%) and the stomatal conductance (20–30%) [34].

CONCLUSION AND FUTURE PROSPECTS

An increasing number of reports highlight biological responses of plants after exposure to HF-EMF at the molecular and the whole plant level. The exposure conditions are, however, far from being standardized and illustrate the diversity of exposure conditions employed. However, future work should avoid exposure in near-field conditions (i.e., in immediate vicinity of the emission antenna) where the field is instable and difficult to characterize. Similarly, the use of communication devices (i.e., cell phones) should be avoided as emission sources since it may be difficult to readily control the exposure conditions because of built-in automation that may overcome the experimental set-up. The use of specialized devices (TEM cells, GTEM cell, waveguides, MSRC, etc.) in which a precise control of exposure condition can be achieved is highly preferable.

Shckorbatov [59] recently reviewed the possible interactions mechanisms of EMF with living organisms. While the classical targets (interaction with membranes, free radicals, and intracellular regulatory systems) have all been observed in plants, a convincing interpretation of the precise mechanism of HF-EMF interaction with living material is still needed. Alternative explanation (i.e., electromagnetic resonance achieved after extremely high frequency stimulation which matches some kind of organ architecture) has also been proposed for very high frequency EMF (several dozen GHz) [60]. However, the reality of this phenomenon in vivo (studied for now only through numerical simulations) and its formal contribution to the regulation of plant development have not yet been experimentally established. Amat et al. [61] proposed that light effects on plants arose not only through chromophores, but also through alternating electric fields which are induced in the medium and able to interact with polar structures through dipole transitions. The possible associated targets (ATP/ADP ratio, ATP synthesis, and Ca^{2+} regulation) are also those affected by exposure to HF-EMF [10]. It could therefore be speculated that HF-EMF may use similar mechanisms. The targeted pathways, especially Ca^{2+}metabolism, are well known to modulate numerous responses of plants to environmental stress. While deeper understanding of plant responses to HF-EMF is still needed, these treatments may initiate a set of molecular responses that may affect plant resistance to environmental stresses, as already demonstrated in wheat for $CaCl_2$ [62] or UV [63] tolerances, and constitute a valuable strategy to increase plant resistance to environmental stressful conditions.

ACKNOWLEDGMENTS

The authors would like to gratefully acknowledge the LRC BIOEM ("Laboratoire de Recherche Conventionné"—LRC no. 002-2011-CEA DAM/Institut Pascal) for its financial support.

REFERENCES

1. A. Balmori, "Electromagnetic pollution from phone masts. Effects on wildlife," Pathophysiology, vol. 16, no. 2-3, pp. 191–199, 2009.
2. H. Kleinlogel, T. Dierks, T. Koenig, H. Lehmann, A. Minder, and R. Berz, "Effects of weak mobile phone—electromagnetic fields (GSM, UMTS) on well-being and resting EEG," Bioelectromagnetics, vol. 29, no. 6, pp. 479–487, 2008.
3. A. Vian, C. Faure, S. Girard et al., "Plants respond to GSM-like radiations," Plant Signaling & Behavior, vol. 2, no. 6, pp. 522–524, 2007.
4. M. Tkalec, K. Malarić, and B. Pevalek-Kozlina, "Exposure to radiofrequency radiation induces oxidative stress in duckweed Lemna minor L.," Science of the Total Environment, vol. 388, no. 1–3, pp. 78–89, 2007.
5. D. Roux, A. Vian, S. Girard et al., "Electromagnetic fields (900 MHz) evoke consistent molecular responses in tomato plants," Physiologia Plantarum, vol. 128, no. 2, pp. 283–288, 2006.
6. É. Beaubois, S. Girard, S. Lallechere et al., "Intercellular communication in plants: evidence for two rapidly transmitted systemic signals generated in response to electromagnetic field stimulation in tomato," Plant, Cell and Environment, vol. 30, no. 7, pp. 834–844, 2007.
7. H. P. Singh, V. P. Sharma, D. R. Batish, and R. K. Kohli, "Cell phone electromagnetic field radiations affect rhizogenesis through impairment of biochemical processes," Environmental Monitoring and Assessment, vol. 184, no. 4, pp. 1813–1821, 2012.
8. A. Vian, D. Roux, S. Girard et al., "Microwave irradiation affects gene expression in plants," Plant Signaling and Behavior, vol. 1, no. 2, pp. 67–69, 2006.
9. M. N. Halgamuge, S. K. Yak, and J. L. Eberhardt, "Reduced growth of soybean seedlings after exposure to weak microwave radiation from GSM 900 mobile phone and base station," Bioelectromagnetics, vol. 36, no. 2, pp. 87–95, 2015.
10. D. Roux, A. Vian, S. Girard et al., "High frequency (900 MHz) low amplitude (5 V m^{-1}) electromagnetic field: a genuine environmental stimulus that affects transcription, translation, calcium and energy charge in tomato," Planta, vol. 227, no. 4, pp. 883–891, 2008.
11. J. A. Stratton, Electromagnetic Theory, McGraw-Hill, New York, NY, USA, 1941.
12. V. P. Sharma, H. P. Singh, R. K. Kohli, and D. R. Batish, "Mobile phone radiation inhibits Vigna radiata(mung bean) root growth by inducing oxidative stress," Science of the Total Environment, vol. 407, no. 21, pp. 5543–5547, 2009.
13. M. Afzal and S. Mansoor, "Effect of mobile phone radiations on morphological and biochemical parameters of Mung bean (Vigna radiata) and wheat (Triticum aestivum) seedlings," Asian Journal of Agricultural Sciences, vol. 4, no. 2, pp. 149–152, 2012.
14. M. D. H. J. Senavirathna and T. Asaeda, "Radio-frequency electromagnetic radiation alters the electric potential of Myriophyllum aquaticum," Biologia Plantarum, vol. 58, no. 2, pp. 355–362, 2014.
15. E. Surducan, V. Surducan, A. Butiuc-Keul, and A. Halgamagy, "Microwaves irradiation experiments on biological samples," Studia Univesitas Babes-Bolyai Biologia, vol. 58, no. 1, pp. 83–98, 2013.
16. P. Jinapang, P. Prakob, P. Wongwattananard, N. E. Islam, and P. Kirawanich, "Growth characteristics of mung beans and water convolvuluses exposed to 425-MHz electromagnetic fields," Bioelectromagnetics, vol. 31, no. 7, pp. 519–527, 2010.
17. M. L. Crawford, "Generation of standard EM fields using TEM transmission cells," IEEE Transactions on Electromagnetic Compatibility, vol. 16, no. 4, pp. 189–195, 1974.
18. M. Tkalec, K. Malarić, and B. Pevalek-Kozlina, "Influence of 400, 900, and 1900 MHz electromagnetic fields on Lemna minor growth and peroxidase activity," Bioelectromagnetics, vol. 26, no. 3, pp. 185–193, 2005.
19. L. M. Liu, F. Garber, and S. F. Cleary, "Investigation of the effects of continuous-wave, pulse- and amplitude-modulated microwaves on single excitable cells of chara corallina," Bioelectromagnetics, vol. 3, no. 2, pp. 203–212, 1982.
20. A. Radzevičius, S. Sakalauskiene, M. Dagys et al., "The effect of strong microwave electric field radiation on: (1) vegetable seed germination and seedling growth rate," Zemdirbyste, vol. 100, no. 2, pp. 179–184, 2013.

21. S. Lalléchère, S. Girard, D. Roux, P. Bonnet, F. Paladian, and A. Vian, "Mode Stirred Reverberation Chamber (MSRC): a large and efficient tool to lead high frequency bioelectromagnetic in vitro experimentation," Progress in Electromagnetics Research B, vol. 26, pp. 257–290, 2010. · ·

22. G. Cogalniceanu, M. Radu, D. Fologea, and A. Brezeanu, "Short high-voltage pulses promote adventitious shoot differentiation from intact tobacco seedlings," Electro- and Magnetobiology, vol. 19, no. 2, pp. 177–187, 2000. ·

23. K. Dymek, P. Dejmek, V. Panarese et al., "Effect of pulsed electric field on the germination of barley seeds," LWT—Food Science and Technology, vol. 47, no. 1, pp. 161–166, 2012.

24. B. Cadilhon, L. Pécastaing, S. Vauchamp, J. Andrieu, V. Bertrand, and M. Lalande, "Improvement of an ultra-wideband antenna for high-power transient applications," IET Microwaves, Antennas and Propagation, vol. 3, no. 7, pp. 1102–1109, 2009.

25. M. Răcuciu, S. Miclăuş, and D. E. Creangă, "Non-thermal, continuous and modulated rf field effects on vegetal tissue developed from exposed seeds," in Electromagnetic Field, Health and Environment, A. Krawczyk, R. Kubacki, S. Wiak, and C. Lemos Antunes, Eds., vol. 29 of Studies in Applied Electromagnetics and Mechanics, pp. 142–148, IOS Press, Amsterdam, The Netherlands, 2008.

26. D. B. Jones, G. P. Bolwell, and G. J. Gilliatt, "Amplification, by pulsed electromagnetic fields, of plant growth regulator induced phenylalanine ammonia-lyase during differentiation in suspension cultured plant cells," Electromagnetic Biology and Medicine, vol. 5, no. 1, pp. 1–12, 1986.

27. V. P. Sharma, H. P. Singh, D. R. Batish, and R. K. Kohli, "Cell phone radiations affect early growth of vigna radiate (Mung Bean) through biochemical alterations," Zeitschrift für Naturforschung C, vol. 65, no. 1-2, pp. 66–72, 2010. ·

28. A. Kumar, H. P. Singh, D. R. Batish, S. Kaur, and R. K. Kohli, "EMF radiations (1800 MHz)-inhibited early seedling growth of maize (Zea mays) involves alterations in starch and sucrose metabolism,"Protoplasma, pp. 1–7, 2015. ·

29. M. Kouzmanova, M. Dimitrova, D. Dragolova, G. Atanasova, and N. Atanasov, "Alterations in enzyme activities in leaves after exposure of Plectranthus Sp. plants to 900 MHZ electromagnetic field,"Biotechnology & Biotechnological Equipment, vol. 23, supplement 1, pp. 611–615, 2009. ·

30. S. Radic, P. Cvjetko, K. Malaric, M. Tkalec, and B. Pevalek-Kozlina, "Radio frequency electromagnetic field (900 MHz) induces oxidative damage to DNA and biomembrane in tobacco shoot cells (Nicotiana tabacum)," in Proceedings of the IEEE/MTT-S International Microwave, pp. 2213–2216, IEEE, Honolulu, Hawaii, USA, June 2007.

31. Y.-P. Chen, J.-F. Jia, and Y.-J. Wang, "Weak microwave can enhance tolerance of wheat seedlings to salt stress," Journal of Plant Growth Regulation, vol. 28, no. 4, pp. 381–385, 2009.

32. V. P. Sharma, G. Singh, and R. K. Kohli, "Effect of mobile phone EMF on biochemical changes in emerging seedlings of Phaseolus aureus Roxb," The Ecoscan, vol. 3, no. 3-4, pp. 211–214, 2009.

33. H. Zare, S. Mohsenzadeh, and A. Moradshahi, "Electromagnetic waves from GSM a mobile phone simulator and abiotic stress in Zea mays L," Journal of Nutrition & Food Sciences, vol. S11, p. 3, 2015. ·

34. M.-L. Soran, M. Stan, Ü. Niinemets, and L. Copolovici, "Influence of microwave frequency electromagnetic radiation on terpene emission and content in aromatic plants," Journal of Plant Physiology, vol. 171, no. 15, pp. 1436–1443, 2014.

35. M. Rammal, F. Jebai, H. Rammal, and W. H. Joumaa, "Effects of long term exposure to RF/MW radiations on the expression of mRNA of stress proteins in Lycopersicon esculentum," WSEAS Transactions on Biology and Biomedicine, vol. 11, pp. 10–14, 2014.

36. J. C. Engelmann, R. Deeken, T. Müller, G. Nimtz, M. R. G. Roelfsema, and R. Hedrich, "Is gene activity in plant cells affected by UMTS-irradiation? A whole genome approach," Advances and Applications in Bioinformatics and Chemistry, vol. 1, pp. 71–83, 2008. ·

37. A. Scialabba and C. Tamburello, "Microwave effects on germination and growth of radish (Raphanus sativus L.) seedlings," Acta Botanica Gallica, vol. 149, no. 2, pp. 113–123, 2002. ·

38. A. Akbal, Y. Kiran, A. Sahin, D. Turgut-Balik, and H. H. Balik, "Effects of electromagnetic waves emitted by mobile phones on germination, root growth, and root tip cell mitotic division of Lens culinarisMedik," Polish Journal of Environmental Studies, vol. 21, no. 1, pp. 23–29, 2012. ·

39. Y. C. Chen and C. Chen, "Effects of mobile phone radiation on germination and early growth of different bean species," Polish Journal of Environmental Studies, vol. 23, no. 6, pp. 1949–1958, 2014. ·

40. M. Răcuciu, C. Iftode, and S. Miclaus, "Inhibitory effects of low thermal radiofrequency radiation on physiological parameters of Zea mays seedlings growth," Romanian Journal of Physics, vol. 60, no. 3-4, pp. 603–612, 2015.

41. L. Ragha, S. Mishra, V. Ramachandran, and M. S. Bhatia, "Effects of low-power microwave fields on seed germination and growth rate," Journal of Electromagnetic Analysis and Applications, vol. 3, no. 5, pp. 165–171, 2011. ·

42. S. Sharma and L. Parihar, "Effect of mobile phone radiation on nodule formation in the leguminous plants," Current World Environment Journal, vol. 9, no. 1, pp. 145–155, 2014. ·

43. A. O. Oluwajobi, O. A. Falusi, and N. A. Zubbair, "Flower bud abscission reduced in hibiscus sabdariffa by radiation from GSM mast," Environment and Pollution, vol. 4, no. 1, pp. 53–57, 2015. ·

44. M. Tafforeau, M.-C. Verdus, V. Norris et al., "Plant sensitivity to low intensity 105 GHz electromagnetic radiation," Bioelectromagnetics, vol. 25, no. 6, pp. 403–

407, 2004.

45. A. Grémiaux, S. Girard, V. Guérin et al., "Low-amplitude, high-frequency electromagnetic field exposure causes delayed and reduced growth in Rosa hybrida," Journal of Plant Physiology, vol. 190, pp. 44–53, 2016.

46. V. J. Berdinas-Torres, Exposure›s systems and dosimetry of large-scale in vivo studies [Ph.D. thesis], Swiss Federal Institute of Technology, Zürich, Switzerland, 2007.

47. M. J. Van Wyk, M. Bingle, and F. U. C. Meyer, "Antenna modeling considerations for accurate SAR calculations in human phantoms in close proximity to GSM cellular base station antennas,"Bioelectromagnetics, vol. 26, no. 6, pp. 502–509, 2005.

48. S. Miclăuş and M. Răcuciu, "A dosimetric study for experimental exposures of vegetal tissues to radiofrequency fields," in Electromagnetic Field, Health and Environment, A. Krawczyk, R. Kubacki, S. Wiak, and C. Lemos Antunes, Eds., vol. 29 of Studies in Applied Electromagnetics and Mechanics, pp. 133–141, IOS Press, Amsterdam, The Netherlands, 2008.

49. J. León, M. C. Castillo, A. Coego, J. Lozano-Juste, and R. Mir, "Diverse functional interactions between nitric oxide and abscisic acid in plant development and responses to stress," Journal of Experimental Botany, vol. 65, no. 4, pp. 907–921, 2014.

50. Y.-P. Chen, J.-F. Jia, and X.-L. Han, "Weak microwave can alleviate water deficit induced by osmotic stress in wheat seedlings," Planta, vol. 229, no. 2, pp. 291–298, 2009. · ·

51. Z.-B. Qiu, J.-L. Guo, M.-M. Zhang, M.-Y. Lei, and Z.-L. Li, "Nitric oxide acts as a signal molecule in microwave pretreatment induced cadmium tolerance in wheat seedlings," Acta Physiologiae Plantarum, vol. 35, no. 1, pp. 65–73, 2013.

52. R. K. Jangid, R. Sharma, Y. Sudarsan, S. Eapen, G. Singh, and A. K. Purohit, "Microwave treatment induced mutations and altered gene expression in Vigna aconitifolia," Biologia Plantarum, vol. 54, no. 4, pp. 703–706, 2010.

53. A. Vian, C. Henry-Vian, and E. Davies, "Rapid and systemic accumulation of chloroplast mRNA-binding protein transcripts after flame stimulus in tomato," Plant Physiology, vol. 121, no. 2, pp. 517–524, 1999. · ·

54. J. A. Teixeira da Silva and J. Dobránszki, "Magnetic fields: how is plant growth and development impacted?" Protoplasma, 2015. ·

55. M. D. H. J. Senavirathna, T. Asaeda, B. L. S. Thilakarathne, and H. Kadono, "Nanometer-scale elongation rate fluctuations in the Myriophyllum aquaticum (Parrot feather) stem were altered by radio-frequency electromagnetic radiation," Plant Signaling and Behavior, vol. 9, no. 4, Article ID e28590, 2014.

56. M. Ursache, G. Mindru, D. E. Creangă, F. M. Tufescu, and C. Goiceanu, "The effects of high frequency electromagnetic waves on the vegetal," Romanian Journal of Physics, vol. 5, no. 1-2, pp. 133–145, 2009.

57. E. A. M. Hamada, "Effects of microwave treatment on growth, photosynthetic pigments and some metabolites of wheat," Biologia Plantarum, vol. 51, no. 2, pp. 343–345, 2007.

58. M.D.H.J.Senavirathna,A.Takashi,andY.Kimura,"Short-duration exposure to radiofrequency electromagnetic radiation alters the chlorophyll fluorescence of duckweeds (Lemna minor),"Electromagnetic Biology and Medicine, vol. 33, no. 4, pp. 327–334, 2014.

59. Y. Shckorbatov, "The main approaches of studying the mechanisms of action of artificial electromagnetic fields on cell," Journal of Electrical & Electronic Systems, vol. 3, no. 2, 2014. ·

60. A. M. Pietak, "Structural evidence for electromagnetic resonance in plant morphogenesis," BioSystems, vol. 109, no. 3, pp. 367–380, 2012.

61. A. Amat, J. Rigau, R. W. Waynant, I. K. Ilev, and J. J. Anders, "The electric field induced by light can explain cellular responses to electromagnetic energy: a hypothesis of mechanism," Journal of Photochemistry and Photobiology B: Biology, vol. 82, no. 2, pp. 152–160, 2006.

62. Z. Qiu, J. Li, Y. Zhang, Z. Bi, and H. Wei, "Microwave pretreatment can enhance tolerance of wheat seedlings to $CdCl_2$ stress," Ecotoxicology and Environmental Safety, vol. 74, no. 4, pp. 820–825, 2011. · ·

63. Y.-P. Chen, "Microwave treatment of eight seconds protects cells of Isatis indigotica from enhanced UV-B radiation lesions," Photochemistry and Photobiology, vol. 82, no. 2, pp. 503–507, 2006.

CHAPTER 16 EVALUATION OF SOME TREE SPECIES TO ABSORB AIR POLLUTANTS IN INDUSTRIAL LOCATIONS

Abida Begum and S. Harikrishna

Keywords: Tree species, Pollution, Industrial area, Bengaluru.

INTRODUCTION

Air pollution can be defined as the human introduction into the atmosphere of chemicals, particulate matter or biological materials that cause harm or discomfort to humans, or other living organism or damage the environment. Air pollution is a major problem arising mainly from industrialization. Air pollutions can directly affect plants via leaves or indirectly via soil acidification. It has also been reported that when exposed to air pollutants, most plant experience physiological changes before exhibiting visible damage to leaves. Even in the kingdom of plants, there is no escaping Darwin's survival of the fittest. Some plants thrive in environments that others would find toxic, these plants can clean-up various sources of man-made pollution; both organic (petrochemical) and inorganic (heavy metal toxins). Trees remove a significant amount of pollution from the atmosphere as part of their normal functioning. They directly increase the quality of the air in the city and its surrounding area and should be considered an integral part of any comprehensive plan aimed at improving overall air quality. The Indian tree flora comprises of nearly 2000 species belonging to about 800 genera and 150 families of flowering plants. This vast diversity ranks 10th in the world and even higher relative to the area[1]. This includes majority of the cultivated and exotic species. Trees provide a large leaf surface onto which particles are deposited and gases are removed. Pollution is removed by nearly all parts of a tree; the soil, roots and vegetative portions of the tree species. Trees respirate and exchange gases through stomates, or holes, on their leaves; these gases include those necessary for the tree's functioning as well as other gaseous air pollutants. Once inside the leaf, gases diffuse into the spaces between the cells of the leaf to be absorbed by water films or chemically altered by plant tissues. Trees also reduce air pollution by intercepting airborne particles and retaining them on the leaf surface, called dry deposition. Some can be absorbed by the leaf surface itself, although most remain on the plant surface[7]. Leaf surfaces are most efficient at removing pollutants that are water soluble including sulfur dioxide, nitrogen dioxide and ozone. Pollutants travel through plants by translocation via the xylem and phloem. Chemical pollutants absorbed by the leaves are translocated to the root areas where they can be broken down

Citation: Abida Begum and S. Harikrishna, "Evaluation of Some Tree Species to Absorb Air Pollutants in Industrial Locations," E-Journal of Chemistry, vol. 7, no. S1, pp. S151-S156, 2010. doi:10.1155/2010/398382.

by microbes in the soil and pollutants absorbed by the roots can be broken down and translocated to the leaves where they are released into the atmosphere.

EXPERIMENTAL

The sampling sites in the present study were three industrial areas (electronic city, bommasandra industrial area and jigini industrial area) of south bengaluru. The traffic density of the locations is moderate to high with peak periods found during morning and evening hours. The durations chosen for sampling the tree species was in summer during 15th March to15th May 2008. Three replicates of fully matured leaves samples from all trees growing in the industrial conditions were selected from the immediate vicinity of the stations for the purpose of determining ATPI index in the leaves. The plant samples[2] were put through a three step washing sequence, air dried, weighed and placed in a dehydrator at approximately 80 °C.

To determine relative leaf water content (RWC) the fresh leaves were weighed and then immersed in water over night, blotted dry and then weighed to get the turgid weight. The leaves were dried overnight in an oven at 55°C and reweighed to obtain the dry weight 5 g of the fresh leaves was homogenized in 10 mL deionised water. This was filtered and the pH of the leaf extract determined after calibrating pH meter using buffer solution.

Ascorbic acid content (AA) (mg/g) was measured using spectrophotemetric method. 1 g of the fresh foliage was put in a test-tube, 4 mL oxalic acid - EDTA extracting solution was added, then 1 mL of orthophosphoric acid and then 1 mL 5% tetraoxosulphate(VI) acid added to this mixture, 2 mL of ammonium molybdate was added and then 3 mL of water. The solution was then allowed to stand for 15 minutes. After which the absorbance at 760 nm was measured with a spectrophotometer. The concentration of ascorbic acid in the sample was then extrapolated from a standard ascorbic acid curve.

To determine total chlorophyll content (TCh) in the leaves, 3 g of fresh leaves were blended and then extracted with 10 mL of 80% acetone and left for 15 minutes for thorough extraction. The liquid portion was decanted into another test-tube and centrifuged at 2,500 rpm for 3 minutes. The supernatant liquid was then collected and the absorbance taken using a spectrophotometer at 645 nm and 665 nm. Calculations were done according to the formula given below.

$$\text{Relative leaf water content} = \frac{\text{Fresh weight - Dry weight}}{\text{Turgid weight} - \text{Dry weight}} \text{ X } 100$$

$$\text{Air pollution tolerance index} = \frac{\text{AA (TCh + pH) + RWC}}{10}$$

$$\text{Chlorophyll 'a'} = \frac{12.7 \text{ Dy x } 663 - 2.69 \text{ Dy x } 645 \text{ x V}}{1000 \text{ W}} \text{ mg/g}$$

$$\text{Chlorophyll 'b'} = \frac{22.9 \text{ Dy x } 645 - 4.68 \text{ Dy x } 663 \text{ x V}}{1000 \text{ W}} \text{ mg/g}$$

Where AA - Ascorbic acid content (mg/g), TCh = Total chlorophyll mg/g, pH = pH of leaf extract and RWC = Relative water content of leaf, % TCh = Chlorophyll a + b mg/g, Dy = Absorbance of the extract at the wavelength y nm. V = Total volume of the chlorophyll solution (mL) and W = Weight of the tissue extracted (g)[3].

RESULTS AND DISCUSSION

Air pollution tolerant index is an index denotes capability of a plant to combat against air pollution. Plants which have higher index value are tolerant to air pollution and can be caused as sink to mitigate pollution, while plants with low index value show less tolerance and can be used to indicate levels of air pollution (Table 2). This study was carried out to assess the impact of industries on the available tree species grown at three different industrial areas (Table 1). In this study, changes in parameters such as ascorbic acid, total chlorophyll, relative water content, pH of leaf extract were used in evaluating the degree of tolerance to air pollution by the plant species. The results are as shown on Table 3-5.

Table 1. Showing the tree species growing in three different stations

S. No.	Tree Species		
	Botonical names	Family	Common Names
1	*Syzygium cumini*	Myrtaceae	Jamun
2	*Michelia champaca*	Magnoliacea	Champa
3	*Acacia melanoxylon*	Fabaceae	Blackwood,
4	*Euculeptus* sp.	graminaea	Euculeptus
5	*Ficus Benghalensis*	Moraceae	Banyan Tree
6	*Delonix regia* Raf.	Fabaceae	flamboyant tree
7	*Morinda pubescens*	Rubiaceae	Indian Mulberry
8	*Millingtonia hortensis*	Bignoniaceae	Jasmine Tree
9	*Leucaena leucocephala*	Fabaceae	Wild Tamarind,
10	*Saraca Indica*	Leguminosae	Ashoka
11	*Caesalpinia pulcherrima*	Fabaceae	Dwarf poinciana
12	*Dalbergia lanceolaria* L.f.	Laguminosae	Dhobin,Satpuria
13	*Ficus religiosa*	Moraceae	Peepal
14	*Azadirachta indica*	Meliaceae	Neem
15	*Pongamia pinnata* (L.)	Fabaceae Dicotyledon	Pongamia
16	*Madhuca latifolia* (Roxb)	Sapotaceae	Mahua
17	*Diploknema butyrace*	Sapotaceaea	Indian butter tree

Table 2. Air pollution index values and plant response to pollution

APTI value	Response[6]
30 to100	Tolerant
29 to17	Intermediate
16 to 1	Sensitive
<1	Very sensitive

Table 3. Air pollution tolerance index (APTI) for tree species from Electronic city of south Bengaluru

S. No.	Tree Species	Total chlorophyll mg/mL	Ascorbic acid, mg/g	pH	Relative water content, %	Air pollution tolerance index, (APTI)
1	*Syzygium cumini*	3.47	0.24	4.7	80.02	16.172
2	*Michelia champaca*	3.41	0.34	4.9	78.56	16.166
3	*Acacia melanoxylon*	3.98	0.15	6.5	77.54	18.234
4	*Euculeptus* sp.	3.43	0.09	6.7	72.53	17.383
5	*Ficus benghalensis*	5.64	0.65	6.4	78.5	19.89
6	*Delonix regia* Raf.	1.56	0.73	5.7	67.35	13.995
7	*Morinda pubescens*	1.54	0.78	5.8	54.65	12.805
8	*Millingtonia hortensis*	1.23	0.99	5.8	60.34	13.064
9	*Leucaena leucocephala*	3.21	0.96	5.7	54.74	14.384
10	*Saraca indica*	2.32	0.54	4.9	51.33	12.353
11	*Caesalpinia pulcherrima*	4.56	0.36	5.1	52.53	14.913
12	*Dalbergia lanceolaria* L.f.	7.44	0.87	6.2	81.6	21.8
13	*Ficus religiosa*	4.56	0.51	5.3	67.45	16.605
14	*Azadirachta indica*	6.03	0.93	5.4	82.5	19.68
15	*Pongamia pinnata* (L.)	7.45	0.64	5.5	85	21.45
16	*Madhuca latifolia* Roxb	4.09	0.57	5.6	65.95	16.285
17	*Diploknema butyrace*	6.48	0.78	5.9	64.35	18.815

Table 4. Air pollution tolerance index (APTI) for tree species Jigini industrial area of south Bengaluru

S. No.	Tree Species	Total chlorophyll mg/mL	Ascorbic acid, mg/g	pH	Relative water content, %	Air pollution tolerance index (APTI)
1	*Syzygium cumini*	4.4	0.44	5.6	86.36	38
2	*Michelia champaca*	4.5	0.56	4.9	61.93	32.6
3	*Acacia melanoxylon*	5.6	0.53	6.5	44.44	28.5
4	*Euculeptus* sp.	4.2	0.4	6.7	55.50	24.2
5	*Ficus benghalensis*	5.64	0.65	6.4	21.47	16.8
6	*Delonix regia* Raf.	5.3	0.73	5.7	18.06	14.5
7	*Morinda pubescens*	3.4	0.78	5.8	41.11	29.5
8	*Millingtonia hortensis*	2.9	0.99	5.8	18.11	15.6
9	*Leucaena leucocephala*	4.5	0.96	5.7	19.30	18.9
10	*Saraca indica*	3.3	0.54	4.9	33.20	14.7
11	*Caesalpinia pulcherrima*	6.2	0.36	5.1	40.31	16.4
12	*Dalbergia lanceolaria* L.f.	7.3	0.87	6.2	27.67	32.5
13	*Ficus religiosa*	4.4	0.51	5.3	37.40	18.5
14	*Azadirachta indica*	6.03	0.93	5.4	33.49	35.6
15	*Pongamia pinnata* (L.)	7.46	0.64	5.5	39.09	32.4
16	*Madhuca latifolia* Roxb	4.09	0.57	5.6	62.64	34.6
17	*Diploknema butyrace*	6.48	0.78	5.9	33.55	32.4

Table 5. Air pollution tolerance index (APTI) of tree species from Bommasandra industrial area of south Bengaluru

S. No.	Tree Species	Total chlorophyll (mg/ml)	Ascorbic acid mg/g	pH	Relative water content, %	Air pollution tolerance index (APTI)
1	*Syzygium cumini*	5.2	0.32	5.9	98.54	35
2	*Michelia champaca*	4.3	0.51	4.7	65.36	30
3	*Acacia melanoxylon*	5.2	0.44	6.2	52.83	26.5
4	*Euculeptus* sp.	4.4	0.5	6.1	42.29	22.2
5	*Ficus benghalensis*	5.9	0.72	6.2	17.45	15.2
6	*Delonix regia* Raf.	5.6	0.82	5.2	15.24	13.5
7	*Morinda pubescens*	4.8	0.72	5.6	35.39	26.5
8	*Millingtonia hortensis*	3.5	0.67	6.2	22.78	14.5
9	*Leucaena leucocephala*	5.4	0.87	6.3	17.88	18.2
10	*Saraca indica*	4.4	0.62	4.9	24.63	14.2
11	*Caesalpinia pulcherrima*	6.2	0.36	5.1	40.31	16.4
12	*Dalbergia lanceolaria* L.f.	7.8	0.87	6.2	27.67	32.5
13	*Ficus religiosa*	4.4	0.51	5.3	37.40	18.5
14	*Azadirachta indica*	6.03	0.93	5.4	33.49	35.6
15	*Pongamia pinnata* (L.)	6.4	0.82	5.6	32.52	32
16	*Madhuca latifolia* Roxb	4.09	0.89	6.4	35.56	33.2
17	*Diploknema butyrace*	6.42	0.74	5.5	35.71	31.5

At Electronic city, the concentration of the ascorbic acid (AA) ranged between 0.09 and 0.99 mg/g of the green weight, chlorophyll (TCh) concentration ranged between 1.32-7.4505 mg/g of the green weight[5,6]. For the tree species, the study showed that the highest concentration in the ascorbic acid (Vitamin C) was in *Millingtonia hortensis* and the least concentration was found in *Euculeptus sp.* The APTI values ranged between 21.8-12.3, the maximum air pollution tolerance index value was found in *Dalbergia lanceolaria* L.f. (21.8) *Pongamia pinnata* (L.) (21.45) *Azadirachta indica* (19.68), *Diploknema butyrace* (18.815) and minimum were in *Saraca indica* (12.3) and *Morinda pubescens* (12.8)

At Jigini industrial area the concentration of the ascorbic acid (AA) ranged between 0.4 and 0.99 mg/l00 g of the green weight, chlorophyll (TCh) concentration ranged between 3.3-7.46 mg/g of the green weight .The Air pollution tolerance index value was found between 38-14.5 the maximum value was in the following tree species *Syzygium cumini* (38) *Pongamia pinnata* (L.) (35.6) *Madhuca latifolia* Roxb (34.6), *Dalbergia lanceolaria* L.f.

(32.5) and minimum was in *Delonix regia* Raf. (14.5) and (14.7) for *Saraca indica.*

At Bommasandra industrial area, the concentration of the ascorbic acid (AA) ranged between 0.32 and 0.93 mg/g of the green weight, chlorophyll (TCh) concentration ranged between 3.5-7.86 mg/g of the green weight. The maximum air pollution tolerance index value was found between 35.6-14.2 the maximum value was in the following tree species *Azadirachta indica* (35.6) *Syzygium cumini* (35) *Madhuca latifolia* Roxb (33.2), *Dalbergia lanceolaria* L.f. (32.5) and minimum was in *Delonix regia* Raf. (13.5) and *Saraca indica* (14.2). The result reveals that out of the seventeen tree species studied, *Syzygium cumini* (Jamun) is most tolerance, followed by *Azadirachta indica, Madhuca latifolia* Roxb and *Madhuca latifolia* Roxb in Bommasandra and Jigini industrial area. For the tree species in all the three station, the study showed that the highest concentration in the ascorbic acid (Vitamin C) was in *Millingtonia hortensis*[8,9] and the least concentration was found in

CONCLUSION

This study has identified the various tree species growing in all the three industrial areas of south Bengaluru. An overview of the entire result obtained from this study reveals that different plants respond differently to air pollution; hence the different indices show that plants growing in apparently polluted environment have higher APTI values compared to moderately polluted environment.

REFERENCES

1. Abida Begum, Ramaiah.M, Harikrishna, Irfanulla Khan and Veena K *E-J Chem.*, 2009, 6(1) 13-22.
2. Brown S, Gillespie A J R and Lugo A E, *Forest Sci.*, 1989, 38, 881–902.
3. Agrawal M, Nandi P K and Rao D N, *Water, Air Soil Pollut.,* 1982, 18, 449-454.
4. Agarwal M, *Natl Acad Sci Lett.,* 2005, 28(3&4), 93-105.
5. Beckett K P, Freer-Smith P H and Taylor G, *Environ Pollut.,* 1998, 99, 347-360.
6. Beg M A A, Yousufzai A H K and Mahmood S N, *Pak J Sci Ind Res.*, 1987, 30, 60-70.
7. Joshi P C and Chauhan A, *Life Sci J.,* 2008, 5(3), 41-45.
8. Joshi P C and Swami A, *Environmentalist,* 2007, 27, 365-374.
9. Darley E F, Dugger W M, Mudd J B and Ordin L, *Arch Environ Health*, 1963, 6, 700-761

CHAPTER 17

BIOCHEMICAL AND ULTRASTRUCTURAL CHANGES IN SIDA CORDIFOLIA L. AND CATHARANTHUS ROSEUS L. TO AUTO POLLUTION

Vijeta Verma and Neelam Chandra

INTRODUCTION

Air pollution, a global problem being faced by both the developed nations and the developing ones, has been aggravated by developments that typically occur as countries become industrialized: growing cities, increased traffic, rapid economic development and industrialization, and high levels of energy consumption. Air pollution is a major problem arising mainly from industrialization [1]. Motor vehicles account for 60–70% of the pollution found in an urban environment [2–4]. It is estimated that vehicles account for 70% of CO, 50% of HC, 30–40% of NOx, 30% of SPM, and 10% of SO_2 of the total pollution load in the major metros of India, of which two-thirds are contributed by two wheelers alone [5, 6].

Besides the human and animal populations, this problem has drastic impacts on the local environment and causes extensive damage to vegetation including crops, fruit trees, medicinal plants, and ornamentals. Plants being directly and constantly exposed to the pollutants (both gaseous and particulates) play a significant role as indicators and in mitigating the problem. They absorb, accumulate, and integrate the pollutants impinging on their foliar surface, acting as the sinks for various pollutants and thus mitigating the problem. The plants do not render this service to mankind without any serious implications; in turn, they suffer from various deformities caused by the integrating pollutants and show diverse morphological, biochemical, anatomical, and physiological responses. Thus, the plants can be used as both passive biomonitors and biomitigators in the urban environment to indicate the environmental quality and to attenuate the pollution level in a locality [7, 8]. Pollutants like SO_2, O_3, and CO_2 have many harmful effects as they affect the physiological activities of the plants. Sulfur dioxide is responsible for the increase in oxidative stress. They enhanced the production of reactive oxygen species like H_2O_2 which is very harmful to the metabolic activities of the plants [9]. Sulfur dioxide polluted regions showed the higher activities of peroxidase [10, 11]. Sulfur dioxide also affects the process of photosynthesis [12]. Vehicular traffic is the main source of particulate air pollution in Lucknow and these are the prime source of air pollution in urban areas [13]. The number of different categories of vehicles registered with RTO (Regional Transport Office), Lucknow, is 12,09,745 on March 31, 2011, which is 9.23% higher over the last year. Charbagh is the most polluted area in Lucknow, says a report published by

Citation: Vijeta Verma and Neelam Chandra, "Biochemical and Ultrastructural Changes in Sida cordifolia L. and Catharanthus roseus L. to Auto Pollutionss," International Scholarly Research Notices, vol. 2014, Article ID 263092, 11 pages, 2014. doi:10.1155/2014/263092.

Indian Institute of Toxicology Research. The study carried out in three locations of the state capital showed that Charbagh has the highest average level of respirable particulate matter (RSPM), sulfur dioxide, and oxides of nitrogen among all the locations. As per the national ambient air quality standards (NAAQS), Charbagh had RSPM of 253.66 with 23.62 and 58.42 micrograms per cubic meter of sulfur dioxide and oxides of nitrogen, respectively. The report cited the addition of 109,773 vehicles on the roads of Lucknow as the chief reason for pollution and called for emergency attention of policy markers, researchers, and regulatory agencies.

Sida cordifolia L. is an erect perennial commonly known as flannel weed. Flannel weed is used in Ayurvedic medicine. It is used in neurological ailments, especially in stroke rehabilitation. It has been investigated as an anti-inflammatory for preventing cell proliferation and for encouraging liver regrowth. Due to its ephedrine content, it possesses psychostimulant properties, affecting the central nervous system and also the heart. The roots and stems contain the alkaloids ephedrine. Catharanthus roseus L., also known as periwinkle, is an important medicinal plant which belongs to family Apocynaceae and is an important source of indole alkaloids which are present in all plant parts. Periwinkle is used for the treatment of diabetes, fever, malaria, throat infections, and chest complaints. It is also used for the regulation of menstrual cycles and as a euphoriant. The physiologically important and antineoplastic alkaloids, namely, Vincristine and Vinblastine, are mainly present in the leaves, whereas antihypertensive alkaloids, such as ajmalicine, serpentine, and reserpine, are reported to be present in the roots. Vincristine and Vinblastine alkaloids are used in the treatment of various types of lymphoma and leukemia. These Catharanthus alkaloids are also used for the treatment of both malignant and nonmalignant diseases and in platelet and platelet associated disorders. Both plants are used as roadside plants as a green belt in cities. In view of plants' role in the indication and abatement of air pollution, this study was carried out to assess the impact of automobile pollution on the plant metabolism and foliar surface configuration of plants growing along the roadsides in the urban locality in order to identify their role as the biomarkers of air pollution.

MATERIALS AND METHODS

For micromorphological and physiological studies, leaves of Sida cordifolia L. (flannel weed) and Catharanthus roseus L. (periwinkle) were collected from the Botany Department, Lucknow University, which was found to be healthy and treated as control (C) and also the plants growing inside the Talkatora (Industrial area P1) which has been taken as polluted site (industrial area, P1), Charbagh, loco workshop, P2, which has been taken as polluted site. For micromorphological evaluation, whole studies were carried out by comparing only one polluted site (P2) with healthy site in the summer. The pieces of leaves were cut ($1 \times 1\ cm^2$) and placed in maceration mixture (conc. $HNO_3 + H_2CrO_3$ in a ratio of 9 : 1). Cuticles were stained in sudan 1 V and mounted in glycerin jelly. Epidermal structures were examined under low and high power of the microscope. Terminology used was the same as suggested by Pant [14], Stace [15], and Dilcher [16]. Scanning electron microscopy (SEM) examination of values of both genera was carried out at the Birbal Sahni Institute of Paleobotany (BSIP), Lucknow, India. For biochemical and physiological studies, both polluted sites (P1 and P2) were taken into consideration in comparison to healthy site in both the summer and the winter. Photosynthetic pigments in leaves were extracted in 80% acetone by the method of Lichtenthaler et al. [17]. The extract was centrifuged and read out at 480 and 510 nm for carotenoid, 645 and 663 nm for chlorophyll, and 655 and 666 for phaeophytin. Catalase was assayed by the method of Euler and Josephson [18]. The reaction mixture for enzyme assay contained 0.005 M hydrogen peroxide in 0.025 potassium phosphate buffer pH 7.0 and this was standardized against 0.1 N $KMnO_4$. Suitable amount of aliquot was added. After 5 minutes, the reaction was stopped by adding 2 N H_2SO_4. Peroxidase was estimated by the modified method of Luck et al. [19]. To 2 mL of 0.1 M phosphate buffer pH 6.0, 0.01% H_2O_2 and 0.5% (w/v) p-phenylenediamine were added. Reaction was stopped by adding 4 N H_2SO_4. The whole reaction was carried out at 25°C and color intensity was read at 485 nm.

Statistical Analysis

The data has been presented in tables and figures and is the mean of the observations made in triplicate (n=3) along with the standard error and the least significant differences between mean values ($P \leq 0.05$). The data has been put to ANOVA.

RESULT AND DISCUSSION

Micromorphological Structures

The use of plants as monitors of air pollution has long been accepted since plants often are the initial indicators of air pollution. Individual plant responds differently to different air pollutants. Plants improve the quality of urban life due to their large leaf areas, relative to the ground on which they stand. Depending on structural properties of their surface, they can act as biological absorbers or filters of pollutants [7]. In this way, they remove huge quantities of gaseous pollutants and airborne particles, thus improving the quality of the environment. Some plant species have been identified to be able to absorb, detoxify, and tolerate high levels of pollution [20]. The foliar surface is the most important receptor of atmospheric pollutants where they cause several structural and biochemical changes. The present studies on the plants growing in loco area indicate that auto exhaust pollution brought appreciable changes in the number of epidermal cells and stomata per unit area. Plants of both genera growing very close to the loco area showed an adverse effect on this number. In the present investigation, it has been observed that the stomatal and epidermal frequencies decline, resulting in the significant fall in stomatal index (Table 1). Size of epidermal cells and stomata was increased in polluted population. Since differentiation of stomata mother cells requires division of epidermal cells, therefore, a decrease in the frequency of stomata should be normally accompanied by an increase in the size of epidermal cells. Our results were in conformity with the findings of [21] which studied the effect of environmental pollution on morphology and leaf epidermis of Commelina bengalensis. Results were the same; stem length, leaf area, flower size, and fruit size showed a marked reduction in growth in the plants from the polluted area. Length and density of trichome increased, while stomatal frequency values and the epidermal frequency decreased, in response to environmental pollution (Table 1). Plants of flannel weed showed changes in their leaf morphology due to environmental stresses. Lower foliar surface showed stomata with board aperture and large size trichome, while slit-like aperture of stomata was observed in polluted site plant in comparison to control (Figure 1). It has been reported that air pollutants increase cell permeability by damaging the membrane integrity [22]. Pollutant induced increased cell permeability may cause a loss of water from the guard cells to make them flaccid, which result in stomatal closure. In Catharanthus roseus L. stomata was observed on either side of mid vein on the upper surface of healthy population [Figure 2(a)], but stomata on one side of mid vein [Figure 2(d)] was observed on the plants collected from highly polluted site. Lower foliar surface showed stomata with thin walled marginal cells (Figure 2(b)) in healthy plants, while, in polluted population, there were thick walled marginal cells with occluded stomata (Figure 1(e)) observed. Morphological characters of plants are very important in determining plant resistance to air pollution. Pal et al. [23] studied sunken stomata and thick cuticles; small and dense cells and suberised cell walls are in favor of reducing pollutant entry into leaves and cells. When the leaf samples of these plants were examined under SEM, the following observations were recorded. SEM micrographs of lower foliar surface of flannel weed showed sunken stomata with occluded stomatal pores. It was also observed that cuticle was ruptured at the edge of stomata to form a crypt-like structure (Figure 3). In SEM micrographs of periwinkle, upper foliar surface showed encrypted and sunken stomata, while lower foliar surface showed stomata with striations arising from subsidiary cells (Figure 3).

Table 1: Effect of air pollution on the micromorphological structures present on the leaves of Sida cordifolia L. and Catharanthus roseus L. collected in the summer

Plant name	Micromorphological structures	Parameters	Lucknow University (nonpolluted area, HS)		Charbagh (polluted site, P2)	
			Abaxial surface	Adaxial surface	Abaxial surface	Adaxial surface
Sida cordifolia L.	Epidermal cells	Frequency	175 ± 1.675	140 ± 1.989	*152 ± 1.543	*118 ± 1.833
		Size in μm	L—12 ± 0.489	L—11 ± 0.324	*L—20 ± 0.571	*L—18 ± 0.110
			B—07 ± 0.421	B—06 ± 0.211	*B—13 ± 0.114	*B—13 ± 0.098
	Stomata	Frequency	20 ± 0.523	15 ± 0.760	*11 ± 0.212	*07 ± 0.311
		Size in μm	L—06 ± 0.214	L—04 ± 0.214	L—08 ± 0.214	L—06 ± 0.189
			B—02 ± 0.115	B—02 ± 0.115	*B—05 ± 0.115	B—04 ± 0.097
		Stomatal index	14 ± 0.117	10 ± 0.387	*06 ± 0.241	*04 ± 0.265
	Trichomes	Frequency	05 ± 0.126	03 ± 0.056	*09 ± 0.162	*07 ± 0.044
		Size in μm	90 ± 0.897	80 ± 1.333	*130 ± 1.112	*95 ± 1.211
Catharanthus roseus L.	Epidermal cells	Frequency	144 ± 1.125	117 ± 1.254	*128 ± 1.105	*100 ± 1.002
		Size in μm	L—20 ± 0.789	L—14 ± 0.324	*L—25 ± 0.771	L—18 ± 0.110
			B—14 ± 0.421	B—09 ± 0.211	*B—17 ± 0.114	B—11 ± 0.098
	Stomata	Frequency	34 ± 0.457	12 ± 0.152	*29 ± 0.356	*08 ± 0.321
		Size in μm	L—20 ± 0.789	L—18 ± 0.324	L—21 ± 0.771	L—20 ± 0.110
			B—12 ± 0.421	B—09 ± 0.211	B—13 ± 0.114	B—10 ± 0.098
		Stomatal index	20 ± 0.459	09 ± 0.103	*15 ± 0.451	*05 ± 0.357

L : length, B: breadth; data represent S.E. (n = 3). * indicates values significant from healthy site ($P \leq 0.05$).

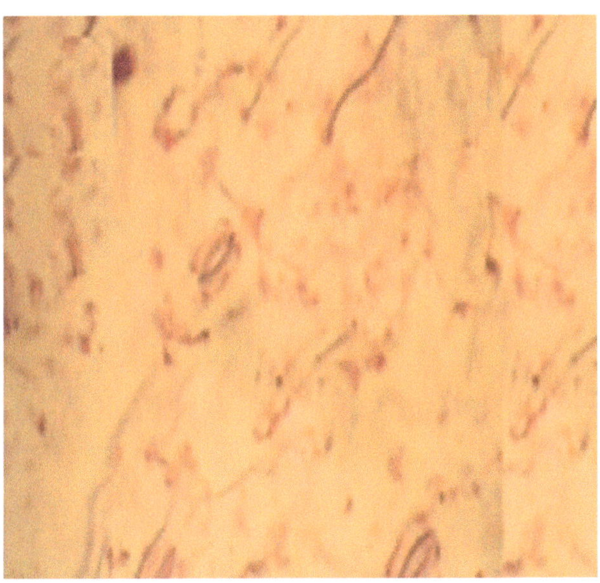

(a)

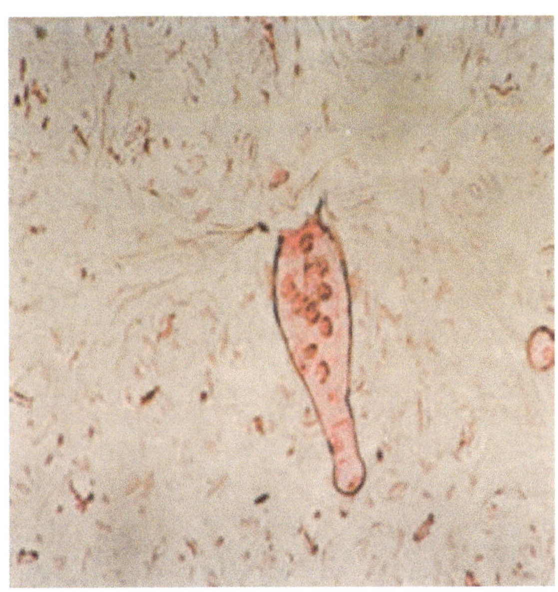

(b)

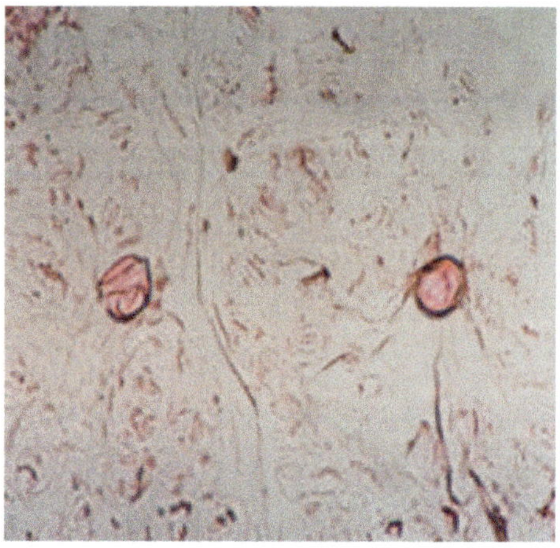

(c)

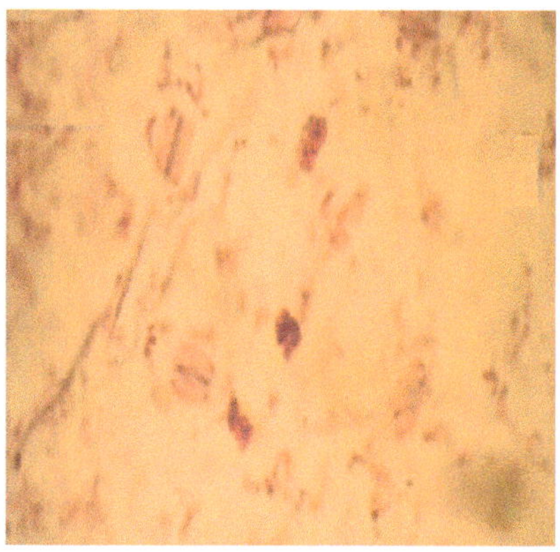

(d)

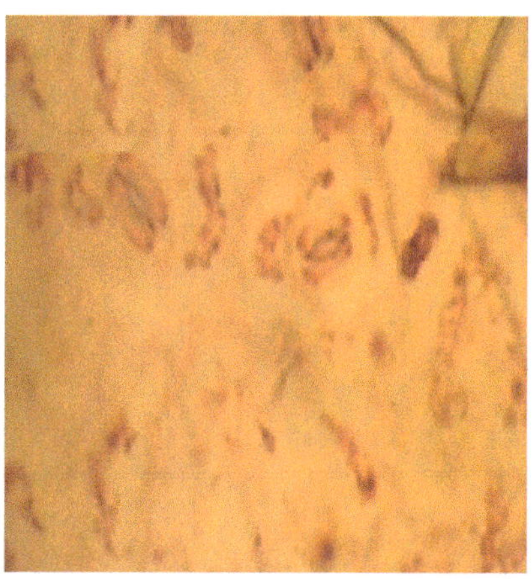

(e)

(f)

Figure 1: Effect of air pollution on micromorphological structures on leaves of Sida cordifolia L. in the summer (left side: healthy site; right side: polluted site 2) ((a) upper foliar surface showing stomata; (b) lower foliar surface showing stomata with striations at the trichome base; (c) lower foliar surface showing distribution of stomata with developing trichome base; (d) upper foliar surface showing slit-like aperture of stomata; (e) lower foliar surface showing distribution of stomata; (f) lower foliar surface with degenerating trichome).

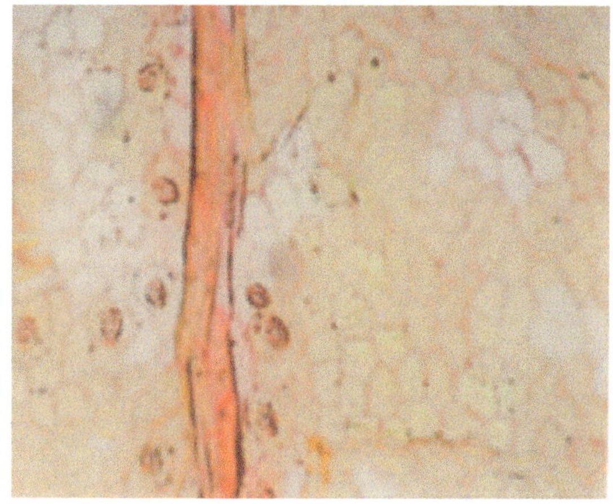

(a)

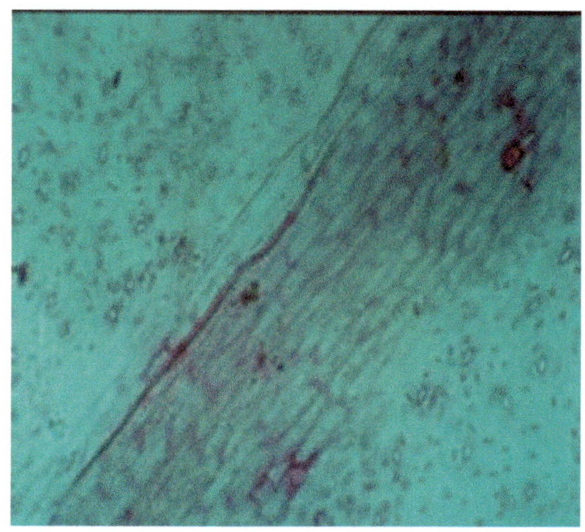

(b)

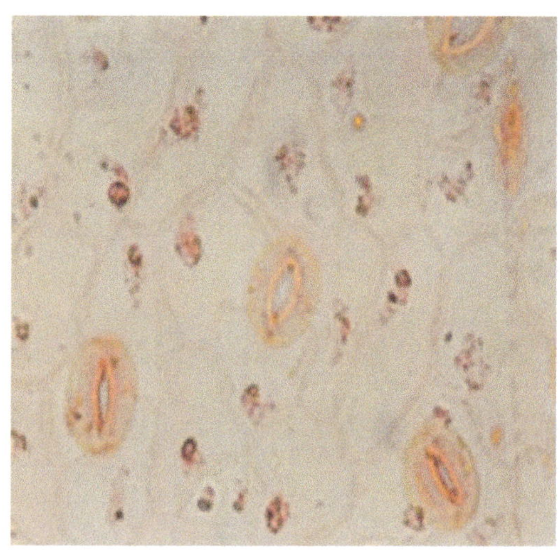

(c)

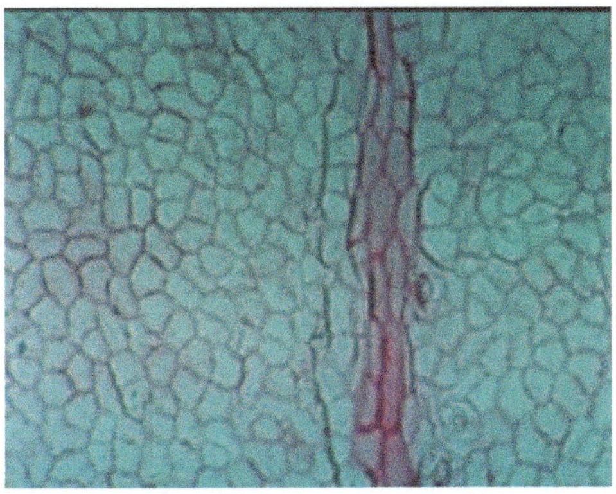

(d)

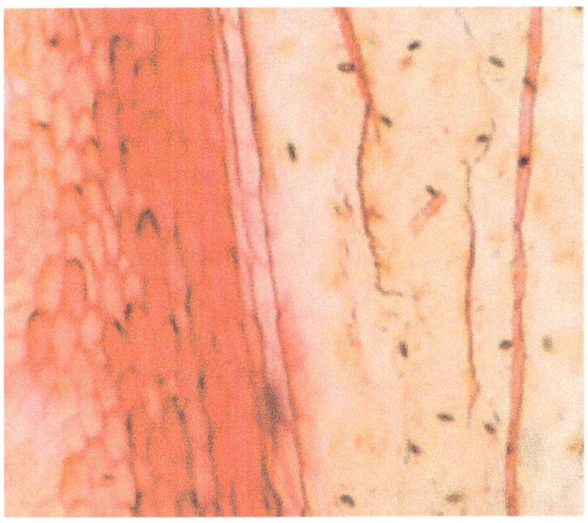

(e)

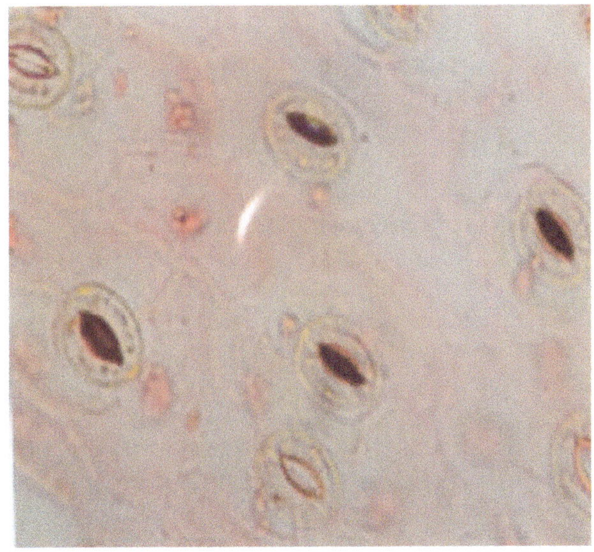

(f)

Figure 2: Effect of air pollution on micromorphological structures on leaves of Catharanthus roseus L. in the summer (left side: control; right side: polluted site 2). ((a) upper foliar surface showing stomata on either side of midvein; (b) lower foliar surface showing stomata with thin walled marginal cells; (c) lower foliar surface showing epidermal cells with stomata; (d) upper foliar surface showing stomata on one side of midvein; (e) lower foliar surface showing occluded stomata with thick walled marginal cells; (f) lower foliar surface showing occluded stomata).

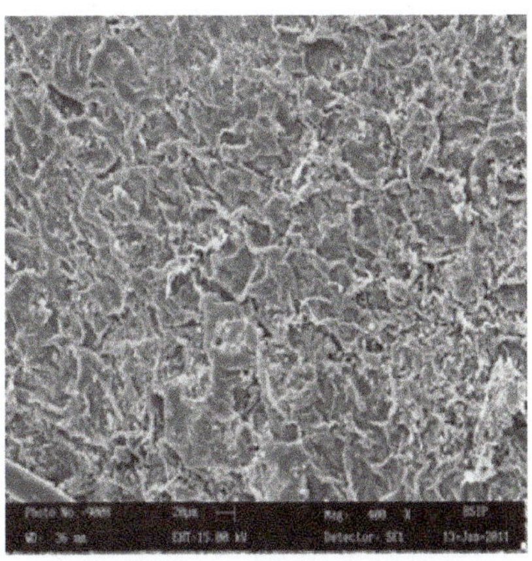

(a)

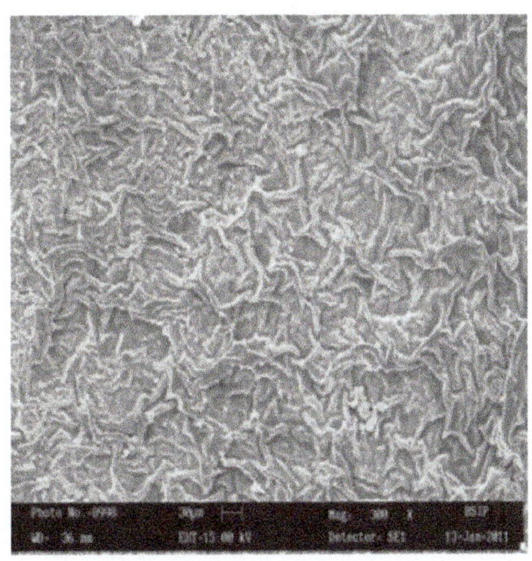

(c)

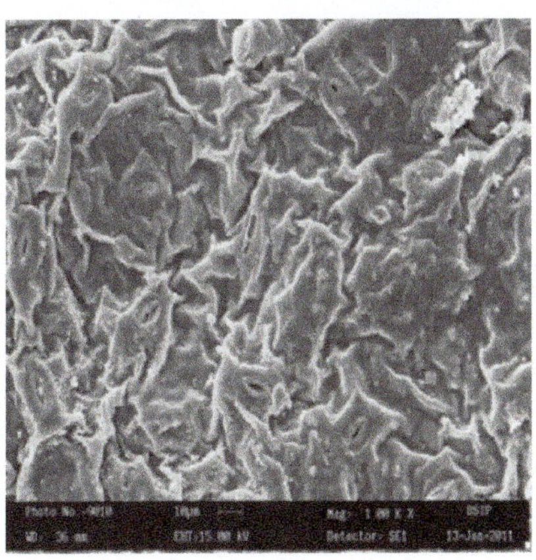

(b)

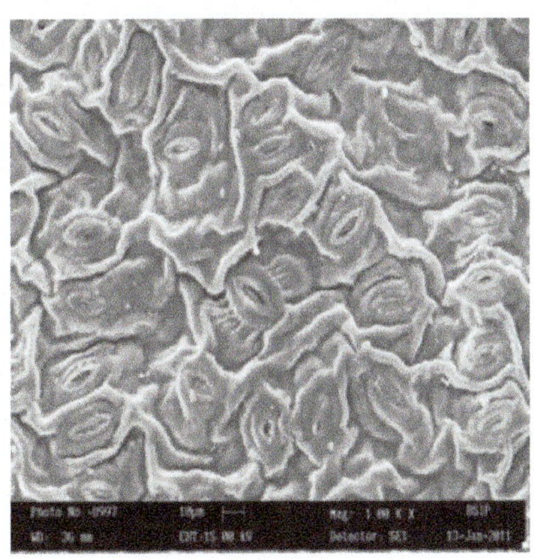

(d)

Figure 3: SEM photographs showing the effect of air pollution on leaf surface of Sida cordifolia L. (left: (a) = lower surface showing sunken stomata; (b) = lower foliar surface showing distribution of stomata with occluded stomatal pore) and Catharanthus roseus L. (right: (c) = upper foliar surface showing encrypted and sunken stomata; (d) = lower foliar surface showing stomata with striations arising from subsidiary cells) collected from Charbagh in the summer.

Chlorophylls

Chlorophylls a and b both were found to be decreased in leaves of both plants collected from polluted site in comparison to plants collected from healthy site (HS) healthy site (Figure 4). The concentration of chl a and chl b content in flannel weed in the summer at polluted site (P1) was decreased up to 30% and 43%, while at polluted site 2 (P2) it was found up to 55% and 71% in comparison to healthy site. In the winter, chlorophyll a was decreased to 43% and 71% at P1 and P2, respectively. In the winter, chlorophyll b also showed a reduction of 52% and 59% at P1 and P2, respectively, in comparison to healthy site. Total chlorophyll was highly reduced at P2 in comparison to P1 in both seasons (Figure 4). Ratio of chl a and chl b was found to be increased in the summer at both sites. But in the winter, ratio of chl a and b was found to be increased (19%) at P1 and decreased (29%) at P2 in comparison to healthy site. In periwinkle, chl a was highly reduced in the winter at both P1 (43%) and P2 (67%) in comparison to healthy site, while in the summer it was decreased up to 15% and 31% at P1 and P2 sites in comparison to healthy site. In periwinkle, total chlorophyll was also reduced in plants collected from P1 and P2 in comparison to healthy site, and reduction was more pronounced in the winter. Ratio of chl a and chl b was increased in both seasons at both sites except in the winter at P1 site where it was found to be decreased (Figure 4). Chlorophyll measurement is an important tool to evaluate the effect of air pollutants on plants as it plays an important role in plant metabolism, and any reduction in chlorophyll content corresponds directly to plant growth [24]. Leaf chlorophyll content, thus, can provide valuable information about physiological status of plants. Pollutants like SO_2 entering the leaf are hydrated to form HSO_3, SO_3, and H+ ions. Sulfite affects the carbon fixation, ribulose biphosphate carboxylase, glycolate oxidase activity, and photophosphorylation [25]. Increasing level of air pollution in Lucknow is a major contributor to the formation of dense fog and low day temperature in winters. Suspended particle matter (SPM) emitted by vehicles and in the air combines with the moisture content in the atmosphere to increase the density of the fog, which lasts for a longer period. The SPM are present in the atmosphere in summers as well but do not pose much problem because moisture content in the atmosphere is very low. In monsoon season, the pollutants are cleared by rainfall. However, in winters, as temperatures come down due to icy winds coming to plains from the snow clad hills, the moisture content is condensed into fog. The presence of SPM expedites the process of fog formation, leading to a drop in day temperatures and chilly weather. The fog thus formed bars penetration of sun rays leading to a fall in temperature during the daytime. A Higher chlorophyll content was found in the monsoon season than in the summer and is the least in the winter; it might be due to the high NO_2concentration which increases the sensitivity of plant to winter stress as observed by Swarnlata et al. [26]. Low-chlorophyll-content winter is due to the high pollution level, temperature stress, low sunlight intensity, and short photoperiod.

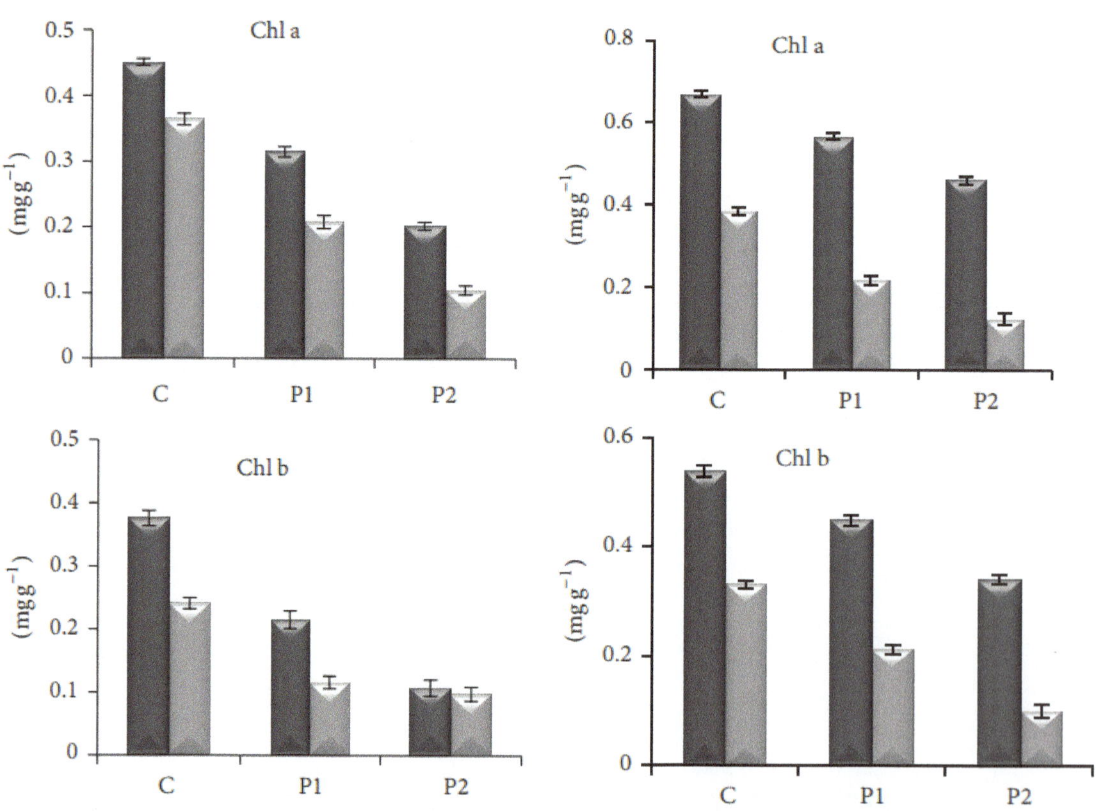

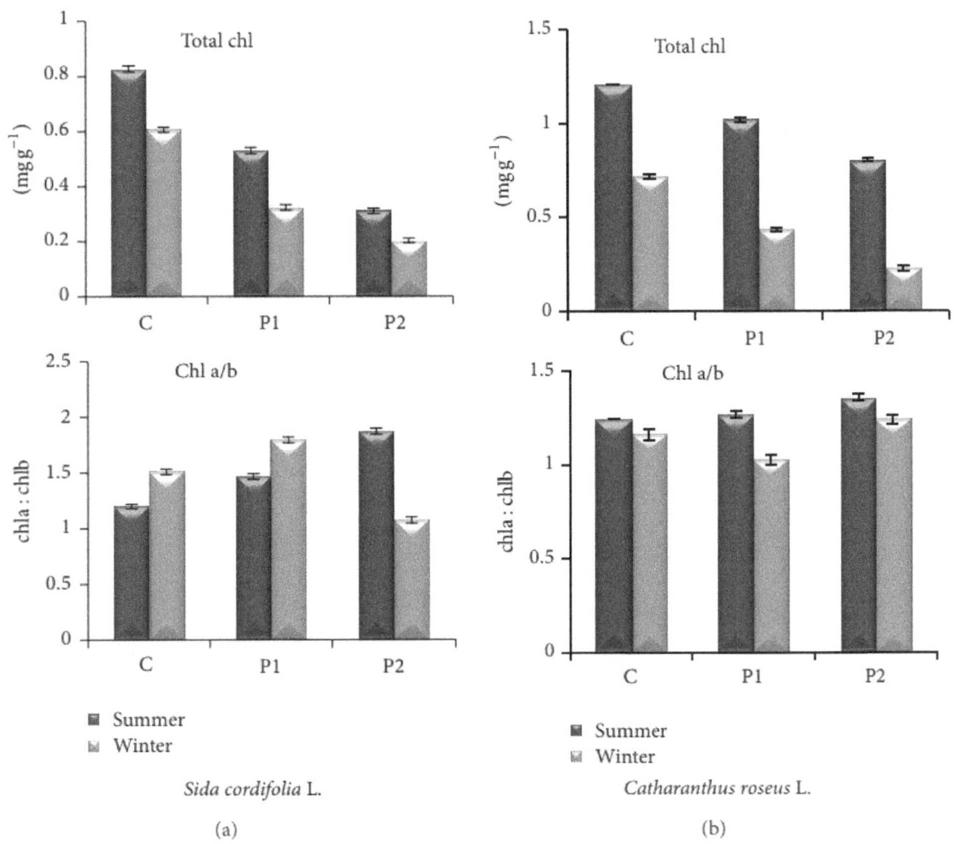

Figure 4: Effect of auto pollution on chlorophylls (chl a, chl b, total chl, and ratio of chl a and chl b) in leaves of Sida cordifolia L. (a) and Catharanthus roseus L. (b) (C = healthy site; P1 = polluted site 1; P2 = polluted site 2). Data represent mean ± S.E. ($P \leq 0.05$).

Present study revealed that chlorophyll content in both plants varies with the pollution status of the area; that is, the higher the pollution level in the form of automobile exhaust the lower the chlorophyll content. It also varies with the tolerance as well as sensitivity of the plant species. Studies [3, 27] also suggest that high levels of automobile pollution decrease chlorophyll content in plants near roadsides. The reason for degradation of chlorophyll pigments can also be attributed to the action of SO_2 and NO_2 on the metabolism of chlorophyll [28]; both of these gases are the constituents of vehicular emissions. The reduction in the concentration of chlorophyll might have also been caused by the increase in chlorophyllase enzyme activities, which, in turn, affects the chlorophyll concentration in plants [29]. Our results are also in agreement with the work of [30].

Carotenoids and Phaeophytin

In both genera, carotenoids were found to be reduced in both seasons at both sites in comparison to healthy site but phaeophytin was found to be increased (Figure 5). In flannel weed, Decrease in carotenoids concentration was more in the winter at healthy site. Percentage decrease at P1 was marginally increased in the summer in comparison to the winter. At P2 carotenoids concentration was highly reduced (up to 74%) and the reduction was more pronounced in the summer. In periwinkle, carotenoids concentration was also reduced in both seasons and both sites. Reduction was significantly high in the winter at P2 in comparison to healthy site (Figure 5). Agarwal and Sharma [31] have reported significant reduction in carotenoids in leaf samples collected from polluted environment. Aquil et al. [32] have also reported a reduction in carotenoids content in the plants of Albizzia lebbek Benth due to being exposed to coal smoke. Carotenoids protect photosynthetic organisms from potentially harmful photooxidative processes and are essential structural components of the photosynthetic antenna and reaction center [33]. The pheophytin concentration of plant leaves of both genera was found to be opposite that of chlorophyll concentration. Higher level of pheophytin was found in the winter in both the plant species when the pollution level was high. In flannel weed and periwinkle, the highest concentration of phaeophytin was observed in the winter at P2 but the increase was more (202%) pronounced in flannel weed (Figure 5). It might be due to the breakdown of chlorophyll to pheophytin as suggested by Rao and Le Blanc [34]. It indicates the sensitivity of plants to the pollutants; the higher the phaeophytin level in the plant is, the greater the sensitivity to air pollution is.

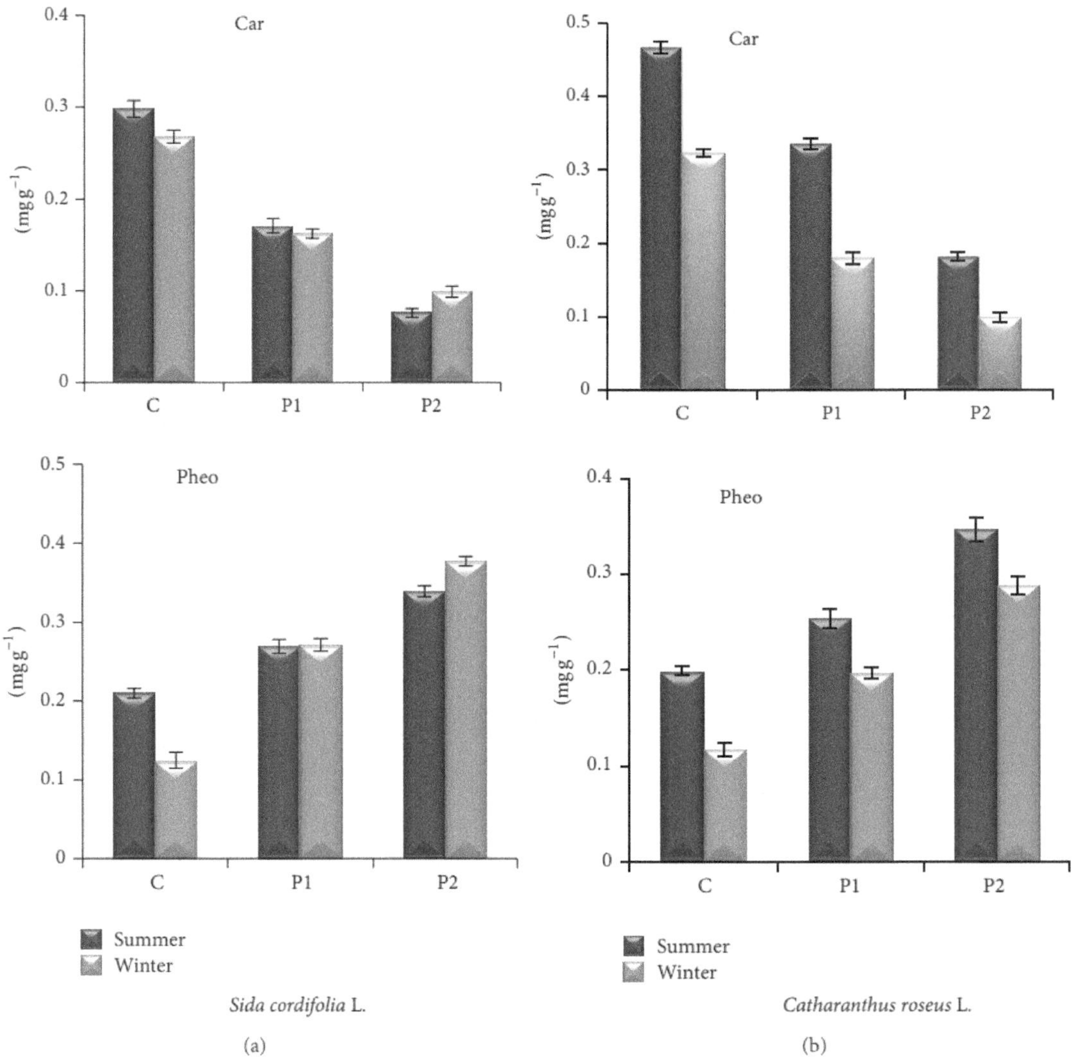

Figure 5: Effect of auto pollution on carotenoids (car) and phaeophytin (Pheo) in leaves of Sida cordifolia L. (a) and Catharanthus roseus L. (b) (C = healthy site; P1 = polluted site 1; P2 = polluted site 2). Data represent mean ± S.E. ($P \leq 0.05$).

CAT, POD, and RWC

The activity of CAT at both sites in both seasons in flannel weed and periwinkle was found to be increased (Figure 6). Maximum increment in CAT activity in flannel weed and periwinkle was 112% and 36%, which was observed at P2 site in the winter. In the summer, maximum (179%) increment in CAT activity was observed at P2 in flannel weed. Peroxidase activity was more in the winter than in the summer at healthy site (Figure 6). In the winter, the percentage increase in POD activity in flannel weed. In periwinkle POD activity was found to be 24% and 26% at P1 and 105% and 81% at P2, respectively. In the summer, the percentage increase in POD activity in flannel weed and periwinkle was found to be 40% and 24% at P1 and 101% and 49% at P2, respectively. Antioxidative enzymes CAT and POD are effective quenchers of reactive oxygen species (ROS) and play an important role in the adaptation and ultimate survival of plants during periods of stress. As plants produce significant amount of antioxidants to prevent oxidative stress caused by photons and oxygen, they represent a potential source of new compounds with antioxidant activity. The activity of CAT and POD increases under the air pollution. Studies of Karjalainen et al. [35] suggested that POD is a specific indication of SO2 and NO2 pollution. In combination, SO2 and NO2 induced strong POD activity in plants. Basically, gaseous pollutants like SO2 are taken up by stomata and solubilized in aqueous phase of the cell wall and can diffuse across the plasma membrane. Sulfur dioxide is oxidized in aqueous phase by producing HSO3 − and H+ ions. Peroxidases have a very important role as they may act as sulfite oxidases in the presence of monophenolics [11].

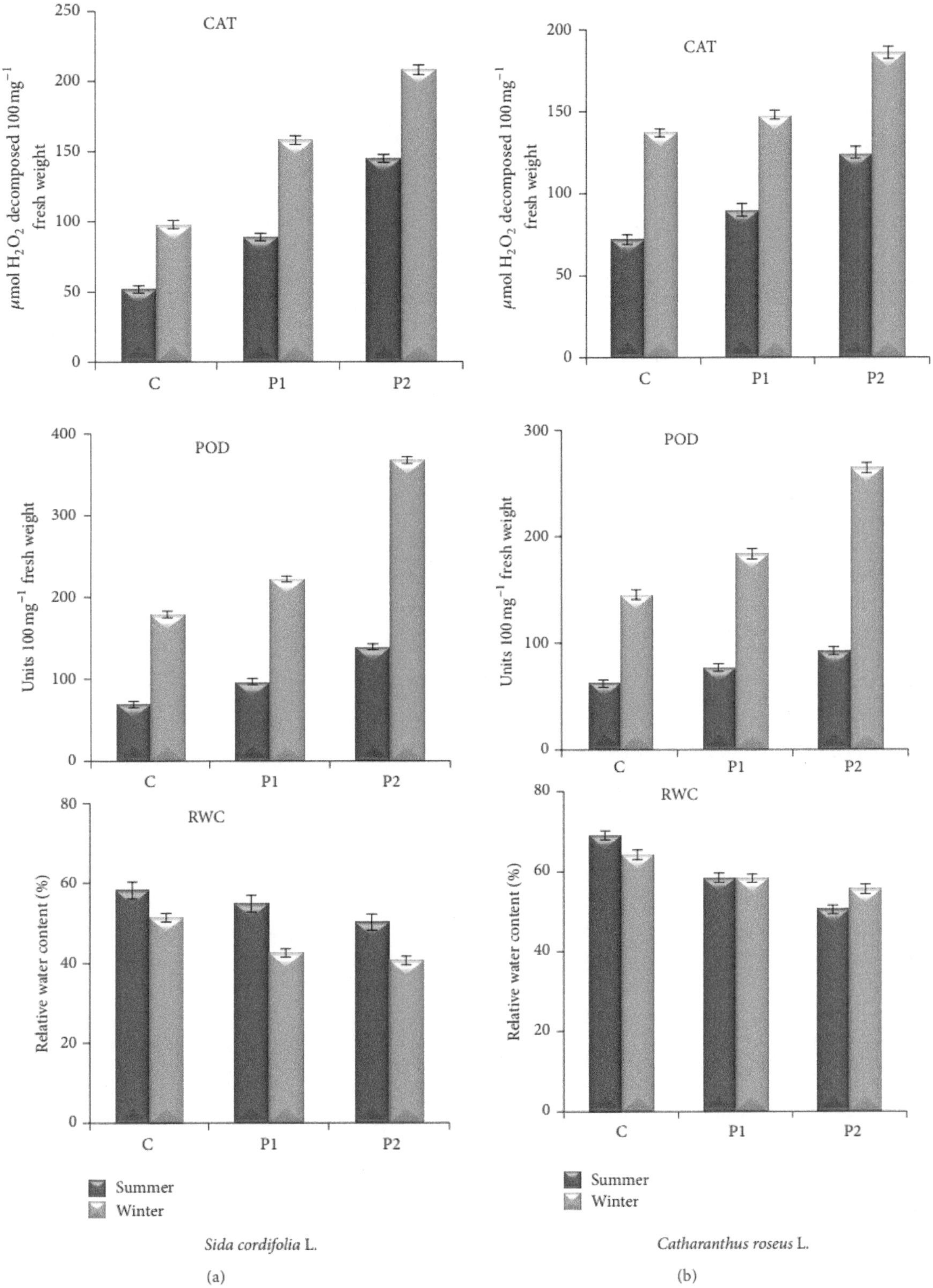

Figure 6: Effect of auto pollution on catalase (CAT), peroxidase (POD), and relative water content (RWC) in leaves of Sida cordifolia L. (a) and Catharanthus roseus L. (b) (C = healthy site; P1 = polluted site 1; P2 = polluted site 2). Data represent mean ± S.E. ($P \leq 0.05$).

Relative water content (RWC) represents a useful indicator of the status of water balance of a plant, essentially because it expresses the absolute amount of water, that the plant requires to reach artificial full saturation (Gonzalez-Vilar, 2001) [36]. Relative water content was found to be low in polluted plants in comparison to healthy plants (Figure 6). Air pollution affects the RWC of leaves very effectively. More RWC of a leaf helps to maintain its physiological balance under stress condition of dust pollutants. Higher RWC also favors drought resistance. The relative water content in the leaves of flannel weed at P1 showed reduction of 6% and 14%, while in periwinkle, the reduction was 15% and 27% at P2 in the summer in comparison to healthy site. Significant reduction was observed in flannel weed (21%) and in periwinkle (14%) in the winter at P2 in comparison to healthy site (Figure 6). It was observed that RWC declines more in sensitive species than that in tolerant ones, as reported by Singh [37]. It is evident from the above discussion that the pollutants such as SPM, SO_2, and NO_2 from automobile exhaust not only cause bad air quality conditions around nearby areas but also cause significant reduction in morphological and physiological parameters.

Present studies clearly indicate that the vehicular induced air pollution reduces the concentration of photosynthetic pigments (chl a, chl b, carotenoid, and phaeophytin) in both the species exposed to roadside pollution. The antioxidative enzymes (catalase and peroxidase) were found to be enhanced while relative water content was reduced. In micromorphological studies, there was a reduction in stomatal frequency, epidermal cell frequency, and stomatal index, while length and breadth of stomata and epidermal cells were found to be increased in both the species. Marked alterations were observed both in the physiological status and in the foliar surface ultrastructural configuration of both periwinkle and flannel weed plants growing at highly polluted site in comparison to healthy site. Therefore, these plant species may be used as biomarkers and mitigators of pollutants coming out of the automobile exhaust. Significant changes were recorded in flannel weed in comparison to periwinkle in the studied parameters. It was noticed from the above result that periwinkle was found to be more tolerant compared to the flannel weed.

ACKNOWLEDGMENT

Authors are thankful to Dr. Naresh Chandra Mehrotra, Director, Birbal Sahni Institute of Palaeobotany, Lucknow, for providing scanning electron microscope facilities.

REFERENCES

1. C. A. Odilara, P. A. Egwaikhide, A. Esekheigbe, and S. A. Emua, "Air pollution tolerance indices (APTI) of some plant species around Ilupeju Industrial Area, Logos," Journal of Engineering Science and Applications, vol. 4, no. 2, pp. 97–101, 2006.
2. N. Singh, M. Yunus, K. Srivastava, S. M. Singh, V. Pandey, and J. Mishra, "Monitoring of auto exhaust pollution by roadside plants," Environmental Monitoring & Assessment, vol. 34, no. 1, pp. 13–25, 1995. · ·
3. A. K. Tripathi and M. Gautam, "Biochemical parameters of plants as indicators of air pollution," Journal of Environmental Biology, vol. 28, no. 1, pp. 127–132, 2007. ·
4. A. K. Dwivedi and B. D. Tripathi, "Effect of ambient air sulphur dioxide on sulphate accumulation in plants," Journal of Environmental Biology, vol. 29, no. 3, pp. 377–379, 2008. ·
5. Centre for Science and Environment, A Special Presentation by CSE to the Committee on Auto Fuel Policy (Mashelkar Committee), New Delhi, India, 2001.
6. Central Pollution Control Board, "Benzene in air and its effects on human health," Parivesh Newsletter, CPCB Ministry of Environment and Forests, New Delhi, India, 2002.
7. K. P. Beckett, P. H. Freer-Smith, and G. Taylor, "Urban woodlands: their role in reducing the effects of particulate pollution," Environmental Pollution, vol. 99, no. 3, pp. 347–360, 1998. · ·
8. R. M. Cox, "The use of passive sampling to monitor forest exposure to O_3, NO_2 and SO_2: a review and some case studies," Environmental Pollution, vol. 126, no. 3, pp. 301–311, 2003. · ·
9. K. Asada and M. Takahashi, "Production and scavenging of active oxygen in photosyntheisis," inPhotoinhibition, D. Kyle, C. Osmond, and C. Arntzen, Eds., pp. 222–287, Elsevier Scientific Publishers BV, Amsterdam, The Netherlands, 1987.
10. S. Huttunen and E. Heiska, "Superoxide dismutase (SOD) activity in Scots pine (Pinus sylvestris L.) and Norway spruce (Picea abies L. Karst.) needles in northern Finland," European Journal of Forest Pathology, vol. 18, no. 6, pp. 343–350, 1988. · ·
11. A. Polle, M. Eiblmeier, and H. Rennenberg, "Sulphate and antioxidants in needles of Scots pine (Pinus sylvestris L.) from three SO_2-polluted field sites in eastern Germany," New Phytologist, vol. 127, no. 3, pp. 571–577, 1994. · ·
12. R. Alscher, J. L. Bower, and W. Zipfel, "The basis for different sensitivities of photosynthesis to SO_2 in two cultivars of pea," Journal of Experimental Botany, vol. 38, no. 186, pp. 99–108, 1987. · ·
13. A. Charron and R. M. Harrison, "Primary particle formation from vehicle emissions during exhaust dilution in the roadside atmosphere," Atmospheric Environment, vol. 37, no. 29, pp. 4109–4119, 2003. · ·
14. D. D. Pant, On the Ontogeny of Stomata and Other Homologous Structures, vol. 1 of Plant Science Series, Indian Uttar Pradesh, Allahabad, India, 1965.
15. C. A. Stace, "Cuticular studies as an aid to plant Taxonomy," Bulletin of the British Museum (Natural History), vol. 4, pp. 1–78, 1965.
16. D. L. Dilcher, "Approaches to the identification of angiosperm leaf remains," The Botanical Review, vol. 40, no. 1, pp. 1–157, 1974. · ·
17. H. K. Lichtenthaler, "Chlorophylls and carotenoids: pigments of photosynthetic biomembranes," inMethods in Enzymology, L. Packer and R. Douce, Eds., pp. 350–382, Academic Press, New York, NY, USA, 1987.
18. H. V. Euler and K. Josephson, "Uber katalase. I," Justus Liebigs Annalen der Chemie, vol. 452, no. 1, pp. 158–181, 1927.
19. M. Luck, "Peroxidase," in Methods of Enzymic Analysis, H. V. Bergmeyer, Ed., pp. 895–897, Academic Press, New York, NY, USA, 1963.
20. S. Y. Ninave, P. R. Chaudhari, D. G. Gajghate, and J. L. Tarar, "Foliar biochemical features of plants as indicators of air pollution," Bulletin of Environmental Contamination and Toxicology, vol. 67, no. 1, pp. 133–140, 2001. · ·
21. L. C. Mishra, "Effect of environmental pollution on the morphology and leaf epidermis of Commelina bengalensis Linn," Environmental Pollution A: Ecological and Biological, vol. 28, no. 4, pp. 281–284, 1982. · ·
22. T. Keller, "The electrical conductivity of Norway spruce needle diffusate as affected by air pollutants,"Tree

Physiology, vol. 1, pp. 85–94, 1986.

23. A. Pal, K. Kulshreshtha, K. J. Ahmad, and H. M. Behl, "Do leaf surface characters play a role in plant resistance to auto-exhaust pollution?" Flora, vol. 197, no. 1, pp. 47–55, 2002. · ·

24. N. D. Wagh, P. V. Shukla, S. B. Tambe, and S. T. Ingle, "Biological monitoring of roadside plants exposed to vehicular pollution in Jalgaon city," Journal of Environmental Biology, vol. 27, no. 2, pp. 419–421, 2006. ·

25. N. R. Madamanchi and R. G. Alscher, "Metabolic bases for differences in sensitivity of two pea cultivars to sulfur dioxide," Plant Physiology, vol. 97, no. 1, pp. 88–93, 1991. · ·

26. I. Swarnlata, B. Samidha, and R. Shashi, "Responses of Ficus relegiosa Exp osed To NO_2," Indian Journal of Environmental and Toxicology, vol. 2, no. 2, pp. 21–26, 1992.

27. A. Q. Mir, T. Yazdani, A. K. K. Narain, and M. Yunus, "Vehicular population and pigment content of certain avenue trees," Pollution Research, vol. 27, no. 1, pp. 59–63, 2008. ·

28. W. K. Lauenroth and J. L. Dodd, "Chlorophyll reduction in western wheatgrass (Agropyron smithii Rydb.) exposed to sulfur dioxide," Water, Air, and Soil Pollution, vol. 15, no. 3, pp. 309–315, 1981. ·

29. M. Mandal and S. Mukherji, "Changes in chlorophyll content, chlorophyllase activity, Hill reaction, photosynthetic CO_2 uptake, sugar and starch contents in five dicotyledonous plants exposed to automobile exhaust pollution," Journal of Environmental Biology, vol. 21, no. 1, pp. 37–41, 2000. ·

30. B. Wali, Mahmooduzzafar, and M. Iqbal, "Plant growth, stomatal response, pigments and photosynthesis of Althea officinalis as affected by SO_2 stress," Indian Journal of Plant Physiology, vol. 3, pp. 224–233, 2004.

31. S. K. Agarwal and A. P. Sharma, "Air pollution induced deranged physiology in plant-1 changes pertaining to chlorophyll, carotene and Dalbergia sissoo Roxb," Acta Ecologia, vol. 6, article 37, 1984.

32. S. Aquil, S. H. Ahmad, Z. A. Reshi, and M. Iqbal, "Physiological and biochemical response of Albizzia lebbeck benth. to coal smoke pollution," Pollution Research, vol. 22, no. 4, pp. 489–493, 2003. ·

33. G. E. Bartley and P. A. Scolnik, "Plant carotenoids: pigments for photoprotection, visual attraction, and human health," Plant Cell, vol. 7, no. 7, pp. 1027–1038, 1995. · ·

34. N. Rao and F. Le Blanc, "Effects of SO_2 pollution on the lichen algae with special reference to chlorophyll," Bryologist, vol. 69, pp. 60–75, 1966.

35. R. Karjalainen, J. Jokinen, E. Laine et al., "Changes in enzyme activity as an indicator of air pollution stress in woody plants," in Proceedings of the 7th International Bioindicators Symposium and Workshop on Environmental Health, S. Roy, L. Karamlamp, and O. Hanninen, Eds., vol. 7, Kuopio University, Kuopio, Finland, 1992.

36. L. Gonzalez and M. Gonzalez-Vilar, "Determination of relative water content," in Handbook of Plant Ecophysiology Techniques, M. J. Reigosa, Ed., pp. 207–211, Kluwer Academic Publisher, Dordrent, The Netherlands, 2001.

37. S. K. Singh, Study of plants in relation to SO_2-pollution abatement [Ph.D. thesis], Banaras Hindu University, Varanasi, India, 1989.

CHAPTER 18

BIOCHEMICAL AND METABOLIC CHANGES IN ARSENIC CONTAMINATED BOEHMERIA NIVEA L.

Hussani Mubarak, Nosheen Mirza, Li-Yuan Chai, Zhi-Hui Yang, Wang Yong, Chong-Jian Tang, Qaisar Mahmood, Arshid Pervez, Umar Farooq, Shah Fahad, Wajid Nasim, and Kadambot H. M. Siddique

INTRODUCTION

Ramie (Boehmeria nivea), commonly known as China grass, is an important fiber crop which has been widely cultivated and distributed in China. It is a perennial plant with at least three harvests per year [1]. The principal end product of B. nivea is textile grade fiber, famous for its fine characteristics in textile industries [2]. The leaf and the root extracts of the plant have antimicrobial, anti-inflammatory, antioxidant, and hepatoprotective properties [3]. Ramie grows in the wild and is known to colonize both active and abandoned metal mine sites. It is capable of accumulating certain amounts of toxins such as Sb, Cd, and Hg [4, 5]. Other studies have shown that ramie can also tolerate certain amounts of heavy metals such as mercury, lead [6], cadmium [7, 8], and arsenic [1]. Across the globe, centuries of unsustainable activities have resulted in severe heavy metal contamination which has damaged aquatic and terrestrial environments [9]. Arsenic (As) is one of these reported toxins and is carcinogenic. It is a ubiquitous element and its toxicity in the environment is a global issue. The increasing As contamination in water, soils, and crops in numerous countries such as India, Bangladesh, Cambodia, Laos, Myanmar, China, Taiwan, United States, Vietnam, Thailand, and Europe is well reported [9, 10]. The health problems associated with chronic exposure to As are diabetes, cancer, poisoning, pathogenic potential of bacteria or fecal coliform, and blood stream infections. In China, high arsenic groundwater has been observed in the Datong basin of Shanxi Province, Hetao basin of Inner Mongolia, Xinjiang and Taiwan Provinces. In these regions, approximately 18.5 million people are at risk of exposure to high arsenic groundwater [11].

Colonization of B. nivea and accumulation of heavy metals such as Sb, Cd, Pb, and Zn in metal-contaminated regions have been reported. B. nivea has a high tolerance to As contamination [12]. The effect of As pollution on the

Citation: Hussani Mubarak, Nosheen Mirza, Li-Yuan Chai, et al., "Biochemical and Metabolic Changes in Arsenic Contaminated Boehmeria nivea L."" BioMed Research International, vol. 2016, Article ID 1423828, 8 pages, 2016. doi:10.1155/2016/1423828.

biology of B. nivea, the potential tolerance of B. nivea to As pollution, and the metabolic changes in B. nivea under the reported ranges of As contamination has not been documented. Based on the above review [12], the objectives of the current study were to evaluate plant growth, metabolic changes, and the ability of B. nivea L. to uptake and accumulate As in an attempt to determine the effects of As on chlorophyll content, antioxidative systems, and lipid membrane peroxidation in order to better understand the cellular basis of As tolerance in B. nivea.

MATERIALS AND METHODS

Plant Materials and Growth Conditions

Young shoot cuttings (14 cm) of B. nivea plants were collected from an active and abandoned mining area in Xikuangshan, Hunan Province, China (29°N, 120°E). The cuttings were planted in sand for root initiation and then transferred to 1/2 strength Hoagland solution until they reached a height of 30 cm (40 days).

Hydroponic Experiment

After 40 days, the plants were transferred to 1/2 strength Hoagland solution (2 L) spiked with 0, 5, 10, 15, or 20 mg L^{-1} of arsenic. Arsenic (As) was applied as $NaAsO_2$ (As-III, 100% purity). Each treatment was replicated three times. After 14 days of As exposure, the roots were immersed in 20 mM Na_2-EDTA for 30 min to remove the As adsorbed to the roots, and the entire plant was rinsed three times with deionized water. The roots, stems, and leaves were separated. For further analysis, some of the fresh top leaves were frozen at −80°C and some were dried at 70°C. The remaining Hoagland solution was filtered and refrigerated for later As analysis. The stress tolerance index of As-contaminated B. nivea was calculated according to Yang et al. and Ismail et al. [1,13].

The stress tolerance index (%) was calculated using the following formula:

$$\text{Stress tolerance index} = \frac{\text{Shoot length of stress plant}}{\text{Shoot length of control plant}} \times 100. \quad (1)$$

Metabolism of B. nivea

Chlorophyll Analysis

The 6th fully expanded leaf from the top of each plant was used to measure chlorophyll a, chlorophyll b, and total chlorophyll (chl (a + b)). Chlorophyll a, chlorophyll b, and chl (a + b) were analyzed according to the methods of Arnon and Huang et al. [14, 15] and estimated using the formulas of Ehsan et al. and Metzner et al. [16, 17]:

$$\text{Chlorophyll a } (\mu g\, mL^{-1}) = 10.3 \times E663 - 0.98 \times E644$$

$$\text{Chlorophyll b } (\mu g\, mL^{-1}) = 19.7 \times E644 - 3.87 \times E663$$

$$\text{Total chlorophyll} = \text{Chlorophyll a} + \text{Chlorophyll b}. \quad (2)$$

Assessment of Antioxidants

Physiological measurements were performed on the 5th leaf from the top of plants growing in well-watered As conditions. Approximately 0.2 g of fresh tissue was homogenized in a precooled mortar with 5 mL of 50 mmol L^{-1} precooled phosphate buffer (pH 7.8). The homogenate was centrifuged at 11,000 g for 20 min at 4°C. The supernatant (i.e., the enzyme extract) was used to determine enzyme activities—superoxide dismutase (SOD), peroxidase (POD), catalase (CAT), and malondialdehyde content (MDA) [18, 19].

Assessment of Electrolyte Leakage and Hydrogen Peroxide Levels

Electrolyte leakages in 5 mm long fragments of fully expanded leaves were determined. The fragment tubes were incubated in a water bath at 32°C for 2 hours with the initial electrical conductivity (EC) of the medium, EC_1, noted. The samples were autoclaved at 121°C for 20 min to discharge the electrolytes and then cooled to 25°C. The final EC_2 was measured [20, 21]. Electrolyte leakage (EL) was calculated using the following formula:

$$EL = \left(\frac{EC_1}{EC_2}\right) \times 100. \quad (3)$$

H_2O_2 contents were assayed colorimetrically as documented by Jana and Choudhuri and Shakoor et al. [22,23]. The hydrogen peroxide was extracted by homogenizing 0.05 g leaf tissues with 3 mL of phosphate buffer (50 mM, pH 6.5). The homogenate was centrifuged at 6,000 g for 25 min. To measure H_2O_2 content, 2.5 mL of the extracted solution was mixed with 1 mL of 0.1% titanium sulfate ($Ti(SO_4)_2$) in 20% (v/v) H_2SO_4. The mixture was centrifuged at 6,000 g for 15 min. The intensity of the yellow color supernatant was analyzed at 410 nm. The H_2O_2 content was calculated by applying an extinction coefficient of 0.28 $mmol^{-1} cm^{-1}$.

Assessment of Relative Water Contents (RWC)

The RWC was determined using the methods of Huang et al. and Yamasaki and Dillenburg [15, 24]. The fresh mass (FM) of leaves was immersed in distilled water for 12 h to determine turgid mass (TM). Leaves were then dried at 70°C for 48 h to determinate dry mass (DM). The RWC was calculated as follows:

$$RWC = \left[\frac{(FM - DM)}{(TM - DM)}\right] \times 100. \quad (4)$$

Arsenic (As) Analysis

The dried plant samples were ground, sieved (1 mm), and digested with HNO_3 : $HClO_4$ (4 : 1, v/v). The As concentration in plant parts was analyzed using Induced Couple Plasma-Optical Emission Spectrometer (ICP-OES) (Perkin Elmer, Precisely, Shelton, CT 06484, USA, Optima™ 5300 DV Spectrometer). For accuracy of the digestion and analytical method, a blank sample (4 mL HNO_3 + 1 mL $HClO_4$) was also run with the samples.

Data Analysis

Analysis of variance (ANOVA) at a significance level of $P < 0.05$ was performed using the General Linear Model (GLM) in the SAS package. The LSD test and *t*-test were employed to compare significant differences between means for the treatments at $P < 0.05$. The results are expressed as means ± SD. Graphical analyses were carried out using Origin Pro 8.5.

RESULTS AND DISCUSSION

Growth of B. nivea

Field surveys have reported the presence of healthy growing B. nivea plants in toxic metal-contaminated areas [8, 25], but only a few studies have reported As resistance in B. nivea. The higher tolerance of B. nivea, compared to other plant species, for toxic [26] and heavy metals [27], has been estimated and documented. The metabolic responses of B. nivea under specific As ranges, say between 10 and 250 mg kg^{-1} soil, have not been assessed.

The As-contamination (hydroponic) treatments inhibited the growth of B. nivea more so as the contamination increased (Figure 1(a)). Plant height decreased with increasing As concentration. At 0 mg L^{-1} As, plant height ranged from 3.8 to 4 cm; the As treatments at 5, 10, 15, and 20 mg L^{-1} reduced plant height by 25, 32, 40, and 73% of the control, respectively. The As tolerance index of B. nivea significantly decreased ($P < 0.05$) as As concentration increased (Figure 1(b)) and ranged from 27 to 77%.

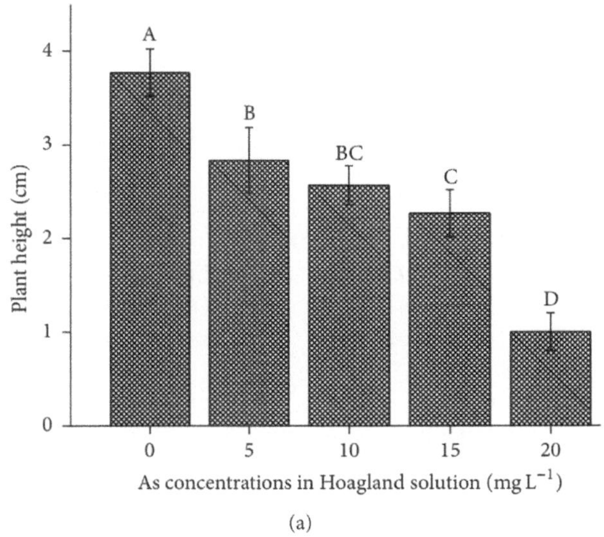

(a)

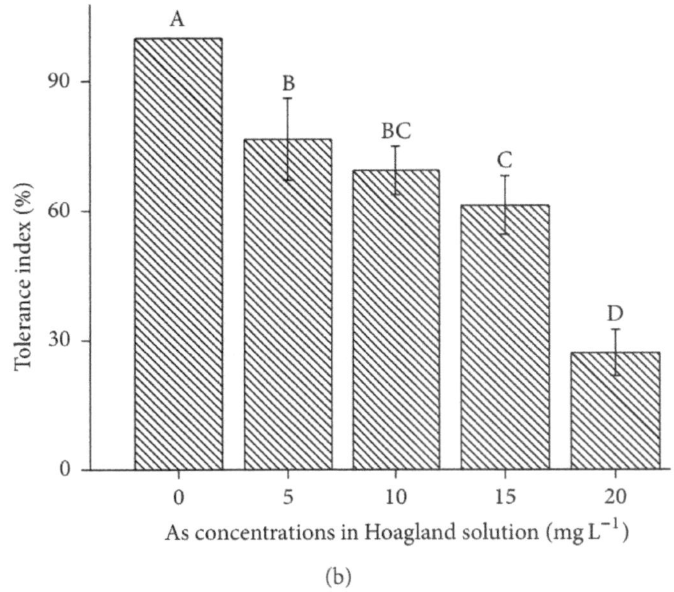

(b)

Figure 1: (a) Effect of As application, on the height of *Boehmeria nivea* L. (b) Effect of As application, on the tolerance index of *Boehmeria nivea*. Values followed by different uppercase letters are significantly different at $P < 0.05$, for treatments. Different letters indicate significant differences between treatments for each parameter, at significant level of 0.05. Values followed by the same letters for each parameter are not significantly different at the 0.05 level (least significant difference). Values in the graph are mean (n=3); error bars are standard deviation (SD).

Metabolism of As-Contaminated B. nivea L.

Chlorophyll Content of B. nivea L.

The effect of applied As on chlorophyll content is presented in Figure 2. As the As concentration increased, all chlorophyll content measurements (chl a, chl b, and chl a + b) significantly decreased ($P < 0.05$) in B. nivea by 11–54%, 22–54% and 14–54%, respectively, relative to the control (Figure 2). Singh et al. [28] reported increased chlorophyll contents in Pteris vittata but decreased chlorophyll contents in Pteris ensiformis under As-induced stress.

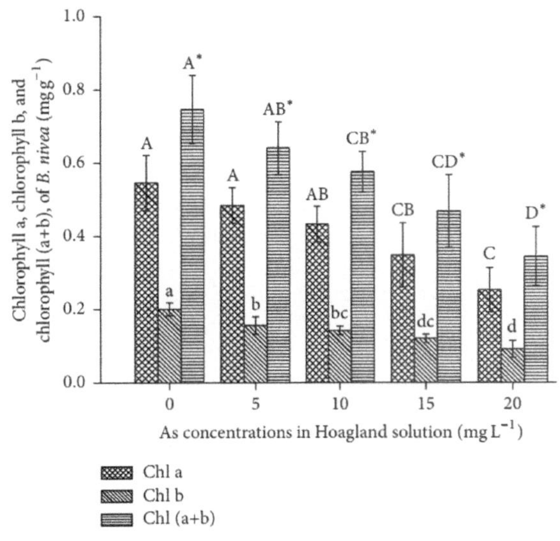

Figure 2: Effect of As application, on the chlorophyll of Boehmeria

nivea L. Values followed by different uppercase letters are significantly different at $P < 0.05$, for treatments. Values followed by different lowercase letters are significantly different at $P < 0.05$, for treatments. Values followed by different uppercase letters* are significantly different at $P < 0.05$, for treatments. Different letters indicate significant differences between treatments for each parameter, at significant level of 0.05. Values followed by the same letters for each parameter are not significantly different at the 0.05 level (least significant difference). Values in the graph are mean (n=3); error bars are standard deviation (SD).

Activities of Antioxidant Enzymes in B. nivea L.

Increasing As concentration had a significant ($P < 0.05$) effect on SOD, CAT, and MDA concentrations in B. nivea (Figures 3(a)–3(d)). Compared with the control, increasing the As concentration significantly decreased ($P < 0.05$) SOD concentration in B. nivea, except at the highest applied As (20 mg L^{-1}), while there were a nonsignificant reduction in POD concentration and a gradual reduction in CAT concentration at 10 mg L^{-1}, but a slight increase at 15 and 20 mg L^{-1}. The MDA content in As-contaminated B. nivea showed a significant ($P <$ 0.05) increasing trend with increasing As concentration.

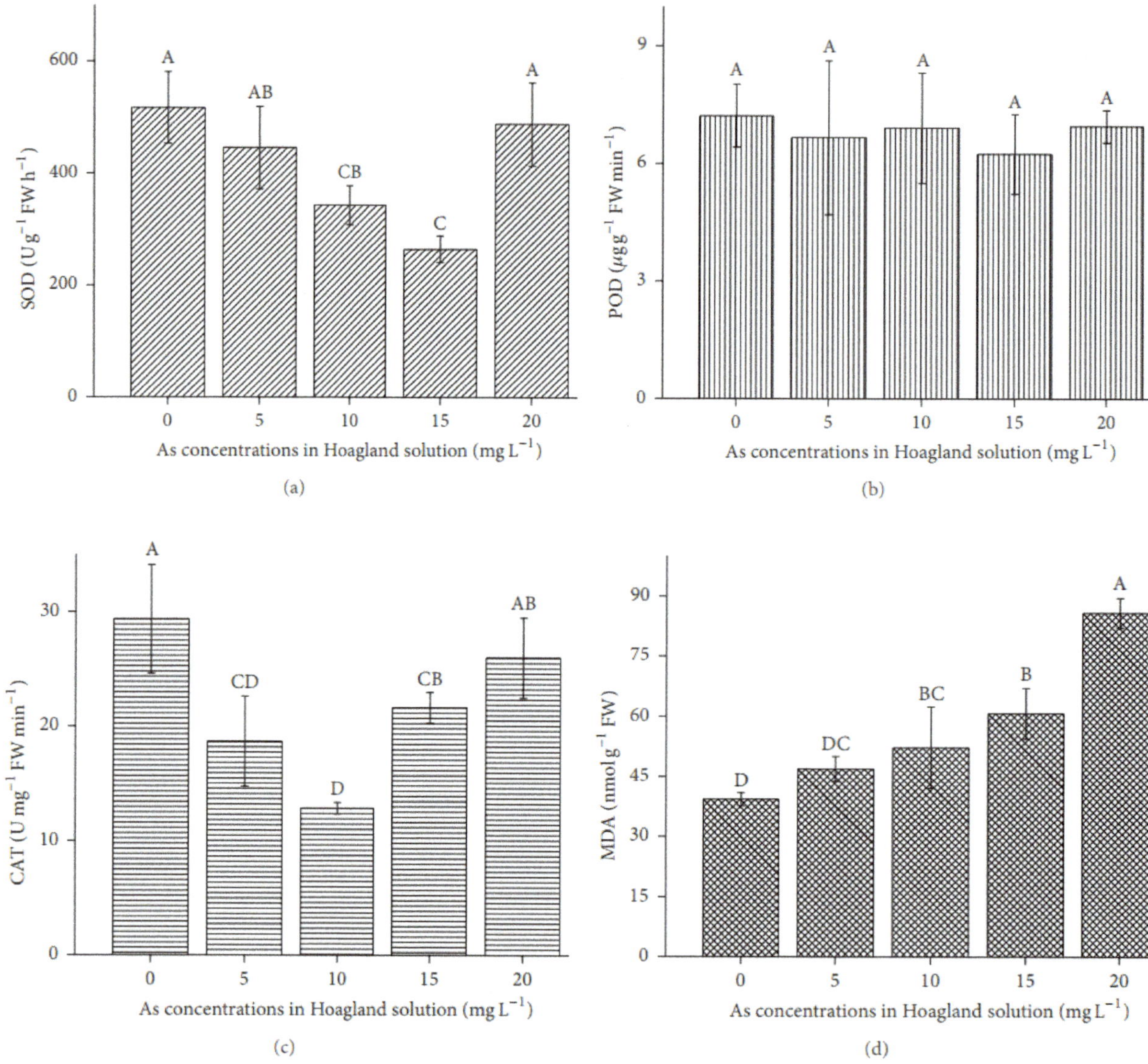

Figure 3: Effect of As application on antioxidant enzymes of Boehmeria nivea. (a): SOD; (b): POD; (c): CAT; (d): MDA. Values followed by different uppercase letters are significantly different at $P < 0.05$, for treatments. Different letters indicate significant differences between treatments for each parameter, at significant level of 0.05. Values followed by the same letters for each parameter are not significantly different at the 0.05 level (least significant difference). Values in the graph are mean ± SD (n=3); error bars are SD.

In plants and living organisms, stress induces the generation of reactive oxygen species (ROS) which may cause oxidative damage to proteins and enzymes. Excessive ROS increases MDA, the last product of membrane liposome peroxidation, which suggests lipid membrane instability [29]. To reduce oxidative damage, plants initiate enzymatic and nonenzymatic antioxidant defense mechanisms, of which the synthesis of SOD, POD, and CAT is the most important. In B. nivea, the greatest increase in MDA, relative to the control, was at 20 mg L^{-1} As (Figures 3(a)–3(c)), demonstrating that POD and CAT are H_2O_2 scavengers in ramie. SOD, POD, and CAT activities had similar suppressive effects and enhanced trends with As addition, except for POD concentration at 10 mg L^{-1} As and CAT at 15 mg L^{-1} As. The critical-stage performance of antioxidants, in B. nivea, was at $20\,mg\,L^{-1}$ As. SOD contents in B. nivea decreased by approximately 1.2, 1.5, 2.0, and 1 times of the control at 5, 10, 15, and $20\,mg\,L^{-1}$ As, respectively (Figure 3(a)). Based on this and the observed fluctuations, we conclude that SOD contributed to the tolerance of B. nivea to As contamination. At 5, 10, 15, and $20\,mg\,L^{-1}$ As, POD content declined by approximately 1.1, 1.0, 1.2, and 1.0 times of the control, respectively (Figure 3(b)). Similar decreasing trends of SOD and POD contents in cadmium-stressed wheat (Triticum durum) and selenium-stressed ryegrass (Lolium perenne) have been reported [19]. A study by Saidi et al. [30] reported suppressed activities of SOD and POD in cadmium-contaminated bean plants, while Huang et al. [31] reported increased SOD and POD contents in hybrid ramie under increased salinity.

The CAT content in B. nivea decreased at 5, 10, 15, and $20\,mg\,L^{-1}$ by 1.6-, 2.3-, 1.4-, and 1.1-fold less than control (Figure 3(c)). Silva et al. and Huang et al. [31, 32] reported decreasing CAT content in aluminum-exposed rye and salinity-stressed ramie. The decreased CAT activity with increasing As contamination confirms the role of CAT in quenching H_2O_2 and preventing oxidative damage in B. nivea.

Increasing the concentration of As in B. nivea increased MDA concentration from 1.12 times greater than control at $5\,mg\,L^{-1}$ to 2.20 times greater than control at $20\,mg\,L^{-1}$ (Figure 3(d)). The increase at $20\,mg\,L^{-1}$ As suggests the role of MDA in lipid peroxidation and the maintenance of homeostasis of B. nivea. Increases in MDA activity below $20\,mg\,L^{-1}$ As inhibited biomass production which is a clear indication of As tolerance ofB. nivea. Our results of enhanced MDA and CAT activities with As addition agree with those of Feng et al. and Huang et al. [29, 31] who reported increased MDA and CAT activities in drought-stressed drought-resistant ramie cultivars, plants (ferns, rice, and maize) and hybrid ramie (B. nivea), respectively. Our results suggest thatB. nivea is capable of alleviating oxidative stress and preventing lipid peroxidation under a specified range (5–15 $mg\,L^{-1}$ As) of As contamination. The highest increase in MDA (2.20 times greater than control) at $20\,mg\,L^{-1}$ confirms lipid peroxidation or damage to the plasma membrane which, in turn, inhibits plant growth.

Electrolyte Leakage and Hydrogen Peroxide (H_2O_2) Levels in B. nivea L.

Solute leakage and H_2O_2 content increased in B. nivea with increasing As contamination (Figures 4(a) and 4 (b)). The increases in electrolyte leakage and H_2O_2 from 5 to $20\,mg\,L^{-1}$ As ranged from 1- to 1.5-fold and 1.13- to 2-fold greater than the control, respectively. The gradual increase in MDA, electrolyte leakage, and H_2O_2 from 5 to $15\,mg\,L^{-1}$ As revealed that As toxicity accelerated the antioxidant defense mechanism [15]. However, $20\,mg\,L^{-1}$ As resulted in oxidative destruction in the plant. Similar trends for electrolyte leakage and lipid peroxidation have been reported in Cu-, Cd-, and Pb-contaminated Brassica napus [16, 20, 23].

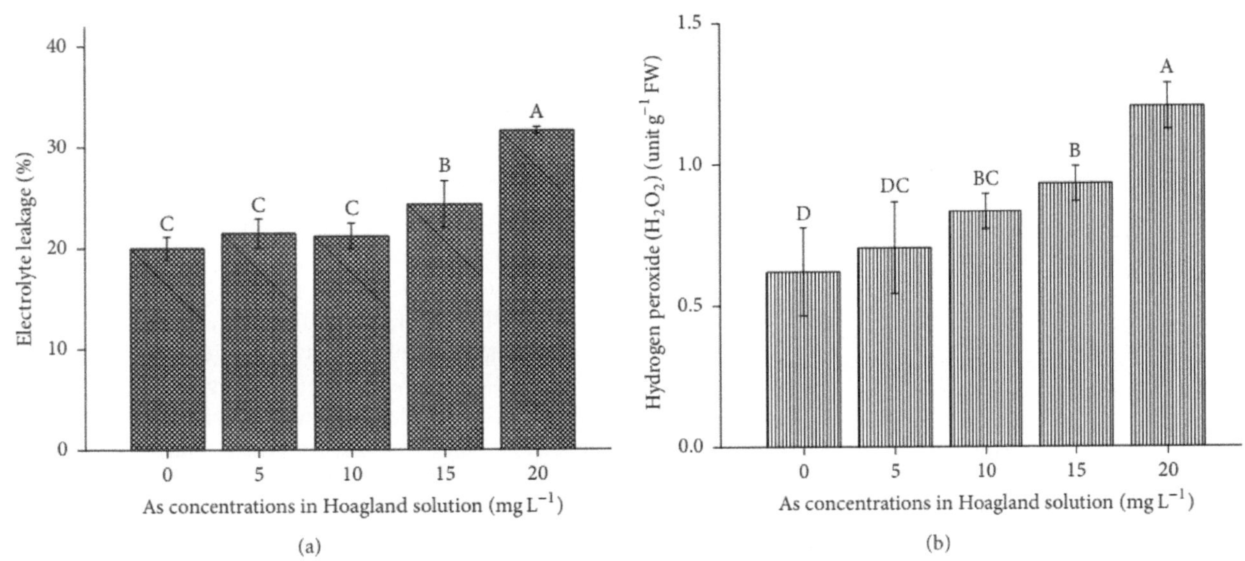

Figure 4: : (a) Effect of As application, on the electrolyte leakage in Boehmeria nivea L. (b) Effect of As application, on the H2O2 in Boehmeria nivea L. Values followed by different uppercase letters are significantly different at $P < 0.05$, for treatments. Different letters indicate significant differences between treatments for each parameter, at significant level of 0.05. Values followed by the same letters for each parameter are not significantly different at the 0.05 level (least significant difference). Values in the graph are mean (n=3); error bars are standard deviation (SD).

Relative Water Contents (RWC) in B. nivea L.

As contamination reduced RWC in ramie, it was not significant ($P > 0.05$) (Figure 5). The reductions in RWC were 1.01–1.04-fold less than the control (1–4%) at 5–20 $mg\,L^{-1}$ As. However, the greatest reduction in RWC (4%) was recorded at $20\,mg\,L^{-1}$ As (Figure 5). The antioxidant defense mechanism enabled B. nivea to maintain tissue water potential and, therefore, cell turgor under the stress conditions [15]. Turgidity maintenance under stress leads to the maintenance of comparatively higher RWC under increasing As contamination (Figure 5). Similar trends for RWC and lipid peroxidation have been reported in Pb-contaminated ramie cultivars [15].

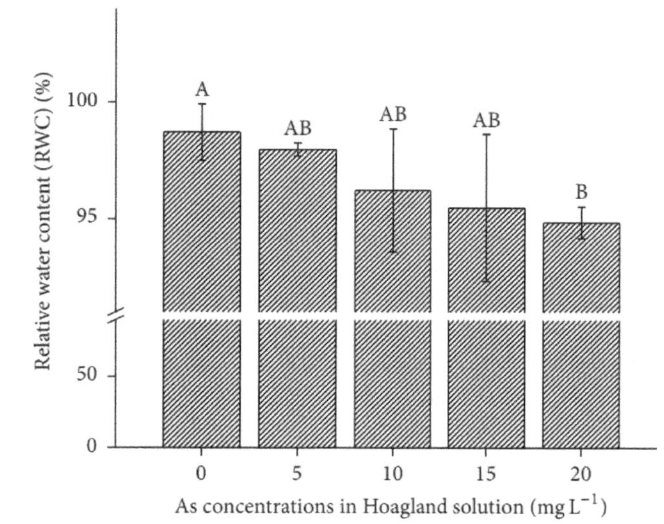

Figure 5: Effect of As application, on the RWC in Boehmeria nivea L. Values followed by different uppercase letters are significantly different at P < 0.05, for treatments. Different letters indicate significant differences between treatments for each parameter, at significant level of 0.05. Values followed by the same letters for each parameter are not significantly different at the 0.05 level (least significant difference). Values in the graph are mean (n=3); error bars are standard deviation (SD).

Arsenic (As) Concentration in B. nivea L.

The As content in dried roots and shoots (leaves plus stem) of B. nivea increased significantly (P < 0.05) with increasing applied As (Figure 6), more so in the shoots than the roots. Arsenic (As) mostly accumulates in the aboveground parts of tolerant plants. The average amount of As remaining in the Hoagland solution at the end of the experiment was 89%.

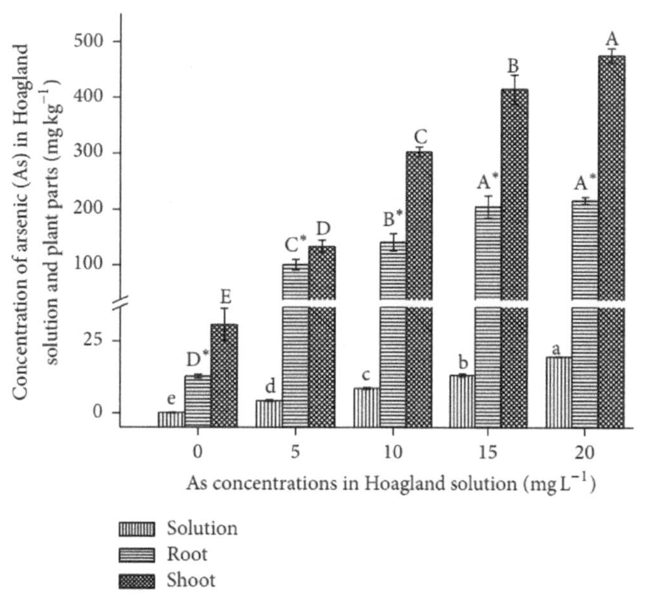

Figure 6: Effect of As application, on the As concentration in Hoagland solution and plant parts of Boehmeria nivea L. Values followed by different uppercase letters are significantly different at P < 0.05, for treatments. Values followed by different lowercase letters are significantly different at P < 0.05, for treatments. Values followed by different uppercase letters* are significantly different at P < 0.05, for treatments. Different letters indicate significant differences between treatments for each parameter, at significant level of 0.05. Values followed by the same letters for each parameter are not significantly different at the 0.05 level (least significant difference). Values in the graph are mean (n=3); error bars are standard deviation (SD).

The concentrations of As in the shoots and roots of B. nivea gradually significantly increased within a certain range, that is, 330–150 mg kg^{-1}, respectively, compared to the control (94–75% > control). Shoot accumulation of As in B. nivea at 5 and 20 mg L^{-1} As was 4.0 and 15.5 times greater than the control, respectively, while in the roots the respective values were 8.0 and 17.0 times greater than the control, respectively. The average amount of As remaining in the Hoagland solution at 5, 10, 15, and 20 mg L^{-1} was 85, 86, 88, and 99% (i.e., 15, 14, 12, and 1% were absorbed by B. nivea), respectively. Thus, the performance of B. nivea improved with increasing addition of As. The mobilization of As from roots to leaves is the greatest threat to the food chain and the survival of life on Earth, but this would not occur in B. nivea because it annually sheds older leaves (which can be collected, removed, and/or recycled) and is a commercial fiber crop; hence, contamination of the food chain is avoided.

In contrast to our results, Otones et al. [33] recommended Agrostis castellana (Boiss. & Reut.), Centaurea jaceaL., Eryngium campestre L., and Scirpus holoschoenus L. for the stabilization of As in abandoned mining areas. According to this study, these plants showed low translocation factors, that is, underground [As] > aboveground [As]. In accordance with our results Helichrysum oligocephalum, Hyoscyamus kurdicus, Nonea persica, Salvia syriaca [34], Rumex acetosella L. [33], Arundo donax [35], Isatis cappadocica, and Hesperis persica[36] reportedly accumulate relatively high As concentrations in their shoots.

CONCLUSIONS

This study reports on the growth of B. nivea in As-contaminated hydroponic cultures up to 20 mg L^{-1}. Arsenic contamination at high concentration, that is, 20 mg L^{-1}, can inhibit growth, chlorophyll content, and SOD, CAT, and POD contents in the plant by inducing electrolyte leakage, lipid peroxidation, and reducing RWC. However, up to 15 mg L^{-1} As resulted in limited cellular oxidative damage in B. nivea. The plant accumulated higher As concentrations in shoots than roots and thus gave higher translocation factors. The ability to accumulate more metals in the stalk and leaves than roots is a positive indicator. The metabolic and biochemical processes in B nivea remain unaffected till 15 mg L^{-1} As, but at 20 mg L^{-1} stress-induced oxidative damage was apparent. This experiment suggests that B. nivea L. may extract a considerable amount of As; however a field-based study is needed to confirm these results.

ACKNOWLEDGMENTS

The financial support of this work by the Science and Technology Program for Public Wellbeing (2012GS430201) and the Key Project of Science and Technology of Hunan Province, China (2012FJ1010), is gratefully acknowledged.

REFERENCES

1. B. Yang, M. Zhou, W. S. Shu et al., "Constitutional tolerance to heavy metals of a fiber crop, ramie (Boehmeria

nivea), and its potential usage," Environmental Pollution, vol. 158, no. 2, pp. 551–558, 2010. · ·

2. R. Kozlowaski, M. Rawluk, and J. Barriga-Bedoya, "Ramie," in Bast and Leaf Fibre Crops, R. R. Frank, Ed., pp. 207–227, Woodhead Publishing, Cambridge, UK; CRC Press, New York, NY, USA, 2005.
3. K.-L. Huang, Y.-K. Lai, C.-C. Lin, and J.-M. Chang, "Involvement of GRP78 in inhibition of HBV secretion by Boehmeria nivea extract in human HepG2 2.2.15 cells," Journal of Viral Hepatitis, vol. 16, no. 5, pp. 367–375, 2009. · ·
4. H. B. Wang, M. H. Wong, C. Y. Lan et al., "Uptake and accumulation of arsenic by 11 Pteris taxa from southern China," Environmental Pollution, vol. 145, no. 1, pp. 225–233, 2007. · ·
5. X. Wang, Y. O. Liu, G. M. Zeng et al., "Subcellular distribution and chemical forms of cadmium inBechmeria nivea (L.) Gaud," Environmental & Experimental Botany, vol. 62, no. 3, pp. 389–395, 2008. · ·
6. M. Lei, Q.-L. Yue, T.-B. Chen et al., "Heavy metal concentrations in soils and plants around Shizuyuan mining area of Hunan Province," Acta Ecologica Sinica, vol. 25, no. 5, pp. 1146–1151, 2005 (Chinese). ·
7. J. Dai, Y. Jie, J. Leng, and Z. Sun, "Distribution of cadmium in different parts of ramie in a polluted environment," Plant Fiber Production, vol. 25, pp. 279–293, 2003 (Chinese).
8. W. She, Y. Jie, H. Xing et al., "Uptake and accumulation of heavy metal by ramie (Boehmeria nivea) growing on antimony mining area in Lengshuijiang City of Hunan Province," Journal of Agro-Environment Science, vol. 29, no. 1, pp. 91–96, 2010 (Chinese, with English abstract).
9. Y. Zheng and J. D. Ayotte, "At the crossroads: hazard assessment and reduction of health risks from arsenic in private well waters of the northeastern United States and Atlantic Canada," Science of the Total Environment, vol. 505, pp. 1237–1247, 2015. · ·
10. R. Singh, S. Singh, P. Parihar, V. P. Singh, and S. M. Prasad, "Arsenic contamination, consequences and remediation techniques: a review," Ecotoxicology and Environmental Safety, vol. 112, pp. 247–270, 2015. · ·
11. H. Guo, D. Wen, Z. Liu, Y. Jia, and Q. Guo, "A review of high arsenic groundwater in Mainland and Taiwan, China: distribution, characteristics and geochemical processes," Applied Geochemistry, vol. 41, pp. 196–217, 2014. · ·
12. Z.E.-C.Huang,T.-B.Chen,M.Lei,Y.-R.Liu,andT.-D.Hu, "Difference of toxicity and accumulation of methylated and inorganic arsenic in arsenic-hyperaccumulating and -hypertolerant plants,"Environmental Science and Technology, vol. 42, no. 14, pp. 5106–5111, 2008. · ·
13. S. Ismail, F. Khan, and M. Z. Iqbal, "Phytoremediation: assessing tolerance of tree species against heavy metal (Pb and Cd) toxicity," Pakistan Journal of Botany, vol. 45, no. 6, pp. 2181–2186, 2013. ·
14. D. T. Arnon, "Copper enzyme in isolated chloroplasts polyphenoloxidase in Beta vulgaris," Plant Physiology, vol. 24, no. 1, pp. 1–15, 1949. ·
15. C. J. Huang, G. Wei, Y. C. Jie et al., "Responses of gas exchange, chlorophyll synthesis and ROS-scavenging systems to salinity stress in two ramie (Boehmeria nivea L.) cultivars," Photosynthetica, vol. 53, no. 3, pp. 455–463, 2015. ·
16. S. Ehsan, S. Ali, S. Noureen et al., "Citric acid assisted phytoremediation of cadmium by Brassica napusL.," Ecotoxicology and Environmental Safety, vol. 106, pp. 164–172, 2014. · ·
17. H. Metzner, H. Rau, and H. Senger, "Untersuchungen zur Synchronisierbarkeit einzelner pigmentmangel-mutanten von Chlorella," Planta, vol. 65, no. 2, pp. 186–194, 1965. ·
18. R. W. Feng and C. Y. Wei, "Antioxidative mechanisms on selenium accumulation in Pteris vittata L., a potential selenium phytoremediation plant," Plant, Soil and Environment, vol. 58, no. 3, pp. 105–110, 2012. ·
19. R. Feng, C. Wei, S. Tu, F. Wu, and L. Yang, "Antimony accumulation and antioxidative responses in four fern plants," Plant and Soil, vol. 317, no. 2, pp. 93–101, 2009. · ·
20. I. E. Zaheer, S. Ali, M. Rizwan et al., "Citric acid assisted phytoremediation of copper by Brassica napusL," Ecotoxicology and Environmental Safety, vol. 120, pp. 310–317, 2015. ·
21. M. L. Dionisio-Sese and S. Tobita, "Antioxidant responses of rice seedlings to salinity stress," Plant Science, vol. 135, no. 1, pp. 1–9, 1998. · ·
22. S. Jana and M. A. Choudhuri, "Glycolate metabolism of three submersed aquatic angiosperms during ageing," Aquatic Botany, vol. 12, no. C, pp. 345–354, 1982. · ·
23. M. B. Shakoor, S. Ali, A. Hameed et al., "Citric acid improves lead (Pb) phytoextraction in Brassica napusL. by mitigating Pb-induced morphological and biochemical damages," Ecotoxicology and Environmental Safety, vol. 109, pp. 38–47, 2014. · ·
24. S. Yamasaki and L. C. Dillenburg, "Measurements of leaf relative water content in Araucaria angustifolia,"Sociedade Brasileira de Fisiologia Vegetal, vol. 11, no. 2, pp. 69–75, 1999.
25. G. Okkenhaug, Y.-G. Zhu, J. He, X. Li, L. Luo, and J. Mulder, "Antimony (Sb) and Arsenic (As) in Sb mining impacted paddy soil from Xikuangshan, China: differences in mechanisms controlling soil sequestration and uptake in Rice," Environmental Science and Technology, vol. 46, no. 6, pp. 3155–3162, 2012. · ·
26. W. She, Y.-C. Jie, H.-C. Xing et al., "Absorption and accumulation of cadmium by ramie (Boehmeria nivea) cultivars: a field study," Acta Agriculturae Scandinavica Section B: Soil & Plant Science, vol. 61, no. 7, pp. 641–647, 2011. · ·
27. W. She, Y.-C. Jie, H.-C. Xing et al., "Tolerance to cadmium in ramie (Boehmeria nivea) genotypes and its evaluation indicators," Acta Agronomica Sinica, vol. 37, no. 2, pp. 348–354, 2011. · ·
28. N. Singh, L. Q. Ma, M. Srivastava, and B. Rathinasabapathi, "Metabolic adaptations to arsenic-

induced oxidative stress in Pteris vittata L and Pteris ensiformis L," Plant Science, vol. 170, no. 2, pp. 274–282, 2006. · ·

29. R. Feng, C. Wei, S. Tu, Y. Ding, R. Wang, and J. Guo, "The uptake and detoxification of antimony by plants: a review," Environmental and Experimental Botany, vol. 96, pp. 28–34, 2013. · ·

30. I. Saidi, M. Ayouni, A. Dhieb, Y. Chtourou, W. Chaïbi, and W. Djebali, "Oxidative damages induced by short-term exposure to cadmium in bean plants: protective role of salicylic acid," South African Journal of Botany, vol. 85, pp. 32–38, 2013. · ·

31. C. Huang, G. Wei, Y. Jie et al., "Effects of concentrations of sodium chloride on photosynthesis, antioxidative enzymes, growth and fiber yield of hybrid ramie," Plant Physiology and Biochemistry, vol. 76, pp. 86–93, 2014. · ·

32. S. Silva, G. Pinto, B. Correia, O. Pinto-Carnide, and C. Santos, "Rye oxidative stress under long term Al exposure," Journal of Plant Physiology, vol. 170, no. 10, pp. 879–889, 2013. · ·

33. V. Otones, E. Álvarez-Ayuso, A. García-Sánchez, I. S. Regina, and A. Murciego, "Mobility and phytoavailability of arsenic in an abandoned mining area," Geoderma, vol. 166, no. 1, pp. 153–161, 2011. · ·

34. N. Karimi, S. M. Ghaderian, and H. Schat, "Arsenic in soil and vegetation of a contaminated area,"International Journal of Environmental Science and Technology, vol. 10, no. 4, pp. 743–752, 2013. · ·

35. N. Mirza, Q. Mahmood, A. Pervez et al., "Phytoremediation potential of Arundo donax in arsenic-contaminated synthetic wastewater," Bioresource Technology, vol. 101, no. 15, pp. 5815–5819, 2010. · ·

36. N. Karimi, S. M. Ghaderian, H. Maroofi, and H. Schat, "Analysis of arsenic in soil and vegetation of a contaminated area in Zarshuran, Iran," International Journal of Phytoremediation, vol. 12, no. 2, pp. 159–173, 2010.

CHAPTER 19 PLANT BIOASSAYS TO ASSESS TOXICITY OF TEXTILE SLUDGE COMPOST

Ademir Sérgio Ferreira Araújo, Regina Teresa Rosim Monteiro

Key words: phytotoxicity, seed germination, solid residue, hydroponics

INTRODUCTION

Industrial waste composting is increasing because of recycling requirements set on organic wastes. Landfilling is also becoming more expensive and industrial companies consequently have to look for cheaper ways of disposing their wastes.

Safety requirements have to be fulfilled if the produced compost is intended for agricultural use and, until now, the harmfulness of solid waste has been estimated mainly on the basis of its chemical composition (Kapanen & Itavaara, 2001). According to the same authors, limit values are set only for heavy metals, and there are no requirements for the ecotoxicity or chemical levels in composts. Pagga et al. (1996) report that more experience is needed in testing the toxicity of composts before most suitable methods can be suggested or limit values established.

The textile sludge contains organic matter, chemical nutrients and also a relative amount of heavy metals and aromatic dyes (Balan & Monteiro, 2001). Therefore, the textile sludge must be further composted in order to decrease or eliminate the toxicity before recycling in soils for crop production. In order to ensure that textile sludge composts will be safe for land use, simple methods to evaluate their toxicity, mainly to plants, are necessary. The bioassays of phytotoxicity have received great attention by environmental agencies of the world. Phytotoxicity is described as an intoxication of living plants by substances present in the growth medium, when these substances are taken up and accumulated in plant tissue (Chang et al., 1992). The phytotoxicity effects produced by organic wastes are the result of a combination of several factors, like presence of heavy metals, ammonia, salts and low molecular weight organic acids (Zucconi et al., 1985). The evaluation of organic waste toxicity by biological testing is therefore extremely important for screening the suitability of wastes for land application (Fuentes et al., 2004).

Seed germination and plant growth bioassays are the most common techniques used to evaluate compost phytotoxicity (Kapanen & Itavaara, 2001). A large number of studies have been carried out with different plant species such cucumber and cress (Helfrich et al., 1998), lettuce and soybean (Gundersson

Citation: Ademir Sérgio Ferreira Araújo; Regina Teresa Rosim Monteiro, "Plant bioassays to assess toxicity of textile sludge compost," http://dx.doi.org/10.1590/S0103-90162005000300013.

et al., 1997), red maple, sugar maple, white pine and pink oak (Maynard, 1998). There are large variations among bioassays and plant species. Itavaara et al. (1997) proposed a lettuce seed germination test in Petri dishes as a potential compost toxicity test and observed that there was a low sensibility of test for compost samples. Fuentes et al. (2004) observed that seed germination has been regarded as a less sensitive method than root length when used as a bioassay for the evaluation of phytotoxicity. Root growth is affected by environmental conditions, such as pH, temperature, salts and metals (Camargo et al., 2004). On the other hand, Kapanen & Itavaara (2001) recommended a standardized plant test for compost and soil samples. The OECD (1984) recommends the use of several plant species including soybean and wheat for bioassays to evaluate phytotoxicity.

In this study, we compared the seed germination and plant growth bioassays to assess the toxicity of a textile sludge compost.

MATERIAL AND METHODS

The compost was produced using sludge, obtained from the wastewater treatment system of a textile mill located at Americana, SP, Brazil, using the Beltsville aerated-pile method (USDA, 1980). The composting process lasted for 90 days, and 20 single samples were collected in several sites of the compost mixture, forming a composite sample. The chemical characteristics of the compost are: pH, 6.8; N, 0.13; K, 0.57; P, 0.22; Ca, 1.39; Mg, 0.43 %; Cu, 110.8; Zn, 397.0; Cd, <0.3; Cr, 73.9; Ni, 30.44; Pb, 33.4; Mo, <4.0 mg kg^{-1}. The heavy metal concentration is lower than the maximum permitted concentration (CETESB, 1999).

Two plant bioassays were used to assess the toxicity of this textile sludge compost, described in experiments I and II. In the Experiment I the extracts of the compost were prepared shaking, for 15 min, fresh samples (75% water content) with distilled water in the following concentrations: 0 g L^{-1} (only water); 19 g L^{-1} (0.57 g of compost plus 30 mL of water); 38 g L^{-1} (1.14 g of compost plus 30 mL of water); 76 g L^{-1} (2.28 g of compost plus 30 mL of water) and 152 g L^{-1} (4.56 g of compost plus 30 mL of water).

The seed germination bioassay was evaluated according to Tam & Tiquia (1994). Seeds of soybean (cv. IAC-foscarin) and wheat (cv. IAC-305) were incubated during five days, in the dark. Seed germination, root elongation and germination index (GI, a factor of relative seed germination and relative root elongation) were determined:

$$\text{Relative seed germination (\%)} = \frac{\text{Number of seeds germinated in the extract}}{\text{Number of seeds germinated in the control}} \times 100$$

$$\text{Relative root elongation (\%)} = \frac{\text{Mean root elongation in the extract}}{\text{Mean root elongation in the control}} \times 100$$

$$GI = \frac{(\text{\% Seed germination}) \times (\text{\% Root elongation})}{100\%}$$

In experiment II the fresh compost was mixed with nutrient solution (Hoagland & Arnon, 1950) in the following concentrations: 0 g L^{-1} (1000 mL of nutrient solution); 19 g L^{-1} (19 g of compost plus 1000 mL nutrient solution); 38 g L^{-1} (38 g of compost plus 1000 mL nutrient solution); 76 g L^{-1} (76 g of compost plus 1000 mL nutrient solution); 152 g L^{-1} (152 g of compost plus 1000 mL nutrient solution).

Soybean (cv. IAC-foscarin) and wheat (cv. IAC-305) seeds were pre-germinated, on distilled water humidified germination paper, during a period of five days in a germination chamber under controlled conditions (temperature of 25°C and photoperiod light and dark of 14 /10 h). Plastic pots with capacity of 2 L were used, each receiving 10 plants distributed in plates of compatible diameter in relation to the superior part of pot. Seedlings were grown, under an aerated hydroponics system, in a greenhouse (temperature of 20°C to 28°C and environmental photoperiod) adding the concentrations of textile sludge compost. Plants were harvested 15 days after transplanting. Shoot and root length were measured using a ruler, and then dried at 65°C for 48 h for total dry matter measurement.

The compost concentrations used in both bioassays were equal to the application of 19, 38, 76 e 152t of compost per ha, corresponding to 1x, 2x, 4x and 8x the amount of nitrogen of compost, respectively, as compared to fertilization of 100 kg ha^{-1} N (CETESB, 1999).

Treatments were replicated four times in a completely randomized design. Statistical analysis was conducted using the SAS System Software (SAS Institute, 2000).

RESULTS AND DISCUSSION

Experiment I

In general, the increase of compost concentration did not harm seed germination. At all concentrations the percentages of seed germination were greater than 90% for soybean and wheat (Figure 1a). Compost application resulted in linear increases of root elongation and quadratic increases of the germination index (GI) as shown in Figures 1b and 1c. The soybean and wheat seeds presented germination indexes (GI) of 100% for all the concentrations of compost extract. However, soybean and wheat seeds presented a higher GI for the concentration of 152 g L^{-1}, as compared to the other treatments.

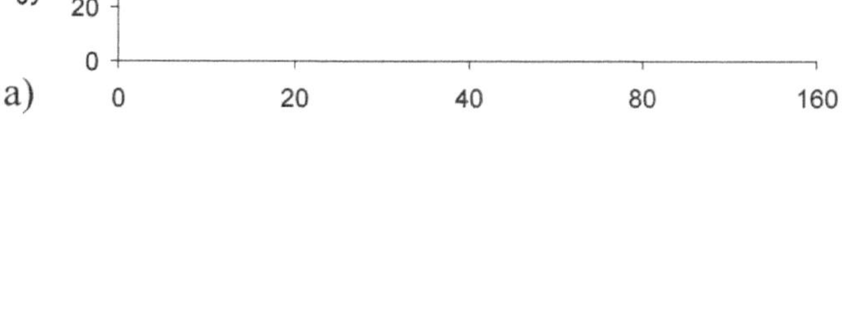

a)

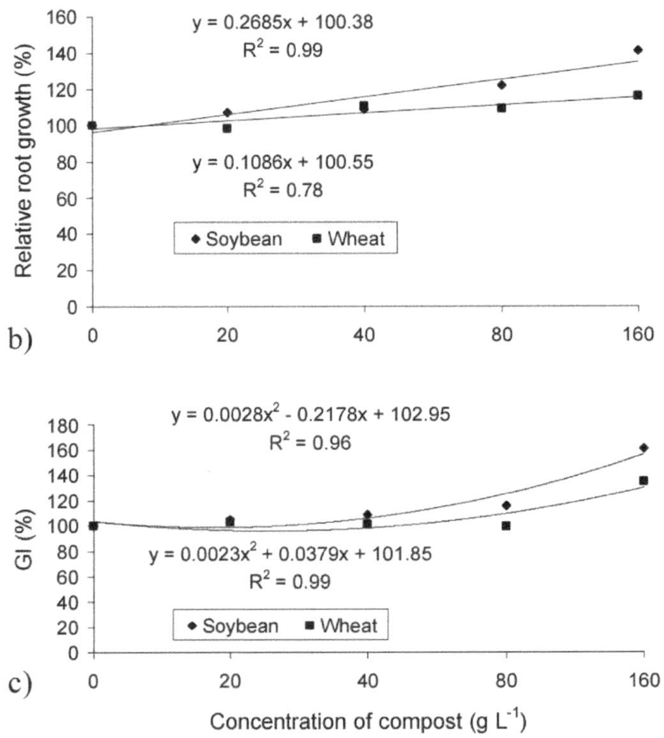

Figure 1: Percentage of seed germination (a), relative root growth (b) and germination index - GI (c) of soybean and wheat for different concentrations of textile sludge compost.

The compost did not show inhibitory effects on soybean and wheat. Similar results were found by Fuentes et al. (2004), who observed a germination percentage of more than 70% for barley and cress seeds in the presence of sewage sludge. In addition, GI values suggest no toxicity of the compost for these plant species. The GI has been proved to be a very sensitive index (Tiquia et al., 1996) indicating, when greater than 80%, the disappearance of the phytotoxicity of the compost.

Experiment II

The increase of concentrations of compost harmed seedling growth. The application of compost in concentrations equal or above 38 g L^{-1} decreases the total dry matter production, shoot and root length of soybean and wheat seedlings (Figures 2 a, b and c). Compared to the control, these decreases for soybean were 68 to76% for total dry matter, 64 to 81% to shoot length and 3 to 68% to root length, and for wheat were 65 to 72% for total dry matter, 84 to 92% for shoot length and 46 to 50% to root length. The application of 19 g L^{-1} of compost did not present negative effects.

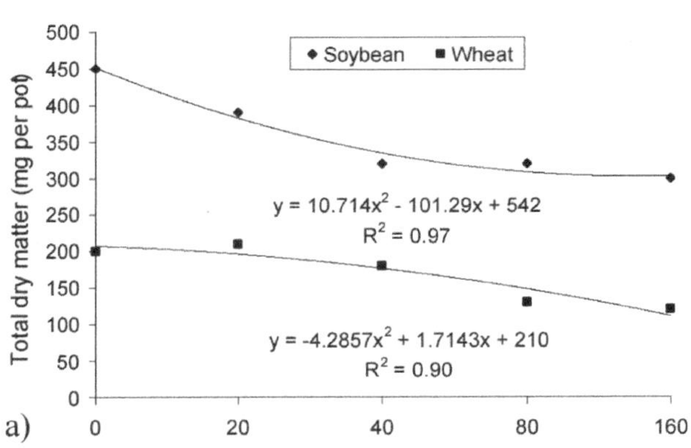

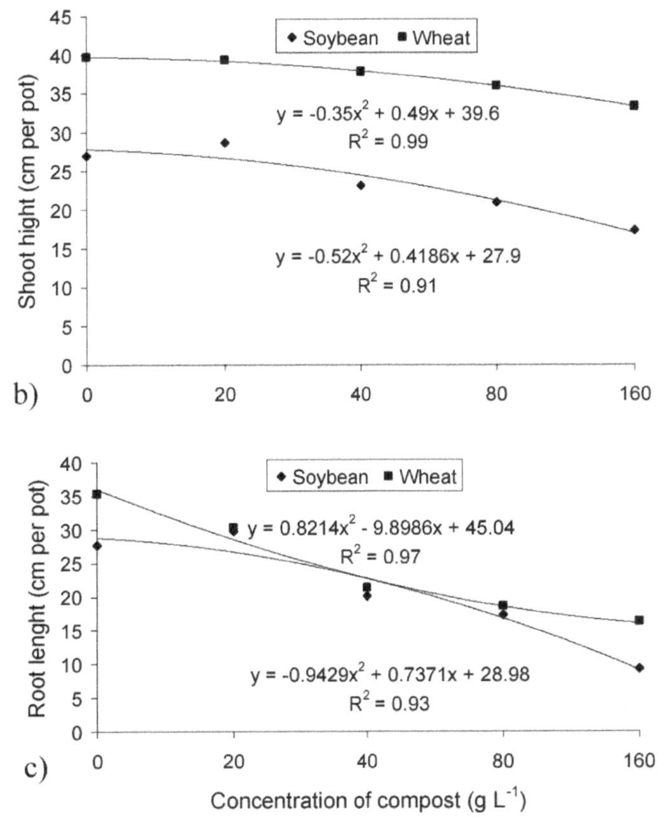

Figure 2: Total dry matter (a), shoot hight (b) and root length (c) of soybean and wheat seedlings in different concentrations of textile sludge compost, under hydroponics.

The results observed for the hydroponics system indicate that the compost in high concentrations is harmful to soybean and wheat seedlings, evaluated mainly through total dry matter. According with Kapustka (1997), the total dry matter provides the best indication of an adverse plant response to toxic substances and the decrease in the total dry matter indicates that the application of compost at concentrations equal or above 38 g L^{-1} leads to toxic effects to plants, probably by heavy metals present in higher concentration and their bioavailability in the hydroponics solution. However, the concentrations of heavy metals in the compost are lower than the maximum heavy metal concentration for soil application permitted by Cetesb (1999). On the other hand, the bioavailability of heavy metals in hydroponics is higher than in soils, due the interactions among heavy metals and edaphic components (Mcbride et al., 1998). Ali et al. (2004) compared conditions of artificial and natural soils and observed that in the artificial soil a maximum bioavailability of heavy metals is present.

Inhibition was also observed on shoot and root lengths, however, the root length were more affected by the concentrations of the compost. The high decrease in the root length of soybean and wheat indicates that they are more sensitive than the shoot. The roots are the responsible for absorption and accumulation of metals (Marques et al., 2002), therefore metal concentrations affect more the roots than the aerial parts of the plant (Oncel et al., 2000).

Plant growth was a more sensitive parameter than the seed germination in relation to concentrations of the compost extracts. Probably, the great quantity of nutrient reserves in soybean and wheat seed would have lower sensitivity to toxicity of the compost. Cheung et al. (1989) reported that seed of root crops, cereals and legumes containing high quantity of reserves, and the sensitivity of a plant species to toxicity depends on the quantity of their reserves.

In addition, Kapustka (1997) reports that the seed germination bioassay is relatively insensitive to many toxic substances. According to the same author, the insensitivity results from two factors: (a) many chemicals may not be absorbed by the seed, and (b) the embryonic plant draws its nutritional requirements internally from seed stored materials and is effectively isolated from the environment.

On the other hand, the plant growth bioassay overcomes deficiencies of the seed germination bioassays. The duration of the test provides for exposure well into the autotrophic stage of plant development, providing a better approximation of field conditions than root elongation bioassay (Kapustka, 1997). Baca et al. (1990) evaluated the phytotoxicity of compost and observed that plant growth bioassay was more sensitive than seed germination.

CONCLUSIONS

Textile sludge compost presented great phytotoxicity under hydroponics condition as compared to seed germination and having a possible short-term response.

Textile sludge compost applied at the concentration of 19 g L^{-1} (19 t ha^{-1}) did not induce toxicity in the evaluated plant species.

Soybean and wheat are sensitive plants for the evaluation of the toxicity of organic wastes in plant growth bioassays.

ACKNOWLEDGEMENT

The authors were supported by personal grant from CNPq.

REFERENCES

1. ALI, N.A.; ATER, M.; SUNAHARA, G.I.; ROBIDOUX, P.Y. Phytotoxicity and bioaccumulation of copper and chromium using barley (Hordeum vulgare L.) in spiked artificial and natural forest soils. Ecotoxicology and Environmental Safety, v.57, p.363-374, 2004.
2. BACA, M.T.; DELGADO, I.C.; SANCHEZ-RAYA, A.J.; GALLARDO-LARA, F. Comparative use of cress seed germination and physiological parameters of *Helianthus annuus* L. to assess compost maturation. Biological Wastes, v.33, p.251-261, 1990.
3. BALAN, D.S.L.; MONTEIRO, R.T.R. Decolorization of textile indigo dye by lignolytic fungi. Journal of Biotechnology, v.89, p.141-145, 2001.
4. CAMARGO, C.E.O.; FERREIRA-FILHO, A.W.P.; SALOMON, M.V. Temperature and pH of nutrient solution on wheat primary root growth. Scientia Agricola, v.61, p.313-318, 2004.
5. CETESB. Aplicação de biossólidos em áreas agrícolas: critérios para projeto e operação. Norma P4230. São Paulo, 1999.
6. CHANG, A.C.; GRANTO, T.C.; PAGE, A.L. A methodology for establishing phytotoxicity criteria for chromium, copper, nickel and zinc in agricultural land application of municipal sewage sludges. Environmental Quality, v.21, p.521-536, 1992.
7. CHEUNG, Y.H.; WONG, M.H.; TAM, N.F.Y. Root and shoot elongation as an assessment of heavy metal toxicity and Zn equivalent value of edible crops. Hydrobiologia, v.188, p.377-383, 1989.
8. FUENTES, A.; LLORENS, M.; SAEZ, J.; AGUILAR, M.I.; ORTUÑO, J.F.; MESEGUER, V.F. Phytotoxicity and heavy metals speciation of stabilized sewage sludges. Journal of Hazardous Materials, v.108, p.161-169, 2004.
9. GUNDERSSON, C.A.; KOSTUK, J.M.; MITCELL, H.G.; NAPOLITANO, G.E.; WICKER, L.F.; RICHMOND, J.E.; STEWART, A.J. Multispecies toxicity assessment of compost produced in bioremediation of an explosives-contaminated sediment. Environmental Toxicology and Chemistry, v.16, p.2529-2537, 1997.
10. HELFRICH, P.; CHEFETZ, B.; HADAR, Y.; CHEN, Y.; SCHNABL, H. A novel method for determining phytotoxicity in compost. Compost Science and Utilization, v.6, p.6-13, 1998.
11. HOAGLAND, D.; ARNON, D.I. The water culture method for growing plants without soil. Berkeley: California Agriculture Experiment Station, 1950.
12. ITAVAARA, M.; VILKMAM, M.; VENELAMPI, O. Windrow composting of biodegradable packaging materials. Compost Science and Utilization, v.5, p.84-92, 1997.
13. KAPANEN, A.; ITAVAARA, M. Ecotoxicity tests for compost applications. Ecotoxicology and Environmental Safety, v.49, p.1-16, 2001.
14. KAPUSTKA, L.A. Selection of phytotoxicity tests for use in ecological risk assessments. In: WANG, W.; GORSUCH, J.W.; HUGHES, D. Plants for environmental studies. New York: CRC Press, 1997. p.516-548.
15. MARQUES, M.O.; MELO, W.J.; MARQUES, T.A. Metais pesados e o uso de biossólidos na agricultura. In: TSUTIYA, M.T. (Ed.) Biossólidos na agricultura. São Paulo: SABESP, 2002. p.365-403.
16. MAYNARD, A.A. Utilization of MSW compost in nursery stock production. Compost Science and Utilization, v.6, p.38-44, 1998.
17. McBRIDE, M.; MARTINEZ, C.E.; SAUVE, S. Copper (II) activity in aged suspensions of geothite and organic matter.Soil Science Society of America Journal, v.62, p.1542-1548, 1998.
18. ONCEL, I.; KELES, Y.; USTUN, A.S. Interactive effects of temperature and heavy metal stress on the growth and some biochemical compounds in wheat seedlings. Environmental Pollution, v.107, p.315-320, 2000.
19. ORGANIZATION FOR ECONOMIC COOPERATION AND DEVELOPMENT. OECD Guidelines for testing of chemicals.Guideline 208: terrestrial plants, growth test. Paris: OECD, 1984.
20. PAGGA, U.; BEIMBORN, D.B.; YAMAMOTO, M. Biodegradability and compostability of polymers: Test methods and criteria for evaluation. Journal of

Environment Polymers Degrading, v.4, p.173-178, 1996.

21. SAS INSTITUTE. System for information. Versão 8.0. Cary: SAS Institute, 2000.

22. TAM, N.F.Y.; TIQUIA, S.M. Assessing toxicity of ‹spent sawdust pig-litter› using seed germination technique. Resource Conservation Recycling, v.11, p.261-274, 1994.

23. TIQUIA, S.M.; TAM, N.F.Y.; HODGKISS, I.J. Effects of composting on phytotoxicity of spent pig-manure sawdust litter. Environmental Pollution, v.93, p.249-256, 1996.

24. ZUCCONI, F.; MONACO, A.; FORTE M.; DE BERTOLDI, M. Phytotoxins during the stabilization of organic matter. In: GASSER, J.K.R. (Ed.) Composting of agricultural and other wastes. London: Elsevier, 1985. p.73-86.

25. USDA. ENVIRONMENTAL PROTETION AGENCY. Manual for composting sewage sludge by the Beltsville aerated-pile method.Washington: EPA, 1980. 65p. (EPA-600/8 80-022).

CHAPTER 20

THE EFFECT OF LEAD ON THE PHYTOCHEMISTRY OF TITHONIA DIVERSIFOLIA EXPOSED TO ROADSIDE AUTOMOTIVE POLLUTION OR GROWN IN POTS OF PB-SUPPLEMENTED SOIL

Elizabeth Olivares

Key words: Asteraceae, metals, Mexican sunflower, phenols, pollution, weeds.

INTRODUCTION

Mexican sunflower, *Tithonia diversifolia* is widely distributed in the tropics, from Mexico to Brazil, and is also found in Asia and Africa, where it has been introduced in view of its multiple uses, for example, as animal forage, green fertilizer, a natural insecticide, a honey-plant and an ornamental shrub (Rios, 1999). The species enhances the availability of P to crops, which led to its recommendation for biomass transfer technologies in Kenya (George et al., 2002 a, b). However, it behaves as an aggressive weed with high invasive capacity in Africa (Ayeni et al.,1997). This species, adapted to a variety of ecological zones, is present in Venezuela, from montane environments in the Coastal and Andean ranges to the savanna areas, and is very common in wasteland, margins of polluted city rivers and roadsides. Studies of biochemical responses to environmental pollution of *T. diversifolia* are important, since this species is recommended in agroforestry for animal production in the tropics and is also a weed capable of inhabiting polluted sites.

García-Miragaya et al. (1981) reported strong Pb pollution in roadside soils at Caracas, Venezuela, due to heavy traffic of motor vehicles and the utilization of leaded gasoline. At present, only motor vehicles bought after the year 2000 use unleaded gasoline in Venezuela and consequently the same or higher levels of Pb contamination prevail.

Lagerwerff and Specht (1970) reported that the concentrations of Cd, Ni, Pb and Zn in roadside soil and grass samples from several locations decreased with the distance from vehicular traffic. Contamination was related to the composition of gasoline, motor oil and automobile tires, and to roadside deposition of residues of these materials. Contamination from motor vehicle emissions within 100 m of the roadway has been reported in vegetation (Motto et al., 1970; Ward et al., 1975), soils (Yassoglou et al., 1987), and spider webs (Hose et al., 2002). Lead added to fuel as tetra ethyl, to act as a lubricant during combustion, is discharged into the environment in exhaust fumes largely as minute particles of inorganic Pb compounds and about 50 % of this falls within a region of 100 m from the road (Mengel and Kirkby, 2001).

Citation: Elizabeth Olivares, "The effect of lead on the phytochemistry of Tithonia diversifolia exposed to roadside automotive pollution or grown in pots of Pb-supplemented soil," ttp://dx.doi.org/10.1590/S1677-04202003000300004.

In the year 2000, the cumulative industrial age anthropogenic global production of Pb was 235 million tonnes (Han et al., 2002). According to Weiss et al. (1999) since the industrial revolution and until 1950, the main sources of Pb in the biosphere have been industrial activity mostly coal combustion and ferrous and non-ferrous smelting, as well as open waste incineration, whereas from 1950 onwards it has been the use of leaded gasoline. The largest Pb emissions into the atmosphere throughout Europe took place between 1950 and 1980 due to automotive exhausts. The marked drop in atmospheric Pb fluxes, recorded since the 1980s, has been attributed to the phasing out of leaded gasoline.

Diverse biochemical changes in green plants in response to Pb have been reported by several authors: decreases in chlorophyll content (Ewais, 1997; Xiong, 1997; Kastori et al., 1998; Fargašová, 2001), carotenoids (Fargašová, 2001), proteins (Kevresan et al., 2001), nitrate reductase activity (Singh et al., 1997; Kevresan et al., 2001), lamina and mesophyll thickness, epidermal cell size and diameter of vessels (Kovačević et al., 1999); increases in chlorophyll *a*/*b*ratio (Fargašová, 2001) and concentration of phenols (Lummerzheim et al., 1995; Lavid et al., 2001 a, b). In the presence of other metals, such as Al, Cu and Cd, lipid peroxidation has been found to increase (Cakmark and Horst, 1991; Gallego et al., 1999; Pereira et al., 1999; Baryla et al., 2000; Hartley-Whitaker et al., 2001; Sandalio et al., 2001; Shah et al., 2001).

Thus, in metal contaminated plants we expect to find senescence symptoms, such as lower concentrations of chlorophyll, carotenoids and proteins, diminished nitrate reductase activity and increased acid phosphatase activity, lipid peroxidation and chlorophyll *a*/*b* ratio compared to plants not stressed by metals. We also expect detoxification through phenols.

The objective of the present study was to evaluate metals in foliar and root material from field samples of *T. diversifolia,* collected at two sampling sites with contrasting pollution levels due to automotive traffic, as well as evaluate foliar compounds sensitive to phytotoxic concentrations of metals. The field study was followed by an experiment using pots of Pb-acetate-supplemented soil to learn whether the parameters affected in the field also change when the plants are exposed to the same air and soil, but to a different concentration of Pb in the soil. This experiment was carried out in order to verify whether the parameters affected in the field were responding to Pb.

MATERIALS AND METHODS

Plant material and sites: Tithonia diversifolia (Hemsl.) Gray (Asteraceae) was sampled in two agricultural fields near satellite towns of Caracas, Venezuela, where the plant grows wild on roadsides. The control site (Site 1) is located at 10° 21› N, 66° 56› W, in a rural area (El Amarillo) close to the town of San Antonio, while the polluted site (Site 2) is located at 10° 22› N, 67° 05› W, where automotive traffic is heavy because it is close to the entrance of the town of San Pedro. Soils from both sites presented a sandy clay loam texture, a pH (KCl) between 5 at Site 1 to 6 at Site 2, together with large concentrations of Fe (36 $mg.g^{-1}$) and Al (17 $mg.g^{-1}$) in similar concentrations at both sites (Olivares et al., 2002a).

At each site, 15 whole plants of approximately 2 m height were collected. The plants were transported to the laboratory with their roots immersed in tap water. In the present study, *T. diversifolia* leaves and roots were washed before analysis, since a previous study conducted on plants exposed to highway dust showed that unwashed plant tissue had a higher Pb content than those analyzed after a 15 min wash with distilled water, indicating that 27 % of the Pb occurred on the surface of the leaves (Isermann, 1977).

Analysis of mineral elements: The leaves and roots were dried in a ventilated oven, for approximately 78 h at 60 °C to constant weight and then ground. Total nitrogen concentration was determined by the Kjeldhal method (Tecator Kjeltec Systems from Foss Tecator, Höganäs, Sweden) after digestion with sulphuric acid of the ground dry plant material.

Phosphorus was measured colorimetrically (Murphy and Riley, 1962) in the digested material by means of an UV/visible spectrophotometer (Ultrospec 2000, Amersham Pharmacia, Cambridge, England).

For the measurement of metals (K, Ca, Mg, Mn, Cd, Co, Pb, Cr, Al, Fe, Cu and Ni) samples were digested in a nitric-perchloric acid mixture (Miller, 1998) and analysed with an atomic absorption spectrometer (SpectrAA 55B, Varian Techtron, Victoria, Australia).

Enzymatic activities: Nitrate reductase activity (NRA) was determined in fresh tissue of leaves and roots from each site by measuring the amount of nitrite produced from nitrate; nitrite was estimated by formation of the diazonium salt after reaction of nitrite with sulphanilamide in acid medium (Lindblad and Guerrero, 1993). Acid phosphatase activity (APA) in fresh tissue of leaves and roots was determined according to Tabatabai and Bremner (1969). Activities were expressed on a dry mass basis, using the fresh mass/dry mass ratio of leaf discs taken from the same leaves used for enzyme activity determinations.

Organic compounds affected by metals and leaf specific mass: For determinations of chlorophyll and carotenoids, fresh tissue of leaves was ground in a mortar with cold 80 % aqueous acetone. The concentration of chlorophyll *a*+*b*and the total concentration of carotenoids were calculated according to Lichtenthaler and Wellburn (1983). The amount of UV-B-absorbing compounds was estimated in fresh leaf tissue by measuring the absorbance at 300 nm in acidified methanol solutions (Day and Demchik, 1996). Total phenols were measured colorimetrically in ethanol extracts of ground dried leaves, using the Folin-Ciocalteu method (Amorim et al., 1977); values were standardized against chlorogenic acid. Lipid peroxidation in fresh tissue of leaves was measured by the formation of a 2-thiobarbituric acid (TBA)-malondial-deyde (MDA) complex (van Hasselt et al., 1996). Water-soluble proteins were determined according to Bradford (1976).

Specific leaf mass (SLM) was calculated from the dry mass expressed per unit area using four leaf discs (1.41 cm^2diameter) dried in a microwave oven for approximately 15 min to constant weight. SLM corresponds to the inverse of specific leaf area (SLA) or ratio of leaf area to leaf dry mass, dimension quotients were introduced by Stocker in 1931 and

Müller-Stoll in 1947 (Medina et al., 1990). *Pot experiment with Pb-supplemented soil:* Twenty young plants were selected in the field according to their uniformity in size (0.5 m height) and were transplanted with their native soil to pots (2 kg soil capacity) and kept in the grounds of the "Instituto Venezolano de Investigaciones Científicas» (I.V.I.C.), a site with low traffic and low Pb (0.41 mg.m^{-3}) in the air (Escalona and Sanhueza, 1981). The soil also presented a low Pb concentration (42 to 64 mg.kg^{-1}). The soil was described in Olivares et al. (2002 b). The pots were supplemented with solutions of analytical grade Pb-acetate at 21 and 29 days after transplantation (1 g.kg^{-1} soil at each time). During the 3 weeks without Pb, the plants become adapted to the pot condition. Lead was added in two progressive sessions in order to increase the probability of Pb incorporation through roots. Pb-acetate was used by Ewais (1997), Xiong (1997) and Singh et al. (1997/98). A gradient was used by Xiong (1997): 0, 0.8, 1.6 and 3.2 g.kg^{-1} Pb. Based on this gradient two treatments were selected, 0 and 2 g.kg^{-1} soil. The plants were watered daily and care was taken that no water was lost from the dishes under the pots. The plants were maintained under natural conditions of light and air temperature and humidity. The maximal radiation in the greenhouse as reported by Suarez (2000) was 1550 ± 370 mmol.m^{-2}.s^{-1}, with daily temperatures of 30.5 ± 3.4 °C during the day and 16.3 ± 2.0 °C during the night and with a relative humidity from 50 to 78 %.

After 34 days of the first metal application measurements were made of the concentration of phenols, APA of leaves and roots and SLM (variables that suffer changes in response to Pb in the field plants) using 5 plants selected randomly from the 10 plants in each treatment.

Statistical analysis: Student's t-test was used for mean comparisons using the software Sigma plot 2001 (V7.0). Regression analysis was used for the APA and Pb data. Two-way ANOVA was carried out to compare the total concentration of mineral elements and enzymatic activities in *T. diversifolia* grown in the field at two sites (with light and heavy automotive traffic) and between two organs (leaves and roots) using the software Statistica 6.0 (Statsoft Inc.). The same analysis was done to compare levels of Pb, phenols, APA and SLM in plants grown in pots between two treatments (with or without Pb-acetate) and two organs. One-way ANOVA was carried out to compare biochemical parameters and SLM in leaves with different foliar Pb content from plants grown in the field.

RESULTS

Analysis of mineral elements in leaves and roots: In leaves, the content of heavy metals (Pb, Cr, Al, Fe, Cu and Ni) was higher in samples from the polluted Site 2 than in those from the control Site 1. The Pb concentration was found to be 15 times higher for Site 2 than Site 1 (table 1). In roots, higher values were also found for Pb at Site 2, while the other metals occurred in similar or lower amounts in samples from Site 2 compared with Site 1. The metals Co and Cd were not detected, the detection limits being 0.05 μg.g^{-1} for Co and 0.02 μg.g^{-1} for Cd.

Roots and leaves from Site 2 showed higher concentrations of P, K and Ca than those from the less polluted Site 1 (table 1). The same relationship between sites held for Mn in leaves, and for Zn, N and Mg in roots. The ANOVA showed differences between sites in the mineral elements except for Fe, Mn and N, and between organs except for K.

Enzymatic activities: The NRA was lower in leaves from Site 2 as compared with those from Site 1 but was similar in roots from both sites (table 1). APA was higher in both leaves and roots of the more polluted Site 2, where higher values of P were also found. The following regressions were found between APA and P concentrations: $y = 0.11x + 1.84$, $r^2 = 0.60$ (roots and leaves); $y = 0.07x + 0.28$, $r^2 = 0.73$ (leaves); and $y = 0.07x - 0.03$, $r^2 = 0.57$ (roots). The ANOVA showed significant differences between sites and organs.

Organic compounds affected by metals and leaf specific mass: Chlorophyll, chlorophyll *a*/*b* ratio, carotenoids, compounds absorbing at 300 nm such as flavonoids and proteins, lipid peroxidation and water-soluble proteins were similar at both sites, while the concentration of phenols was higher and SLM was lower for Site 2 (table 2). The t-test and one way ANOVA showed similar results.

Pot experiment with Pb-supplemented soil: Young plants transplanted to pots with Pb-acetate in the soil showed a highly significant increase in phenol concentration in roots compared to the controls without Pb-acetate (table 3). The concentration of lead in roots was very high and higher than in leaves, whereas in the field, where the main source of Pb would be from the emissions of motor vehicles, a higher concentration was observed in leaves than in roots. For the plants from the greenhouse experiment there were no significant differences in SLM and the APA did not increase in Pb-contaminated plants in contrast to the field plants, where these responses may depend on multiple factors rather than Pb alone. However, a very highly significant increase of phenols was observed in Pb-contaminated roots.

The ANOVA showed differences between treatments and organs for Pb. The difference in phenol concentration in roots was highly significant for both treatments but not in leaves. The two-way ANOVA therefore did not show differences between treatments because it includes both organs.

DISCUSSION

Analysis of mineral elements in leaves and roots: The concentration of Pb in leaves from heavy traffic roadside sites found in this study is higher than those reported by Ross (1994) as typical values for Pb-contaminated plants (30-300mg.g^{-1}). Mengel and Kikby (2001) reported values of 50 mg.g^{-1} Pb in vegetation at the side of the road in motorway areas, with roots being very little affected, but only 2 to 3 mg.g^{-1} Pb at a distance of 150 m from the motorway.

The concentrations of Cr, Cu and Ni in leaves from polluted Site 2 were at the lower limit of the range typical of contaminated plants, whereas at Site 1 leaves showed concentrations of Cr, Cu and Ni below this range.

T. diversifolia from Site 2 had a higher concentration of Pb in leaves than in roots. This may be explained by the fact that there are two major pathways for the incorporation of heavy metals into plant tissues: soil uptake and the atmospheric deposition from settling dust (Fodor, 2002). On the other hand, similar concentrations of Cr, Al, Fe, Cu and Ni were found in both organs at Site 2, while roots contained larger amounts of these metals than leaves collected at control Site 1. Usually, roots have a higher concentration of Pb than leaves (Wierzbicka, 1999). Higher metal concentrations in roots than in shoots are frequently observed in non-hyperaccumulator plants (McGrath et al., 2001). Generally the majority of Pb is attached to the cell walls in the roots and competes with Ca in the adsorption of cations by the carboxylic groups of the matrix substances, galacturonic and glucuronic acids. However, despite the retention of Pb in the root, there are plants in which a significant amount of the metal is transported to the shoot (Cseh, 2002).

Compartmentalization and sequestration with organic acids have been proposed as mechanisms of tolerance to metals in Asteraceae. In sunflower, trichomes play a role in the tolerance to Mn toxicity (Blamey et al., 1986).

T. diversifolia also possesses trichomes in the leaves but the compartmentalization of Pb in the organs was not evaluated in the present study. Nevertheless, the localization of metals in a metabolically inactive form in trichomes is a possible explanation for the unusually higher concentration of Pb found in leaves compared to roots of *T. diversifolia* field plants.

Although the leaves were washed, the hair and cuticle can trap contaminants. Metals, such as Fe and Al, can be used as markers for soil contamination in plant analyses. The metals with higher concentration in the leaves were Al, Fe and Cu (table 1) and these were 3 or 4 times higher for the polluted site compared to Site 1, whereas Pb was 15 times higher, indicating uptake of this metal.

In general, toxic metals may cause deficiency of other elements essential for plants (Fodor, 2002; Krupa et al., 2002). However, this was not observed in the data of table 1. At Site 2 agricultural and urban development is higher than for Site 1, and consequently the input of nutrients through irrigation by streams is larger.

Enzymatic activities: The NRA in leaves diminished at the polluted site but no changes were observed in roots (table 1). Klobus et al. (2002) showed different examples of inhibition of nitrate reductase in plants under heavy metal stress.

Singh et al. (1997) reported that nitrate reduction was inhibited drastically in roots by Pb but in leaves different effects have been observed with various cultivars. Leaf NRA in *Vigna radiata* increased with increasing Pb (Singh et al., 1997/98), while Kevresan et al. (2001) found a significant decrease in the NRA in the aboveground parts of young pea plants.

The APA increased at Site 2, where leaves had a higher concentration of P. The exposure of roots to Ni markedly stimulated phosphatase activity in *Alyssum bertolonii* (Gabbrielli et al., 1989). The deposition of Pb pyrophosphate along cell walls was proposed as the main process responsible for Pb accumulation in root tissues. A favourable P regime reduces the effects of Pb toxicity, since Pb can form insoluble phosphates (Kabata Pendias and Pendias, 1984).

Organic compounds affected by metals and leaf specific mass: Plants from Site 2 were exposed to daily traffic during their entire lifespan (long time exposure). The plant samples from the field analysed here did not produce responses such as depletion of chlorophyll, carotenoids and proteins, nor increases in lipid peroxidation, reported in the literature for leaves of plantlets under greenhouse conditions.

Changes of the foliar structure in response to Pb have been reported by Kovacevic et al. (1999), who found smaller lamina and mesophyll thickness, epidermal cell size and diameter of vessels. Differences in SLM, related to leaf thickness, were expected for both sites. Detoxification through phenols was also expected (Giertych et al., 1999).

An increase in phenol concentration and a highly significant drop in SLM was observed in Site 2 (table 2). However, phenol concentration in leaves and SLM were not affected, in spite of a significant increase (13.6 times) in foliar Pb observed between the treatments in the pot experiment (see below). On the other hand, Pb concentration in contaminated leaves in the field (table 1) was 6 times higher than in the +Pb treatment in the pot experiment (table 3) presumably because there was an input of Pb from the air due to the traffic, absent in the pot experiment.

Pot experiment with Pb-supplemented soil: APA of roots was not affected when comparing both treatments despite the Pb concentration being different by a factor of 62. In leaves a decrease of activity was observed. According to these result the increase of APA observed in the field for contaminated plants is probably not a response to Pb concentration. In this experiment plants were grown in the same soil with or without the Pb supplement.

The highly significant increase of phenols in response to Pb observed for roots has been reported by several authors for other plants in response to metals.

Ruso et al. (2001) found a good correlation between the induced synthesis of phenols, mainly in root tissue, and the tolerance and accumulation of heavy metals in sunflower. It was proposed the selection of lines for their constitutive and induced phenol content to be used for phytoremediation.

Giertych et al. (1999) found higher phenol contents in plants at a polluted site compared with a control site. Total phenols include condensed and hydrolysable tannins, flavonoids, phenolic glucosides and chlorogenic acids. The hydroxyl radicals of tannins have a strong affinity for metals and are used in industry as metal chelators. For example, the high concentrations of tannins in *Psychotria douarrei* leaves might function as a detoxification mechanism for elevated cytoplasmic metal concentrations, in addition to providing defensive benefits (Davis et al., 2001). Lavid et al. (2001 a,b) demonstrated the involvement of phenols in Cd accumulation in aquatic plants. Polyphenols in *Nymphaea* chelate Pb, Cr and Hg *in vitro* (Lavid et al., 2001b). The phenolics are generally thought to prevent oxidative damage by scavenging active oxygen species and by breaking the radical chain reactions during lipid peroxidation. These antioxidative effects require the reduced form of phenolics. In the oxidized form (unstable under physiological conditions but stabilized by Al) act as prooxidants (Sakihama and Yamasaki, 2002) .

As a conclusion *T. diversifola* growing wild on roadsides with heavy urban traffic presented levels of Pb in leaves and roots considered as typical of contaminated plants, but without visible symptoms of damage to the leaves or depletion of chlorophyll, carotenoids, proteins nor increases in lipid peroxidation. These are responses reported in the literature for plants under Pb stress. On the other hand, phenols increased and this was also observed in roots with high Pb concentration after Pb-acetate application. The identification and role of phenols in the tolerance to Pb in this species deserves the attention of future research.

ACKNOWLEDGEMENTS

The author thanks Eder Peña for carrying out statistical and mineral analyses. Guillermina Aguiar performed the measurements with the atomic absorption spectrophotometer. This work was partially sponsored by CONICIT, project S1-98003716.

REFERENCES

1. Amorim HV, Dougall DK, Sharp WR (1977) The effect of carbohydrate and nitrogen concentration on phenol synthesis in Paul's Scarlet Rose cells grown in tissue culture. Physiol. Plant. 39:91-95.
2. Ayeni AO, Loordbanjou DT, Majek BA (1997) *Tithonia diversifolia* (Mexican sunflower) in South-western Nigeria: Occurrence and growth habit. Weed Res.37:443-449.
3. Baryla A, Laborde C, Montillet JL, Triantaphylidès C, Chagvardieff P (2000) Evaluation of lipid peroxidation as a toxicity bioassay for plants exposed to copper. Environ. Pollut. 109:131-135.
4. Blamey FPC, Joyce DC, Edwards DG, Asher CJ (1986) Role of trichomes in sunflower tolerance to manganese toxicity. Plant Soil 91:171-180.
5. Bradford MM (1976) A rapid and sensitive method for the quantitation of microgram quantities of protein utilizing the principle of protein-dye binding. Anal. Biochem. 72:248-254.
6. Cakmak I, Horst WJ (1991) Effect of aluminium on lipid peroxidation, superoxide dismutase, catalase, and peroxidase activities in root tips of soybean (*Glycine max*). Physiol. Plant. 83:463-468.
7. Cseh E (2002) Metal permeability, transport and efflux in plants. In: Prasad MNV, Strzalka K (eds), Physiology and Biochemistry of Metal Toxicity and Tolerance in Plants, pp.1-36. Kluwer Academic Publishers, Dordrecht, The Netherlands.
8. Davis MA, Pritchard SG, Boyd RS, Prior SA (2001) Developmental and induced responses of nickel-based and organic defences of the nickel-hyperaccumulating shrub, *Psichotria douarrei*. New Phytol.150:49-58.
9. Day TA, Demchik SM (1996) Influence of enhanced UV-B radiation on biomass allocation and pigment concentrations in leaves and reproductive structures of greenhouse-grown *Brassica rapa*. Vegetatio 127:109-116.
10. Escalona L, Sanhueza E (1981) Elemental analysis of the total suspended matter in the air in downtown Caracas. Atmospheric Environ. 15:61-64.
11. Ewais EA (1997) Effects of cadmium, nickel and lead on growth, chlorophyll content and proteins of weeds. Biol. Plant. 39:403-410.
12. Fargašová A (2001) Phytotoxic effects of Cd, Zn, Pb, Cu and Fe on *Sinapis alba* L. seedlings and their accumulation in roots and shoots. Biol. Plant. 44:471-473.
13. Fodor F (2002) Physiological responses of vascular plants to heavy metals. In: Prasad MNV, Strzalka K (eds), Physiology and Biochemistry of Metal Toxicity and Tolerance in Plants, pp.149-177. Kluwer Academic Publishers, Dordrecht, The Netherlands.
14. Gallego SM, Benavides MP, Tomaro ML (1999) Effect of cadmium ions on antioxidant defense system in sunflower cotyledons. Biol. Plant. 42:49-55.
15. García-Miragaya J, Castro S, Paolini J (1981) Lead and zinc levels and chemical fractionation in road-side soils of Caracas, Venezuela. Water Air Soil Pollut. 15:285-297.
16. George TS, Gregory PJ, Robinson JS, Buresh RJ, Jama B (2002 a) Utilisation of soil organic P by agroforestry and crop species in the field, western Kenya. Plant Soil 246:53-63.
17. George TS, Gregory TJ, Robinson JS, Buresh RJ (2002 b) Changes in phosphorus concentrations and pH in the rhizosphere of some agroforestry and crop species. Plant Soil 246:65-73.
18. Gabbrielli R, Grosi L, Vergnano O (1989) The effects of nickel, calcium and magnesium on the acid phosphatase activity of two Alyssum species. New Phytol.111:631-636.
19. Giertych MJ, Karolewski P, De Temmerman LO (1999) Foliage age and pollution alter content of phenolic compounds and chemical elements in *Pinus nigra* needles. Water Air Soil Pollut. 110:363-377.
20. Han FX, Banin A, Su Y, Monts DL, Plodinec MJ, Kingery WL, Triplett GE (2002) Industrial age anthropogenic inputs of heavy metals into the pedosphere. Naturwissenschaften 89:497-504.
21. Hartley-Whitaker J, Ainsworth G, Meharg AA (2001) Copper- and arsenate-induced oxidative stress in *Holcus lanatus*L. clones with differential sensitivity. Plant Cell Environ. 24:713-722.
22. Hose GC, James JM, Gray MR (2002) Spider webs as environmental indicators. Environ. Pollut. 120:725-733.
23. Isermann K (1977) A method to reduce contamination and uptake of lead by plants from car exhaust gases. Environ. Pollut. 12:199-203.
24. Kabata-Pendias A, Pendias H (1984) Trace Elements in Soils and Plants. CRC Press, Boca Raton, U.S.A.
25. Kastori R, Plesnicar M, Sakac Z, Pancovic D, Arsenijevic Maksimovic I (1998) Effect of excess lead on sunflower growth and photosynthesis. J. Plant Nutr.21:75-85.
26. Kevresan S, Petrovic N, Popovic M, Kandrac J (2001) Nitrogen and protein metabolism in young pea plants as affected by different concentrations of nickel, cadmium, lead, and molybdenum. J. Plant Nutr. 24:1633-1644.
27. Klobus G, Burzynsky M, Buczek J (2002) Heavy metals and nitrogen metabolism. In: Prasad MNV, Strzalka K (eds), Physiology and Biochemistry of Metal Toxicity and Tolerance in Plants, pp.325-355. Kluwer Academic Publishers, Dordrecht, The Netherlands.
28. Kovacevic G, Kastori R, Merkulov LJ (1999) Dry matter

and leaf structure in young wheat plants as affected by cadmium, lead, and nickel. Biol. Plant. 42:119-123.

29. Krupa Z, Siedlecka A, Skórzynska-Polit E, Maksymiec W (2002) Heavy metal interactions with plant nutrients. In: Prasad MNV, StrzaBka K (eds), Physiology and Biochemistry of Metal Toxicity and Tolerance in Plants, pp.287-301. Kluwer Academic Publishers, Dordrecht, The Netherlands.

30. Lagerwerff JV, Specht AW (1970) Contamination of roadside soil and vegetation with cadmium, nickel, lead, and zinc. Environ. Sci. Technol.4:583-586.

31. Lavid N, Schwartz A, Lewinsohn E, Tel-Or E (2001 a) Phenols and phenol oxidases are involved in cadmium accumulation in the water plants *Nymphoides peltata* (Menyanthaceae) and *Nymphaeae* (Nymphaeaceae). Planta 214:189-195.

32. Lavid N, Schwartz A , Yarden O, Tel-Or E (2001 b) The involvement of polyphenols and peroxidase activities in heavy-metal accumulation by epidermal glands of the waterlily (Nymphaeaceae). Planta 212:323-331.

33. Lichtenthaler HK, Wellburn AR (1983) Determinations of total carotenoids and chlorophylls *a* and *b* of leaf extracts in different solvents. Biochem. Soc. Transactions11:591-592.

34. Lindblad P, Guerrero MG (1993) Nitrogen fixation and nitrate reduction. In: Hall DO, Scurlock JMO, Bolhar-Nordenkampf HR, Leegood RC, Long SP (eds), Photosynthesis and Production in a Changing Environment: a Field and Laboratory Manual, pp. 299-312. Chapman and Hall, London, UK.

35. Lummerzheim M, Sandroni M, Castresana C, De Oliveira D, van Montagu M, Roby D, Timmerman B (1995) Comparative microscopic and enzymatic characterization of the leaf necrosis induced in *Arabidopsis thaliana* by lead nitrate and by *Xanthomonas campestris* pv. *campestris* after foliar spray. Plant Cell Environ.18:499-509.

36. McGrath SP, Zhao FJ, Lombi E (2001) Plant and rhizosphere processes involved in phytoremediation of metal-contaminated soils. Plant Soil 232:207-214.

37. Medina E, García V, Cuevas E (1990) Sclerophylly and oligotrophic environments: relationships between leaf structure, mineral nutrient content, and drought resistance in tropical rain forests of the upper rio Negro region. Biotropica 22:51-64.

38. Mengel K, Kirkby E (2001) Principles of Plant Nutrition. Kluwer Academic Publishers, Dordrecht, The Netherlands.

39. Miller RO (1998) Nitric-perchloric acid wet digestion in an open vessel. In: Kalra YP (ed), Handbook of Reference Methods for Plant Analysis, pp.57-61. CRC Press, Boca Raton, U.S.A.

40. Motto HL, Daines RH, Chilko DM, Motto CK (1970) Lead in soils and plants: its relationship to traffic volume and proximity to highways. Environ. Sci. Technol.4:231-237.

41. Murphy J, Riley J (1962) A modified single solution method for determination of phosphate in natural waters. Anal. Chem. Acta 27:31-36.

42. Olivares E, Peña E, Aguiar G (2002a) Nutrición mineral y estrés oxidativo por metales en espinaca y lechuga, en comparación con dos malezas asociadas, en cultivo semi-urbanos. Interciencia 27:454-464.

43. Olivares E, Peña E, Aguiar G (2002b) Metals and oxalate in *Tithonia diversifolia* (Asteraceae): concentrations in plants growing in contrasting soils, and Al induction of oxalate exudation by roots. J. Plant Physiol.159:743-749.

44. Pereira PH, Cambraia J, Sant›Anna R, Mosquim PR, Moreira MA (1999) Aluminum effects on lipid peroxidation and on the activities of enzymes of oxidative metabolism in sorghum. Rev. Bras. Fisiol. Veg.11:137-143.

45. Rios CI (1999) *Tithonia diversifolia* una planta con potencial para la producción sostenible en el trópico. In: Sánchez MD, Rosales M (eds), Agroforestería para la Producción Animal en América Latina. Estudio FAO Producción y Sanidad Animal 143, pp. 311-325. Roma, Italy.

46. Ross SM (1994) Sources and forms of potentially toxic metals in soil-plant systems. In: Ross SM (ed), Toxic Metals in Soil-Plant System, pp.3-25. John Wiley and Sons, Chichester, UK.

47. Ruso J, Zapata J, Hernández M, Ojeda MA, Benlloch M, Prats-Pérez E, Tena M, López-Valbuena R, Jorrin JV (2001) Toxic metals accumulation and total soluble phenolics in sunflower and tobacco plants. Minerva Biotecnol.13:93-95.

48. Sakihama Y, Yamasaki H (2002) Lipid peroxidation induced by phenolics in conjunction with aluminium ions. Biol. Plant. 45:249-254.

49. Sandalio, LM, Dalurzo HC, Gómez M, Romero-Puertas MC, del Río LA (2001) Cadmium-induced changes in the growth and oxidative metabolism of pea plants. J. Exp. Bot.52:2115-2126.

50. Shah K, Kumar RG, Verma S, Dubey RS (2001) Effect of cadmium on lipid peroxidation, superoxide anion generation and activities of antioxidant enzymes in growing rice seedlings. Plant Sci. 161:1135-1144.

51. Singh RP, Dabas S, Choudhary A, Maheshwari R (1997/98) Effect of lead on nitrate reductase activity and alleviation of lead toxicity by inorganic salts and 6-benzyla-minopurine. Biol. Plant. 40:399-404.

52. Singh RP, Tripathi RD, Sinha SK, Maheshwari R, Srivastava HS (1997) Response of higher plants to lead contaminated environment. Chemosphere 34:2467-2493.

53. Suarez N (2000) Mecanismos de regulación salina bajo condiciones constantes y fluctuantes de salinidad en *Avicennia germinans* L. Caracas, Instituto Venezolano de Investigaciones Científicas. PhD thesis

54. Tabatabai MA, Bremner JM (1969) Use of p-nitrophenylphosphate for assay of soil phosphatase activity. Soil Biol. Biochem.1:301-307.

55. van Hasselt PR, Chow WS, Anderson JM (1996) Short-term treatment of pea leaves with supplementary UV-B at different oxygen concentrations: impacts on

chloroplast and plasma membrane bound processes. Plant Sci. 120:1-9.

56. Ward NI, Reeves RD, Brooks RR (1975) Lead in soil and vegetation along a New Zealand state highway with low traffic volume. Environ. Pollut.9:243-351.

57. Weiss D, Shotyk W, Kempf O (1999) Archives of atmospheric lead pollution. Naturwissenschaften 86:262-275.

58. Wierzbicka M (1999) Comparison of lead tolerance in *Allium cepa* with other plant species. Environ. Pollut. 104:41-52.

59. Xiong ZT (1997) Bioaccumulation and physiological effects of excess lead in a roadside pioneer species *Sonchus oleraceus* L. Environ. Pollut. 97:275-279.

60. Yassoglou N, Kosmas C, Asimakopolus J, Kallianou C (1987) Heavy metal contamination of roadside soils in the Greater Athens area. Environ. Pollut. 47:293-304.

INDEX

P

R

S

T

U

V

W

X